ACS SYMPOSIUM SERIES **413**

Probing Bioactive Mechanisms

Philip S. Magee, EDITOR
BIOSAR Research Project

Douglas R. Henry, EDITOR
Molecular Design Limited

John H. Block, EDITOR
Oregon State University

Developed from a symposium sponsored
by the Division of Agrochemicals
at the 196th National Meeting
of the American Chemical Society,
Los Angeles, California,
September 25–30, 1988

American Chemical Society, Washington, DC 1989

Library of Congress Cataloging-in-Publication Data

Probing bioactive mechanisms
   Philip S. Magee, editor, Douglas R. Henry, editor,
John H. Block, editor

   Developed from a symposium sponsored by the Division of Agrochemicals at the 196th National Meeting of the American Chemical Society, Los Angeles, California, September 25–30, 1988.

   p.   cm.—(ACS Symposium Series, 0097–6156; 413).
Bibliography: p.

   Includes index.

   ISBN 0–8412–1702–5
   1. Structure–activity relationships (Biochemistry)—Congresses. 2. Pharmacology—Methodology—Congresses.
3. Agricultural chemistry—Methodology—Congresses. 4. Toxicology—Methodology—Congresses.

   I. Magee, Philip, 1926–   .   II. Henry, Douglas R.
III. Block, John H.   IV. American Chemical Society.
Division of Agrochemicals. V. American Chemical Society. Meeting (196th: 1988: Los Angeles, Calif.)
VI. Series.

QP517.S85P76   1989
574.19'283—dc20   89–17998
                                                      CIP

Copyright © 1989

American Chemical Society

All Rights Reserved. The appearance of the code at the bottom of the first page of each chapter in this volume indicates the copyright owner's consent that reprographic copies of the chapter may be made for personal or internal use or for the personal or internal use of specific clients. This consent is given on the condition, however, that the copier pay the stated per-copy fee through the Copyright Clearance Center, Inc., 27 Congress Street, Salem, MA 01970, for copying beyond that permitted by Sections 107 or 108 of the U.S. Copyright Law. This consent does not extend to copying or transmission by any means—graphic or electronic—for any other purpose, such as for general distribution, for advertising or promotional purposes, for creating a new collective work, for resale, or for information storage and retrieval systems. The copying fee for each chapter is indicated in the code at the bottom of the first page of the chapter.

The citation of trade names and/or names of manufacturers in this publication is not to be construed as an endorsement or as approval by ACS of the commercial products or services referenced herein; nor should the mere reference herein to any drawing, specification, chemical process, or other data be regarded as a license or as a conveyance of any right or permission to the holder, reader, or any other person or corporation, to manufacture, reproduce, use, or sell any patented invention or copyrighted work that may in any way be related thereto. Registered names, trademarks, etc., used in this publication, even without specific indication thereof, are not to be considered unprotected by law.

PRINTED IN THE UNITED STATES OF AMERICA

ACS Symposium Series

M. Joan Comstock, *Series Editor*

1989 ACS Books Advisory Board

Paul S. Anderson
Merck Sharp & Dohme Research Laboratories

Alexis T. Bell
University of California—Berkeley

Harvey W. Blanch
University of California—Berkeley

Malcolm H. Chisholm
Indiana University

Alan Elzerman
Clemson University

John W. Finley
Nabisco Brands, Inc.

Natalie Foster
Lehigh University

Marye Anne Fox
The University of Texas—Austin

G. Wayne Ivie
U.S. Department of Agriculture, Agricultural Research Service

Mary A. Kaiser
E. I. du Pont de Nemours and Company

Michael R. Ladisch
Purdue University

John L. Massingill
Dow Chemical Company

Daniel M. Quinn
University of Iowa

James C. Randall
Exxon Chemical Company

Elsa Reichmanis
AT&T Bell Laboratories

C. M. Roland
U.S. Naval Research Laboratory

Stephen A. Szabo
Conoco Inc.

Wendy A. Warr
Imperial Chemical Industries

Robert A. Weiss
University of Connecticut

Foreword

The ACS SYMPOSIUM SERIES was founded in 1974 to provide a medium for publishing symposia quickly in book form. The format of the Series parallels that of the continuing ADVANCES IN CHEMISTRY SERIES except that, in order to save time, the papers are not typeset but are reproduced as they are submitted by the authors in camera-ready form. Papers are reviewed under the supervision of the Editors with the assistance of the Series Advisory Board and are selected to maintain the integrity of the symposia; however, verbatim reproductions of previously published papers are not accepted. Both reviews and reports of research are acceptable, because symposia may embrace both types of presentation.

Contents

Preface ix

VIEWS OF THE FIELD

1. Progress in the Design of Bioactive Molecules 2
   John H. Block

2. Predicting Mechanism and Activity: The Trendy
   Computational Soothsayer 26
   Douglas R. Henry

3. Interfacing Statistics, Quantum Chemistry, and Molecular
   Modeling 37
   Philip S. Magee

WAYS AND MEANS

4. Second-Generation Computer-Assisted Inhibitor
   Design Method 60
   Renee L. DesJarlais, George L. Seibel, and
   Irwin D. Kuntz, Jr.

5. Statistical Modeling of Molecular Shape, Similarity,
   and Mechanism 70
   Douglas R. Henry and A. Morrie Craig

6. New Tool for the Study of Structure–Activity Relationships
   in Three Dimensions: The Hypothetical Active-Site Lattice 82
   Arthur M. Doweyko

7. Finding Lead Structures from Amino Acid Sequence
   Similarities of Target Proteins 105
   Takaaki Nishioka, Kazuo Sumi, and Jun'ichi Oda

8. Application of Micellar Liquid Chromatography to Modeling of Organic Compounds by Quantitative Structure–Activity Relationships 123
   Barry K. Lavine, Anthony J. I. Ward, Jian Hwa Han, and Orla Donoghue

AGROCHEMICAL MECHANISMS

9. Inhibition of Susceptible and Resistant Green Rice Leafhopper Acetylcholinesterase by N-Methylcarbamate and Oxadiazolone Insecticides 136
   Hiroki Ohta, Noburo Kyomura, Yoji Takahashi, and Philip S. Magee

10. Critical Differences in the Binding of Aryl Phosphate and Carbamate Inhibitors of Acetylcholinesterases 147
    Philip S. Magee

11. Contribution of Quantitative Agrochemical Design Strategies to Mechanism-of-Action Studies 157
    E. L. Plummer, J. A. Dixson, and R. M. Kral

12. Quantitative Structure–Activity Relationship Study of Aromatic Trifluoromethyl Ketones: In Vitro Inhibitors of Insect Juvenile Hormone Esterase 169
    András Székács, Barna Bordás, György Matolcsy, and Bruce D. Hammock

13. Conformational Analysis of Fenvalerate and an Ether-Type Pyrethroid 183
    Yasuyuki Kurita, Kazunori Tsushima, and Chiyozo Takayama

14. Molecular Design and Target Site Analysis in Fungicide Development 198
    Hugh D. Sisler and Nancy N. Ragsdale

15. Modeling of Photosystem II Inhibitors of the Herbicide-Binding Protein: Inhibitory Pattern, Quantitative Structure–Activity Relationships, and Quantum Mechanical Calculations of New Hydroxyquinoline Derivatives 215
    W. Draber, B. Pittel, and A. Trebst

DRUG MECHANISMS

16. A₁ Adenosine Receptors in the Heart: Functional and Biochemical Consequences of Activation 232
    T. F. Murray, T. A. Blair, M. Leid, P. H. Franklin, and J. F. Siebenaller

17. Therapeutic Potential for Parathyroid Hormone Antagonists 243
    Mark E. Goldman and Michael Rosenblatt

18. Interaction of Phenylisopropylamines with Central 5-HT2 Receptors: Analysis by Quantitative Structure–Activity Relationships 264
    Richard A. Glennon and Mark R. Seggel

19. Analysis of Binding at 4-Aminobutyric Acid Receptor Sites by Structure–Activity Relationships 281
    Philip S. Magee and James W. King

20. Quantitative Structure–Activity Relationships for the Cytotoxicity of Substituted Aniline Mustards in Tissue Culture 291
    William A. Denny, William R. Wilson, and Brian D. Palmer

21. Quantitative Structure–Activity Relationships of Antibacterial Compounds Based on the Nalidixic Acid Structure 301
    John H. Block, Yupei Yu, James W. King, and Arie Verloop

TOXICITY MECHANISMS

22. Structurally Specific Interaction of Halogenated Dioxin and Biphenyl Derivatives with Iodothyronine-5'-deiodinase in Rat Liver 354
    U. Rickenbacher, S. Jordan, and J. D. McKinney

23. Base-Line Toxicity Predicted by Quantitative Structure–Activity Relationships as a Probe for Molecular Mechanism of Toxicity 366
    Robert L. Lipnick

24. Correlations and Mechanisms of Chemical Toxicity in Animals 390
   Philip S. Magee and James W. King

INDEXES

Author Index 402

Affiliation Index 402

Subject Index 403

Preface

SEVERAL YEARS AGO, ONE OF US (P.S.M.) participated in a 1984 ACS Symposium Series book that described quantitative structure–activity relationship (QSAR) research (*Pesticide Synthesis Through Rational Approaches*, ACS Symposium Series 255, edited by Philip S. Magee, Gustave K. Kohn, and Julius J. Menn). The prevalent theme at that time was drug and pesticide design, and our book fell into this category. Now it is increasingly clear that structure–activity relationship studies in the hands of the best investigators are also addressing mechanisms in both simple and complex biochemical systems. The classical approaches of experimental kinetic, equilibrium, and isotopic studies still form the basis of mechanistic proof that is acceptable to the scientific community. However, the data developed in these and related studies can be extended by statistical and modeling techniques to provide mechanistic inferences beyond experimental results. Such inferences are probabilistic in nature and do not enjoy the same standing as experimental proof. What they do provide are clear working hypotheses that point the way to new experimental designs. This book addresses the inferential approach to extending mechanistic insight by developing relations from raw, experimental data and molecular structure.

Among other changes in recent years is the lowering of conceptual barriers among medicinal, agrochemical, and environmental problems. These problems address analogous underlying mechanisms where cuticle penetration, transport, active-site binding, and irreversible processes can be described by a common model. Moreover, scientists in each area employ related experimental methods and identical computer-assisted methods for developing structure–activity relationship models. This book reflects the merging of these apparently diverse fields under a canopy of common mechanistic events.

The editors of this volume are typical of many in the QSAR field, working alone, without immediately close colleagues. Each of us practices and teaches within the general fields of modeling and statistical QSAR. Though physically separated, we converse by letter, telephone, and modem. We like to call ourselves Research Triangle–West.

PHILP S. MAGEE
BIOSAR Research Project
Vallejo, CA 94591
and
School of Medicine
University of California
San Francisco, CA 94143

DOUGLAS R. HENRY
Molecular Design Limited
2132 Farallon Drive
San Leandro, CA 94577

JOHN H. BLOCK
College of Pharmacy
Oregon State University
Corvallis, OR 97331–3507

June 29, 1989

VIEWS OF THE FIELD

Chapter 1

Progress in the Design of Bioactive Molecules

John H. Block

College of Pharmacy, Oregon State University, Corvallis, OR 97331-3507

The traditional search for drugs and agricultural chemicals has been based on observing the use of plants by human populations, dietary habits, general pharmacological screens, and chance observations. The discovery of many biologically active chemicals began with their isolation from natural products. These substances became the prototype molecules from which modifications were made. Other drug discoveries have been based on what could be called the shotgun approach and *chance favors the prepared mind*. For the former, a large number of compounds are synthesized or plant extracts are isolated and then subjected either to general or specific pharmacological screens. In today's economic and regulatory climate, this latter approach is very expensive because there are relatively few commercially successful products obtained from the thousands of compounds tested. Increasingly, a productive search for biologically active molecules requires a fundamental understanding of the disease for which the drug is targeted. Statistical techniques, conformational analysis, and receptor characterization will provide valuable information for the synthetic chemist to better tailor the molecular structure required for desired activity.

Over the long history in the development of bioactive molecules, there have been many approaches used. These range from rational, carefully thought out hypotheses, to general and specific pharmacological screens used to identify compounds with desired biological activity, to serendipity. Listed below are different ways that commercially useful bioactive molecules have been discovered. Because of the author's background, there will be more examples from human medicine than from agrochemicals. At the same time, the principles discussed for humans will apply generally to mammals including commercial livestock. In the broad sense, the development of chemicals used to maximize crop yields use the same approaches.

NATURAL PRODUCTS.

This has been and probably will remain one of the main means of discovering biologically active molecules. It is realized that there is a tremendous number of terrestrial plants that have never been screened and, unfortunately, may never be.(1) There is concern that, with the loss of the tropical rain forests, many plant species will become extinct before there is a chance to evaluate their chemical constituents. On the other hand, the examination of natural sources now includes marine species. The search for active compounds from natural product sources will continue because so little is known about the etiology of so many diseases that it is difficult to design potentially active molecules for these conditions. Some classic examples of successful drugs and agrochemicals derived from plant extracts and the results from attempts at chemically modifying their structures include the following.

<u>Cardiac glycosides:</u> These are the drugs of choice in the treatment of congestive heart failure. The synthetic medicinal chemist has not produced a product superior to cardiac glycosides such as digoxin.

<u>Atropine:</u> This is a classic example of the prototype drug from which the anticholinergic class of agents are derived. In contrast with digitalis, a wide variety of anticholinergics, chemicals which block the cholinergic receptors, which are superior to the parent alkaloid have been made and continue to be introduced into medicine.

<u>Cocaine:</u> This drug, which has become such a pariah in our society, is the prototype for the local anesthetics. It is one of the success stories as evidenced by the wide variety of local anesthetics that have been made free of any abuse potential. Indeed, the structure of procaine, which is a simple benzoic acid ester, illustrates how the pharmacophore moiety can be abstracted from a more complex natural product.

<u>Penicillins, cephalosporins, tetracyclines, actinomycins:</u> These are examples of classes of antibiotics each from a microorganism producing cytotoxic agents. The first three classes are selectively cytotoxic to bacteria, and the fourth cytotoxic to mammalian cancer cells, unfortunately with poor selective toxicity. The penicillins and cephalosporins inhibit bacterial cell wall synthesis, and tetracyclines selectively block protein synthesis at the bacterial ribosome. The actinomycins intercalate in a relative nonselective manner the

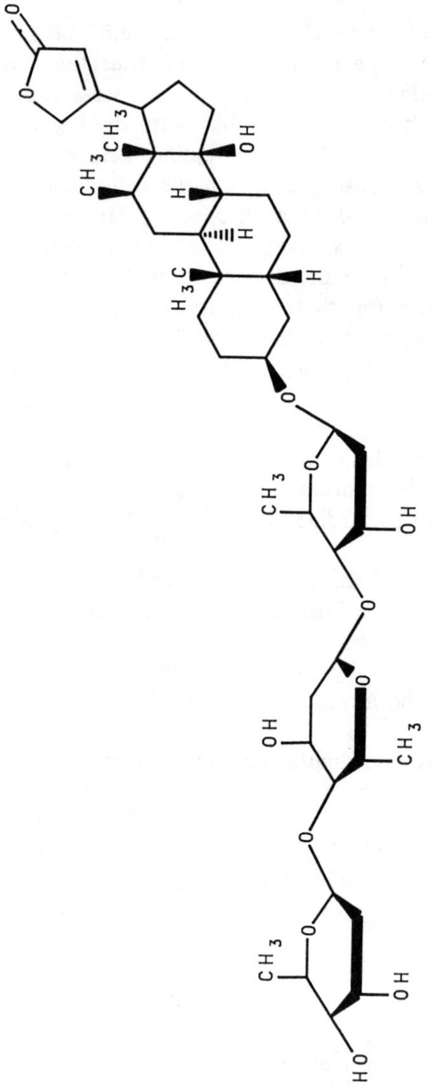

Digoxin

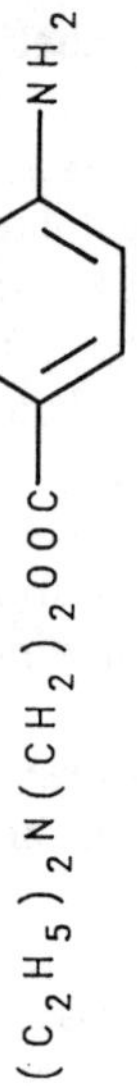
(C2H5)2N(CH2)2OOC
NH2

Procaine

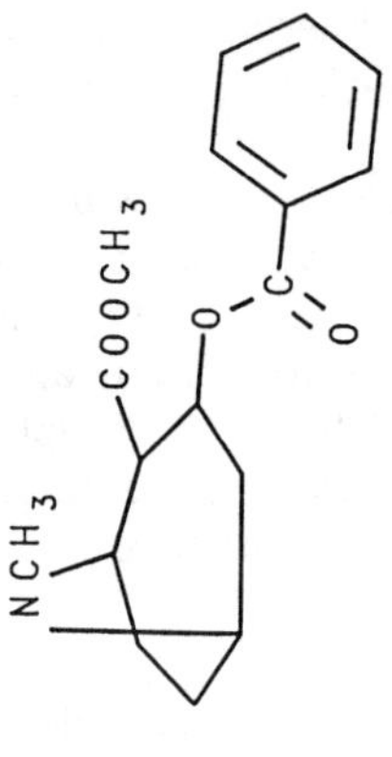

Cocaine

chromosomal DNA in both malignant and benign cells. They were initially screened in the 1940's for their antibacterial activity and, while active against bacteria, were found to be too toxic. Several years later they were found to be successful against selected cancers. While synthetic and semi-synthetic analogues of the antibiotics are continually being synthesized, evaluated and marketed, microorganisms are still being actively screened for new leads.

Pyrethrins: This family of insecticides from plants of the genus *Chrysanthemum* were first used in the early 1800's and continue to be widely used to the present, particularly in household insecticide products because of their relatively low toxicity in humans. Classical compounds like the pyrethrins illustrate one of the more frustrating aspects in the design of bioactive molecules. Even though this chemical class has been used for nearly 200 years, little can be said regarding their mechanism of action. Further, it is doubtful that much basic research will be reported in the immediate future because there is little economic incentive in the private sector, and the government funding agencies likely will continue to award grants for projects that investigate more novel chemistry.

Cyclosporin: The history of this drug, isolated from the fungus *Cylindrocarpon ludidium* and *Trichoderma polysporum*, shows how perseverance of an individual scientist, Dr. Jean-Francoise Borel (Sandoz), has led to the marketing of a drug which has increased significantly the prognosis of patients receiving organ transplants. In 1978, the one year survival rate for was only 66 percent for transplanted hearts and 65 percent for transplanted kidneys. The most recent figures show the one year survival rates now to be 80 percent for hearts and 91 percent for kidneys.(2) The search for additional immunosuppressant drugs has led to a new natural product, FK-506, isolated from *Streptomyces tsukubaensis* No. 9993.(3) It appears that both compounds, as structurally diverse as they appear, possibly have the same mechanism of action. Using the newer techniques of conformational analysis, it will be interesting to determine if there is a common pharmacophore.

The list of commercial products from natural sources could fill volumes. It must be emphasized that their discoveries range from systematic searches, to tradition to *chance favors the prepared mind*. The latter is reemphasized in the discovery of the alkaloids from the periwinkle plant (vincristine and vinblastine) which were first screened for their hypoglycemic activity based on reports of their use by local groups in Madagascar. While the hypoglycemic response could not be confirmed under controlled laboratory conditions, an immunosuppressive effect due to

Dactinomycin (Actinomycin-D)

Pyrethrins

Cyclosporin

FK-506

drastic reduction of white cells was seen. This led to the introduction of two effective agents used to treat leukemias and lymphomas.(4,5)

BIOCHEMICALLY ACTIVE MOLECULES:

The neurotransmitters and hormones are good examples based on the approach of starting with a biochemically active substance as the prototype molecule. This group of compounds combine with specific receptors, and therefore, provide the basic structure for synthetic modification in order to obtain more specific activity or even antagonistic response. The following will serve as examples of this approach at developing bioactive molecules.

Acetylcholine: Both anticholinergics (antagonists) that block the cholinergic receptor and acetylcholinesterase inhibitors (potentiate acetylcholine) are in use today and are based on the acetylcholine structure. Because of the efficiency of acetylcholinesterase, it has proved more productive to inhibit the enzyme that hydrolyzes this neurotransmitter rather than develop cholinergic agonists.

Acetylcholinesterase inhibitors has been a productive approach in the design of insecticides starting with the phosphate esters such as malathion and continuing on to carbamate esters such as carbaryl. For the latter, leads came from two carbamate reversible cholinesterase inhibitors used in medicine, the natural product physostigmine and the synthetic derivative, neostigmine.

The history of the development of the organophosphate acetylcholinesterase inhibitors as insecticides is a classic example of examining the early literature and systematically synthesizing a large group of compounds. Contrary to popular opinion, this group of insecticides was not a spin off from the development of nerve gases. Indeed the research apparently began prior to the nerve gas research when industrial chemists in Germany began a search for synthetic chemicals to replace the insecticides nicotine, pyrethrum (see below) and rotenone which had to be imported into Germany. Based on earlier published work, a large number of organophosphate chemicals were synthesized from which active acetylcholinesterase inhibitors were obtained. It was only later, after the toxicity of the organophosphates was realized, that research on nerve gas began in earnest.

Histamine: At least two responses are attributed to this neurotransmitter derived from the amino acid histidine. It is part of the allergic response (H_1 receptor) and, in the stomach, stimulates the release of gastric hydrochloric acid (H_2 receptor). Mild allergic responses, such as hay fever, have been treated for years with antihistamines. Their development pretty much followed the classical approach where

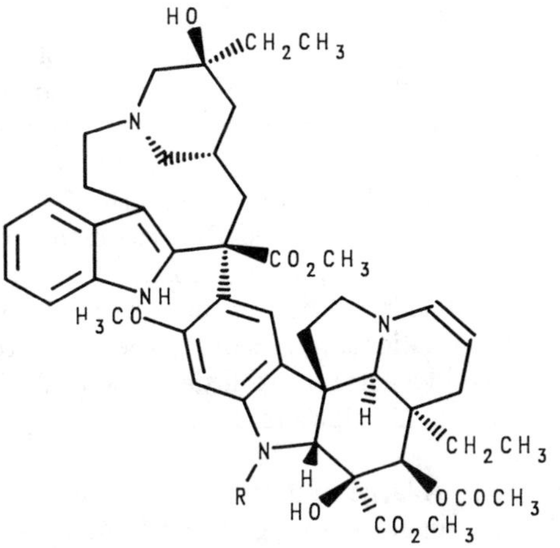

Vinblastine: R = −CH₃;   Vincristine: R = −CHO

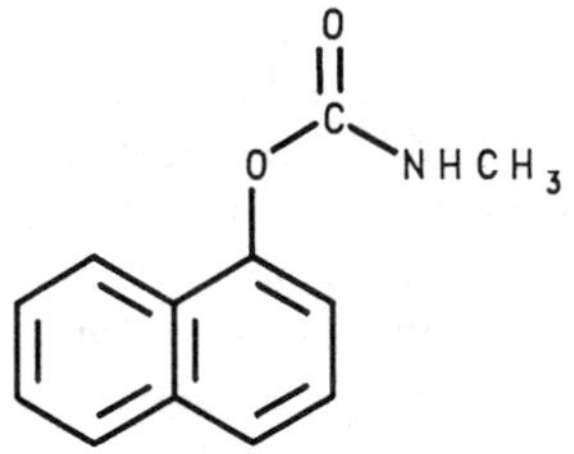

Carbaryl

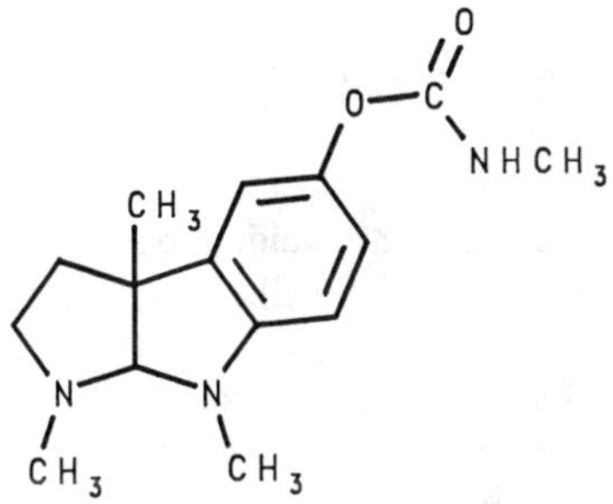

Physostigmine

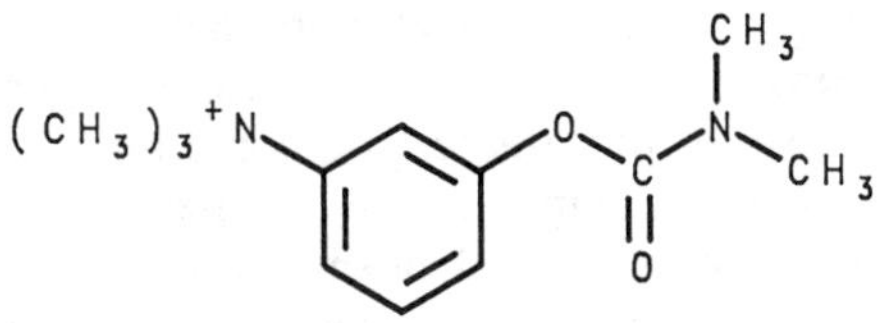

Neostigmine

histamine was first realized to be a crucial component of the allergic response. There already were known compounds which had antihistamine activity, but they were too toxic. The first clinically useful antihistamines were reported in 1942 to be followed by hundreds of useful compounds.([5](#)) Co-recipient of the 1988 Nobel Prize in Medicine was Sir James Black who observed that antihistamines did not work against ulcers and postulated that there must be a second histamine receptor, now called the H_2 receptor. This led to a new class of histamine antagonists (e.g. cimetidine and rantidine) which has greatly altered the treatment of peptic ulcers. No longer are patients with peptic ulcers dependent on dosing themselves with antacids. What is interesting to note from the structure activity relationship (SAR) aspect is the traditional antihistamines (H_1 blockers) bear little structural resemblance to histamine whereas the H_2 blockers do.

Cortisone/Hydrocortisone: A large number of steroid analogues have been made in order to separate the glucocorticoid (antiinflammatory) response from the mineralcorticoid (fluid retention) with a fair degree of success. The result has been a large number of synthetic corticosteroids used systemically and topically for their antiinflammatory activity.

Phenoxyacetic acids: The development of the 2,4-dichloro- and 2,4,5-trichlorophenoxyacetic acids (R = H and Cl, respectively) was an outgrowth of work based on the plant hormone activity (auxin) of indole-3-acetic acid. Based on the earlier concepts of rigid receptors, there was little likelihood that the chlorinated phenoxyacetic acids would show auxin activity. When it was realized that the margin of safety between induction of healthy root growth and the induction of excessive root thickening was too small for this group to be used as growth simulators, their use as herbicides developed.([6](#)) Of course, today it is realized that substituted benzene rings can be used as bioisoteric replacements for heterocyclic rings such as indole.

The opportunities for developing new approaches for attacking medical and agricultural problems is limited only by the complexity of the biochemical milieu of interest, i.e. the greater the complexity, the greater the number of opportunities. An example of exciting new approaches are the chapters in this book discussing the adenosine receptor in the heart which has potential use in developing new cardiotonic and antiarrhythmic agents, a group of potent parathyroid hormone antagonists that may prove useful in the treatment of hypercalcemia, and the serotonin receptor, the first of which are just coming onto the market.

SELECTIVE TOXICITY.

The ideal bioactive molecule acts solely on the target organ, organism, or receptor. The antibacterial agents in use today are excellent examples of this approach in human medicine. In general the key is to identify a metabolic reaction unique to the microorganism or an enzyme used by the microorganism or agricultural pest that is so physically or chemically different that the drug will have no significant effect on the patient's metabolism. Professor Adrian Albert takes a broader view of the term *selective toxicity*.(7) He considers antagonists as being toxic in the sense that they occupy receptors preventing the binding of the normal ligand. Thus an antihistamine shows selectivity by combining mostly with the H_1 receptor and ignoring the H_2 receptor. This class of drugs is not completely selective as evidenced by their centrally acting depressant actions leading to sedation and the anticholinergic response. The latter can range from a nuisance in some individuals to potentially harmful in asthmatics who may have trouble expectorating fluids from their lungs. The more restricted approach to *selective toxicity* will be used in this chapter and in the following examples.

Trimethoprim/Methotrexate: Both drugs inhibit dihydrofolate reductase, but trimethoprim (developed by George Hitchings and Gertrude Elion, also 1988 co-recipients of the Nobel Prize in Medicine) is selective for the bacterial dihydrofolate reductase while methotrexate is an inhibitor of the mammalian enzyme and is used in cancer chemotherapy. The latter drug cannot distinguish between the enzyme in malignant and normal cells with the result that it is a very toxic drug. Many times the antidote, calcium leucovorin which is the calcium salt of one of the forms of tetrahydrofolic acid, is administered to the patient following a course of intense methotrexate therapy. Because the pteridine ring is already reduced, calcium leucovorin does not require conversion by active dihydrofolate reductase into an active form.

    The discovery of methotrexate again shows how alert scientists exploit what first appeared to be a puzzling observation. It was found that administration of folates to patients with acute leukemia hastened the progress of the disease. Positive results from crude folate antagonists that were available at the time led eventually to the synthesis of methotrexate.(5)

Tetracyclines: This very successful class of antibacterials act selectively on the bacterial ribosome inhibiting protein synthesis. They do not bind with mammalian ribosomes in the cytoplasm and, therefore, do not have a direct effect on the patient's metabolism. Like any drug, they are not free of potentially harmful side effects. They complex calcium and can interfere with development of the permanent teeth prior to their erupting through the gums As with the sulfonamides, this class

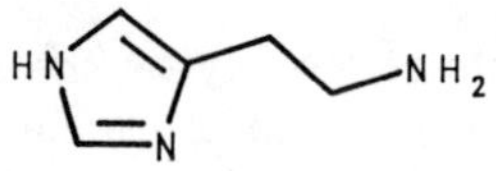

Histamine

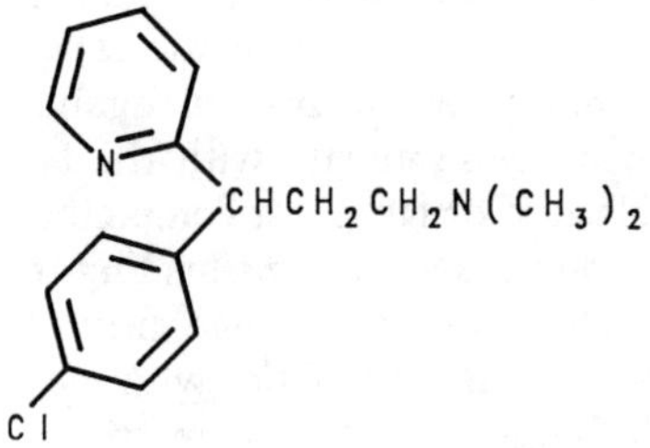

Chlorpheniramine

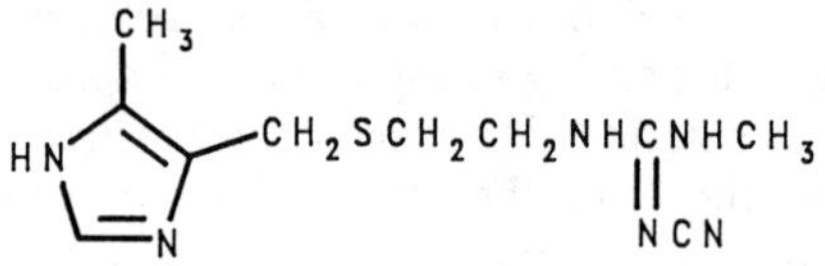

Cimetidine

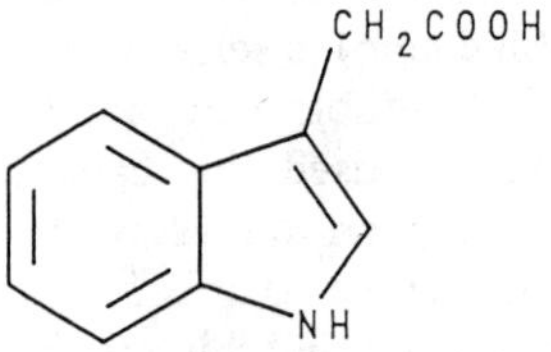

Indoleacetic Acid

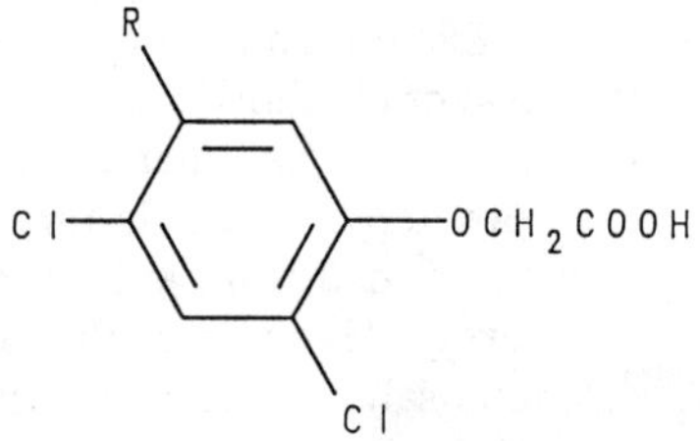

Chlorinated Phenoxyacetic Acids

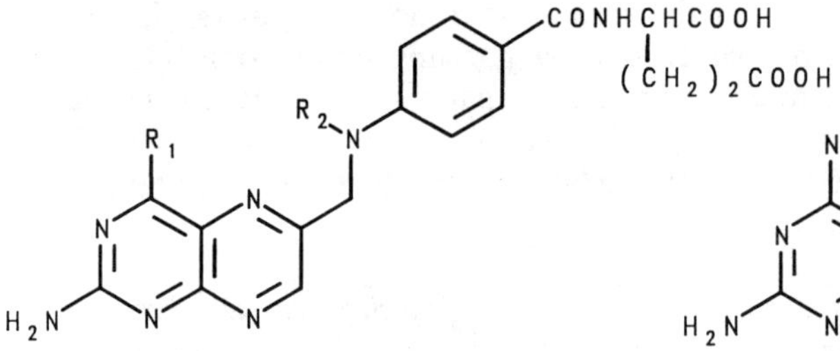

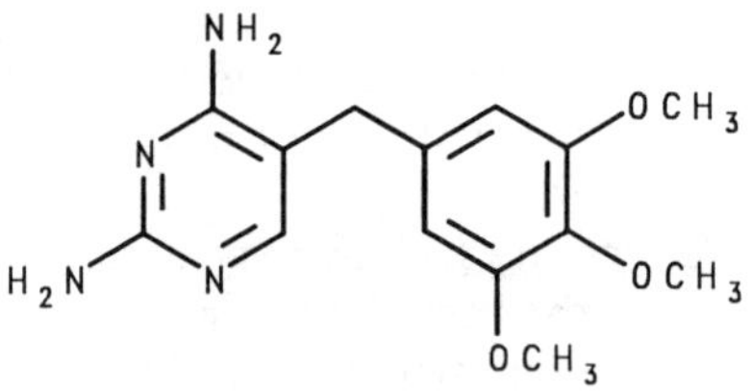

R_1 = OH; R_2 = H: Folic Acid

R_1 = NH_2; R_2 = CH_3: Methotrexate

Trimethoprim

of drugs was discovered as the result of an antibacterial screen of bacterial extracts.

<u>Acyclovir:</u> This agent is quite effective against Herpes Simplex II (genital herpes) if the patient follows the regimen carefully. This drug is selective for viral thymidine kinase which converts acyclovir to the nucleotide triphosphate whereas the host cell kinase does not. Thus the drug remains in the inactive prodrug form in noninfected cells. The acyclovir triphosphate, which lacks a 3'-OH, now inhibits viral DNA polymerase preventing the synthesis of the new viral DNA needed for herpes virus reproduction.

<u>Azidothymidine:</u> This drug shows a reasonable degree of selectivity for the viral RNA dependent DNA polymerase (reverse transcriptase), an enzyme found only in retroviruses. Because the Human Immunodeficiency Virus-I (HIV-1), the cause of Acquired Immunodeficiency Syndrome (AIDS), requires this enzyme to reproduce itself, azidothymidine slows the progress of this tragic, largely preventable disease. But bone marrow depression is a common complication indicating that it is inhibiting cell division in the patient.

<u>Antifungal Agents:</u> It has been very difficult to design agents effective against fungal infections whether they are in humans, livestock or plants. In theory it should be possible to control fungal infections/infestations with the appropriate chemical because they do have unique biochemistry which can be exploited. For example, the imidazole class of antifungal agents (miconazole, ketoconazole) are selective for the incorporation of acetate into ergosterol, a route not found in humans. In practice, fungal infections can be very difficult to control. The structure activity relationships of the agents used for the fungal caused diseases are very diverse. Besides the imidazoles, there is the antibiotic griseofulvin which is active against both plant and animal fungal infections and binds preferentially to fungal RNA and the polyene antibiotics (nystatin, amphotericin B) which bind preferentially to the ergosterol in fungal membranes relative to cholesterol in mammalian cell membranes.(8) For more information, please see the chapter in this book describing approaches for developing fungicides used in agriculture based on taking advantage of the metabolic differences between the fungus and its host.

Ideally, selective toxicity is the one of the best approaches to use in the design of biologically active molecules. It is very expensive and requires a considerable investment of time and capital because the metabolism of the pathogen must be elucidated in order to locate the unique transformations,

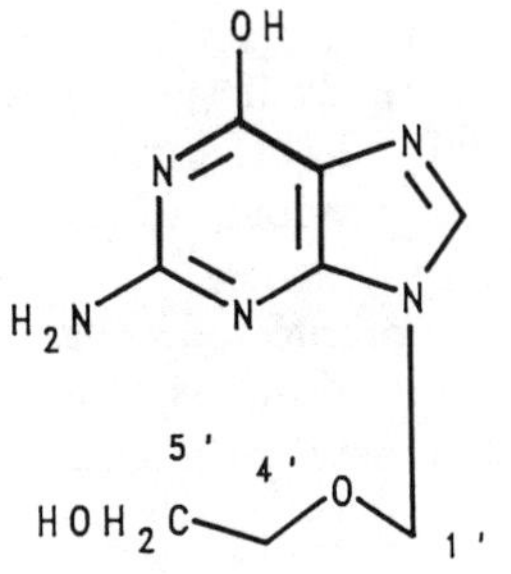

Acyclovir

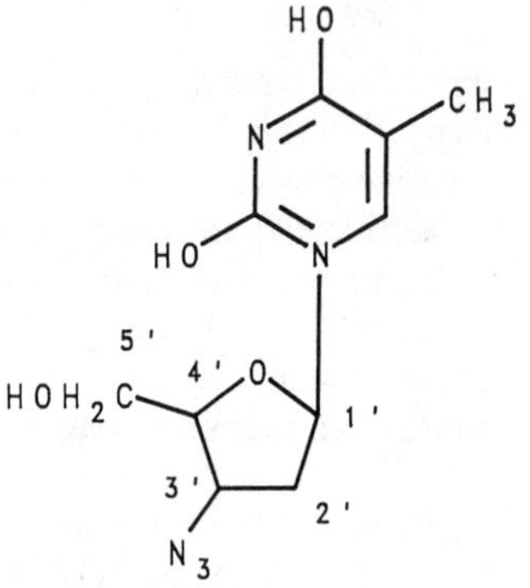

Zidovudine (azidothymidine)

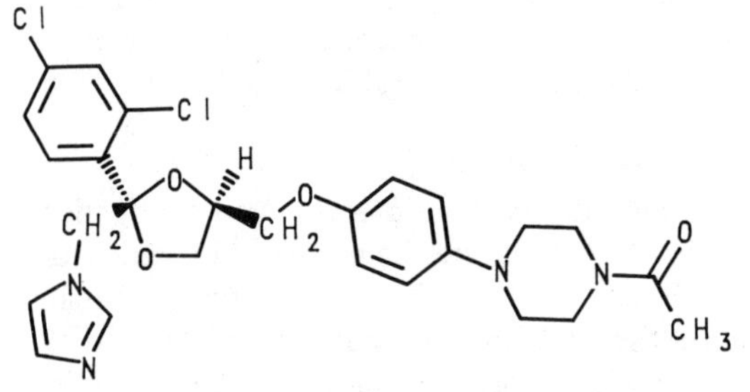

Ketoconazole

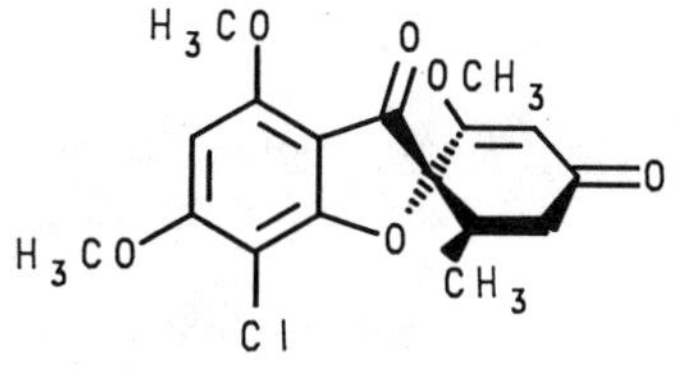

Griseofulvin

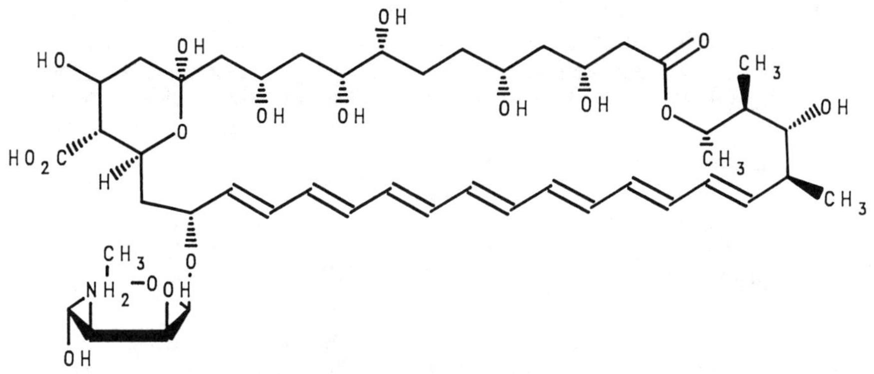

Amphotericin B

enzymes or structures that can be the target of chemical intervention. In practice, it has worked best for treatment of bacterial diseases as there is a reasonable chance that there is some key metabolic difference between the pathogen and the host. Bacteria have cell walls rather than membranes (selective site for penicillin) and unique ribosomes (selective site of the tetracyclines). Bacterial infections are not feared in developed economies due to the large armamentarium of antibiotics available to the medical and veterinary professions. Fungal infections have proved to be more of a problem in both human and plant diseases. Fortunately, many fungi are found on the surface of the skin or on the plant and can be treated topically. This permits the use of some fairly toxic agents which, as long as they are not absorbed, cause little harm to the host and can be washed off the plant.

Parasites and insects can have such complex metabolism and life histories that use of chemical agents has had only limited success when the offending organism is a pathogen for humans or livestock. First, it has been difficult to find chemicals that are selective for only the offending species. Many go through various changes as they move from egg to larva to mature adult. The organism's susceptibility will change with the stage in its life. If the organism cannot be stopped outside of the animal or plant host, a bioactive chemical will have to be introduced into the patient or plant in the form of a systemic agent. These can be very toxic to humans and animals requiring repeated applications until the patient is free of the organism. Compliance in humans is a problem due to the harsh side effects of these drugs. Use of these chemicals is complicated further in livestock or plants because it is more difficult to remove the chemical prior to or during food processing. Obviously it is easier to wash off a chemical from the surface of the plant or a dipped animal. At the same time it must be realized that since plants do not have nervous systems, very toxic pesticides can be applied without harm to plants. In other words, the principle of selective toxicity works very well in terms of protecting the plants against a variety of insects.

Viruses have proved to be a real dilemma. First, they can only reproduce inside the cell. This means that the chemical agent must enter the cell in order to reach the virus. Compare this with the previously described pathogens in which the bioactive agent intercepts the organism before it penetrates the patient's cells. The alternative is to take an antiviral drug prophylactically. To date there is only one agent in the U. S. market, amantadine (see below), which is a very effective prophylactic agent against the Type A influenza virus. At the same time, most patients prefer vaccination because it is simpler and much less expensive.

Most antiviral agents are nucleotide antimetabolites similar in structure to the nucleotide antimetabolites used in cancer chemotherapy. The viruses that lend themselves to drug therapy are those with complicated genomes. The herpes simplex virus has over 120 genes making it likely that there are unique enzymes, such as the viral thymidine kinase, required for its reproduction. Similarly, the HIV-I, a retrovirus with the unique reverse transcriptase enzyme, is potentially vulnerable to rationally designed molecules.

METABOLISM OF FOREIGN MOLECULES (XENOBIOTICS).

A drug should be considered a xenobiotic. All mammals have the capabilities to transform these molecules. The *drug metabolizing enzymes* are a misnomer because their natural substrates are part of normal metabolism. Among other things, these enzyme systems hydroxylate steroids, degrade the porphyrin rings from aged erythrocytes, and conjugate the bile pigment, bilirubin, with glucuronic acid. Fortunately, the substrate specificities do not appear to be very strict with this diverse group of enzymes. The net result is that most drugs can be administered to humans and livestock with the correct assumption that, in most cases, they will be transformed and excreted. In other words, the elimination of administered drugs from the tissues of commercial livestock is due to these diverse group of enzymes. Indeed, the measurement of the biological half-life and determination of the metabolic fate are part of any application requesting permission to market a drug.

It is now realized that many drugs are converted to active metabolites. Indeed, the parent drug molecule may largely be inactive. This has led to a systematic approach called prodrug design.(9) While there have been elegant approaches published for getting drugs to specific organs by the prodrug approach, the rigors of obtaining approval of new chemical entities has restricted this technique largely to the use of simple esters. This can be illustrated with Vitamin E or α-tocopherol. The acetate ester produces an oil soluble vitamin while the hemisuccinate yields a water soluble derivative. The new hypotensive drug, enalapril (see below) is marketed as the ethyl ester because the free acid, enalaprilic acid, is poorly absorbed orally. All of these examples are hydrolyzed in the patient to the active form. Examine the label on a container of multivitamins and notice how many of the vitamins are in a chemically more stable precursor form: retinol acetate or palmitate (retinal), pyridoxine (pyridoxal), and pantothenol (pantothenic acid).

A study of a drug's metabolism can lead to the design of better compounds. This is illustrated by the popular local anesthetic, lidocaine, which also is an excellent drug for the treatment of arrhythmias. It has one drawback. It is so rapidly N-dealkylated by the hepatic cytochrome P450 enzyme system followed by hydrolysis that it cannot be given orally.(10) The initial N-deethylated product, monoethylglycinexylide, has excellent antiarrhythmic activity. This information led to the development of tocainide, an orally active antiarrhythmic drug. It can be considered the α-methyl analog of glycinexylide.

α-Tocopherol (R = H)
Acetate ester (R = CH₃CO)
Hemisuccinate ester (R = Na⁺ ⁻OCH₂CH₂CO)

R = COOH: Pantothenic Acid
R = CH₂OH: Pantothenol

Lidocaine → CH₃CHO → Active Metabolite

Tocainide

The situation for chemicals used on crops is somewhat different although the goal is the same. By the time of harvest, there should be no residues left. Rather than rely on the plant's enzymes, it is more common to time the application to minimize uptake into the plant's edible portion. There is increasing interest in designing molecules that will break down chemically upon exposure to the environment. Complicating the situation with agents applied to large fields of plantings is the concern of these chemicals migrating into the ground water. Again, chemicals that are degradable is one solution.

EXPLOITATION OF SECONDARY EFFECTS/SERENDIPITY.

This approach could be subtitled *Chance Favors the Prepared Mind*. Several examples of this have been described in the previous discussion. What follows are agents that were synthesized in search of a specific response. In some cases marketable products were obtained and, as a result of their use, new applications were found. In other cases a side effect was exploited.

<u>Oral Antidiabetic Agents:</u> The oral hypoglycemics were developed from the observation that sulfonamides had hypoglycemic activity.

<u>Amantadine:</u> The antiviral agent, amantadine, which is used as a prophylactic for Type A influenza, largely in elderly populations, was found to reduce tremors in Parkinsonian patients in these populations and was subsequently developed for treatment of Parkinsonism.

<u>DDT:</u> This classical insecticide was discovered from the observation that wool dyed with Mitin Green FF showed resistance to moths. A careful dissection of the dye's chemical structure showed that the insecticidal activity was not from the dye, itself, but from an impurity. Using the impurity's structure as a prototype, a search of related chemicals uncovered DDT (synthesized in 1874 by condensing chloral or chloral hydrate with chlorobenzene) which showed activity when screened. Fortunately, the screen was against insects rather than mites as DDT is ineffective against this group of pests because of its high lipophilicity. In contrast, the more polar kelthane (R = OH) controls mites but not insects.(<u>11</u>)

<u>Bordeaux Mixture:</u> This common product got its start with growers in the Bordeaux region of France spraying the outside rows of grapes with "awful blue-colored chemical ($CuSO_4$ and lime)" to discourage theft. Then it was observed that the treated vines did not get powdery mildew.

RECEPTOR MAPPING.

In one sense, this has been the approach in any broad, synthetic approach where a systematic replacement of substituents is done on a reference molecule. Assumptions are made that aliphatic moieties fit into hydrophobic pockets, anionic residues bind by an ionic bond to a positive charged group such as the ε-amino of lysine, cationic residues bind by a salt linkage to the anionic residues of aspartic or glutamic acids, alcohols and amides hydrogen bond to electron rich moieties such as oxygen, nitrogen and sulfur functions.

Today, receptor mapping refers to isolation of the receptor and chemically characterizing it. From a detailed knowledge of its chemical structure, attempts are made to develop three-dimensional pictures of what the drug molecules *sees* when approaching the receptor.(12) To be successful, it requires considerable work and carefully thought out assumptions. Isolation of the receptor is the first hurdle, but it is increasingly being accomplished using such techniques as affinity chromatography. Of course, it helps when the receptor is a free standing enzyme such as acetylcholinesterase or dihydrofolate reductase. Because most receptors are membrane bound, it is nearly impossible to obtain them in a crystalline form. This means that the shape must be calculated using such computational chemistry methods as molecular mechanics and molecular dynamics. This requires very powerful, high speed computer capabilities. Perhaps, more important, it requires that the parameters used for each atom be correct. Today, there is considerable research and discussion over those parameters.

Because of the difficulty in obtaining an accurate representation of the receptor, an alternate and productive approach is to determine the conformation of the ligand for the receptor. Because these molecules are simpler, it is easier to obtain a crystalline form or calculate the shape using molecular or quantum mechanics. Efficient computers and accurate parameters are still needed, but due to the years of results obtained from conformational analyses of small molecules, peptides and many proteins, it is possible to know when reasonable, as opposed to spurious, results are obtained. The results from this strategy complement the synthetic approach of mapping the receptor.

A good example is the development of the angiotensin converting enzyme (ACE) inhibitors successfully used in the treatment of hypertension. A careful study of the natural substrate and peptide inhibitors of this enzyme led first to captopril followed by enalapril (see above). The former is considered an analogue of a proline dipeptide and the latter a proline tripeptide analogue.

On the horizon is the exciting work with the opioid receptors for pain control. In one sense, the development of analgesics based on the morphine structure has to be considered a failure if the criteria of success is the separation of addiction liability from analgesia. Literally thousands

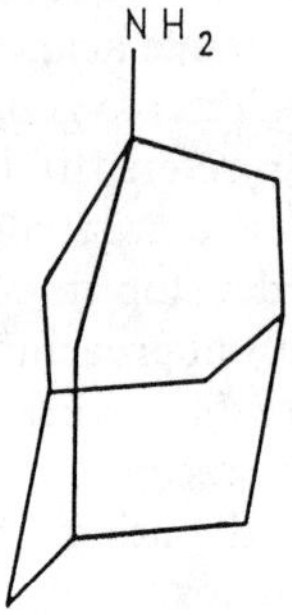

Amantadine

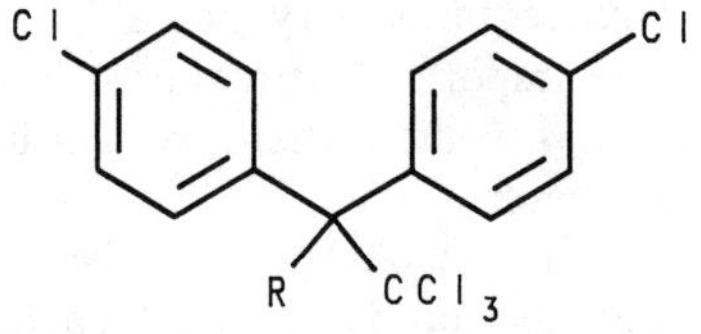

R = H: DDT;   R = OH: Kelthane

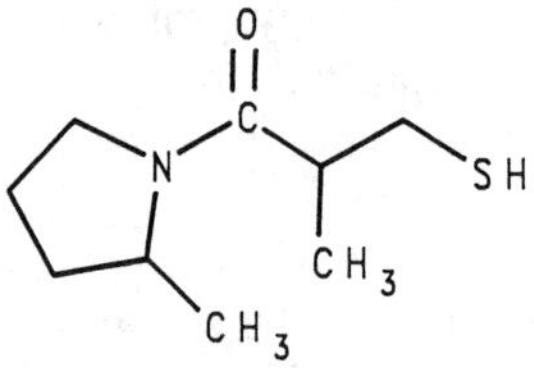

Captopril

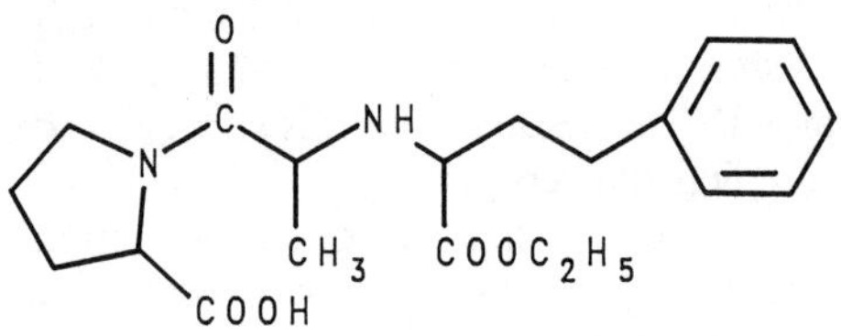

Enalapril

of compounds have been synthesized and tested and millions of dollars have been spent. Finally, in 1967, it was hypothesized that there was more than one type of receptor. Then in 1975 two endogenous pentapeptides with opioid activity were identified.(13) The emphasis has now shifted from refining the latest structure activity relationships (SAR) based on morphine and its congeners to determine the role of each of the opioid receptors on pain control and addiction liability coupled with the conformation of the natural ligands.

It is obvious that receptor mapping is computer intensive and has been limited by the availability of large scale computing and appropriate software. Both of these limitations are being met in today's research environment. First, the increasing power of the microcomputer now permits the calculation of the shape of the smaller molecules. The graphics work stations allow excellent visualization of both the receptor combined with ligands. Several software vendors have produced commercial programs that can carry out energy minimization routines. A relatively complete system for computer aided design of bioactive molecules will include software for statistical analysis, pattern recognition, molecular and quantum mechanics, data and structural file capabilities both for storage and interfacing with the chemical literature such as **Chemical Abstracts**, a high speed graphics terminal capable of three dimensional drawings which can be manipulated, and appropriate output devices.

For further current information and applications, the reader is referred to Part 2 of this book where a variety of topics is presented. These include the calculation of electrostatic potentials using quantum mechanics, modeling of molecular shape, similarity analysis and a method of molecular comparison by developing a hypothetical lattice to represent points on the molecule.

QUANTITATIVE STRUCTURE ACTIVITY RELATIONSHIPS (QSAR).

QSAR studies tend to be either retrospective or provide an experimental design approach which permits obtaining maximum information from a minimum number of compounds. Retrospective studies usually are based on studies from previously published reports involving biological results of a series of compounds. The chapter in this book on nalidixic acid analogues is an example of such a study. QSAR, whether it is the linear free energy relationship model (LFER) developed by Hansch and his co-workers or the *do novo* model of Free and Wilson, provides a means for the synthetic medicinal chemist to develop a proposed list of compounds that will measure the role of lipophilicity, steric influence, and electronic parameters independent of each other or the influence of specific moieties at each position independent of any accidental combination of substituents.(14,15) A properly designed test set of compounds will provide maximum information from a relatively small number of compounds concerning what structural attributes are important determinants of activity. There are many compilations of such studies from pharmaceutical and agrochemical laboratories.(16,17,18) In addition there are several QSAR papers in this book divided by agrochemical mechanisms (Part 3) and drug mechanisms (Part 4). Included in the former are discussions on acetylcholinesterase inhibitors, fungicides and photosynthesis inhibitors. Topics from the section on drug mechanisms include aniline mustards, nalidixic acid analogues, and binding at the GABA receptor. What many people don't realize is that QSAR can be used to analyze the important structural components that elicit a toxic response. For examples the reader is referred to Part 5 Toxicity Mechanisms of this book.

GENERAL SCREENING.

It still must be stated that so little is known about the disease process, whether in humans, livestock, or plants, that general pharmacological screens commonly are used to discover new and novel biologically active compounds. This will continue to be used by the natural product chemists as they screen extracts looking for leads. It will be used looking for more effective, less toxic drug treatments for cancer. Finally, the large chemical companies have thousands of compounds *sitting the shelf* any one of which may turn out to be the next big discovery. Azidothymidine, first synthesized and evaluated in a cancer screen with negative results, was brought back out years later in a general screen for agents effective against the HIV-I virus. Only then was it realized that it is a reverse transcriptase inhibitor. The enzyme wasn't even known at the time this compound was first evaluated.

Why would a company use the *shotgun* approach rather than a carefully thought out hypothesis based on years of fundamental research on the disease process which would lead to the rationale design of new agents? General screening is significantly less expensive. In the 1960's DuPont screened its chemicals and discovered amantadine which, has already been noted, is effective as a prophylaxis against the Type A Influenza virus (found by screening) and reduces the Parkinsonian tremors (serendipity). Based on current knowledge, there is no rationale for a medicinal chemist to postulate that the amantadine structure would show any antiviral activity or be of benefit to Parkinsonian patients. Indeed, serendipity played a role in the discovery of its antiviral role as the drug is not effective against other viruses including the Type B influenza virus group.

At the same time, modern receptor isolation technology is claimed to make general screening faster. Instead of a pharmacological, bacterial or viral screen, the compound's binding properties to isolated receptors is evaluated. Note the following quote from the **Wall Street Journal**.

For more than a century, Kodak has collected chemicals related to its photography and other business. The current inventory numbers a half million compounds, stored in bottles on chemists' shelves or logged as formulas in computers.

Researchers are emboldened to do more screening because of advances in laboratory technique. The automation of a wide number of tests makes it easier to assess a chemical before proceeding with expensive and time-consuming trials in animals and people.(19)

SUMMARY.

The discovery of biologically active molecules occurs by a variety of means ranging from a carefully thought out research program starting with an understanding to the disease process for which the agents are being sought to just plain luck. The current research being conducted on treatment of the acquired immunodeficiency syndrome (AIDS) and elimination of human immunodeficiency virus (HIV-I) provides an excellent example of a multifaceted attack on a disease. First, there was the general screening which led to the discovery of the already synthesized azidothymidine. The history, function and structure of T lymphocytes are being examined carefully leading to the discovery of the CD4 receptor and cloning of its gene. The shape and chemical composition of the virus is being subjected to intense scrutiny. Its genes have been mapped and their protein products studied. It is exhilarating to see front page articles in the **Wall Street Journal** describing the race between companies as to which one published the first paper in a peer reviewed journal.(20) Indeed, these are exciting times. The successful companies and laboratories are those that have combined the best instrumentation and facilities with highly educated, imaginative scientists who work in an atmosphere where inquiry and risk taking is expected.

ACKNOWLEDGMENTS

Dr. James Witt, Professor of Agricultural Chemistry and Extension Specialist in Chemistry and Toxicology, Oregon State University, has provided much of the historical background information regarding biologically active agrochemicals.

LITERATURE CITED

1. Roberts, L. Science 1988, **241**, 1759.
2. *Cyclosporin Turns Five*, Science 1988, **242**, 198.

3. *Powerful Immunosuppressant Synthesized*, Chemical and Engineering News 1989, 67, 29.

4. Montgomery, S. Omni October 1988, 11(1), 42.

5. Sneader, W. Drug Discovery: The Evolution of Modern Medicines; John Wiley: New York, 1985, Chapter 9, 14, 16.

6. Green, M. B.; Hartley, G. S.; West, T. F. Chemicals for Crop Protection and Pest Control; Pergamon Press: New York, 1977; p 144.

7. Albert, A. Selective Toxicity; Chapman and Hall: New York, 1985.

8. Weinberg, E. D. In Principles of Medicinal Chemistry; Foye, W. O., Ed.; Lea and Febiger: Philadelphia, 1989; Chapter 35.

9. Bodor, N. and Kaminski, J. J. In Annual Reports in Medicinal Chemistry; Bailey, D. M., Ed.; Academic Press: New York, 1987; Vol. 22, p. 303.

10. Pieper, J. A.; Rodman, J. H. In Applied Pharmacokinetics, Principles of Therapeutic Drug Monitoring, 2nd Ed; Evans, W. E., Schentag, J. J., Jusko, W. J., Eds.; Applied Therapeutics: Spokane, 1986; Chapter 20.

11. Witt, J. M. In Chemistry, Biochemistry and Toxicology of Pesticides; Witt, J. M., Ed.; Oregon State University Extension Service: Corvallis, OR, 1988; pp. 7-15.

12. Hom, A. S. and DeRanter, C. J. X-ray Crystallography and Drug Action; Clarendon Press: Oxford, 1984.

13. Paterson, S. J.; Robson, L. E.; and Kosterlitz, H. W. in The Peptides; Udenfriend, S. and Meienhofer, J., Eds.; Academic Press: New York, 1984; Vol. 6 Opioid Peptides: Biology, Chemistry, and Genetics, Chapter 5.

14. Martin, Y. C. Quantitative Drug Design; Dekker: New York, 1978.

15. Franke, R. Theoretical Drug Design Methods; Elsevier: New York, 1984.

16. Topliss, J. G. Quantitative Structure-Activity Relationships of Drugs; Academic Press: New York, 1983.

17. Seydel, J. K. QSAR and Strategies in the Design of Bioactive Compounds; VCH: Deerfield Beach, FL, 1985.

18. Hadzi, D. and Jerman-Blazic, B. QSAR in Drug Design and Toxicology; Elsevier: New York, 1986.

19. Koenig, R. and Ansberry, C. The Wall Street Journal July 7, 1988, **119**(4), 26.

20. Chase, M. The Wall Street Journal March 3, 1989, **120**(43), A1.

RECEIVED May 23, 1989

Chapter 2

Predicting Mechanism and Activity

The Trendy Computational Soothsayer

Douglas R. Henry

Molecular Design Limited, 2132 Farallon Drive, San Leandro, CA 94577

>The techniques of computer-aided molecular design have been in use for over 25 years. During this period, trends in hardware, software, and methodology have become evident, and they strongly influence the direction of research into bioactive mechanism. Topics which are covered in this overview include the role of the computational chemist in the design of new pharmaceutical and agricultural agents, along with descriptions of current trends in the field. Particular emphasis is placed on minicomputer graphics workstations, 3D structural databases, and the merging of statistical and molecular modeling techniques to predict activity and mechanism.

The role of computational chemistry in the design and study of drug and agricultural chemicals has grown considerably since the first applications of computer-aided Structure-Activity Relationships (SAR) over a quarter-century ago (1,2). The initial motivation for the use of SAR was to empirically define and quantify the effects of steric, electronic, and lipophilic properties of substituents on biological activity. Not surprisingly, the retrospective success of early studies in predicting bioactivity attracted industrial attention, taking Computer-Aided Molecular Design (CAMD) out of the academic realm and into the pharmaceutical, agricultural, and polymer chemistry companies.

In industry, the motivation for the use of CAMD was and still is, largely commercial, with the hope that these techniques can lead to new and more effective agents with less synthesis and testing effort. It presently costs $50M to $100M and some five to ten years of testing to take a new drug or agricultural entity from its first conception to final marketing. As Figure 1 schematically shows, each stage in the development process is characterized by a relatively low yield of successful compounds. For each new compound that enters the market, over 10,000 may need to be synthesized and tested, at a cost of hundreds to thousands of dollars each. In 1988, about 20 new drugs appeared on the market (3), which translates to over 200,000 new compounds which did not. It might seem that reducing the effort at the lowest level of the process - design and synthesis - would give a directly proportional reduction in the total effort and cost required. In fact, if a safe and effective agent is obtained earlier in the process, a greater-than-proportional saving could in theory be obtained (Figure 2). For this reason, most major pharmaceutical

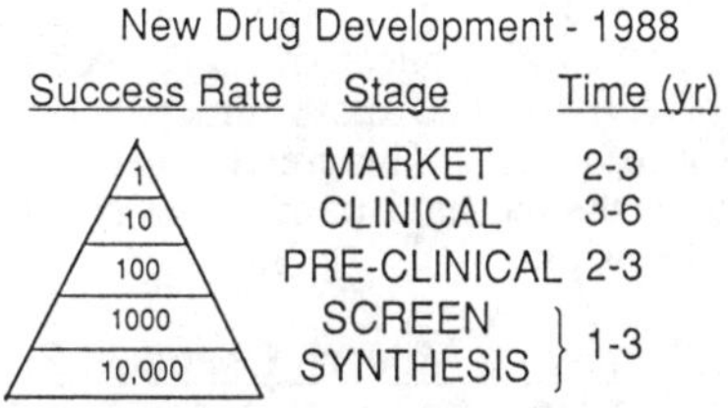

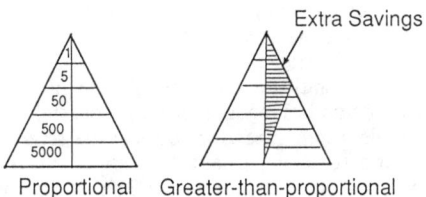

Figure 1. Likelihood of success at any given stage in the drug development process. A similar pyramid can be drawn for agricultural chemicals.

Figure 2. The potential benefit of applying CAMD techniques at the lowest level of development - design and synthesis of new agents - can be more than proportional if an effective agent is arrived at early in the process.

companies, and many agricultural and polymer chemical companies have hired scientists who specialize in the use of CAMD methods, which include SAR and molecular modeling.

In fairness to older, more traditional techniques of molecular design (serendipity, intuition, brute-force, etc.)(4), one should note that the success ratio mentioned above has not in fact improved over the years as CAMD methods have come into use. In light of increasing competition in the field, stricter government regulation, and the much larger number of compounds on the market, it is perhaps gratifying that the ratio has not gotten much worse.

Although the motivations for SAR research in academia and industry may differ, the techniques are largely the same. This is because the usual route for software development in the field is from academia to industry, with (increasingly) a commercial software vendor as a middleman. Over the years, CAMD methods have become much more sophisticated. At the same time, and largely because of commercial software, they can be much more easily and routinely applied. As a result, many published industrial applications have become less oriented towards the empirical prediction of activity, and more directed towards the study of receptor function and mechanism of action. This trend is certain to continue, since knowing the shape of a receptor or the mechanism of action of a particular compound makes it much easier to design new ones.

The Computational Chemist - a Modern Chemical Soothsayer. A central player in the design of new molecules is the computational chemist, who works as part of a team which may include synthetic chemists, biologists or pharmacologists, and perhaps a statistician. Until recently, a typical 'computational chemist' in industry was most likely a synthetic organic chemist who just happened to have an interest in computers (or perhaps was not particularly good in the lab). Although courses in theoretical chemistry and survey courses in molecular design have been around for years, academic programs which aim at training industrial chemists in the use of modern CAMD techniques have only recently become available (5). The popularity of such courses is likely to grow as ads for hiring computational chemists become a regular feature of chemical journals.

The role of a computational chemist may be likened to that of an ancient soothsayer. Whereas the soothsayer looked into a crystal ball to foretell the future, the computational chemist often gazes for hours at the screen of a graphics terminal, not to foretell the future (except perhaps in company profits), but to predict the structure or activity of a new compound. The soothsayer of old resorted to incantations and compendia of spells, while the computational chemist refers to the Command Language Reference Manual. The soothsayer invoked deities with strange-sounding names - the computational chemist calls upon MNDO, MM2, SIMCA, FRODO, and a host of other modeling and display programs. Finally, the soothsayer was often treated as an outcast, with an aura of mystique - which is exactly how some computational chemists are treated!

Trends in Computer-Aided Molecular Design

As one could expect over a 25-year period, several trends have surfaced in the CAMD field. Many of these were a result of the computer hardware that was available at the time, and CAMD studies have matured along with hardware and software developments. Early studies were primarily statistical in nature, using regression analysis, analysis of variance, and some multivariate techniques such as discriminant analysis and principal components analysis (6). The emphasis of these studies was on finding the highest correlations with biological activity, and finding the descriptors which were most significant statistically. Many of the early studies were run in batch mode, using in-house or commercial statistics packages, and graphical display of results and interaction with the data were minimal.

As CAMD matured, analyses became more interactive, more specific, and more structure and mechanism-oriented. The 1970's might be considered a period of interactive computing, as programs moved from the card-deck and batch-oriented mode of the 1960's to computer-terminal mode, using either conversational (question and answer) or command language operation. In the 1970's too, the first low-cost storage display graphics terminals appeared, allowing scientists to view static, monochrome structure and data representations.

The 1980's can be characterized as a period when graphical SAR (GSAR) came into prominence. Graphic displays have improved steadily, from 8-color 256x256 pixel low resolution displays on 8-bit microcomputers to modern 1280x1024 pixel million-color displays on minisupercomputer 3D graphics workstations. Faster computers have allowed researchers to work with larger molecules and to use more sophisticated calculations.

Accompanying the improvements in hardware, there have been corresponding shifts in the emphasis of CAMD studies. As scientists have been given the opportunity to view models of structures and receptors, there has been a trend away from what could be termed _extensive_ SAR studies. In these studies, knowledge about the receptor is weak or lacking, and the amount of information per-structure is small. Consequently, a relatively large number of compounds must be studied (10 to 100 or more), and statistical modeling of the data is typically used. This is the basis for Hansch and Free-Wilson analysis (7). In the area of drug design, recent studies have tended to be more _intensive_, utilizing more information about the structure of ligands and receptors, so fewer compounds need to be studied. The techniques of force-field molecular modeling and conformational analysis (8), semi-empirical quantum mechanics (9), molecular dynamics (10), and Monte Carlo simulation are typically used to find low-energy comformations of structures and receptor-ligand complexes.

The total information available to the scientist, in terms of chemical structure and biological activity, may be about the same in each type of study. The methods that are appropriate to use will depend very much on the amount of specific information that is available, either about ligand or receptor structure or about biological mechanism. Thus, statistical and data-analytic methods are most appropriate when specific information is lacking. Molecular graphics and structure-analytic methods are more appropriate when specific information is available. Recently, some excellent books have appeared, which discuss the present state of SAR and CAMD (11,12,13).

In a field that is changing so rapidly, it is difficult to make specific predictions about the direction of future developments (14). The remainder of this section will discuss in general some areas where trends in hardware, software, and methodology are having significant impact on CAMD studies. Hopefully, this will provide a useful perspective to scientists who are involved in the field, or who anticipate getting into CAMD.

Hardware Trends

The most significant advances over the last few years have been in computer hardware design and function. The earliest SAR studies were performed on mainframe computers with perhaps 64K to 512K of memory, and running at 1/10 the speed of todays desktop microcomputers. The last five years, especially, have seen the growth of powerful personal computers and minicomputer graphics workstations, which are transforming the way CAMD is studied.

3D Graphics Workstations. Several measures of performance have been defined for graphics workstations (15). These include raw processor speed (MIPS, or Million Instructions per Second - the number of assembly language instructions that the CPU can process per second), floating point processor speed (MFLOPS or Million FLoating point OPerations per Second - applicable to processors which have specialized hardware for doing noninteger arithmetic), vector display rate (v/s the number of vector line segments that can be drawn on the screen - which may depend on the length of the segment), and polygon display rate (p/s, the number of colored, shaded polygons that can be displayed per second). Each of these measures is somewhat subjective, and hardware vendors will always report the values obtained under the most optimum conditions. They nevertheless provide the main basis for comparing graphics workstation performance.

Three levels of graphics workstation have come into use over the last few years. The <u>low-end</u> systems typically have 2 to 5 MIPS of processor performance, numeric computing performance of 0.25 to 1 MFLOPS, and a drawing performance of 200,000 v/s and 2,000 to 5,000 p/s. Many of these systems were developed around the Motorola 68020 processor with a floating point processor. Since they are created from off-the-shelf components, they are relatively inexpensive ($10,000 to $30,000). They are adequate when used for small molecule (250 atoms or less) calculations and interactive display.

The <u>mid-range</u> workstations offer 10 to 15 MIPS performance, with numeric processing at 1 to 2 MFLOPS, drawing rates of 200,000 to 400,000 v/s and 20,000 p/s. Their higher performance arises from so-called RISC architecture (Reduced Instruction Set CPU), which allow the computer to perform fewer tasks per CPU instruction. They also utilize faster, proprietary graphics display processors and larger display memory, which allows more colors and multiple windows. These units cost between $30,000 and $70,000, and they probably make up the bulk of recent CAMD workstation purchases. They are suitable for solid model display and manipulation of small molecules, and wireframe and dot-surface display of macromolecules. Molecular mechanics and dynamics calcluations on small molecules and ensembles can be run in batch mode on these machines, and the results can be displayed and manipulated interactively.

At the <u>high-end</u> of graphics workstation technology are multiprocessor machines with peak processor performance of 40 to 80 MIPS, 40 to 60 MFLOPS, and display speeds of 400,000 to 600,000 v/s and 20,000 to 100,000 p/s. They use multiple, large, general purpose vector CPU processors, which are capable of performing rapid, parallel calculations on arrays of numbers (the processors in low-end workstations are scalar in design, operating on each element of an array in sequence). The high-end machines differ from each other in the degree to which they utilize the central processors to generate and manipulate the graphics image, as opposed to using separate graphics processors. They cost typically $70,000 to $100,000 or more, although recent competition has led to price reductions. These workstations are capable of the most sophisticated CAMD tasks, and are suitable for real-time molecular mechanics and dynamics calculations on small molecules, and batch processing of macromolecule calculations.

In the next few years, one can expect workstation technology to lead to faster, cheaper machines, just as microcomputer technology has done in the last few years. Most of the workstations on the market are UNIX-based RISC processor machines. This will continue, leading to a downward shift of each of the three categories listed above, by one level in the next 2 to 5 years. It is not unreasonable to expect machines which today cost $1000/MIPS to drop to $200/MIPS or less, in this period. Display enhancements, such as 3D liquid crystal shutters, anti-aliasing, shading and depth-cueing are already available for most of the workstations, making them extremely attractive for CAMD purposes.

The introduction of faster microcomputer chips, like the recently announced Intel 80860 superchip (a 64 bit processor, with 33 MIP, 80 MFLOP performance, capable of 50,000 p/s), will make microcomputers the new low-end graphics workstations of the 1990's. However, it will be several years before they can catch up to the RISC machines, which are optimized for graphics display purposes. Finally, the problem of hardcopy output from 3D graphics display is still an unsolved one. The recent introduction of color laser printers will help some, as will emerging standards like color Postscript. Prices on these output devices will likely be initially high, so being able to obtain high-quality color hardcopy from graphics workstations will continue to lag somewhat behind the workstations themselves.

Optical Disc and Fiber Optic Technology. A brief mention of the potential influence of read/write compact optical discs (CD), and fiber optic communications on CAMD is in order (16). One of the major problems of modern computational chemistry research is the sheer volume of data that is generated. It is easily possible to generate tens of megabytes of numeric information when running a molecule simulation, either by molecular dynamics, Monte Carlo simulation, or ab-initio quantum mechanics calculations. It is not always necessary to search this data, as in database applications, so storage on a high-capacity, relatively slow storage medium is feasible. The emerging optical disc technology, which typically allows 600 Mb of information to be stored on a single disc, offers an attractive alternative to faster magnetic hard disk storage. Even if searching is necessary, as in retrieving a particular set of conformations, it is possible to use up to half the optical disc to index virtually every piece of information in the data portion of the disk. This allows very fast searching and retrieval. Current uses of optical discs are mainly for read-only access, but as writeable discs become more reliable and affordable, this will change. Use of optical discs in structure databases and in text and chemical information processing is certain to increase, as well.

A related, emerging technology involves the use of fiber optic cables to replace twisted wire and coaxial connectors between computers. Several standard protocols are in use today for network communication (RS232, TCP/IP, DECnet, NFS, etc.). Most systems utilize Ethernet cables and controllers, which can in theory transmit 10 Mbit/sec, but which in practice usually attain only about 1 Mbit performance. Fiber optic standards (such as the Los Alamos HFC, and FDDI) are available or under development, which will allow up to 100 Mbyte/sec transmission. The hardware for these is expensive now, but as it becomes cheaper, these should allow much faster data and program interchange, and make the terms 'distributed systems' and 'distributed database access' take on new meaning. Especially attractive are distributed computing systems such as Apollo Computer's Network Computer System. Here, parts of a single application can be distributed to run at the same time on several computers in a network, while still appearing to the workstation user as a single program running on his machine. This has obvious applications in CAMD and molecular graphics, where calculations on parts of a molecule or display can be carried out independently, if there is no interaction with other parts.

Software Trends

By and large, software development in CAMD has not kept pace with new hardware developments. Several reasons for this can be imagined. As mentioned previously, CAMD software and methodology usually originates in a university setting, and most often in a chemistry group. Chemists are usually not computer science professionals, and they rarely have the luxury of hiring, or even consulting computer scientists when developing software. The resulting programs are often specific in their purpose, and they may lack the polish and sophistication that is found in commercial software. All of this is understandable, given the financial, hardware, and software constraints of most computational chemistry groups. It does pose problems for industrial scientists, and even other academics who wish to use a given program or system. It is often necessary to invest considerable time and effort, to get a program running on a computer or graphics system that differs from the one used for development. Help files and documentation for computational chemistry programs are usually aimed at practiced users of the program, who are the least likely to read them!

As a result, there has been a trend over the last several years, for molecular modeling and SAR software to be marketed by commercial software vendors, who add value in the form of graphics support, a convenient user interface, documentation and customer support. Most commercial modeling programs had their origin in this manner. The trend is, if anything, increasing, prompted largely by the highly competitive and lucrative market for modeling software, and by the demands from university administrations for overhead funds and support money. This is unfortunate for academics who wish to use a program that has been commercialized, but cannot afford the 'academic discount' price, let alone the industrial price for many systems. An alternative solution, which has been available for some time in the case of MM2 (17), is to distribute the program to academic users through the Quantum Chemistry Program Exchange (QCPE, University of Indiana, Bloomington, IN 47405), at a modest charge.

There are numerous current trends in the software and methodology of CAMD. Two of the most important are 1) the growing use of 3D structural databases, and 2) the merging of statistical modeling and molecular modeling techniques in finding solutions to CAMD problems.

3D Structural Databases. The 'grandfather' of all 3D structural databases is, of course, the Cambridge Crystallographic Structure Database (CSD)(18). This serves as the repository for about 70,000 published X-ray crystal structures, and it is widely used in industry and academia. The macromolecular analog of CSD is the Brookhaven Protein Databank (PDB), which contains about 400 protein structures (19). Other 3D databases exist and are occasionally referenced in the literature (20). Some problems with CSD and PDB have prompted the development of alternative in-house and commercial systems. First, the data in CSD and PDB is public. Although data can be submitted and held before general users are allowed access, it will eventually become public. The addition of proprietary data to in-house copies of CSD and PDB is not a trivial matter. Secondly, these systems are nongraphical and batch-oriented in operation, which limits their usefulness for modeling chemists who are accustomed to graphical interaction with their structures and data. For CSD, each of these limitations is being corrected by software development, but this will take a couple more years to complete.

In-house alternatives to CSD have been used by drug and chemical companies for some time. In addition, commercial vendors of 2D structural databases have recently announced 3D enhancements. The THOR database system (Daylight Chemical Systems, Inc.) is a tree-structured chemical database which uses the linear SMILES notation to represent a chemical structure (21). Different conformations and orientations of a structure can be stored in THOR, along with data such as energy or partial charges on the atoms. THOR software provides substructure searching in the 2D domain. An add-on program, ALADDIN, provides 3D pharmacophore searching by creating geometric objects and constraints, using SMILES notation and a control language (22).

The most recent development in 3D structural databases is the 3D module of MACCS-II (Molecular Design Limited). Users can build 3D databases to parallel existing corporate databases, and store any model, atom, or atom-pair data. Search queries can be constructed graphically, and they can contain any combination of 2D substructures, 3D substructural fragments, geometric objects and constraints, and atom or atom-pair data constraints. Figure 3 shows a 3D query for the dopamine D2 receptor. This query consists of a 2D substructure, containing the phenol in one fragment, and the tertiary nitrogen in the other. The geometric objects which are defined include the plane of the ring, the centroid of the ring atoms, the normal to the ring, and a point on the normal. The geometric constraints which are defined consist of two angles, a dihedral angle, and the distance between the nitrogen and the ring centroid. Figure 4 shows the superposition of the 3D query on apomorphine, a known D2 agonist, obtained by searching a database of about 50,000 3D structures. The search time depends on the complexity of the 3D query, and on the composition of the database. This example required about one minute of CPU time on a DEC Vax 8820 computer. By varying the values in a geometric or data constraint, or by changing the tolerance for matching fixed atoms in the query, the user can selectively refine portions of the pharmacophore.

The most obvious application of 3D structural databases is for pharmacophore or toxicophore searching. More diverse applications include using the database as a source of structures to fit a known receptor (23), and using the database as a source of 3D fragments to rapidly build molecular models (24). An extremely important development in the generation of 3D structural databases is the emergence of programs for the rapid, approximate modeling of structures. These programs, which include Pearlman's CONCORD (Tripos Associates, St. Louis, MO 63144) and Dolata's WIZARD program (25), rely on rules and heuristics to build the acyclic portions of molecules, and use either template libraries or simplified force-field calculations to generate rings. These programs are typically 100 to 1000 times faster than ordinary force-field modeling programs, generating an average structure in one or two CPU seconds. The speed is of course offset by the approximateness of the model. In most cases, it is an acceptable tradeoff, allowing conversion of a typical corporate database of 100,000 structures in the course of a weekend. Some problems which still exist with these approaches include stereochemical centers at ring fusion atoms, specific nonbonded interactions in distant parts of the molecule, and disconnected fragments, most of which typically require force-field approaches for solution.

The Merging of Statistical and Molecular Modeling. As mentioned, the trend in CAMD methodology shifted during the early 1980's from statistical data-analytic solutions to graphical structure-analytic ones. Whenever appropriate, these are the methods of choice for studying SAR. Recently, papers have appeared which describe a merging of statistical modeling, in the form of either regression analysis or principal components modeling of data, with molecular modeling and molecular graphics analysis of structures.

In one review, Hansch and Klein described the application of an artificial intelligence approach, coupled with a database of QSAR equations and data, to aid in the physical description of a receptor site (26). Using heuristics derived from the QSAR, it was possible to rationalize features of the receptor using coefficients in the statistical models. In a unique application of Partial Least Squares (PLS) analysis, Cramer has described a method for COMparative molecular Field Analysis (COMFA) of 3D structures (27). This approach is based on the premise that early recognition of a ligand by a receptor is based primarily on the electrostatic field surrounding the ligand molecule, rather than its topological or even topographical structure. The field is computed over a 3D grid of points surrounding the molecule,

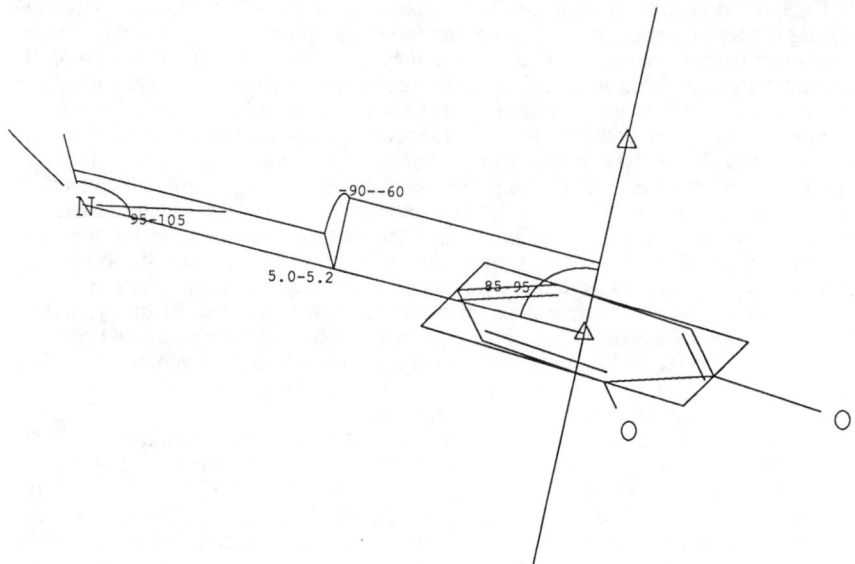

Figure 3. 3D substructure query for the dopamine D2 pharmacophore. The substructure has been rotated to show the geometric objects (plane, centroid, extra point) and constraints (two angles, a dihedral, and a distance). The search constraint values are shown.

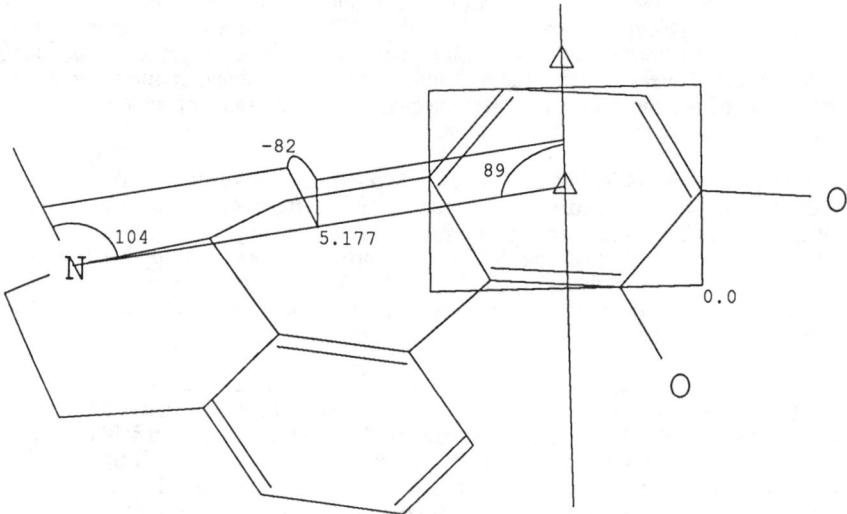

Figure 4. Apomorphine, with the D2 pharmacophore superimposed. Here, measured values for the molecule are shown.

and by comparing several structures, one can deduce the regions of space which favor activity, and those which do not. As with the Hansch and Klein work, the coefficients in a linear statistical model are used to deduce these regions.

The merging of statistical and molecular modeling techniques is a welcome trend. Over the years, two quite separate camps have emerged among CAMD researchers - the statistical modellers and the molecular modellers. As more examples of combined techniques appear in the literature, these separate groups should become more united again, and begin to appreciate the value of each other's approaches.

Summary and Conclusions. This overview has briefly covered some current trends in CAMD which are likely to have significant impact on the science in the near future. There are many areas which were not covered, such as calculations for genetic and protein engineering, including protein structure determination; distance geometry calculations on small and large structures; newer methods for quantum and molecular mechanics calculations; the role of supercomputers in CAMD; and the slowly growing importance of artificial intelligence techniques, including neural networks. By and large, these are for the experts. The issues which have been discussed are ones which will have impact on all scientists in the field. Though they seem mundane, hardware and database aspects of CAMD will have a profound and lasting influence on the tools and the techniques that computational chemists use to study structure and mechanism.

Literature Cited

1. Hansch, C; Muir, R. M.; Fujita, T.; Maloney, P. P.; Geiger, F.; Steich, M. J. Am. Chem. Soc. 1963, 85, 2817-2824.
2. Free, S. M.; Wilson, J. W. J. Med. Chem. 1964, 7, 395.
3. Hussar, D. A. Amer. Phar. 1989, NS29, 25-52.
4. Franke, R Theoretical Drug Design Methods; Elsevier: New York, 1984; p 12-14.
5. Bowen, J. P. Chemical Design Automation News 1988, 3, 4.
6. Blankley, J. In Quantitative Structure-Activity Relationships of Drugs; Topliss, J. G., Ed.; Academic: New York, 1983; Chapter 1.
7. Martin, Y. C. Quantitative Drug Design; Marcel Dekker: New York, 1978.
8. Kollman, P. Ann. Rev. Phys. Chem. 1987, 38, 303-316.
9. Schleyer, P. J. Computer-Aided Molec. Design 1988, 2, 223-224.
10. Wong, C.; McCammon, A. J. Israel J. Chem. 1986, 27, 211-215.
11. Martin, Y. C.; Kutter, E.; Austel, V. Eds. Modern Drug Research; Marcel Dekker: New York, 1989.
12. Fauchere, J. L. Ed. QSAR: Quantitative Structure-Activity Relationships in Drug Design; Alan R. Liss: New York, 1989.
13. Perun, T. J.; Propst, C. L. Eds. Computer-Aided Drug Design - Methods and Applications; Marcel Dekker: New York, 1989.
14. Unger, S. H. Drug Inf. J. 1987, 21, 267-275.
15. Kinnucan, P. Comput. Graph. Rev. January 1989, 20-32.
16. Borman, S. C. & E. News May 29, 1989, pp 22-25.
17. Sprague, J. T.; Tai, J. C.; Yuh, Young; Allinger, N. L. J. Comp. Chem. 1987, 8, 581-603.
18. Taylor, R.; Kennard, O. J. Chem. Inf. Comput. Sci. 1986, 26, 28-32.
19. Bernstein, F. C.; Koetzle, T.; Williams, G. J. B.; Meyer, E. F. Jr.; Brice, M. D.; Rodgers, J. R.;, Kennard, O.; Shimanouchi, T.; Tasumi, M. J. Mol. Biol. 1977, 112, 535-542.

20. Bergerhoff, G. In Crystallographic Computing 3: Data Collection, Structure Determination, Proteins, and Databases; Sheldrick, G.; Krueger, C.; Goddard, R. Eds.; Clarendon: Oxford, 1985; pp 85-95.
21. Martin, Y. C.; Danaher, E. B.; May, C. S.; Weininger, D. J. Computer-Aided Molec. Design 1988, 2, 15-29.
22. Martin, Y. C.; Danaher, E. B.; May, C. S.; Weininger, D.; Van Drie, J. H. In QSAR: Quantitative Structure-Activity Relationships in Drug Design; Fauchere, J. L. Ed.; Alan Liss: New York, 1989, pp 177-181.
23. DesJarlais, R. L.; Sheridan, R. P.; Seibel, G. L.; Dixon, J. S.; Kuntz, I. D.;, Venkataraghavan, R. J. Med. Chem. 1988, 31, 722.
24. Wipke, W. T.; Hahn, M. A. In Artificial Intelligence Applications in Chemistry; Pierce, T. H.; Hohne, B. A. Eds.; ACS Symposium Series No. 306; American Chemical Society: Washington, DC, 1986; pp 136-146.
25. Dolata, D. P.; Leach, A. R.; Prout, K. J. Computer-Aided Molec. Design 1987, 1, 73-85.
26. Hansch, C.; Klein, T. Acc. Chem. Res. 1986, 19, 392-400.
27. Cramer, R. D.; Patterson, D. E.; Bunce, J. D. J. Am. Chem. Soc. 1988, 110, 5959-5967.

RECEIVED August 2, 1989

Chapter 3

Interfacing Statistics, Quantum Chemistry, and Molecular Modeling

Philip S. Magee

BIOSAR Research Project, Vallejo, CA 94591 and
School of Medicine, University of California, San Francisco, CA 94143

> Quantum chemistry, classical modeling and statistical
> approaches come together in the special arena of
> specific binding events at the molecular level. Such
> problems are treatable by a variety of precise tech-
> niques that tend to complement each other in giving
> different views of the same picture. This review
> considers the overlap of techniques in some detail
> and presents a new complementary method best described
> as a statistical docking experiment. This method
> permits the partial mapping of the active site in
> terms of energetics and mechanism at each strongly
> binding atom. It has the unique advantage of re-
> quiring no three-dimensional knowledge or assumptions
> about the active site prior to analysis. Thus, a
> clear view of the binding event becomes possible
> even when no visual experiment can be performed.

One of our objectives in this book is to bring agricultural and medicinal sciences together over common ground. There is already considerable overlap in the areas of disease control where many compounds have good to excellent activity in both realms. But even in outlying areas as diverse as herbicidal weed control and cardiovascular agents, there is common mechanistic ground. Most agrochemicals and drugs share both physical and mechanistic problems in formulation, skin irritation, acute toxicity, metabolism, cuticle penetration, transport through living tissue, active-site binding and reactivity in irreversible inhibition. Identical techniques are used to resolve experimental data into predictive models and mechanistic insights in both fields.
 We view structure-activity relations (SAR) as a means to correlate experimental data and support mechanistic hypotheses. Mechanisms are proven, within reasonable doubt, by direct experimental observations with skilled inferences from the expressed data. SAR extends these observations by developing relations in terms of

mechanistic descriptors that point the way to extrapolated inferences and the design of new experiments. Experimentation provides the facts; SAR provides the network of inferred relations.

Methodology for SAR is diverse, perhaps as diverse as the investigators themselves. Insights from quantum chemistry (QC), classical modeling, and statistical inference all have the common goal of understanding features concealed by the molecular structure and raw experimental data. Many specialists believe that modern QC calculations and precise modeling techniques will replace the "old" statistical methods. This is unlikely as each approach has areas of unique strength, and where they overlap in treating the same problem, the output is independent and complimentary. The interrelation of statistics with classical and QC modeling is the subject at hand.

Let us now restrict the experimental field where the merging of techniques can be observed. Problems involving passive transport and distribution of chemicals remain bound to statistical interpretation despite some attempts to describe the key descriptor, logP (octanol/water), with QC data (1,2). Soil adsorption (3), fish toxicity (4-6) and percutaneous absorption (7) are examples of many problems best handled through partition coefficients. Charton has clearly demonstrated the composite nature of logP in amino acid studies (8), in clear support of its complex constitutional nature. As all live plant and animal studies incorporate the distributive process, they are currently assailable only through statistical analysis of experimental data. These restrictions focus the area of modeling and statistical overlap clearly at the _in vitro_ stage of reversible and irreversible inhibition, areas where all techniques enter the arena.

One further set of restrictions concerns the statistical approach to SAR analysis. While all statistical approaches have merit in particular problems, the various forms of multiple regression analysis or principal component analysis coupled with regression to dissect the components are methods leading directly to mechanistic insight. Classification methods provide inherently weaker insights to mechanism as the goal of classification lies at a lower level than direct prediction of activity. Finally, correlations with descriptors having no clear basis for mechanistic insight, however strong the correlation, must be excluded from this discussion. Examples are the popular connectivity indices of Kier and Hall (9,10) and the very interesting shape descriptors recently developed by Kier (11).

Insights from Energy-Minimized Structures

Both classical methods based on molecular force fields (12,13) and QC methods (MNDO, MINDO/3, CNDO/2, PCILO) as contained in CHEMLAB II (14) and MOPAC (15) are capable of minimizing the potential energy of a molecular structure. Both are capable of thermodynamic description and dipole moment calculation but only the QC methods can estimate atomic sigma charges, HOMO and LUMO energy levels and a wealth of other electronic descriptors having potential value in SAR (16,17). Both are capable of mapping the potential energy response to group rotations or molecular distortion.

The concept of a receptor-inhibitor recognition point (18) is based on the idea that weak long-range interactions are sufficient to initiate reception. At this point, the receptor "recognizes" matching key features in the drug and begins the orientation. Right or not so right, this idea has stimulated an enormous volume of work on drug and pesticide ground states. Energy-minimization is now a highly refined art through a variety of available techniques. The value of this work is questionable in terms of mechanism as it seems unlikely that the receptor site "sees" anything before significant exchanges of ion-paring or Debye, Keesom and London forces promote both partners into activated states. However, these studies have allowed the measurement of known pharmacophores and many interatomic distances between supposed binding points of natural substrates, drugs and pesticides (19-23,31,36). These critical distances based on a working hypothesis of mechanism have been invaluable in the design of new bioactives and, in this sense, confirm the proposed mechanism. In working with new heterocyclic phosphates at Chevron Chemical, we achieved a 50% success rate by assuming the need for 5.2 Angstroms spacing between the phosphorus atom and a branched group on the heterocycle. The 5.2 Angstroms was based on our own model of acetylcholine in the extended form (24).

Ground-state modeling is also of value when the receptor site is undescribed but very active substrates are in hand. By modeling the active substrates and various experimental or conceptual candidates for activity, superpositioning for visual or least squares fit can be achieved with programs like COMPAR (25). Low energy rotations are permissible and the degree of fit in selecting experimental candidates is a judgement call, receptor fit being only one factor determining activity. Superposition has been used in this manner (26-42) and also in matching transition state analogs to TS models as illustrated by the rearrangement of chorismate to prephenate (43-46).

Where the receptor site is described, docking experiments with enzyme and DNA segments may be carried out computer-graphically (47-49). Moreover, receptor-inhibitor complexes can be modeled and studied as single entities (47,50-55). The availability of many important enzymes and enzyme-inhibitor complexes as 3D-models (X-Ray crystal structure) has greatly facilitated these studies (56). Minimized conformations are ideal standard states for assessing physical properties (shape, interatomic distances, dihedral angles, non-bonded interaction, etc.) and for comparing candidate inhibitors with known drugs and pesticides. It is clear that many planer and caged drugs (sulfonamides, barbiturates, morphines, etc.) and pesticides (triazines, diphenyl ethers, captans, etc.) must bind to receptors in or near their minimum energy conformations. That this is not universally true is explored in the next section.

Activation in the Bound State

Measured or inferred binding energies are considerable. A single hydrogen-bond of 4-8 Kcal/mol has more than sufficient energy to overcome the average axial-equatorial preference (0.2-2.6 Kcal/mol) or chair-boat inversion (5.5 Kcal/mol) (57). In principle then, a favorably placed H-bond can convert an energy-minimized chair-form into a twist-boat with energy left over. A combination of more ordinary forces can also exceed the conversion energy. In mathematical experiments at 3.5 Angstroms separation, McFarland found 990 cal/mol for two opposed carbonyls (Keesom), 230 cal/mol for a carbonyl group over a disulfide linkage (Debye), and 600 cal/mol for a bromine and sulfur atom (London)(58). In more recent studies of the binding energies of 200 drugs, Andrews and co-workers have evaluated functional group contributions to the process (59). Intrinsic binding energies in Kcal/mol run from 0.7-3.4 for neutral atoms and groups (C,N,O,S,halogen,C=O,OH) to 8.2-11.5 for charged groups (N^+, CO_2-, OPO_3-). Over 60% of the drugs studied had binding energies of more than 10 Kcal/mol, sufficient to distort any flexible compound and many proteins. A simple calculation shows the free energy of binding for good to excellent substrates ($K_d=10^{-5}$ to 10^{-10}) is 6.8-13.6 Kcal/mol at 25°C. For flexible molecules, binding research clearly requires a facility for the study of distorted conformations to effect an experimental approach to the minimum energy conformation of the receptor-inhibitor complex. An excellent example of this approach is provided in the conformational analysis of some flexible antidepressants by Andrews and co-workers (60). The overall mobility of two tricyclic antidepressants (imiprine, amitriptyline) was explored in a step-wise manner to describe a multitude of torsional conformations at and just above the ground state.

The Cambridge Crystallographic Data File (61) is often used as a source for valid molecular models to compare with experimental QC and classical MM2 models. Available on line through the NIH-EPA Chem Info System, this data base provides the largest single repository for 3D structures (ca. 60,000). How reliably are X-Ray crystal structures as minimum energy models likely to reflect solution behavior? It requires little thought to realize that all crystals exist in bound states of appreciable energetics. The crystal lattice energy is basically intermolecular binding energy with a direct relation to the crystal melting point. The melting point, in fact, is a valid descriptor in solubility correlations involving solids of varying crystal stabilities (62). Typical heats of fusion for simple aromatics with melting points from 110-189°C (resorcinol, diphenyl, p-aminobenzoic acid) are near 5 Kcal/mol. This level of binding energy is clearly responsible for the dihedral variance of diphenyl (45° by electron diffraction [63]; coplanar by X-Ray crystal [64]). In large molecules of moderate melting point (tristearin, m.p. 73°C), the heat of fusion exceeds 40 Kcal/mol. At the same specific heat of fusion (45.63 cal/g), triacetin (m.p. 4.1°C) has slightly less than 10 Kcal/mol of crystal lattice energy. The remaining 30 Kcal/mol is the cumulative effect of London forces in the stearyl chains, which can reach 1 Kcal/CH_2(s → 1) in tightly packed chains (154). Statements that crystal conformations are usually within 2-3 Kcal/mol of the lowest energy (65), are simply not correct. Moreover, the remarkable torsional flexibility of alicyclic drugs like morphine at minimum ±5 Kcal/mol (65, Fig. 8), cloud the usefulness of X-Ray structures. While X-Ray structures often fall within the range of potential minima for multiring structures (32, Fig. 7), measurements on open-chain structures can be grossly misleading. As one example, mescaline hydrochloride has an extended conformation (66), while the hydrobromide has a gauche conformation (67). In summary, the use of X-Ray crystal structures as models needs to be qualified by an understanding of the lattice energetics and its probable consequences.

Ideally, detailed studies of receptor complexes can be made by classical and QC modeling approaches (47,50-55), though this information is rarely available. Studies based on the active analog approach by classical or QC modeling (30,68) can be powerfully enhanced by concurrent studies of molecular electrostatic potentials (70-74). Reference 74, though short, provides a particularly clear introduction to MEP's for the non-expert. Static in nature, these MEP maps provide critical insight to the approach of polarizing reactants, especially electrophiles, providing credibility to electrostatic recognition of the pharmacophore. MEP's are both calculable and experimentally accessible. Usually presented as 2-D contour maps with a Kcal/mol energy scale, it is now possible to generate MEP's on a van der Waals molecular surface, enhanced by color graphics coding (75). Electrostatic potentials have been calculated for cholinergic agents (76-78), dopamine and 5-HT agonists/antagonists (79-81), MAO substrates (82), DHFR ligands (83) and insect juvenile hormone mimics (84) as a sampling of examples. Reference 74 describes MEP studies of nucleic acid bases, 5-HT receptor agents, beta-adrenergic blocking agents and some examples from chemical carcinogenesis. MEP's are coming into frequent use, providing more reliable predictions than point charges (74).

From a mechanistic point of view, the major weakness of MEP's is the inability to interpret the dynamic events that occur on close approach (polarization, charge transfer, steric repulsion). Other methods based on mathematical modeling are needed to understand the actual binding process at the docking stage.

The Need For Mathematical Modeling

Most receptor sites in medicine and agrochemistry are undescribed. The availability of the Brookhaven file (56) and studies of receptor-inhibitor complexes (47,50-55) has been helpful in specific studies, but does not reflect the common situation. All major drug and pesticide companies screen thousands of experimental compounds each year against 30-50 or more complex targets, most of which are undescribed. As one example, topical or microspray application of experimental insecticides to the common housefly (Musca domestica) addresses four acetylcolinesterase isozymes in the head and three others from the thorax (85). Each, on isolation, responds differently to standard organophosphates as measured by K_m and V_{max}. None have been sequenced and defined by X-Ray crystallography. The output of drug and pesticide screening is basically serial dilution versus biological response in terms of % kill/control or some form of rated score. For analysis of bioresponse in live organisms, representing the majority of drug and pesticide problems, the only successful approach has been some form of mathematical modeling based on multivariate statistics. Those who believe that statistical analysis of response data is archaic and soon to be replaced by molecular modeling and computer graphics are unaware of the nature of the problem.

Modeling and computer graphics do intersect with statistical research at the in vitro level of bioresponse in a most interesting and synergistic way. Let us explore this critical region of overlap.

Many valid approaches have been taken in multivariate modeling from classification techniques (86) to mathematical models containing a linear combination of weighted variables (87-89). In terms of mechanistic insight, the most successful analyses have been based on some form of multiple regression analysis, either directly or through pre-processing by principal component analysis (90). Although a well-established technique, factor analysis (91) does not address mechanism directly as the orthogonal factors are of unknown composition. While many variations of regression exist, our discussion is limited to ordinary and multiple regression analysis in order to simplify the treatment.

Measurement of binding constants (K_d) or 50% inhibition concentrations (I50) for in vitro events usually provides data of far higher precision than that of in vivo studies. When carried out on a series of structurally related compounds, the data set ($log1/K_d$ or pI50) can be analyzed in terms of a linear combination of relevant descriptors to provide a mathematical model of the event. The number of descriptors the model will support depends partly on the size and data span of the set and partly on the strength of statistical measures (r^2, F, T's) associated with the analysis. Most important is the selection of descriptors that relate directly to the microscopic steps of the event. These mechanistic descriptors fall in two major classes: 1. Derived by physicochemical measurement, including fragment-based descriptors, and 2. Estimated by quantum chemical procedures. The two descriptor sets are fundamentally different and require separate discussion.

Descriptors used in the classical Hansch approach are selected to model transport (1opP, Pi), active-site binding (logP, Pi, MR, dipole moment) and reactivity in irreversible inhibition (electronic and steric descriptors). These are supplemented by indicator variables to represent hydrogen-bonding by some substituents (HB) (92) and a variety of other discontinuous structural features within the data set (I[X]=1.0 or 0.0). Indicator variables correct for parallel behavior in the Free-Wilson sense when a unique characteristic provides a constant increment or decrement to the bioresponse. Unfortunately, the selection of electronic and steric descriptors is quite large and there is wide disagreement on the best set for analysis. Moreover, there is widespread misunderstanding of the distinction between bulk descriptors (MW, MR, V_w) which model London forces (93) and steric descriptors (E_s, υ, L, B1, B5) which measure projection in terms of van der Waals radii (94). To add to the general confusion, there are now over 40 electronic (sigma-type) descriptors to choose from (95). This variety of selection has led to statistical concerns for the probability of accidental correlation when too many descriptors are tested (96). The situation is less serious in actual practice by the best investigators. They differ in selection of the fundamental set of descriptors (most alternatives are highly colinear), but they are internally consistent in their own version of the linear model for transport, binding and reactivity. In my opinion, Charton's intermolecular force equation (IMF) is the best model covering all physicochemical and physicobiochemical events (97), but it is not in general use. Hansch (98), Fujita (99) and Verloop (100) all use internally consistent variations in their own research. By any consistent approach, accidental correlations are of little concern in the analysis of statistically large (n>30) sets of well measured binding data. Even smaller sets can reliably extract the major mechanistic components provided overdescription is not attempted (less than 4 data points/variable).

Understanding the nature of the experimental descriptors is important in their use to probe mechanism. All except Bondi's Volume and Surface Area (V_w, A_w) (101) and Verloop's Sterimol decriptors (B1, B5, L) (100) are based on experimental measurements in solution as rate or equilibrium constants. Thus LogP and the derived Pi values for substituents are determined by a shake-flask method (102), Hammett's sigma is based on benzoic acid ionization (103) and Taft's steric constants (E_s) on the kinetics of acid-catalyzed ester hydrolysis (104). Each of the other available descriptors are similarly based on measurement of a simple, standard event that we presume to understand so well that most mechanistic questions are answered. In using these descriptors, we correlate more complex events with simpler standard events. We "understand" what a positive or negative coefficient of a sigma constant "means" in terms of electron flow at a complex reactive center. Thus we correlate by analogy from well-understood model systems to complex problems and phrase our conclusions in mechanistic terms. That these conclusions and the SAR equations are probabilistic in nature supports a level of understanding more consistent with a working hypothesis than with proof. However, some correlation phenomena such as logP correlations in transport and non-specific binding, have been seen so many times that proof is virtually assured. One final point about experimental or inferential descriptors is the associated error of measurement. This violates the mathematical dictum of no error in the independent variables (105). Enforcing this dictum strictly would bring the Hansch approach to an end. Fortunately, the viewpoint held by everyone in the field simply requires the descriptor error to be substantially less than that of the test set. This seems to be an adequate concept as SAR analysis of well-measured enzyme reactions commonly show very high correlations (r=0.95-0.98) (106-108).

Quantum chemically-derived descriptors are fundamentally different from those experimentally measured, though there is some natural overlap. For example, dipole moments can be measured in solution (109) and also estimated from the sigma charges and conformations calculated by various QC programs. The numbers typically differ and are sensitive to conformation, as only the experimental approach addresses solvation-minimized conformations (110). In addition, sigma charges commonly described by QC calculations can now be estimated by a classical iterative approach based on orbital electronegativities (Gasteiger charges) (111,112). We have found these charges to be highly colinear with those calculated by extended Huckel and MOPAC on a 17-atom bicyclic containing F, Cl, Br and two types each of C, N, O, S (113). Unlike experimental descriptors,

there is no statistical error in the QC calculations. There is error
in the absolute sense of necessary assumptions to facilitate the calculations. In most cases the direction of error is known but the
degree is not. An exception is the estimation of known thermochemi-
cal quantities (121-123). In using QC based descriptors with a
series of related compounds, the calculation error is considered to
be approximately constant throughout. Common QC descriptors are
listed in Table 5.5 by Franke (16); a more extensive listing by
Kikuchi (114). The number of descriptors listed by Kikuchi (n=54)
pose a clear problem of accidental correlation if used without reference to a mechanistic model (96). One weakness of QC descriptors is
the failure to directly address steric and bulk effects.

In principle, QC descriptors and derived quantities like dipole
moments should address mechanism at the primary level through direct
understanding of the correlating factors. Time will tell if QC descriptors will match or surpass the success of experimental descriptors based on model systems. That they are in current use in statistical correlations with valuable results is beyond question. Lien
and co-workers have shown the utility of dipole moments as parameters
in drug-receptor interactions and, while they used measured values,
the extension is obvious (115-119). In a recent study of a fungicide
series with eight local minima, the calculated dipole moments correlated with activity for only one minima, thereby identifying the
crucial binding conformation. In the same study, activity correlated
with Gasteiger and MOPAC charges at only one carbon where we suspec-
ted a single-electron transfer (120). All other conformations and
charged positions gave random results. In terms of simple direct
correlations of physico-chemical and biochemical data, there are
many documented examples (124-129). More complex relations supported
by multiple regression analysis are considered in the next section.

The mathematical model based on mechanism-related descriptors
from classical and QC sources directly addresses the energetics of
the binding process, not recognition or close approach, but actual
contact. The energetics of the event are encoded in the bioresponse
along with experimental error. No other technique sorts random
error from fact as clearly as multivariate analysis. Direct access
to the microscopic event is the "raison d'etre" of statistics that
provides it with a clear path to the future of mechanism. New techniques can provide powerful synergism, but cannot supercede it as
long as physico-chemical and bioresponse data are generated.

Binding as a Soft Reaction

Binding can be viewed as a low-energy reaction in which translation
is lost, rotations are frozen, an activated conformation established,
weak bonds are formed and a variety of forces are exchanged (London,
Keesom, Debye, CT) (130). There are six possible forces among ions,
dipoles and induced dipoles. Single electron charge transfer,
hydrogen bonding and steric repulsion provide three more. Overlay-
ing these forces on the complexity or organic structural variation
results in a finite, but enormous variety of reversible binding
events. A subsequent hard reaction leading to irreversible inhibition adds further levels of complexity. Thus, we rarely see the

"same" binding mechanism twice. Each binding event is unique and, without mechanistic insight, we would have a neverending story. However, the individual binding forces can each be addressed by correlation analysis and the near-infinite complexity is reducible to relatively simple linear combinations in the regression model. We now consider four examples each of binding events successfully modeled with experimental and quantum chemical descriptors. The modeling technique is independent of the descriptor type and provides a common ground for merging two very different approaches.

Binding Events Modeled with Physico-Chemical Descriptors. The binding of a diverse set of pyridines to silica gel during thin-layer chromatography was recently modeled with exceptional clarity (131). A negative coefficient for $\Sigma\pi$ means that only polar substituents (-Pi) increase R_M while lipophilic groups are repelled, logical as there are no lipophilic regions on silica. The negative coefficient of $\Sigma\sigma$ indicates nucleophilic binding through the nitrogen electron-pair (donating groups assist). This is supported by the negative

$R_M = \text{Log}(1/R_F-1)$ - proportional to binding energy
$R_M = -0.120 \Sigma\pi -0.539 \Sigma\sigma -0.184 \upsilon_{2,6} -0.027$
n = 25   r = 0.923   F = 40.26

*-95% confidence limits omitted to simplify examination of equations. All terms are highly significant.

coefficient for Charton's steric descriptor (positively scaled to van der Waals radii) (132). Thus, nucleophilic binding is partially blocked by large 2,6-substituents. As the major electrophilic centers on silica are the silicon atom, this equation points to a fairly clear working hypothesis, namely, formation of a weak N-Si covalent bond.

In the binding of meta-substituted N-methyl arylcarbamates to bovine erythrocyte AChE, it is necessary to factor substituents into electron donor and withdrawing groups and further, to describe those capable of hydrogen-bond formation. This becomes clear in the derived binding model (133). A change of mechanism from acid-catalyzed carbonyl protonation (negative rho) to direct serine attack (positve rho) occurs for the highly electropositive groups (NO_2, CN, etc.). Also clear is the bioresponse enhancement by those groups able to form a hydrogen-bond with the meta-position receptor site (HB = 1.0 for OR, NO_2, CN, NMe_2).

$\text{Log}(1/K_d) = 1.515 \pi + 0.827 \sigma° \text{ (+ groups)} - 2.393 \sigma° \text{ (-groups)}$
$+ 1.344\text{HB} + 0.025$
n = 21   r = 0.984   s = 0.179

In a related study, ortho-substituted N-methyl arylcarbamates were assayed for binding to AChE derived from suseptible (S) and resistant (R) green rice leafhoppers (135). The same series of compounds were explored on each enzyme using identical assay methods.

The results are striking in showing a much larger favorable steric effect for the susceptible insect and a different rate-determining electronic effect. The susceptible enzyme is sensitive to the resonance component of sigma (negative rho), while the resistant enzyme responds strongly to the inductive component (positive rho). Thus, both spatial changes and the timing of the carbamate cleavage by serine oxyl have changed in the genetic modification.

$$pI50(S) = 0.34 \pi + 1.65 \upsilon - 0.74 \sigma_R + 3.74$$
$$n = 19 \quad r = 0.900 \quad F = 21.62$$
$$pI50(R) = 0.53 \pi + 0.71 \upsilon + 1.30 \sigma_I + 2.84$$
$$n = 20 \quad r = 0.876 \quad F = 17.67$$

Phenols inhibiting oxidative phosphorylation of ADP to ATP in yeast were studied by Dedeken (136) and reexamined by Hansch and co-workers (137). Their results showing dependence on lipophilicity and phenolic pKa were recast by Magee in equivalent sigma terms (138). It is clear that activity depends on optimal lipophilicity and the population of phenolate ion. Despite 2,6-disubstitution with ortho-groups as large as isopropyl, there is no steric effect and hence no direct reactivity with phenolate oxygen. A shuttle-type mechanism

$$Log(1/C) = 1.14 \Sigma\pi - 0.372(\Sigma\pi)^2 + 1.14 \sigma^- + 7.73$$
$$n = 14 \quad r = 0.968 \quad F = 49.0$$

in which acidic uncouplers work as protonophores across the inner membrane is supported by recent work (151). A later study of phenol activity against the fungus, Aspergillis niger, takes on a familiar form and leaves little doubt that uncoupling of oxidative phosphorylation is the mechanism of death. The coefficients of the _in vivo_ study are nearly identical to the above _in vitro_ model. Moreover, this set of phenols possess ortho-groups of majestic dimensions (t-butyl, phenyl, cyclohexyl) with no detectable steric effect (138).

$$Log(1/C) = 1.39 \Sigma\pi - 0.205 (\Sigma\pi)^2 + 1.00 \sigma + 2.18$$
$$n = 18 \quad r = 0.978 \quad F = 104.0$$

Binding Events Modeled with Quantum Chemical Descriptors. Inhibition of the Hill reaction (spinach chloroplasts) correlates with the energy level of the highest occupied MO on amide nitrogen for a series of piperidinoacetanilides (139). Electron density on amide nitrogen also correlates well with E_{HOMO} ($r = 0.909$) and suggests the probability of charge-transfer interactions with the receptor site.

$$Log(1/C) = 20.69 E_{HOMO} - 10.57$$
$$n = 7 \quad r = 0.992 \quad s = 0.047$$

The binding of 7-substituted-5-hydroxytryptamine derivatives to an LSD receptor site correlates with the frontier electron density (f_1) and charge ($q_1\pi$) of the ring nitrogen (position 1), modified by the hydrophobicity of groups at ring position 7 (140). These results show a strong dipolar involvement of the indole nitrogen in the

$$\text{Log}(1/C) = 18.90 f_1 - 74.77 q_1\pi + 1.82 \pi (7) - 13.06$$
$$n = 15 \quad r = 0.962 \quad S = 0.288$$

binding process. This would be difficult to assess by physicochemical descriptors as the ring-NH is part of an invariant parent structure.

Molecular shape descriptors based on common overlap steric volume with a reference compound (S_o) are modified by residual charge density (Qc)(CNDO/2) on the SP^2 carbon and the lipophilicity of substituents in N-methyl oxime carbamates inhibiting electric eel AChE (141). The correlation indicates both optimum receptor fit and optimum imino-carbon charge. The latter is reasonable as charge

$$pI50 = 1.887 S_o - 0.0138 (S_o)^2 + 175.26 Qc - 567.49 (QC)^2 - 0.0865 \Sigma\pi - 67.90$$
$$n = 20 \quad r = 0.929 \quad s = 0.631$$

correlates with leaving-group efficiency in the carbamoylation step, with excessive efficiency leading to hydrolytic instability. Similar conclusions would be reached through the use of classical steric (υ, E_s) and electronic (σ) descriptors.

Protein binding and several other SAR's of phenols are correlated by summed electrophilic superdelocalizabilities [$\Sigma S_i(E)$] and the electron density of the hydroxyl oxygen atom divided by MW (q_o/MW) (142). Descriptor estimations are based on HMO calculations. The involvement of molecular electrophilic character with specific involvment of phenolic OH suggests general dipolar binding with probable H-bond formation. As in the previous case, the same conclusions would likely be reached with classical descriptors, though it is rare for quantum chemists to cross-validate in this way.

$$\text{Log}(1/C) = 0.148 \; \Sigma Si(E) - 84.249 q_o/MW + 4.176$$
$$n = 10 \quad r = 0.970 \quad s = 0.11$$

The approximate nature of QC calculations has led Sklenar and co-workers to suggest the use of generalized Rank correlation methods where exact values of data and descriptors are unimportant (143). The use of Ranked methods is an outstanding idea whenever data and/or descriptors lack precision (144,145). We have had exceptional successes in using Rank regression with herbicidal score data (146,147). In this method, bioresponse data and descriptors are uniformly ranked from 1 to n with actual or perceived ties having the same value. Very high, statistically strong correlations can be achieved with this non-parametric, distribution-free method.

A New Statistical Method of Mapping Binding Sites

There have been many attempts to map receptor sites by exploring structural variation and binding response (pK_d, pI50). The early, extensive work of Baker (148) provides a protocol for later approaches such as Distance Geometry (149) and HASL (The Hypothetical Active Site Lattice) (150). A completely different approach, best described as a statistical docking experiment, has been recently developed (113) and applied (147). This method, as outlined below, has the advantage of identifying each location and type of binding energy exchange. The results are fully complimentary to a visual docking experiment by computer graphics (47-49).

Factoring the binding process into different positional contributions is based on the hypermolecule approach as developed by Tipker and Verloop in their study of phytotoxic benzonitriles and nitrophenols (152). A hypermolecule is the lowest common structure that will accommodate every occupied position of a related series of compounds. By careful study of the expressed activities of each member (pK_d, pI50), it is possible to assign qualitative values to each position, namely, favorable, unfavorable or indifferent. For our purposes, this translates into binding, steric or ionic repulsion, and weakly involved, possibly not even touching. By using a special set of atomic descriptors to model London forces, lipophilicity, electronic and steric behavior, it is possible to identify the key positions and describe the nature of the energy exchange. Now, if binding is a whole molecule process and all binding positions are important as descriptors like LogP suggest, then the approach will fail in most cases unless n is very large. Experience, so far, suggests that a relatively few positions contribute most of the binding energy exchange. Correlations of the data matrix by multiple regression provide equations that are statistically comparable to the same data treated by a standard Hansch approach. The information, however, is much different.

Briefly, the method requires well-measured pK_d or pI50 data for a large set of related compounds (n>20, preferably >30). Each position in the hypermolecule is identified and described in one of two ways. If the position has no variety (all C, all S, etc.), a simple indicator variable (P6 = 1.0 if occupied, 0.0 if empty) is used. For an invariant position, a significant positive or negative coefficient is interpreted as binding or repulsion of unknown type. Where the occupied position has atomic variety, a more analytical approach is taken. Each position is described by actual lipophilicity (f), polarizability (MR), electronic nature (X), and steric size (r_v).

As the developed matrix is quite large, backward selection is used to generate a best model with final retesting of all discarded descriptors.

Currently, only saturated positions are analyzable as the lipophilic (f) and electronic descriptors (χ) lose meaning for delocalized structures. Electronegativity correlates nicely with sigma charges only in localized structures. Eventual extension of the method to delocalized structures will, of course, use sigma charges as descriptors. Two examples of the method follow.

Meta-Substituted N-Methyl Arylcarbamates vs. Housefly Head AChE.
Inhibitors of HFAChE have been studied by Metcalf and Fukuto over an
extended period of time, as summarized by Hansch and co-workers (153).
The 36 members of the meta-substituted carbamates fit nicely into an
11-position hypermolecule. Position 1 has exceptional variation,
allowing this position to be tested for a range of possible effects.
The other positions are all hydrocarbyl in nature and are analyzed
for non-specific energetic effects based on simple occupancy.

Data = pI50 against
housefly head AChE

meta-X = H, F, Cl, Br, I, OR, SR, NR_2, PR_2, SiR_3

Descriptors Tested   P1 - F, MR, X, R_v
                      P2-P11 - 1.0/0.0 (C, CH, CH_2, CH_3)

pI50 = -0.835 X(1) + 0.693P2 + 0.299P5 + 7.18

n = 36   r = 0.841   s = 0.390   F = 25.82

The derived model indicates important binding effects at only
three positions (P1, P2, P5), the other eight positions being too
weakly involved to register statistically. Position 1 correlates
with electronegativity. As Chi (X) is negatively correlated with
charge, pI50 is increased by electropositive atoms, indicating that
P1 is located on or near the anionic carboxylate site. Positions P2
and P5 also interact strongly, though by unknown forces. Despite
the simplicity of the model, the explained variance ($100r^2$ = 71%)
and other statistical measures (s,F) are comparable to a standard
Hansch treatment of the same data (153). The remaining 29% of
variance is distributed among 8 other positions and the experimental
error.

Ortho-Substituted N-Methyl Arylcarbamates vs. Housefly Head AChE.
Data from the same source is available for a comparative study of
the ortho-position (153). The hypermolecule is identical (11 positions) and the variety at position 1 is similar. As the carbamate
group is fixed to the esteratic site, it is not possible for the
ortho-substituents to bind in exactly the same region as the meta-
groups. This is clearly supported by the different behavior at P1

and a more complex distribution of effects at P3, P5, P7 and P9.
Position 1 no longer depends on partial charge but reflects a mix of
lipophilicity and polarizability. Position 2 is indifferent while
P3, P5 and P9 provide a band of uniform binding equidistant from P1.

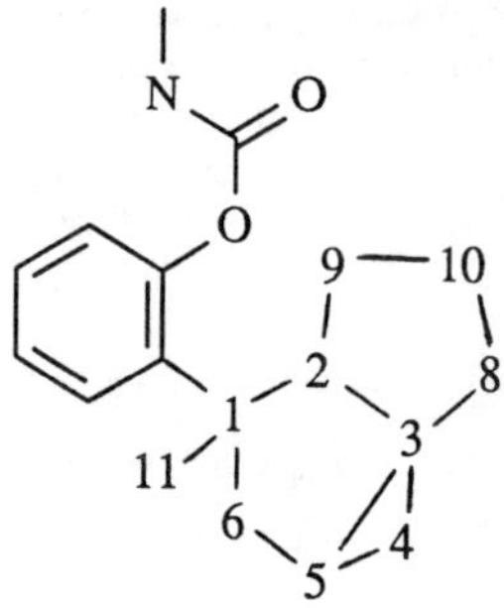

Data = pI50 against housefly head AChE

ortho-X = H, F, Cl, Br, I, Alkyl, OR, SR, NR_2, SiR_3

Descriptors Tested   P1 - F, MR, X, R_v

                            P2-P11 - 1.0/0.0 (C, CH, CH_2, CH_3)

pI50 = -0.547F(1) + 0.199MR(1) + 0.399P3 + 0.588P5 - 0.908P7

        + 0.512P9 - 4.33

n = 46   r = 0.829   s = 0.485   F = 14.24

Position 7 is no longer tolerated, having clearly located a strong
region of steric repulsion. Again, the statistical measures are
comparable to those derived by the standard analysis against physicochemical descriptors (153).

    Many other examples of this method are now in hand and in the
process of publication (113,147).

A Brief SAR Philosophy

Very little has been written about the interconnections of different
SAR approaches, probably because most of us feel hesitant to discuss
the less familiar areas. My own bias toward statistical research
and lack of expertise in modeling and quantum chemistry must be
clear to my colleagues in these fields of study. My purpose in
writing this overview was not to emphasize my bias toward statistics,
but to make an honest attempt to show relations in the different
roads to Rome. They may not all lead directly to Rome but they all
go to Italy.
    It needs to be clearly understood that statistical methods and
QC/modeling methods are basically non-competitive. Modeling and
quantum chemical approaches will not replace statistical analysis of

bioresponse data. More likely, they will continue to enhance it by providing important mechanistic descriptors for multivariate methods. The statistical docking experiments described in the previous section may finally provide the energetic response lacking in the visual analog. Of most importance, this technique provides a view of the binding event even when no visual experiment is possible (receptor site undescribed).

SAR has a brilliant future due to its capacity to build on itself by knowledge feedback. There are many different players in the field with personalized methods and mindsets. There will be many more and the variety of approaches will increase with time, as the need to tackle problems of increasing complexity will demand. However diverse the method, the objective remains the same for all, mechanistic insight. This alone guarantees than all approaches, however different, will have a common goal directed vector.

Literature Cited

1. Cammarata, A.; Rogers, K. S. J. Med. Chem. 1971, 14, 269; 1969, 12, 692. Rogers, K. S.; Cammarata, A. Biochim. Biophys. Acta, 1969, 193, 22.
2. Klopman, G.; Iroff, L. D. J. Comput. Chem. 1981, 2, 157.
3. Briggs, G. G. J. Agric. Food Chem. 1981, 29, 1050.
4. Lipnick, R. L. et al, Environ. Toxicol. Chem. 1985, 4, 281.
5. Call, D. J. et al, Environ. Toxicol. Chem. 1985, 4, 335.
6. Charton, M. Environ. Health Perspec. 1985, 61, 229.
7. Wester, R. C.; Maibach, H. I. In Percutaneous Absorption, Bronaugh, R. L.; Maibach, H. I., Eds., Marcel Dekker, New York/Basel, 1985, pp. 107-123.
8. Charton, M.; Charton, B. I. J. Theoret. Biol. 1982, 99, 629.
9. Kier, L. B.; Hall, L. H. Molecular Connectivity in Chemistry and Drug Research, Academic Press, New York, 1976.
10. Kier, L. B., Hall, L. H. Molecular Connectivity in Structure-Activity Analysis, Research Studies Press, Letchworth, Hertfordshire, England, 1986.
11. Kier, L. B. Quant. Struct.-Act. Relat. 1986, 5, 1,7; 1987, 6, 8.
12. Burkert, U.; Allinger, N. L. Molecular Mechanics, ACS Monograph 177, American Chemical Society, Washington, DC, 1982.
13. Ermer, O. In Structure and Bonding, Volume 27, Springer-Verlag, Berlin 1976, pp. 161-211.
14. Hopfinger, A. J.; Pearlstein, R. A.; Malhotra, D. Developed for Molecular Design Ltd.
15. Written by Stewart, J. P.; Seiler, F. J. (QCPE 455).
16. Franke, R. Theoretical Drug Design Methods, Akademie-Verlag, Berlin 1984, 116.
17. Kikuchi, O. Quant. Struct.-Act. Relat. 1987, 6, 179.
18. Reference 16, 316-322.
19. Borea, P. A.; Bertolasi, V.; Gilli, G. Arzneim.-Forsch. 1986, 36, 895.
20. Kier, L. B.; Holtje, H.-D. J. Theor. Biol. 1975, 49, 401.
21. Kier, L. B. J. Pharm. Sci. 1968, 57, 1188.
22. Kier, L. B. J. Pharm. Pharmac. 1969, 21, 93.
23. Kier, L. B. Mol. Pharmacol. 1968, 4, 70; 1967, 3, 487.
24. Magee, P. S. Unpublished Studies, Chevron Chemical Company 1980.

25. Available through Molecular Design Ltd., 2132 Farallon Drive, San Leandro, CA 94577.
26. McKinney, J. D. et al, Environ. Health Perspec. 1985, 61, 41.
27. Duax, W. L. et al, Environ. Health Perspec. 1985, 61, 111.
28. Weinstein, H. et al, Environ. Health Perspec. 1985, 61, 147.
29. Iwamura, H.; Nishimura, K.; Fujita, T. Environ. Health Perspec. 1985, 61, 307.
30. Marshall, G. R. et al, In Computer-Assisted Drug Design, Olson, E. C.; Christoffersen, R. E., Eds.; ACS Symposium Series 112, 1979, 205-226.
31. Humber, L. G. et al, In Computer-Assisted Drug Design, cited in 30; 227-241.
32. Rohrer, D. C. et al, In Computer-Assisted Drug Design, cited in 30; 259-279.
33. Nichols, D. E. In Dopamine Receptors, Kaiser, C.; Kebabian, J. W., Eds.; ACS Symposium Series 224, 1983, 201-218.
34. Olson, G. L.; Cheung, H.-C.; Chiang, E.; Berger, L. In Dopamine Receptors, cited in 33, 251-274.
35. Motoc, I.; Dammkoehler, R. A.; Mayer, D.; Labanowski, J. Quant. Struct.-Act. Relat. 1986, 5, 99.
36. Naruto, S.; Motoc, I.; Marshall, G. R. J. Med. Chem. 1985, 20, 529.
37. Haviv, F. et al, J. Med. Chem. 1987, 30, 254.
38. Schneider, C. S.; Mieran, J. J. Med. Chem. 1987, 30, 494.
39. Lyon, R. A.; Titeler, M.; McKenney, J. D.; Magee, P. S.; Glennon, R. A. J. Med. Chem. 1986, 29, 630.
40. DiMaio, J.; Bayly, C. I.; Villeneuve, G.; Michel, A. J. Med. Chem. 1986, 29, 1658.
41. Kocjan, D.; Hodoscek, M.; Hadzi, D. J. Med. Chem. 1986, 29, 1418.
42. Andrews, P. R.; Craik, D. J.; Munro, S. L. Quant. Struct.-Act. Relat. 1987, 6, 97.
43. Andrews, P. R.; Smith, G. D.; Young, I. G. Biochem. 1973, 12, 3492.
44. Andrews, P. R.; Cain, E. N.; Rizzardo, E.; Smith, G. D. Biochem. 1977, 16, 4848.
45. Andrews, P. R.; Haddon, R. C. Aust. J. Chem. 1979, 32, 1921.
46. Andrews, P. R. In Computer-Assisted Drug Design, cited in 30, 149-159.
47. Cole, G. M.; Meyer, E. F. Jr.; Swanson, S. M.; White, W. G. In Computer-Assisted Drug Design, cited in 30, 189-204.
48. Roth, B. et al, J. Med. Chem. 1987, 30, 348.
49. DesJarlais, R. L.; Sheridan, R. P.; Dixon, J. S.; Kuntz, I. D.; Venkataraghavan, R. J. Med. Chem. 1986, 29, 2149.
50. Remers, W. A.; Mabilia, M.; Hopfinger, A. J. J. Med. Chem. 1986, 29, 2492.
51. Andrews, P. R.; Iskander, M. N.; Issa, J.; Reiss, J. A. Quant. Struct.-Act. Relat. 1988, 7, 1.
52. Labanowski, J.; Motoc, I.; Naylor, C. B.; Mayer, D.; Dammkoehler, R. A. Quant. Struct.-Act. Relat. 1986, 5, 138.
53. Fujita, T. Quant. Struct.-Act. Relat. 1987, 6, 54.
54. Hansch, C.; Langridge, R. et al, Quant. Struct.-Act. Relat. 1982, 1, 1.

55. Cody, V.; Zakrzewski, S. F. J. Med. Chem. 1982, 25, 427.
56. Bernstein, F. C. et al, J. Mol. Biol. 1977, 112, 535.
57. Eliel, E. L. Stereochemistry of Carbon Compounds, McGraw-Hill, New York, 1962, 206, 236.
58. McFarland, J. W. Prog. Drug Res. 1971, 15, 123.
59. Andrews, P. R.; Craik, D. J.; Martin, J. L. J. Med. Chem. 1984, 27, 1648.
60. Munro, S. L.; Craik, D. J.; Andrews, P. R. Quant. Struct.-Act. Relat. 1987, 6, 104.
61. Wilson, S. R.; Huffman, J. C. J. Org. Chem. 1980, 45, 560; Bergerhoff, G.; Hundt, R.; Sievers, R.; Brown, I. D. J. Chem. Inf. Comput. Sci., 1983, 23, 66.
62. Yalkowski, S. H.; Valvani, S. C. J. Pharm. Sci. 1980, 69, 912.
63. Karle, I. L.; Brockway, L.O. J. Am. Chem. Soc. 1944, 66, 1974.
64. Dhar, J. Indian J. Phys. 1932, 7, 43.
65. Duchamp, D. J. In Computer-Assisted Drug Design, cited in 30, 87, 95.
66. Tsoucaris, D. et al, Cryst. Struct. Commun. 1973, 2, 193.
67. Ernst, S. R.; Cagle, F. W. Jr. Acta Cryst. 1973, B29, 1543.
68. Grunewald, G. L.; Creese, M. W.; Walters, D. E. In Computer-Assisted Drug Design, cited in 30, 439-487.
69. Small, D. M. Pure Appl. Chem. 1981, 53, 2095.
70. Scrocco, E.; Tomasi, J. Top. Curr. Chem. 1973, 42, 95.
71. Scrocco, E.; Tomasi, J. In Advances in Quantum Chemistry, Vol. 11; Lowdin, Per-Olov, Ed., Academic Press, 1978, 155-193.
72. Reference 16, 335-340.
73. Politzer P.; Truhlar, D. G., Eds.; Chemical Applications of Atomic and Molecular Electrostatic Potentials, Plenum Publishing Corporation, New York, 1981.
74. Politzer, P.; Laurence, P. L.; Jayasuriya, K. Environ. Health Perspec. 1985, 61, 191.
75. Weiner, P. K.; Langridge, R.; Blaney, J. M.; Schaefer, R.; Kollman, P. A. Proc. Natl. Acad. Sci. 1982, 79, 3754.
76. Weinstein, H. et al, Mol. Pharmacol. 1973, 9, 820; 1975, 11, 671.
77. Froimowitz, M.; Gans, P. J. J. Am. Chem. Soc. 1972, 94, 8021.
78. Radna, R. J.; Beveridge, D. L.; Bender, A. L. J. Am. Chem. Soc. 1973, 95, 3831.
79. Liebman, M. N. J. Mol. Graphics 1986, 4, 61.
80. Baldwin, S.; Kier, L. B.; Shillady, D. Mol. Pharmacol. 1980, 18, 455.
81. Pepe, G.; Reboul, J.-P.; Cristau, B.; Oddon, Y. Eur. J. Med. Chem. 1986, 21, 339.
82. Sanz, F.; Martin, M.; Lapena, F.; Manaut, F. Quant. Struct.-Act. Relat. 1986, 5, 54.
83. Andrews, P. R.; Sadek, M.; Spark, M. J.; Winkler, D. A. J. Med. Chem. 1986, 29, 698.
84. Nakayama, A.; Richards, W. G. Quant. Struct.-Act. Relat. 1987, 6, 153.
85. Tripathi, R. K.; O'Brien, R. D. Pest. Biochem. Physiol. 1973, 2, 418.
86. Stuper, A. J.; Brugger, W. E.; Jurs, P. C. Computer-Assisted Studies of Chemical Structure and Biological Function, John Wiley & Sons, New York, 1979.

87. Van Valkenberg, W. Biological Correlations-The Hansch Approach, Advances in Chemistry Series 114, American Chemical Society, Washington, DC, 1972.
88. Martin, Y. C. Quantitative Drug Design, Marcel Dekker, New York, 1978, Chapters 2, 4, 7, 9, 12.
89. Quantitative Structure-Activity Relationships of Drugs, Topliss, J. G., Ed.; Academic Press, New York, 1983.
90. Reference 16, Chapter 9.
91. Malinowski, E. R.; Howery, D. G. Factor Analysis in Chemistry, John Wiley & Sons, New York, 1980.
92. Fujita, T.; Nishioka, T.; Nakajima, M. J. Med. Chem. 1977, $\underline{20}$, 1071.
93. Charton, M. In Topics in Current Chemistry, Vol. 114; Charton, M.; Motoc, I. Eds.; Springer-Verlag, Berlin, 1983, 107-118.
94. Charton, M. J. Am. Chem. Soc. 1969, $\underline{91}$, 615.
95. Swain, C. G.; Lupton, E. C. Jr. J. Am. Chem. Soc. 1968, $\underline{90}$, 4328.
96. Topliss, J. G.; Edwards, R. P. In Computer-Assisted Drug Design, cited in 30, 131-145.
97. Charton, M.; Charton, B. I. J. Theoret. Biol. 1982, $\underline{99}$, 629.
98. Hansch, C. et al, J. Med. Chem. 1973, $\underline{16}$, 1207.
99. Fujita, T.; Takayama, C.; Nakajima, M. J. Org. Chem. 1973, $\underline{38}$, 1623.
100. Verloop, A.; Hoogenstraaten, W.; Tipker, J. In Drug Design; Ariens, E. J., Ed.; Academic Press, New York, 1976, Vol. VII, 165-207.
101. Bondi, A. J. Phys. Chem. 1964, $\underline{68}$, 441.
102. Leo, A.; Hansch, C.; Elkins, D. Chem. Rev. 1971, $\underline{71}$, 525.
103. Hammett, L. P. Physical Organic Chemistry, McGraw-Hill, New York, 1940, Chapter 7.
104. Taft, R. W. Jr. In Steric Effects in Organic Chemistry, Newman, M. S., Ed.; John Wiley & Sons, New York, 1956, Chapter 13.
105. Daniel, C.; Wood, F. S. Fitting Equations to Data, John Wiley & Sons, New York, 1980, 32.
106. Zimmerman, J. J.; Goyan, J. E. J. Med. Chem. 1970, $\underline{13}$, 492; 1971, $\underline{14}$, 1206.
107. Hansch, C.; Deutsch, E. W.; Smith, R. N. J. Am. Chem. Soc. 1965, $\underline{87}$, 2738.
108. Hansch, C.; Coats, E. J. Pharm. Sci. 1970, $\underline{59}$, 731.
109. McClellan, A. L. Tables of Experimental Dipole Moments, Vol. 1, W. H. Freeman and Co., San Francisco, 1963; Vol. 2, Rahara Enterprises, El Cerrito, CA, 1974.
110. Exner, O. Dipole Moments in Organic Chemistry, Georg Thieme Publishers Stuttgart, 1975, 20-22, 44-47, 92-93 (Figures 5-2 to 5-5).
111. Marsili, M.; Gasteiger, J. Croatica Chem. Acta 1980, $\underline{53}$, 601.
112. Gasteiger, J.; Marsili, M. Tetrahedron 1980, $\underline{36}$, 3219.
113. Magee, P. S. submitted to Quant. Struct.-Act. Relat. 1989.
114. Kikuchi, O. Quant. Struct.-Act. Relat. 1987, $\underline{6}$, 179.
115. Lien, E. J.; Guo, Z.-R.; Li, R.-L.; Su, C.-T. J. Pharm. Sci. 1982, $\underline{71}$, 641.
116. Lien, E. J.; Liao, R. C. H.; Shinouda, H. G. J. Pharm. Sci. 1979, $\underline{68}$, 463.

117. Hussain, M. H.; Lien, E. J. J. Med. Chem. 1971, 14, 138.
118. Lien, E. J.; Kumler, W. D. J. Med. Chem. 1968, 11, 214.
119. Lien, E. J. J. Med. Chem. 1970, 13, 1189.
120. Magee, P. S.; Ohta, H. confidential studies for Mitsubishi Kasei Corporation 1988.
121. Dewar, M. J. S. Science 1975, 187, 1037.
122. Harris, J. M.; Shafer, S. G.; Worley, S. D. J. Comput. Chem. 1982, 3, 208.
123. McManus, S. P.; Smith, M. R.; Shafer, S. G. J. Comput. Chem. 1982, 3, 229.
124. Kier, L. B. Molecular Orbital Theory in Drug Research, Academic Press, New York, 1971, pp. 75, 83, 86, 89, 117, 218.
125. Dewar, M. J. S.; Dougherty, R. C. The PMO Theory of Organic Chemistry, Plenum Press, New York, 1975, pp. 142, 143, 156, 261, 293, 319, 382.
126. Hintsche, R. et al, In Quantitative Structure-Activity Analysis; Franke, R.; Oehme, P., Eds.; Akademie-Verlag, Berlin, 1978, 221-225.
127. Vorpagel, E. R.; Steitwieser, A. Jr.; Alexandratos, S. D. J. Am. Chem. Soc. 1981, 103, 3777.
128. Gerhards, J.; Mehler, E. L. In QSAR and Strategies in the Design of Bioactive Compounds, Seydel, J. K., Ed.; VCH Verlagsgesellschaft, Weinheim, 1985, 153-161.
129. De Benedetti, P. G. et al, Quant. Struct.-Act. Relat. 1987, 6, 51.
130. Magee, P. S. In Insecticide Mode of Action,; Coats, J. R., Ed.; Academic Press, 1982, Chapter 5, 115-117.
131. Magee, P. S. Quant. Struct.-Act. Relat. 1986, 5, 158.
132. Charton, M. Topics Current Chem. 1983, 114, 58.
133. Nishioka, T.; Fujita, T.; Kamoshita, K.; Nakajima, M. Pest. Biochem. Physiol. 1977, 7, 107.
134. Kamoshita, K.; Ohno, I.; Fujita, T.; Nishioka, T.; Nakajima, M. Pest. Biochem. Physiol. 1979, 11, 83.
135. Magee, P. S.; Kyomura, N.; Takahashi, Y.; Ohta, H.; Yamamoto, I. Mitsubishi Chemical R & D, 1988, Rev. 2, 65.
136. DeDeken, R. H. Biochim. Biophys. Acta 1955, 17, 494.
137. Hansch, C.; Kiehs, K.; Lawrence, G. L. J. Am. Chem. Soc. 1965, 87, 5770.
138. Magee, P. S. unpublished study 1976, PM84 and PM103.
139. Franke, R. et al unpublished studies, cited in Reference 16, p. 120.
140. Johnson, C. L.; Green, J.-P. Int. J. Quantum Chem. 1974, QBS 1, 159.
141. Walters, D. E.; Hopfinger, A. J. J. Mol. Struct. (Theochem.) 1986, 134, 317.
142. Waisser, K.; Rubacek, F.; Vlcek, J.; Celadnik, M. In Quantitative Structure-Activity Analysis, Proceedings of the Symposium on Quantitative Approaches, Suhl, 1976; Franke, R.; Oehme, P., Eds.; Akademie-Verlag, Berlin, 1978, 209-214.
143. Sklenar, H.; Jager, J.; Sussmilch, R., Reference 142, pp. 239-250.
144. Pleiss, M. A. In QSAR in Design of Bioactive Compounds; Kuchar, M., Ed.; J. R. Prous International Publishers, Barcelona, 1984, 403-424.

145. Iman, R. L.; Connover, W. J. Technometrics 1979, 21, 499.
146. Magee, P. S. confidential studies for Mitsubishi Kasei Corporation, 1986-1988.
147. Bell, A. R.; Covey, R. A.; Relyea, D. I.; Magee, P. S. Proc. Br. Crop Prot. Conf.-Weeds 1987, (1), 249.
148. Baker, B. R. Design of Active-Site-Directed Irreversible Enzyme Inhibitors, John Wiley & Sons, New York, 1967.
149. Crippen, G. M. J. Med. Chem. 1979, 22, 988; 1980, 23, 599.
150. Doweyko, A. M. J. Med. Chem. 1988, 31, 1396.
151. Fujita, T. et al, Biochim, Biophys. Acta 1987, 891, 194, 293; 1988, 935, 312.
152. Tipker, J.; Verloop, A. In Pesticide Synthesis Through Rational Approaches; Magee, P. S.; Kohn, G. K.; Menn, J. J., Eds.; ACS Symposium Series 255, 1984, 279-296.
153. Goldblum, A.; Yoshimoto, M.; Hansch, C. J. Agric, Food Chem. 1981, 29, 277.
154. Small, D. M. Pure Appl. Chem. 1981, 53, 2095.

RECEIVED June 19, 1989

WAYS AND MEANS

Ways and Means: Introduction

In science, as in art, three aspects must meet to produce a worthwhile endeavor: the craftsman (or scientist), the tools of the trade (laboratory instrumentation and computer hardware and software), and the working material (the data). This section deals with tools for the computer-aided study of biomechanism, and with new descriptors for structure-activity analysis. The practical application of computer techniques to study mechanism and activity began a quarter-century ago with the work of Hansch, finding 'simple' correlations of bioactivity with the bulk steric, electronic, and lipophilic properties of molecules, often determined experimentally. As computers have grown in performance and importance, the science has matured, and applications have become increasingly complex and specific, relying more on computer-generated models and descriptors. In the hands of drug design scientists, there has been a shift away from data-analytic techniques, such as statistics and QSAR, towards structure-analytic ones, such as molecular modeling and graphics. In the agricultural field, where knowledge about specific receptors and mechanism of action is more sparse, data-analytic tools are still very appropriate. The combining of statistical and molecular modeling techniques is a welcome recent trend.

In the first paper in this section, "A Second-Generation Computer-Assisted Inhibitor Design Method", DesJarlais and coworkers describe an elegant method of combining 3D structural database searching with a unique means of describing a receptor site. This has obvious applications in molecular design. Since the mechanism of action of most commercial agents involves either reversible or irreversible enzyme inhibition, it also has promise for the study of biomechanism.

The next paper deals with the statistical description of molecular shape. Conventional methods of molecular shape analysis are largely nonspecific and nondirectional. As molecular modeling techniques become increasingly capable of simulating the 'reality' of molecular structure, we must use more specific methods to quantify and compare molecular shape. Henry and Craig describe such an approach in the "Statistical Modeling of Molecular Shape, Similarity, and Mechanism".

The first paper describes the use of the receptor structure to find ligands which fit, which is analogous to finding a foot to fit a given shoe (the hand-in-glove analogy is so old...). An alternative approach, especially applicable to agricultural problems, is to describine the receptor from the shape of molecules which bind to it. Doweyko describes the HASL approach to SAR, which generates a hypothetical active site as well as quantitative predictions of activity. This approach also sidesteps a persistent problem in molecular comparison, which is orientation of the structures.

Throughout most of its history, QSAR has been concerned with lead optimization. Only fairly recently have researchers approached the more challenging problem of lead generation. If we could 'think' like an enzyme, we would realize that molecular recognition by a receptor proceeds in stages, relying first on the interaction between the ligand electrostatic field with that of the enzyme, then relying more and more on complementarity of steric and lipophilic forces as binding takes place. Nishioka and colleagues, in "Lead-Structure Finding Based on Similarity of Amino Acid Sequences of Target Proteins", show how similarity and complementarity with the primary structure of a protein can be exploited for lead design.

Lipophilicity has been the cornerstone descriptor of QSAR studies. But measured and calculated octanol/water log P values have known shortcomings, not the least of which are that 1) highly lipophilic structures often cannot be measured or computed accurately, and 2) the partition coefficient models an equilibrium distribution of the solute, which is far from the dynamic nature of the cell membrane. To help overcome these limitations, and to provide researchers with a new measure of lipophilicity, Lavine and coworkers have created a novel micellar liquid chromatographic system, which they describe in "Application of Micellar Liquid Chromatography to QSAR Modelling of Organic Compounds".

One participant of the Symposium, George Purvis, was in transition from academia to industry, and was unfortunately not able to provide a paper for this chapter. His recent work on the display of electorstatic potentials is described in the March, 1989 issue of the *Journal of Molecular Graphics*.

So, there you have it - a toolbox, if you will, of programs and methods for the computer-aided study of structures and mechanism. These articles do not address specific modes of action directly; in most cases, the methods are so new they have not been applied to very many problems in the field. Their potential is considerable, and in the hands of imaginative researchers, they should help greatly in the probing of biomechanism.

DOUGLAS R. HENRY
Molecular Design Limited
2132 Farallon Drive
San Leandro, CA 94577

Chapter 4

Second-Generation Computer-Assisted Inhibitor Design Method

Renee L. DesJarlais, George L. Seibel, and Irwin D. Kuntz, Jr.

Department of Pharmaceutical Chemistry, University of California, San Francisco, CA 94143

We present a second generation computer-assisted method to provide novel candidates for drug design. The method utilizes a rapid and automatic method of locating sterically reasonable orientations of small molecules in a receptor site of known three-dimensional structure combined with a scoring scheme that ranks the orientations by how well they fit the site (DesJarlais, et al., J. Med. Chem., 1988, 31 722). This docking procedure is the first step in a two step process to provide candidates for novel enzyme inhibitors. A large database of small molecule structures is searched for those molecules with shapes complementary to the receptor structure. This ensures good van der Waals interactions between the receptor and the top scoring molecules, but it is unlikely that any of these molecules will have the appropriate electrostatic and hydrogen bonding properties to interact favorably with the receptor. To the docking procedure, we have added a second step that examines the electrostatic and hydrogen bonding properties of the receptor site. These properties are displayed with computer graphics models and are used to suggest chemical modifications to the correctly shaped skeleton structures in order to provide chemical complementarity as well as shape complementarity to the receptor site. At the present time, evaluation of the chemical properties is performed by the chemist. It is our long-term goal to use what is learned from this interactive design to automate the process of modifying the molecular frameworks found in the shape search.

The identification of molecules with a specific biological activity is a necessary first step in developing a new drug. One requirement for biological activity is that the compound bind to a specific receptor. In theory, one could use the three-dimensional structure of a receptor to design molecules that would have binding affinity for the receptor. Structures of some medicinally interesting receptors are now available. We present a semi-automatic approach to the problem of designing molecules complementary to a specific receptor structure. Our procedure uses the DOCK package of computer programs to find molecules with shape complementary to that of a particular receptor site. (1,2) The package characterizes the shape of a potential binding site and searches a database of small molecules whose structures were obtained from the Cambridge Crystallographic Database. (3) This search method finds molecules with good van der Waals interactions at the receptor site and eliminates those that would overlap the receptor atoms severely but ignores other chemical properties of the receptor and the small molecule. The next step is to examine the electrostatic and hydrogen bonding properties of the receptor and to decide which of the molecules from the database search can be made to interact favorably with appropriate chemical modification. The receptor properties are displayed using the molecular graphics package MIDAS. (4)

The method is applicable to any system where a three-dimensional structure of the receptor is available. The steps of the method are described below and illustrated using the enzyme penicillopepsin, an aspartyl protease.

METHODS

The design procedure begins by using the DOCK programs (1,2) to find a set of molecules that fit a receptor site. First, the shape of the receptor site is characterized. The molecular surface as described by Richards (5) is the basis for this characterization. The program MS (6-8) is used to generate a dot representation of this surface. Spheres are generated that touch this surface at two points, do not overlap the receptor atoms, and have their center along the surface normal of one of the points. These spheres are of various sizes. The smallest spheres are the size of the spherical probe used to calculate the molecular surface and the largest spheres are limited to 5Å in radius. The number of spheres is also reduced by retaining only the largest sphere associated with a particular atom. Finally the spheres are grouped into sets in which spheres overlap each other. Each of these sets of spheres characterizes a depression in the receptor surface. While any of these sets may be used in the rest of the calculation, the set with the largest number of spheres is typically the one associated with an enzyme active site.

Next, a database of small molecules is screened for those whose shape complements the shape of the active site best. The database that we have used is a subset of the Cambridge Structural Database consisting of approximately 10,000 molecules. These molecules have been selected to have a wide variety of shapes (Seibel, manuscript in preparation). Various orientations of each small molecule in the site are found by matching atom-atom distances from the small molecule to sphere-sphere distances from the receptor. If a distance between receptor spheres ($d_{A,B}$) is equal to an interatomic distance from the small molecule ($d_{a,b}$) within a user-defined tolerance, then a third sphere (C) and atom (c) are sought such that distance $d_{A,C}$ equals $d_{a,c}$ and $d_{B,C}$ equals $d_{b,c}$ again within the user defined tolerance. Atom centers and sphere centers are added to the set in this way until no more can be found. At least four atom-sphere pairs must be assigned to proceed, but the user may require a larger number. A rotation/translation matrix is then determined that will best superimpose the small molecule atoms onto their paired sphere centers. This rotation/translation is then applied to the entire small molecule to orient it in the site. Several hundred sets of atom-sphere pairs are found and thus many different orientations for each small molecule are explored. The orientations are scored based on a simple scoring function shown in Equation 1 where $d_{i,a}$ is the distance between a particular receptor atom, i, and a particular ligand atom, a. The other terms are defined in Table I. This function approximates a soft van der Waals potential.

$$\begin{aligned}&\text{if any } d_{i,a} < concut \text{ then}\\&\quad score = -999.0\\&\text{else}\\&\quad score = \sum_{i=1}^{\text{receptor atoms}} \sum_{a=1}^{\text{ligand atoms}} F_{i,a}\end{aligned} \quad (1)$$

$$F_{i,a} = \begin{cases} 1.0 & \text{if } concut \leq d_{i,a} < dmin \\ \exp[-(d_{i,a} - dmin)^2] & \text{if } dmin \leq d_{i,a} \leq discut \\ 0.0 & \text{if } discut < d_{i,a} \end{cases}$$

The top scoring orientation of each small molecule is compared to the top scoring of the other molecules, and the best are saved to be viewed graphically.

At this point in the design procedure, the molecules that have been found fit the site in a geometric sense but do not necessarily match in detailed chemistry. Electrostatic and hydrogen bonding features of the receptor are now examined. In this calculation, each atom of the small molecule is used to define a location at which the electrostatic potential from the protein atoms is evaluated (Equation 2).

$$E_j = \sum_{i=1}^{\text{receptor atoms}} \frac{331.5 \, q_i}{\varepsilon \, R_{i,j}} \qquad (2)$$

This potential is calculated using the partial charges in the AMBER united atom force field. (9) Next, potential hydrogen bonds are identified. Oxygen and nitrogen atoms in the protein are located, and if there is a small molecule atom within hydrogen bonding distance of one of the protein oxygens or nitrogens it is labeled as a potential hydrogen bond. The design process now becomes interactive. Properties of the receptor are highlighted as the user views the molecules in the receptor site using molecular graphics. The electrostatic potential at each small molecule is indicated by the color of the atom, and the potential hydrogen bonds are displayed as lines drawn from the oxygens and nitrogens in the protein to atoms in the small molecule that are within hydrogen bonding distance. It is useful to display the potential hydrogen bonds as lines, because this emphasizes the angular component of the hydrogen bonding interaction.

RESULTS

The design procedure described above was applied to the protein penicillopepsin, an aspartyl protease from *Penicillium janthinellum*. The coordinates for the protein were obtained from the Protein Data Bank (10) entry 2APP. The crystal structure was determined by James and Sielecki (11) at 1.8Å resolution.

The DOCK package of programs was used to obtain molecules that have a good geometric fit to the receptor. Pertinent parameters used in the docking step are listed in Table I.

Table I. Description of variables used in the docking procedure

variable	value	description
dislim	2.0Å	error limit for distance matching
nodlim	8	minimum number of sphere-atom pairs in a match
concut	2.3Å	scoring parameter--(Equation 1)
dmin	3.5Å	scoring parameter--(Equation 1)
discut	5.0Å	scoring parameter--(Equation 1)

One of the molecules found in the shape search will be discussed in detail to explain the evaluation of the electrostatic and hydrogen bonding interactions with the receptor.

Molecule I, 2-(2-quinolyl) cyclohexane phenylhydrazone, (<u>12</u>) was ranked fifteenth in the shape fit. It is more interesting than the higher scoring molecules because its framework is fairly simple from a synthetic standpoint. Molecule I fits tightly into the site with its quinolyl end in a deep channel that extends into the protein (see Figure 1a). The cyclohexane ring gives the molecule an overall "L" shape, and the phenyl portion lies along the active site groove near the active site aspartyls.

I

2-(2-QUINOLYL)CYCLOHEXANE PHENYLHYDRAZONE

Figure 1b shows the electrostatic potential at the small molecule. The potential ranges from -100 kcal/mole per unit electron charge (red) to 0 kcal/mole per unit electron charge (green) with intermediate values shaded intermediate colors (e.g., yellow at about -50 kcal/mole per unit electron charge). The potential indicates that there are two positions where positive charge would be desirable: position 16 and position 23. There are many

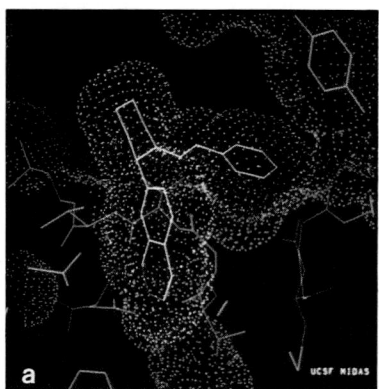
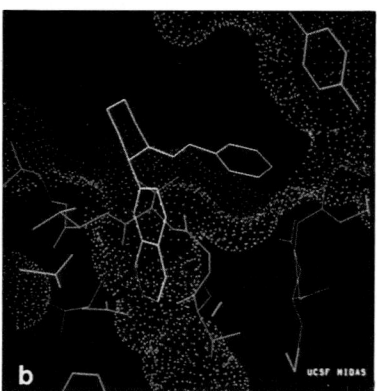
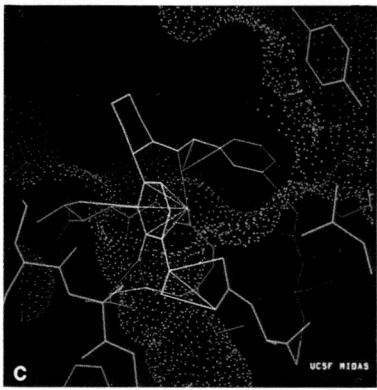

Figure 1. Molecule **I** is shown in the active site of Penicillopepsin. The enzyme and its molecular surface are shown in blue with the active site aspartates in red. (a) The van der Waals surface of molecule **I** (orange) is shown to illustrate the good fit. (b) Molecule **I** with its atoms colored by the value of the electrostatic potential from the protein. Red indicates a potential of -100 kcal/mole per unit electron charge and green indicates a potential of 0 kcal/mole per unit electron charge. The color of intermediate values of the potential are scaled accordingly. (c) Molecule **I** with potential hydrogen bonds to the protein shown as lines between the protein atoms and small molecule atoms. Pink lines indicate that the protein atom is a carbonyl oxygen. Green lines indicate that the protein atom is a hydroxyl oxygen. Purple lines indicate that the protein atom is a nitrogen.

ways that these electrostatic requirements could be satisfied. One approach would be to eliminate the atoms at positions 17 and 18 opening the ring to make 16 a primary amine, and to change the phenyl ring to a piperidinyl ring with the nitrogen at position 23. While changing the phenyl ring to an aliphatic ring changes the shape somewhat, the active site is wide at this point and should be able to accommodate the change.

Receptor polar atoms that might require hydrogen bonds when molecule **I** is docked with the proposed geometry are shown in Figure 1c, with lines drawn from them to nearby ligand atoms. Some of these are in the regions that we have satisfied with charged groups, and we do not need to consider these further. There are two areas that are of concern, however. There is a threonine hydroxyl group near position 3 of the small molecule. Position 3 however is already a nitrogen and is in a reasonable hydrogen bonding geometry, so this requirement is fulfilled. There is a carbonyl below the quinolyl ring that is problematic. It is close to four of the atoms of the quinolyl ring, but the hydrogen bond angles in this receptor-ligand geometry would be unacceptable. There is not room to add other atoms to the ring. The same carbonyl is close to atom position 2 of the small molecule. The hydrazone will not be protonated though, so a change is necessary. If the hydrazone were changed to an amide with its nitrogen at position 2 and its carbonyl at position 1, the nitrogen would be a hydrogen bond donor and the overall geometry of the molecule would be about the same. There is room for the extra atom at position 1. Some other potential hydrogen bonds are shown, but they are not completely shielded from solvent by the ligand. Thus, a proposed target based on molecule **I** is molecule **II**. The next steps in the design process are to evaluate synthetic feasibility and to synthesize and test compounds. This is currently underway.

II

DISCUSSION

The DOCK package of programs is able to find small molecules with shape complementarity to a given receptor site. The question we address in this paper is how to turn these molecular frameworks into molecules that will not only have favorable van der Waals interactions with the receptor but also the appropriate electrostatic and hydrogen bond properties. The use of interactive molecular graphics to display potential hydrogen bonds and the electrostatic potential has proven to be an effective guide in determining what modifications might be made to increase chemical complementarity with the receptor site. Considerable amounts of time are required to do this, however. For the molecules that look promising, it may take several hours at an interactive graphics device to design a target molecule. Moreover, it often takes significant effort to decide that a molecule cannot complement the chemistry of the active site adequately and is not a good candidate.

The design of hydrogen bonding interactions is a particular problem. Because hydrogen bonds are directional, the position of the small molecule atom with respect to the protein atom is of more concern than for electrostatic interactions. We are working on a procedure for assessing how well each small molecule might be able to satisfy the hydrogen bond requirements of the protein atoms that it is near. We expect that this will allow identification of the most promising small molecules early in the screening and reduce the amount of time spent examining molecules that cannot make the necessary interactions with the receptor.

Beyond the problem of whether a particular interaction will be favorable is the question of how many potential interactions must be made for a ligand to bind tightly. Experimental data indicate that a single hydrogen bond can be worth 0.5-1.5 kcal/mole in binding energy. (13) It would be desirable to design as many of these interactions as the protein would seem to require, but it may not be necessary to make them all. X-ray crystallographic data on the binding of antiviral compounds to human rhinovirus (14) shows that although the compound is virtually surrounded by protein, only one hydrogen bond is made leaving some protein atoms and some atoms from the antiviral compound without hydrogen bond partners. Clearly, one would like to make as many favorable interactions as possible for the best binding affinity, but one must also be aware of the potential for designing in interactions that cannot all be made at the same time.

Synthesis of penicillopepsin inhibitors as well as inhibitors of other enzyme systems is on going and testing will begin when molecules are in hand. These studies will involve crystallographic examination as well as conventional binding studies. We expect that the results of these experiments will allow us to improve our design method and better understand intermolecular interactions in general.

Acknowledgments

We thank Dale L. Bodian and Brian K. Shoichet for their help. We thank the Computer Graphics Laboratory (supported by National Institutes of Health RR-01081) for use of their facilities for parts of this project. Partial support was received from the National Institutes of Health GM-31497 (IDK) GM-39552 (IDK), GM-07175 (GLS), and GM-29072 to Peter Kollman. The Floating Point Systems 264 array processor was purchased with grant support from National Science Foundation DMB-84-13762 and National Institutes of Health RR-02441. Assistance from the Macromolecular Workbench supported by the Defense Advanced Research Projects Agency under contract N00014-86-K0757 administered by the Office of Naval Research is also appreciated.

REFERENCES

1. Kuntz, I. D.; Blaney J. M.; Oatley S. J.; Langridge R.; Ferrin, T. E. J. Mol. Biol. 1982, 161, 269.
2. DesJarlais, R. L.; Sheridan, R. P.; Seibel, George L.; Dixon, J. S.; Kuntz, I. D.; Venkataraghavan, R. J. Med. Chem. 1988, 31, 722
3. Allen, F. H.; Kennard, O.; Motherwell, W. D. S.; Town, W. G.; Watson, D. G. J. Chem. Doc. 1973, 13, 119.
4. Ferrin, T. E. and Langridge, R. Comp. Graphics 1980, 13, 320.
5. Richards, F. M. Annu. Rev. Biophys. Bioeng. 1977, 6, 151.
6. Connolly, M. L. Ph.D. Thesis, University of California, Berkeley,1981.
7. Connolly, M. L. J. Appl. Crystallogr. 1983, 16, 548.
8. Connolly, M. L. Science 1983, 221, 709.
9. Weiner, S. J.; Kollman, P. A.; Case, D. A.; Singh, U. C.; Ghio, C.; Alagona, G.; Profeta, S.; Weiner, P. J. Am. Chem. Soc. 1984, 106, 765.
10. Bernstein, F. C.; Koetzle, T. F.; Williams, G. J. B.; Meyer, E. F. Jr.; Brice, M. D.; Rodgers, J. R.; Kennard, O.; Shimanouchi, T.; Tasumi, M. J. Mol. Biol. 1977, 112, 535.
11. James, M. N. G., and Sielecki, A. R. J. Mol. Biol. 1983, 163, 299.
12. Bocelli, G.; Tosi, G; Cardellini, L. Acta Cryst. 1984, C40, 1952.

13. Fersht, A. R.; Shi, J.-P.; Knill-Jones, J.; Lowe, D. M.; Wilkinson, A. J.; Blow, D. M.; Brick, P.; Carter, P.; Waye, M. M. Y.; Winter, G. *Nature* 1985, *314*, 235.
14. Smith, T. J.; Kremer, M. J.; Luo, M.; Vriend, G.; Arnold, E.; Kamer, G.; Rossmann, M. G.; McKinlay, M. A.; Diana, G. D.; Otto, M. J. *Science* 1986, *233*, 1286.

RECEIVED May 23, 1989

Chapter 5

Statistical Modeling of Molecular Shape, Similarity, and Mechanism

Douglas R. Henry[1] and A. Morrie Craig[2]

[1]Molecular Design Limited, 2132 Farallon Drive, San Leandro, CA 94577
[2]College of Veterinary Medicine, Oregon State University, Corvallis, OR 97331

> Current methods of molecular shape description and comparison have two major shortcomings: they are relatively nondirectional, and the superposition of molecules is usually a subjective procedure. A new method of shape description and comparison is described, which uses a binding moment, derived from fragment binding constants, to orient the molecules, followed by SIMCA modeling of the shape and similarity of the structures. This method is compared with other molecular shape descriptors of varying dimensionality. The shape descriptors are correlated with acute and chronic lung and liver toxicities of a set of pyrrolizidine alkaloids. The overall results, although not precise enough for quantitative prediction, demonstrate the value of using more specific and directional shape descriptors, coupled with a more rational method for molecule orientation and superposition.

The interaction of a small molecule with an enzyme receptor, which is the basis for most therapeutic and many agricultural agents, proceeds in several stages (1). The first stage is a long-range through-solvent recognition and reorientation of the ligand, caused by interaction between the electrostatic fields of the ligand and the enzyme. In the next stage, the ligand approaches the receptor and becomes reoriented spatially and conformationally as it contacts and conforms to the surface of the active site. Finally, the ligand binds to the receptor, with possible perturbation of the receptor structure. This binding may be either a nonbonded interaction (reversible inhibition - the basis of action for many drugs), or it may involve formation of covalent bonds (irreversible and suicide inhibition - the basis for many agricultural agents).

Depending on the stage of interaction, different aspects of a molecule's structure become important. At a distance, electrostatic field effects and the dipole moment of the structure are important; at first contact, steric effects play a role, and finally during the binding phase, more specific electronic and lipophilic properties come into play. Molecular shape plays an important part at all steps in the process. In solution, shape helps determine the dynamics of the molecule's movement and its interaction with solvent molecules. At first contact, shape is of paramount importance in limiting or enabling the interaction. In the final binding stage, the shape of one part of the molecule may sterically interfere with a crucial rotation necessary to being a functional group into proximity with a specific region of the receptor.

A precise and relevant description of the 'shape' of a molecule is difficult to calculate. Although several descriptors of molecular shape have been proposed, most fall short of being ideal. One way of viewing shape is from the standpoint of the dimensionality of the information the descriptor encodes. Kier has described a number of shape descriptors which are purely topological in nature (2). These do not depend on the 3D or even the 2D structure of the molecule, and they might thus be considered zero-dimensional measures of shape. Simple whole-molecule measures of size and bulk, such as the volume, the surface area, or any single dimension of the structure, might be considered one-dimensional measures of shape. Two-dimensional measures of shape include the cross-sectional shadow descriptors of Rohrbaugh and Jurs (3).

Spherical harmonics provide a parameterization of three-dimensional shape which is especially useful for protein structure description (4). Overlap volume comparisons are the basis for the Molecular Shape Analysis (MSA) method of Hopfinger (5,6). This technique has been extended to include a quantification of the steric and electrostatic fields surrounding a molecule (7). A further refinement of field analysis, which merges statistical and molecular modeling techniques, is the COMparative Molecular Field Analysis method (COMFA) of Cramer (8). These latter approaches seek to encode information about more than just steric bulk or form. They express multivariate information about the structure, so they might be considered multidimensional shape descriptors.

A couple of problems exist with most of the shape descriptors mentioned above. The first is their nonspecific and nondirectional nature. In a single value, one cannot simultaneously encode information about both the direction and the form of the space the molecule occupies. Thus, a parameterized or at least a multivariate, description of shape is necessary. This is a feature of some of the descriptors mentioned, but it is usually not easy to interpret.

A second problem deals with the orientation and superposition of molecules for shape comparison. Although methods have been published to automatically generate the best geometric fit to a set of matched points (9), in shape analysis this is almost always done subjectively, and it works best for structures which share a common parent substructure of atoms which can be superimposed in an obvious manner. There are many examples showing that chemical intuition can be misleading when orienting structures for comparison, especially if only 2D structures are considered. A classic example is the flip of the pteridine ring between the binding of dihydrofolate and the binding of methotrexate to DHFR (10).

To describe and compare molecular shape, and to correlate shape with biological activity, we sought a methodology that would not be limited by orientation and directionality shortcomings. We also wanted the method to be be implementable with existing modeling and data analysis software. The compounds we selected for analysis were a representative sample of toxic pyrrolizidine alkaloids, which are derivatives of the structures in Figure 1. These compounds produce lung and liver toxicity in horses and cattle which ingest the tansy ragwort (*Senecio jacobaea*). This problem has a multimillion dollar economic impact on agriculture in the Pacific Northwest, where the plant is abundant.

Methodology

The structures and biological data were taken from an article by Culvenor and coworkers (11). There, 62 pyrrolizidine compounds were tested for acute and chronic lung and liver toxicity in young rats. The dose consisted of a single i.p.

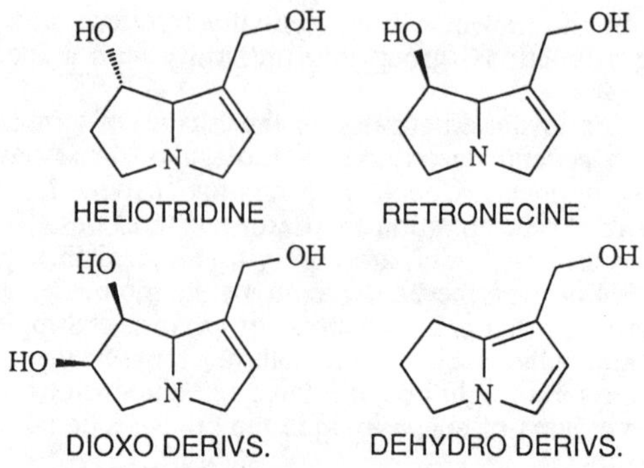

Figure 1. Toxic pyrrolozidine structures.

injection of the hydrochloride salt of the compound in neutral aqueous solution. Eight dose levels were used (0.025 mmol/kg to 3.2 mmol/kg). After four weeks, the rats were sacrificed, and microscopic examination was used to determine the level of tissue damage. The authors defined five types of biological response: 1) peracute morbidity (death in 1 day or less); 2) acute morbidity (death in 1-7 days); 3) acute liver toxicity (centrilobular necrosis); 4) chronic liver toxicity (parenchymal megalocytosis); and 5) chronic lung lesions (intravascular and interstitial accumulation of mononuclear cells). The biological response was recorded as the lowest dose level producing the given lesions. A complication was that each of the compounds was not tested at all dosage levels.

The structures were quite diverse, ranging from those with simple alkyl ether attachments, to complex cyclized derivatives. The entire set could not be modeled and analyzed, so to obtain a representative sampling of the compounds, several molecular connectivity descriptors (12) were computed using the ADAPT program (13). These relate to the degree of branching in the structures, and they provide a convenient means of describing 2D structural diversity. The connectivity descriptors were combined with the biological data and a hierarchical cluster analysis was performed on range-scaled values (14). After analysis of the resulting dendrogram, 21 of the structures were selected (Figure 2). For the 21 structures, lung toxicity showed a fairly high correlation with acute death and liver toxicity. Otherwise, the biodata are fairly uncorrelated (Table I; P=peracute, A=acute, C=chronic).

The 21 structures were modeled using the CHEMLAB-II system of Hopfinger and Pearlstein (15) and Allinger's MMP2 program (16). A rigorous conformational analysis was not performed for each structure; several low-energy conformers were identified for representative members of the set, and in the case of the macrocycles, their structures were compared with crystal structures for jacobine

Table I. Biodata Correlation Matrix

	Death(P)	Death(A)	Liver(A)	Liver(C)
Death(A)	0.106			
Liver(A)	-.228	0.385		
Liver(C)	-.119	0.294	0.360	
Lung	-.022	0.840	0.784	0.790

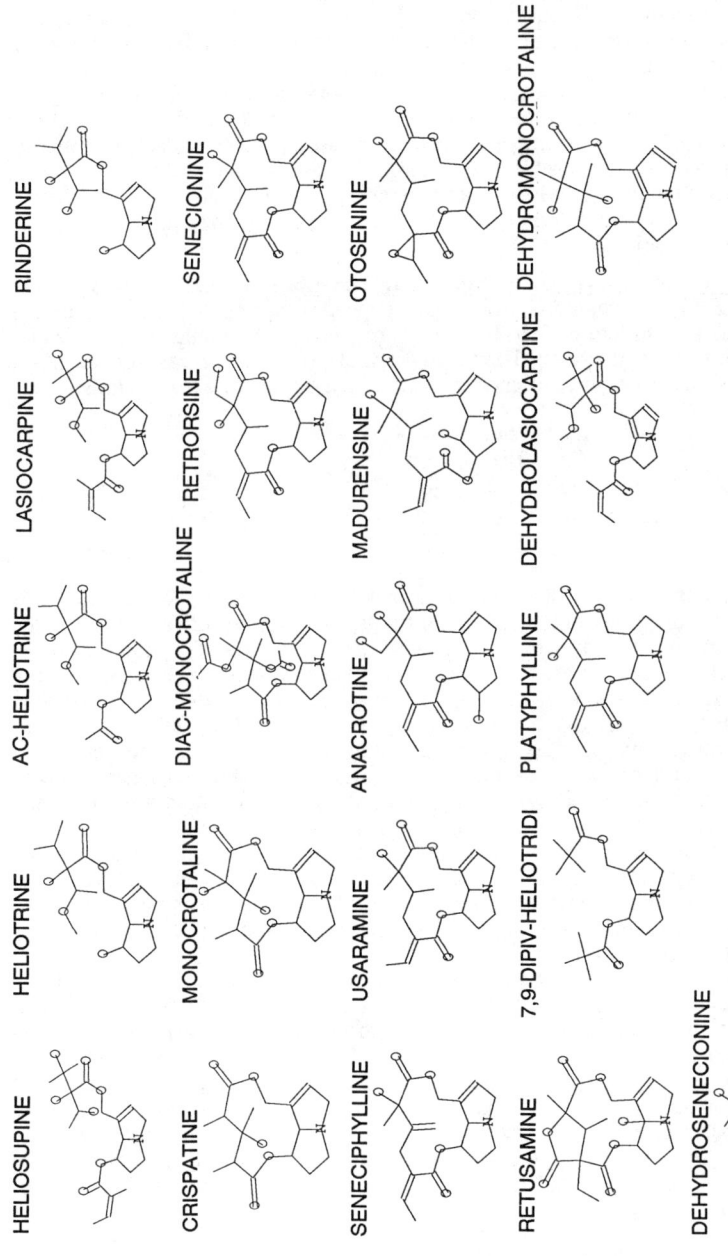

Figure 2. Structures used in the study.

(17) and retrorsine (18). The conformer matching most closely was used in the analysis, and for building related structures.

Standard molecular shape descriptors which were calculated included the following: Kier's $^2\kappa$ index, the Jurs shadow descriptors (6 descriptors) and length/breadth descriptors (2 descriptors), and Hopfinger's MSA volume descriptors (3 descriptors, using each structure in turn as a reference). The default conditions of the descriptor generating routines were used in each case.

For comparison, some simple correlations were obtained between biological activity and the octanol/water partition coefficients, as calculated by the CLOGP program of Leo (19). An example is seen in Figure 3, which shows an apparent parabolic relationship between the log(1/minimum dose for acute death) and log P. The two outlier structures may have poorly calculated log P values (the epoxide ring and the conjugated system are special features), or they may undergo special metabolic transformation.

A Binding Moment Approach to Molecular Orientation. Except for the Kier index, each of the conventional molecular shape descriptors requires a standard orientation of the structure. In the case of the shadow and length/breadth descriptors, this is along principal axes of the molecule. In the case of the MSA descriptors, the structures were aligned by matching the following three atoms:

As an alternative to specific atom matching, a moment of binding was calculated using the fragment binding constants of Andrews (20). The binding moment was computed in a manner similar to the calculation of a dipole moment from partial residual atomic charges. The partial charge values for the nonhydrogen atoms in the structure were replaced by binding constant values. In the case of carbonyl groups, the binding constant value was positioned on the oxygen atom. Alternatives would have been to split the binding constant value, or to generate a pseudoatom at the centroid of the atoms in the fragment. The calculated moment was scaled to unit length and stored as pseudoatoms with the structure. Figure 4 shows a plot of the MMP2-modeled structures of senecionine and dehydrosenecionine, with circles representing the fragment binding constants, and the binding moment displayed as a vector. Also shown are circular profiles of the biological activity. The length of any given spoke in the profile is proportional to the level of activity, and here one sees that the compounds have quite different spectra of biological activity.

To superimpose structures for comparison, the corresponding ends of the binding moment vector, plus the ring nitrogen atom, were used as reference points. In many cases, this gave superpositions similar to those obtained using the nitrogen and the two oxygen atoms, in the MSA analysis. In some cases, as shown in Figure 5, there was a reorientation of the structure being superimposed, sometimes amounting to an almost 90 degree rotation about the vertical axis. It was interesting to note that these differences in orientation often occurred for structures which had quite different toxicity spectra. Of course, other moments, such as lipophilicity and dipole moments, could be matched for superposition as well, though this was not done in this case.

SIMCA Modeling of Molecular Shape. To model the shapes of the molecules in a manner that would express direction as well as bulk, and allow comparison of

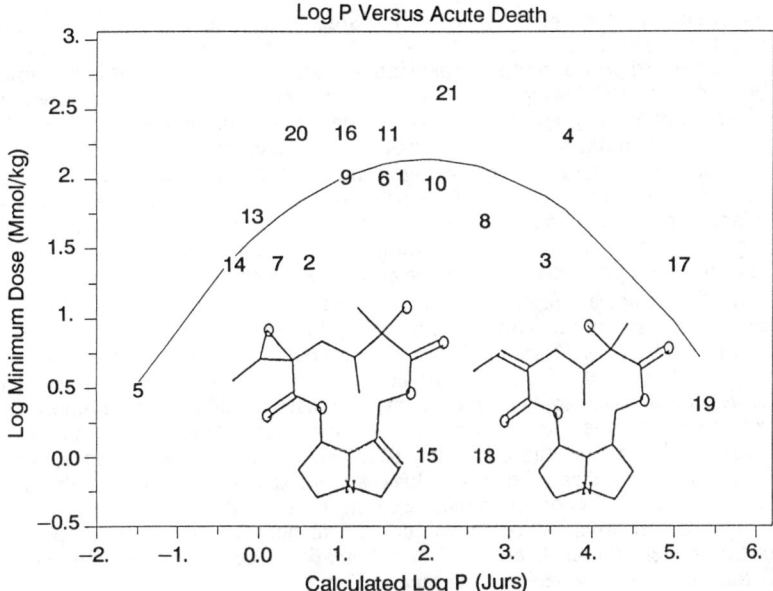

Figure 3. Correlation of calculated log P (octanol/water) values with acute morbidity of pyrrolizidine alkaloids. Structures of two outlying compounds are shown.

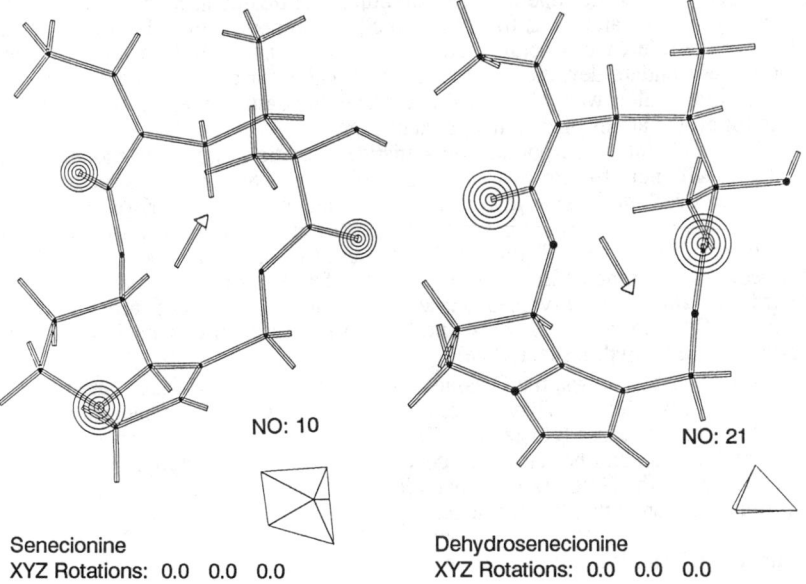

Figure 4. Modeled structures of senecionine and dehydrosenecionine, showing circles for fragment binding constants and the binding moment vector. Circular profiles of biological activity are also shown.

different structures, we chose the SIMCA (Soft Independent Modeling by Class Analogy) technique (21). This method generates disjoint principal component models of clustered points in multidimensional space. With SIMCA, it is possible to describe and parameterize spherical (zero components), linear (one component), planar (two components), and box-shaped regions (three and higher components) of space. The usual limitations apply in terms of the statistical degrees of freedom. In this case, the points were the nonhydrogen atoms in the structure. The variables were simply the Cartesian coordinates of the atoms. It is also possible to include other atomic properties, such as electrostatic and lipophilic characteristics, if necessary, to generate an extended definition of molecular shape. This was not done here, since we were primarily interested in conventional shape description.

SIMCA is not a clustering techniqe. Thus, it was first necessary to cluster the atoms in each structure. This was done using hierarchical clustering in Cartesian space, with ADAPT. A sample cluster dendrogram for senecionine is shown in Figure 6. The number of clusters for each structure was manually determined, though we could have applied one of many cluster validation techniques (22). Between three and five clusters were chosen for each structure. As one would expect for bonded atoms, the clustering followed bonding patterns, since bonded atoms are closest together. Once the clusters were selected, SIMCA analysis was performed on unscaled coordinate data, treating each cluster as a separate class of points. Default cross-validation techniques led to between one and two principal component models for each cluster. A schematic representation of the SIMCA models for the atoms in senecionine is seen in Figure 7.

Comparison of Structures. To compare the structures and generate shape descriptors, the reference compound approach was used. Each compound was used in turn as a reference structure. Every other structure was superimposed on the reference using the binding moment technique described previously. Then, each nonhydrogen atom in the superimposed structure was treated as a new "observation" in the analysis. The atom was fit to the principal component models of each cluster of atoms in the reference structure. The atom was 'assigned' to the cluster it fit best, and both the standard deviation value and the F-value for the fit was recorded. The final descriptor value was taken as the average standard deviation or the average F-value for all the atoms placed in a given cluster.

This technique incorporates the statistical variation of a given model, so that a structure will not 'fit' its own models perfectly. The fit of a structure to its models, however, defines a lower range of values for comparison with other structures. As values increase beyond this lower range, it indicates a poorer shape comparison. The fact that disjoint models are used means that directionality can be expressed. When senecionine is used as the reference structure, one obtains the standard deviation and F-values shown in Table II. Values are shown for senecionine fitted to its own models, and for a similar structure (seneciphylline) and a dissimilar one (dehydrosenecionine).

As this table shows, the F-values are larger and show a wider variation than the standard deviations. They generally correlated with the standard deviation values, but in some cases (cluster 2 in Table II), there were discrepancies. Because of their wider span, and because they generally accorded with graphical comparison of the structures, the F-values were used in the correlations with biological activity and for comparison with other shape descriptors.

Results and Discussion

In general, the various different shape descriptors in this study were not intercorrelated (R values 0.7 or less). Simple correlations of the SIMCA F-value descriptors with the various biological response variables yielded rather poor results.

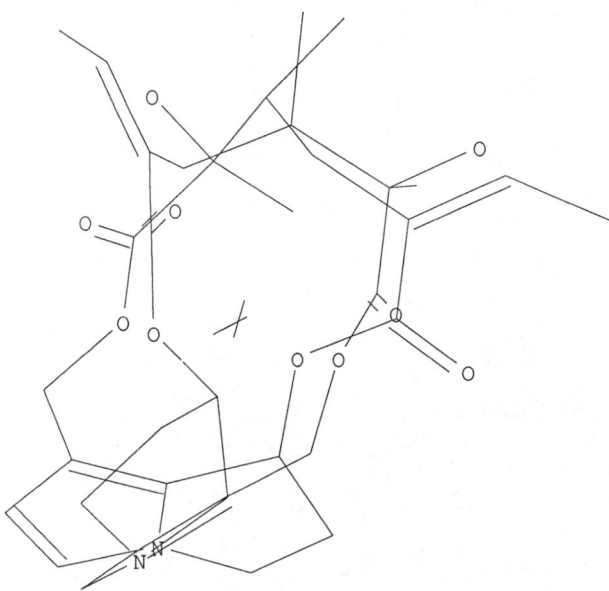

Figure 5. Superposition of senecionine and dehydrosenecionine obtained by matching binding moment vectors and ring nitrogen atoms. The dehydro compound is rotated almost 90 degrees relative to senecionine.

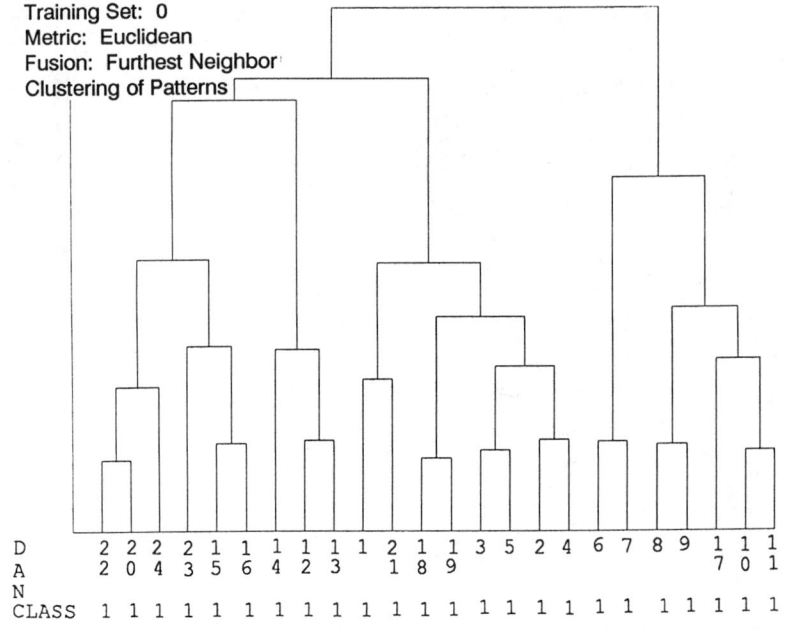

Figure 6. Dendrogram showing atom clusters for senecionine. Three clusters were selected to represent the structure.

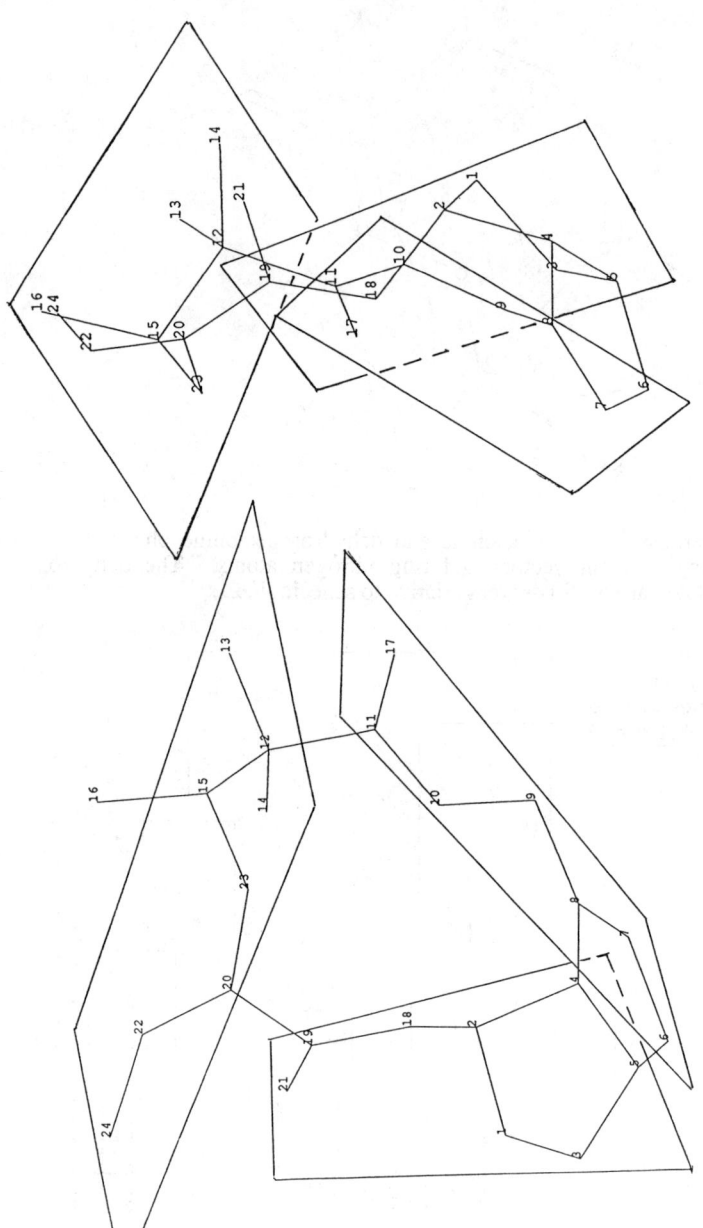

Figure 7. Orthogonal views of a schematic representation of the SIMCA models for the atom clusters in senecionine.

Table II. Average and Maximum Standard Deviation and F-value Descriptors with Senecionine as Reference (maximum in parentheses)

Cluster		Senecionine	Fitted Compound Seneciphylline	Dehydro-senecionine
1	(sd)	0.218 (0.510)	0.643 (1.313)	0.873 (1.799)
1	(F)	0.469 (1.921)	3.590 (13.43)	5.020 (20.45)
2	(sd)	0.557 (1.390)	0.907 (1.610)	0.505 (1.322)
2	(F)	0.625 (2.350)	1.680 (4.030)	5.570 (26.24)
3	(sd)	0.230 (0.530)	0.663 (1.245)	0.958 (2.458)
3	(F)	0.547 (1.928)	1.500 (3.501)	4.436 (11.56)

Correlation coefficients ranged from 0.6 or lower to 0.7, depending on the reference compound that was used. With only a few exceptions, similar results were obtained for the other descriptors. Since the biodata and the shape descriptors in this study are both multivariate, canonical correlation analysis was selected to provide a single overall measure of correlation between molecular shape and biological activity, for comparison of the various shape descriptors. In canonical correlation analysis, the combination of the predictor variables is found, which correlates highest with any possible combination of the response variables (23). A similar approach is taken in Partial Least Squares (PLS) analysis (24).

Table III shows, for the 21 structures in the analysis, the first canonical correlation coefficients relating the shape descriptors with the five measures of biological activity. In the case of a single descriptor (Kier's $^2\kappa$ index, for example) the canonical correlation coefficient is the same as the simple correlation coefficient, so univariate and multivariate correlations can be compared directly.

The correlations in this table are the highest that were observed in the analysis. The activities in the last column are the biological response variables that were most highly associated with the first canonical variable of the biological data.

The best correlation was observed with Jurs' shadow descriptors. This may be partly a result of the larger number of descriptors used, since the simplest topological and one-dimensional descriptors show the lowest levels of correlation. The MSA and SIMCA descriptors are directly comparable. The SIMCA descriptors have somewhat higher correlations with activity, though the differences may not be statistically significant. They are computed with about the same amount of effort as the MSA descriptors; in addition, the SIMCA descriptors have directionality, which could allow a researcher to determine which part of the molecule is responsible for the shape differences and presumably, the differences in activity. The quality of the biodata was not really high enough to associate particular types of activity with certain regions of the molecules. In the future, we hope to be able to accomplish this with newer *in vitro* assays of pyrrolizidine toxicity.

Conclusions

As the technology for computer simulation of 3D molecular structures improves, we are able to generate ever more accurate models of structures and activity. A recurrent trend in QSAR science has been the shift from the more general to the

Table III.   Canonical Correlation Results

Descriptor Set	Vars.	Canon. Corr.	Highest Correlating Activity
Kier $^2\kappa$	1	0.50	Liver (C)
Jur's Shadow	6	0.86	Death (P)
Jurs' L/B	2	0.42	Death (P)
MSA			
Senecionine	3	0.72	Liver (A)
Usaramine	3	0.74	Liver (C)
Madurensine	3	0.64	Lung
Average		0.66	
SIMCA			
Retrorsine	3	0.80	Liver (A)
Rinderine	3	0.75	Liver (C)
Madurensine	4	0.78	Lung
Average		0.72	

more specific (as in proceeding from bulk log P to π values, and in moving from topological to geometrical descriptors). More accurate models of structure require more specific shape descriptors to adequately encode relevant information for correlation with biological activity. As we have shown, there are advantages to multivariate descriptions of molecular shape based on statistical modeling of the structure. Such models are capable of expressing both the amount and the direction of shape differences. We are presently investigating the use of lipophilic and electrostatic extensions of our definition of molecular shape.

Literature Cited

1. Franke, R. Theoretical Drug Design Methods; Elsevier: New York, 1989; pp 316-322.
2. Kier, Lemont B. Quant. Struct.-Act. Relat. 1986, 5, 11-12.
3. Rohrbaugh, R. H.; Jurs, P. C. Anal. Chem. 1987, 59, 1048-1054.
4. Max, Nelson; Getzoff, Elizabeth D. IEEE Comput. Graph. Appl. July 1988; pp 42-50.
5. Hopfinger, A. J. J. Am. Chem. Soc. 1980, 120, 7196-7206.
6. Hopfinger, A. J.; Burke, Benjamin J. In QSAR: Quantitative Structure-Activity Relationships in Drug Design; Fauchere, J. L. Ed.; Alan R. Liss: New York, 1989; pp 151-159.
7. Hopfinger, A. J. J. Med. Chem. 1983, 26, 990-996.
8. Cramer, R. D.; Patterson, D. E.; Bunce, J. D. J. Am. Chem. Soc. 1988, 110, 5959-5967.
9. Danziger, D. J.; Dean, P. M. J. Theor. Biol. 1985, 116, 215-224.
10. Kuyper, Lee F. In Computer-Aided Drug Design; Perun, Thomas J; Propst, C. L. Eds.; Marcel Dekker: New York, 1989; pp 337-338.
11. Culvenor, C. C. J.; Edgar, J. A.; Jago, M. V.; Outteridge, A.; Peterson, J. E.; Smith, L. W. Chem.-Biol. Interactions 1976, 12, 299-324.
12. Kier, Lemont B.; Hall, Lowell H. Molecular Connectivity in Drug Research; Academic: New York, 1976.

13. Stuper, Andrew J.; Brugger, William E.; Jurs, Peter C. Computer Assisted Studies of Chemical Structure and Biological Function; Wiley: New York, 1979.
14. Romesburg, H. C. Cluster Analysis for Researchers; Lifetime Learning Press: Redwood City, CA, 1984.
15. Pearlstein, R. A. In Chemlab-II Reference Manual; Molecular Design Limited: San Leandro, CA, 1988.
16. Sprague, Joseph T.; Tai, Julia C.; Yuh, Young; Allinger, Normal L. J. Comp. Chem. 1987, 8, 581-603.
17. Rohrer, D. C.; Karchesy, J.; Deinzer, M. Acta. Cryst. 1984, C40, 1449.
18. Coleman, P. C.; Coucourakis, E. D.; Pretorious, J. A. S. Afr. J. Chem. 1980, 33, 116.
19. Leo, A. In Medchem Software Manual - Release 3.52; Daylight Chemical Information Systems: Irvine, CA., 1987; Chapter 14.
20. Andrews, P. R.; Craik, D. J.; Martin, J. L. J. Med. Chem. 1984, 27, 1648-1657.
21. Wold, Svante Pattern Recognition 1976, 8, 127-139.
22. Milligan, Glenn W.; Cooper, Martha C. Psychometrica 1985, 50, 159-179.
23. Morrison, Donald F. Multivariate Statistical Methods; McGraw-Hill: New York, 1976; pp 259-263.
24. Wold, Svante; Geladi, Paul; Ebensen, Kim; Ohman, Jerker J. Chemometrics 1987, 1, 41-56.

RECEIVED August 2, 1989

Chapter 6

New Tool for the Study of Structure–Activity Relationships in Three Dimensions

The Hypothetical Active-Site Lattice

Arthur M. Doweyko[1]

Uniroyal Chemical Company, Inc., World Headquarters, Middlebury, CT 06749

A three-dimensional computer-based approach has been developed which predictively models receptor or enzyme binding for molecules of widely different structural types. The method is based upon the intermediate conversion of molecules to points in space which embody both atomic character and partial binding values. The merging of data for a series of known inhibitors results in the construction of a HASL (Hypothetical Active Site Lattice) which serves to quantitatively and predictively model enzyme-inhibitor interaction. Details of the HASL methodology are discussed and the approach illustrated using _E. coli_ dihydrofolate reductase inhibitors.

The creative use of computers has led to a variety of ways to generate structure-activity relationships for sets of molecules which bind to receptors or active sites (1-12). These techniques often involve statistical treatment of data in a largely retrospective manner which is further limited by the arbitrary choice of common substructures, pharmacophores, binding points, and molecular overlays. In classical QSAR the limitations can be quite severe in that regressions are confined to a series of closely related structures often differing in relatively minor ways. More recent innovations involve molecular shape analysis (13) and distance geometry methods (14). These techniques represent significant steps toward a true three-dimensional SAR with some predictive

[1]Current address: Ciba–Geigy Corporation, Environmental Health Center, 400 Farmington Avenue, Farmington, CT 06032

modelling capabilities. The present investigation represents an extension of both molecular shape and distance geometry approaches in handling molecule-to-molecule comparisons which permits the construction of an active site model capable of predicting molecular binding based on inhibition data. Structural comparisons between molecules, without arbitrary molecular superpositions, is made possible through the use of a three-dimensional molecular lattice of points. In addition, a working multi-dimensional model of a binding site, a hypothetical active site lattice (HASL), can be built from binding data, e.g. Ki, and used to test molecules with novel structures (15). Unlike other 3D QSAR approaches, the molecules under consideration need not be of the same structural class.

Creation of a Molecular Lattice

In order to minimize the number of calculations needed for three dimensional molecular manipulations a computationally simple method for molecular representation was sought. This concern is particularly significant when using desk-top microcomputers having limited speed, even when equipped with numeric coprocessors. The construction of a computationally facile molecular lattice is illustrated in Figure 1.

The Cartesian coordinates of a molecule are converted to a set of equidistant points arranged orthogonally to each other in three dimensions (x,y,z directions). These points are separated by a distance referred to as the resolution and are all located within the van der Waal's radii of the atoms constituting the molecule (steps A-C, Figure 1). The resulting framework of points will be referred to as the molecular lattice. The number of such points is dependent upon the size of the molecule, the resolution chosen, and the nature of the atoms.

In order for the molecular lattice to accurately represent the molecule, it is necessary to add information to each lattice point reflecting the atom within which it is found. This process is illustrated in Figure 1 (steps D-E) wherein a portion of the molecule is shaded to represent different types of atoms (e.g. electron-rich, electron-poor). The correspondingly shaded lattice points reflect the atom type and thus can be considered a fourth dimensional extension of the three dimensional molecular lattice.

A simple indicator variable was adopted to serve as a physiochemical descriptor and is referred to as H, the HASL type. Loosely based on the quantitative assessment of hydrophobicity derived from a variety of atom types reported for dihydrofolate reductase inhibitors (16), the values for H are integers equal to -1, 0, and +1, roughly corresponding to atoms having low, medium or high electron density, respectively. The H values are defined by MM2 atom type in Table I. Thus, most aliphatic carbons, silane and hydrogens are considered equivalent neutral atoms (H=0), while halogens, oxygen, and nitrogen (H=+1) are electron-rich and

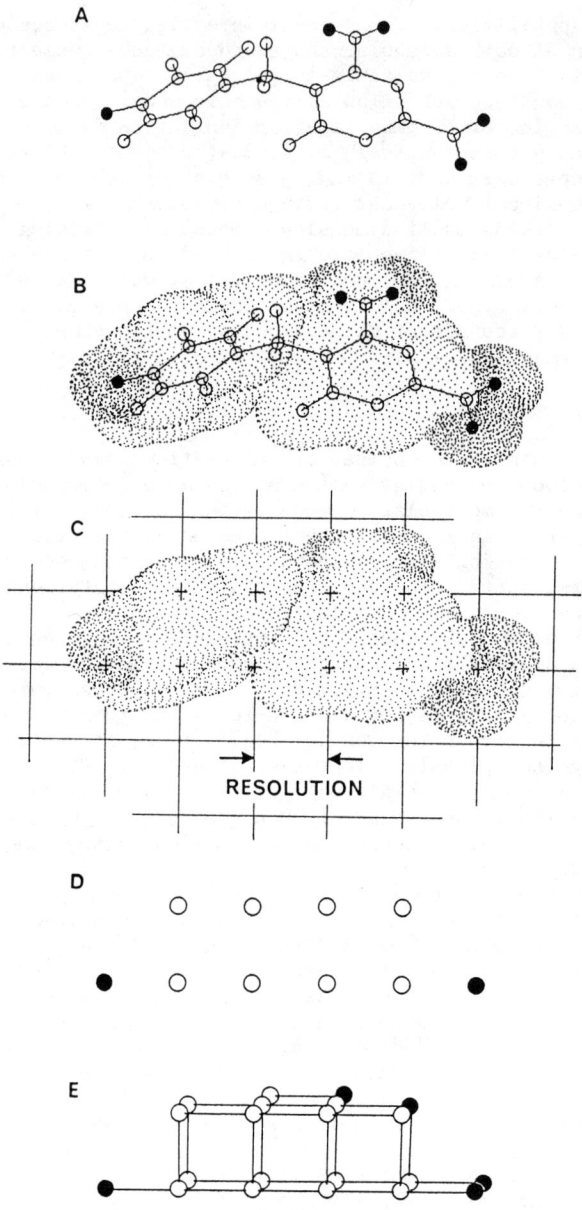

Figure 1. A schematic representation of 4D HASL construction: [A] molecular structure in 3D space, containing different atoms (black and white), [B] superposition of van der Waal's molecular volume, [C] identification of molecular lattice points imbedded within the molecular volume, [D] lattice points shown in two dimensions, [E] a three- and four-dimensional view of these lattice points.

carbonyl carbons, sulfonium and phosphine atoms (H=-1) are electron-poor. This distillation of the atom types to three classes is likely an over-simplification, but appears to serve well in HASL construction in that it is the key to a consistent alignment of molecules by virtue of their atomic makeup. Molecular lattices calculated for p-aminobenzoic acid at several resolution values are illustrated in Figure 2.

Table I. HASL Type (H) Definitions

MM2	H	Atom	Type	MM2	H	Atom	Type
1	0	C	sp alkane	15	+1	S	sulfide
2	0	C	sp alkene	16	-1	S+	sulfonium
3	-1	C	sp carbonyl	17	-1	S	sulfoxide
4	0	C	sp acetylene	18	-1	S	sulfone
5	0	H	hydrogen	19	0	Si	silane
6	+1	O	C(O)H, C(O)C	21	-1	H	O(H)
7	+1	O	C=(O)	22	0	C	cyclopropyl
8	+1	N	sp	23	-1	H	N(H)
9	+1	N	sp	24	-1	H	COO(H)
10	+1	N	sp	25	+1	P	phosphine
11	+1	F	fluoride	26	-1	B	>B-
12	+1	Cl	chloride	27	0	B	>B<
13	+1	Br	bromide	28	-1	H	C=C(H)
14	+1	I	iodide	37	+1	N	C=(N)

Comparing Molecular Lattices

Molecules can be quantitatively compared to one another through the use of molecular lattices. Since each lattice represents a molecule as a finite number of points in space, and each point also contains atomic character information (such as H), it becomes a simple matter to overlay the lattice description of one molecule over that of another, and compute the number of common points found between the two. First, one molecule and its lattice can act as a stationary reference. A second molecule is then centered on the first and its lattice is generated (using the same resolution). After the lattices are checked for commonalties, the second molecule is stepped through a series of translational and rotational movements with a lattice generated at each step. The degree of matching, or FIT, can be computed in a number of ways. A preferred definition of FIT is shown in equation 1. The coordinates of the second molecule that generated the best overall FIT are then taken to represent the optimum overlap on the stationary molecule. Since each

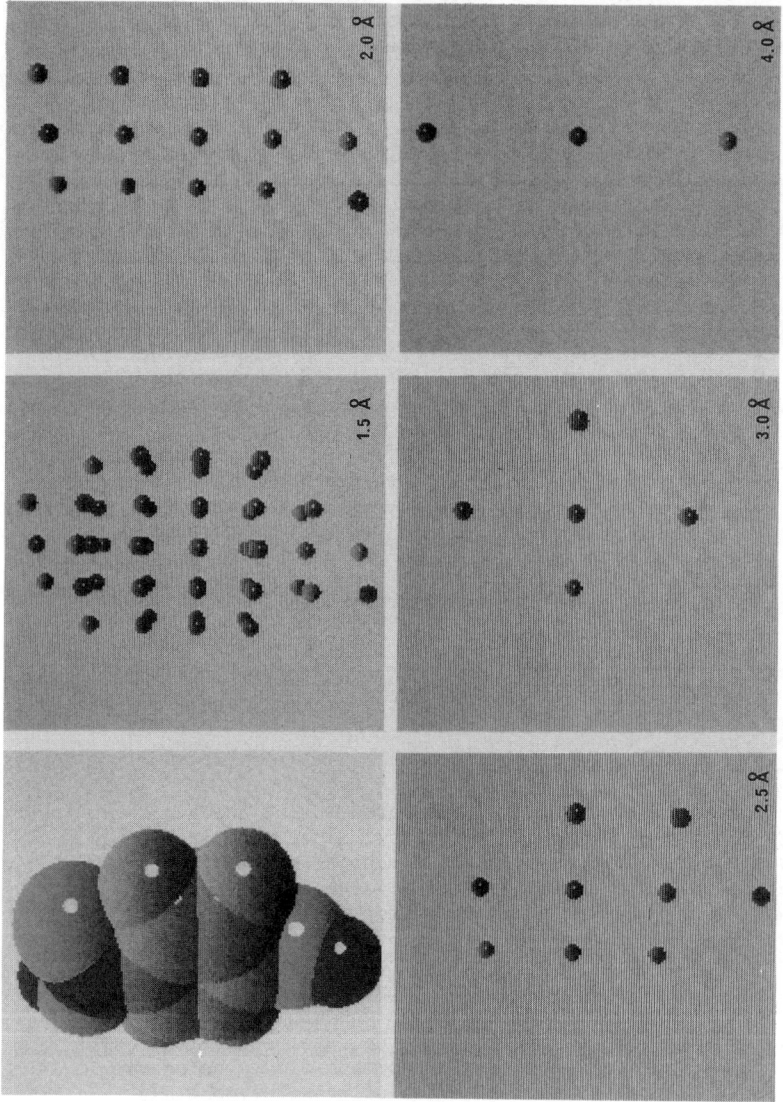

Figure 2. An illustration of lattices at several different resolutions using p-aminobenzoic acid as an example molecule.

lattice point is four-dimensional, a necessary condition for commonality is that two lattice points have identical H values as well as three-dimensional coordinates.

$$\text{FIT} = \frac{L(\text{common})}{L(\text{ref})} + \frac{L(\text{common})}{L(\text{molecule})} \qquad (1)$$

where L (common) = number of lattice points in common between two molecules
L (ref) = number of lattice points belonging to the stationary molecule
L (molecule) = number of lattice points belonging to the moving molecule

The comparison of molecules to one another using lattices for quantitation is illustrated for two simple cases in Figure 3. Benzene overlaid upon itself and rotated about its central axis produces a undulating line reflecting FIT values which peak every 60 degrees corresponding to overlapping carbon-hydrogen nodes. In the case of toluene on toluene, a similar rotational pattern is observed with a maximum FIT achieved as the staggered methyl groups align at the 300 degree mark. The fitting routine conducted at smaller resolution values, as shown in the benzene/benzene example, results in a more sensitive assessment of overlap, reflected by the response/degree in the value of FIT.

After fitting one molecular lattice to another, it is possible to merge the information contained in both to form a composite lattice. This lattice would then reflect the spatial and atomic requirements of both molecules simultaneously. The fitting and merging cycle can be repeated for each additional molecule, continually building up the information content of the resulting composite lattice. Thus, the composite lattice of points represents the spatial and atomic character requirements of all molecules used in its construction.

The Partial pKi Distribution

Enzyme inhibition data is commonly expressed as I(50) or Ki. This data can be made a part of the composite lattice. In this way, not only can the lattice act to assess FIT of novel structures, but it can also provide an estimate of Ki for the novel structures. In order to clearly focus on active site binding, the present investigation is limited to the consideration of competitive enzyme inhibitors. However, in principle, the HASL methodology would lend itself to applications involving less definitive binding as is often encountered in studies where only I(50) values are available, in cases of unspecified binding to a receptor, or where in vivo data, e.g. percent growth inhibition, are considered.

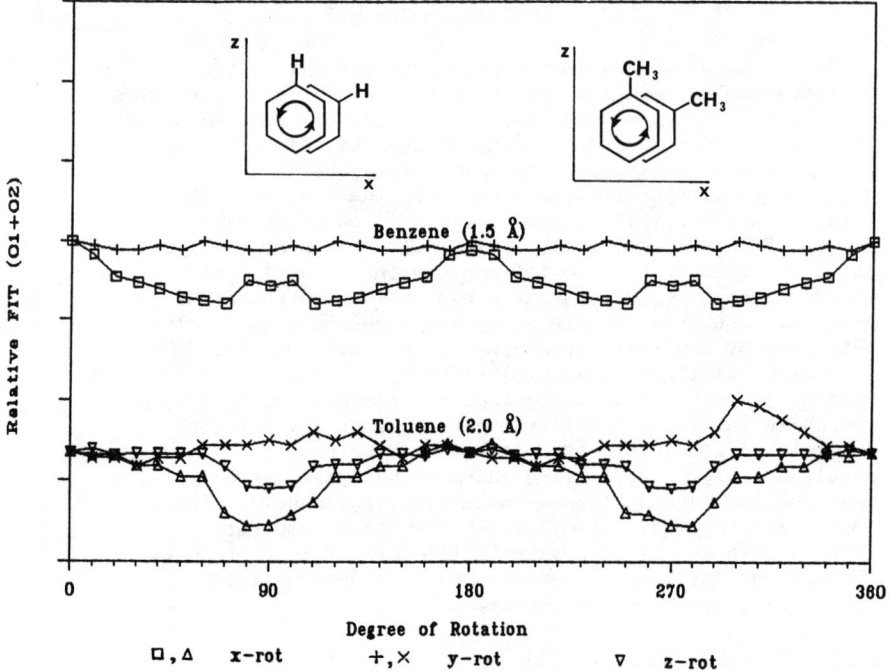

Figure 3. The quantitation of molecular overlap through the intermediacy of molecular lattices. Examples illustrated are the rotational superpositioning of benzene on benzene and toluene on toluene. O1 = L(common)/L(ref) and O2 = L(common)/L(molecule).

Enzyme inhibition data is most conveniently expressed as -log Ki or pKi. In this form, the binding values are directly proportional to the free energy of binding and, therefore, can be dissected into smaller, additive components. For example, as a first approximation, the total pKi of an inhibitor can be divided evenly among its lattice points. These lattice points are presumed to account for the binding of every part of the molecule. Such an equal distribution of partial pKi values is clearly simplistic, since it is likely that a molecule contains portions (typically functional groups) that bind more strongly than other portions.

The key to solving the partial pKi distribution problem is found in the body of inhibition data. The pKi of each molecule (or molecular lattice) that was used to construct the composite lattice needs to be incorporated into the lattice in such a way that refitting an inhibitor molecule to this lattice and adding the partial pKi terms at each lattice point representing that molecule would result in a predicted pKi identical to the original value for the fitted molecule. In this way, a self-consistent mathematical model of active site binding is produced.

A method was found that distributes the partial pKi values in the required predictive manner. This method is illustrated in Figure 4. In the example given, molecules A and B, having pKi's of 3.00 and 6.00, respectively, are used in an attempt to gain information about an active site whose lattice (HASL) consists of four points with partial pKi values of 3.00, 2.00, 1.00, and 0.00. This partial pKi distribution is to be considered as the target of the A/B analysis.

Initially, both molecular lattices reflect an even distribution of pKi. The fitting and merging of molecules A and B results in a HASL with partial pKi values averaged at each point. It is apparent that refitting either molecule onto the present HASL would result in poor pKi predictivity. Fitting molecule A onto the "averaged" HASL provides a prediction of 4.00 (actual pKi = 3.00), while molecule B is predicted to have a pKi of 5.00 (actual pKi = 6.00). It is through a subsequent reiterative technique, reminiscent of the solution of simultaneous equations, that a truly predictive HASL is obtained. Using the "averaged" HASL as a starting point, the fitting of molecule A gives rise to a set of corrections referred to as IN and OUT, whose values are dependent upon the overall error in predicted pKi (ERROR). IN corrections are applied only to those HASL points common to both molecule A and the HASL, while OUT corrections are made to HASL points not used. Step $\underline{i}$ illustrates this process wherein predicted pKi ERROR = -1.00, IN = -0.33 and OUT = 1.00, giving rise to a corrected HASL (3.00, 1.17, 1.17, 0.67). The procedure is repeated with every molecule that was used to create the HASL. In the A/B case a single iterative cycle would be considered as the fitting of molecules A and B, each with corrections applied. The

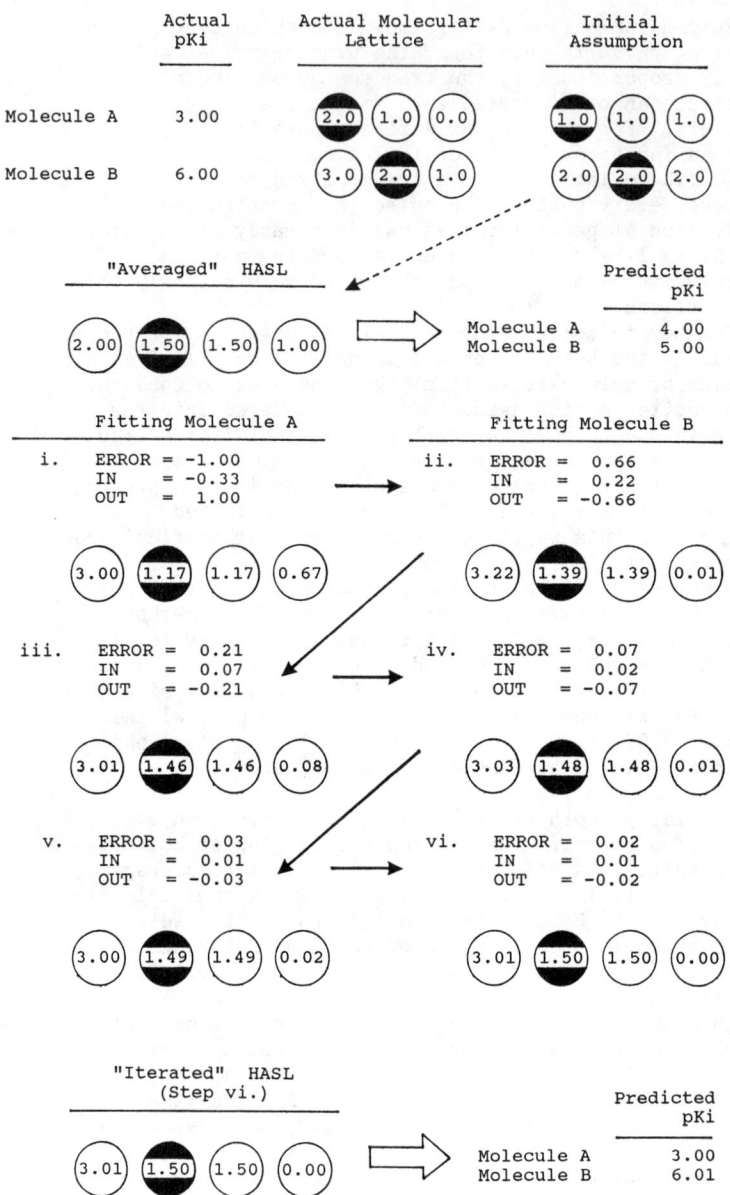

Figure 4. Partial pKi estimates made by carrying out an iterative method on the degree of fitting for molecules A and B. ERROR = actual pKi - predicted pKi. IN is the correction applied to each common lattice point, IN = ERROR/NI, where NI = number of lattice points in common. OUT is the correction applied to each point outside the overlap, OUT = -ERROR/NO, where NO = number of lattice points outside the overlap.

iterative cycle can be repeated until either the error in prediction is removed or minimized. In the present A/B example, six such cycles resulted in near perfect predictivity.

Effect of Resolution

Resolution, or the orthogonal lattice point spacing, can play an important role in the capability of a HASL to effectively predict binding. There exist two logical extremes: (1) very small spacing which results in a large number of lattice points, or (2) very wide spacing which results in very few lattice points. There are disadvantages at either extreme.

As the resolution becomes smaller, the number of lattice points increases. This effect is illustrated for 2,4-diamino-5-methylphenylpyrimidine (D01) in Figure 5A. When the resolution is less than the average atomic van der Waal's radius (ca. 1.8-2.0 Å), the number of resulting points increases with the cube of the spacing. This situation leads to an over-description of the molecule. In addition to an obvious increase in computational time for the assessment of fitting, a subtle but important drawback arises: the over-description of a molecule will result in more points for partial pKi distribution. Essentially, these points represent degrees of freedom to the iterative solution for the partial pKi distribution problem, and therefore, increase the likelihood of obtaining misleading solutions which are not unique. This effect would be expected to compromise HASL predictivity.

The second extreme of wide spacing resulting from a large resolution value would represent an under-description of a molecule. The presence of different atom types may not be accurately assessed when using a large resolution. This effect is illustrated in Figure 5B for the same pyrimidine discussed above. As the resolution is increased beyond 3 angstroms, the observed percent composition of the three HASL types (-1,0,+1) is found to deviate significantly from the theoretical values calculated from van der Waal's volumes. In addition, the solution of the partial pKi distribution problem is affected once again. Without sufficient points to load with partial pKi estimates, predictivity is neccesarily limited to some minimal error regardless of the number of iterative cycles performed.

From the above considerations, it would appear that a resolution choice of 2-3 angstroms would be optimal. This intuitive assessment was tested using several model systems.

Five substituted benzenes were chosen to create a QSAR. These compounds are listed in Table II along with their corresponding substituent physiochemical parameters pi (logP) and MR (molar refractivity). A QSAR set with real parameter values was used in this study in order to obtain a corresponding set of realistic "actual" pKi values.

Table II. The Five Compound QSAR Set Used to Test HASL Predictivity as a Function of Resolution

	pi	MR	pKi
C₆H₅-H	0.00	1.03	3.48
C₆H₅-I	1.12	13.94	2.68
C₆H₅-NH₂	-1.23	5.42	1.05
C₆H₅-CO₂H	-0.32	6.93	2.52
C₆H₅-NHCH₃	-0.47	10.33	1.97

Arbitrary regression equation (2) provided the "actual" pKi values for this compound set:

$$pKi = pi - \frac{pi^2}{2} - 0.1 \, MR + 3.58 \qquad (2)$$

Each compound in the set was fitted to a HASL constructed of the other four. Thus, at each resolution studied, it was possible to obtain five predicted pKi values and compare them to "actual" values. Resolution values ranging from 1.5 to 4.0 angstroms (in increments of 0.125 Å) were each used to construct a four-compound HASL which yielded five tests of predictivity. The average error (actual-predicted pKi) was determined at each resolution value. The results of this analysis are plotted in Figure 6,

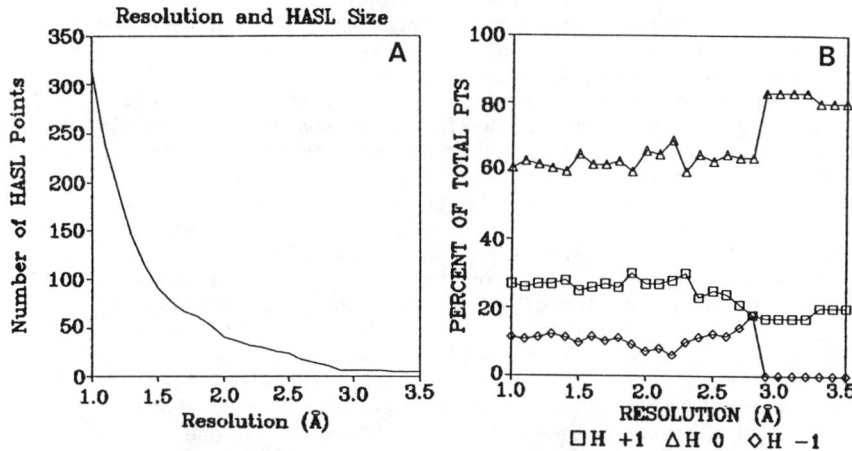

Figure 5. The effects of resolution choice on molecular lattice construction using D01 as an example. [A] A plot of the total number of lattice points as a function of resolution, and [B] a plot of the H distribution as a function of resolution, indicating that H-distribution values begin to approach theoretical levels at resolutions less than or equal to 2.8 angstroms. (Reproduced from Ref. 15. Copyright 1988 American Chemical Society.)

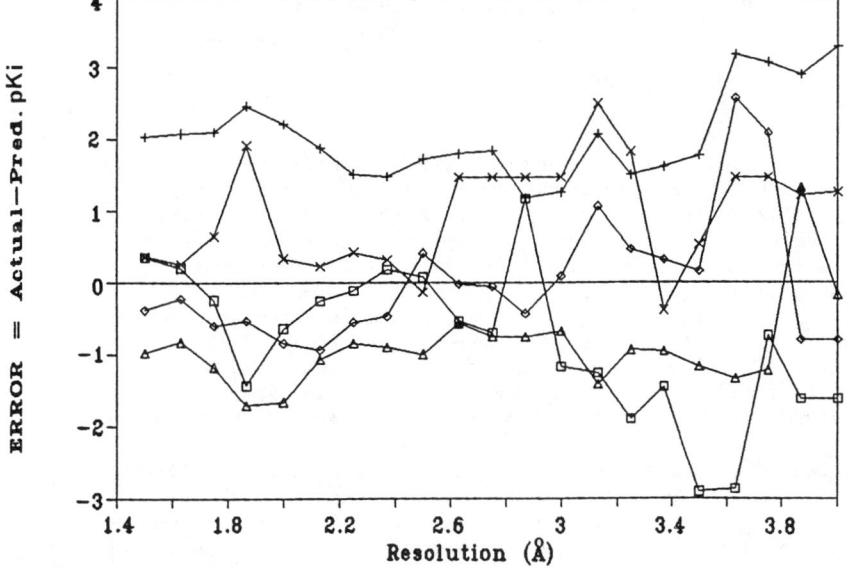

Figure 6. The effect of resolution choice on predictivity. Each point represents an error in the predicted pKi of one of five substituted benzenes (taken from Table II) using a HASL made from the other four compounds.

and indicate that the best predictivity was observed when the resolution was in the range of 2-3 angstroms.

A second test was developed to examine the effect of HASL resolution on the estimates of partial pKi within that HASL. Figure 7 illustrates the four compounds and pKi values comprising this test. p-Aminobenzoic acid is shown to have a pKi of 7.00 and is further detailed to indicate that the carboxyl group, amino group, and phenyl ring binding occurs with partial pKi's of 4.00, 2.00, and 1.00, respectively. Similar detailing of partial pKi values is shown for the other three compounds consistent with the p-aminobenzoic acid structure. Construction of a HASL from the four compounds is then followed by the fitting of p-aminobenzoic acid. From the fitting results it is possible to compute the partial pKi estimates made by the HASL for each of the three moieties under consideration, i.e., the carboxyl, amino, and phenyl groups and compare these estimates with the actual ones. This is done by examining which HASL points correspond to each of these moieties and adding their partial pKi terms. This procedure was carried out over a resolution range of 1.7 to 3.4 angstroms (in increments of 0.1 Å). The results are listed in Table III and illustrated in Figure 8. The absolute average error in HASL predictivity of partial pKi among the three functionalities is found to undergo a minimum roughly in the range of 1.9 to 2.5 angstroms.

Two independent tests designed to examine resolution effects on HASL predictivity have confirmed that the best results are achieved when HASL resolution lies within a 2-3 angstrom range. A more demanding test of HASL predictivity is in its application to an actual inhibitor set using an enzyme whose active site is well known.

Dihydrofolate Reductase

E. Coli dihydrofolate reductase (DHFR) inhibitors were chosen to assess the HASL methodology since a large number of competitive inhibitors are known which encompass a variety of structures. In addition, the crystal structure of the enzyme and the active site orientation of a strong inhibitor, methotrexate (MTX), are also known (17). The set of 72 inhibitor structures (18-22) used for this analysis is listed in Table 4. All structures were energy minimized using MM2 (QCPE MM2 was adapted for use on the IBM-PC by Kevin E. Gilbert and Joseph J. Gajewski). As done in classical QSAR methodology, the DHFR inhibitor set was divided into two groups: a learning set and a test set. The learning set was selected by using parameters such as the number of different atoms, number of rings, and molecular weights in a repetitive version of the algorithm of Wooton (23-24), which yielded 37 compounds that differ from one another as much as possible in terms of these three parameters. These compounds are indicated with asterisks in Table IV.

The learning set was used to generate HASL descriptions at resolutions of 2.8, 3.2, 3.5, and 4.0 angstroms. Partial

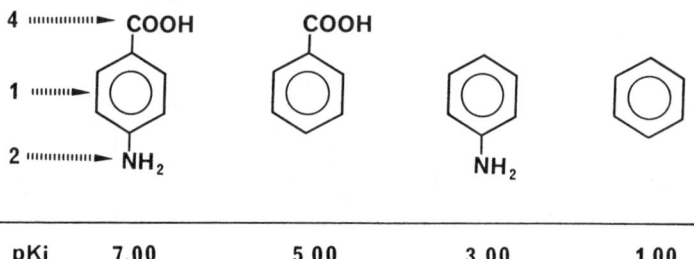

| pKi | 7.00 | 5.00 | 3.00 | 1.00 |

Figure 7. The partial pKi values arbitrarily set for p-aminobenzoic acid and three related molecules used to test the HASL capability to estimate partial pKi values at different resolutions.

Table III.   p-Aminobenzoic Acid Partial pKi Estimations

	COOH	NH2	Phenyl	Total
Partial pKi (Actual)	4.00	2.00	1.00	7.00

Resolution (Angstroms)	Estimated Partial pKi				Abs. Ave. Error
	COOH	NH2	Phenyl	Total	
1.7	2.72	2.48	1.95	7.15	0.90
1.8	2.42	2.33	2.18	6.93	1.03
1.9	3.30	2.28	1.46	7.05	0.48
2.0	3.94	2.10	1.05	7.09	0.07
2.1	4.37	1.85	0.93	7.15	0.20
2.2	4.50	1.66	1.03	7.19	0.29
2.3	2.88	2.47	1.63	6.98	0.74
2.4	3.19	2.10	1.71	7.00	0.54
2.5	3.30	2.04	1.70	7.04	0.48
2.6	3.08	2.05	1.91	7.04	0.63
2.7	2.86	2.05	2.12	7.03	0.77
2.8	2.48	2.65	2.03	7.16	1.07
2.9	2.69	2.70	1.79	7.18	0.93
3.0	1.91	3.94	1.25	7.10	1.09
3.1	1.91	3.94	1.25	7.10	1.09
3.2	1.21	3.94	1.99	7.14	1.91
3.3	1.65	3.79	1.73	7.17	1.62

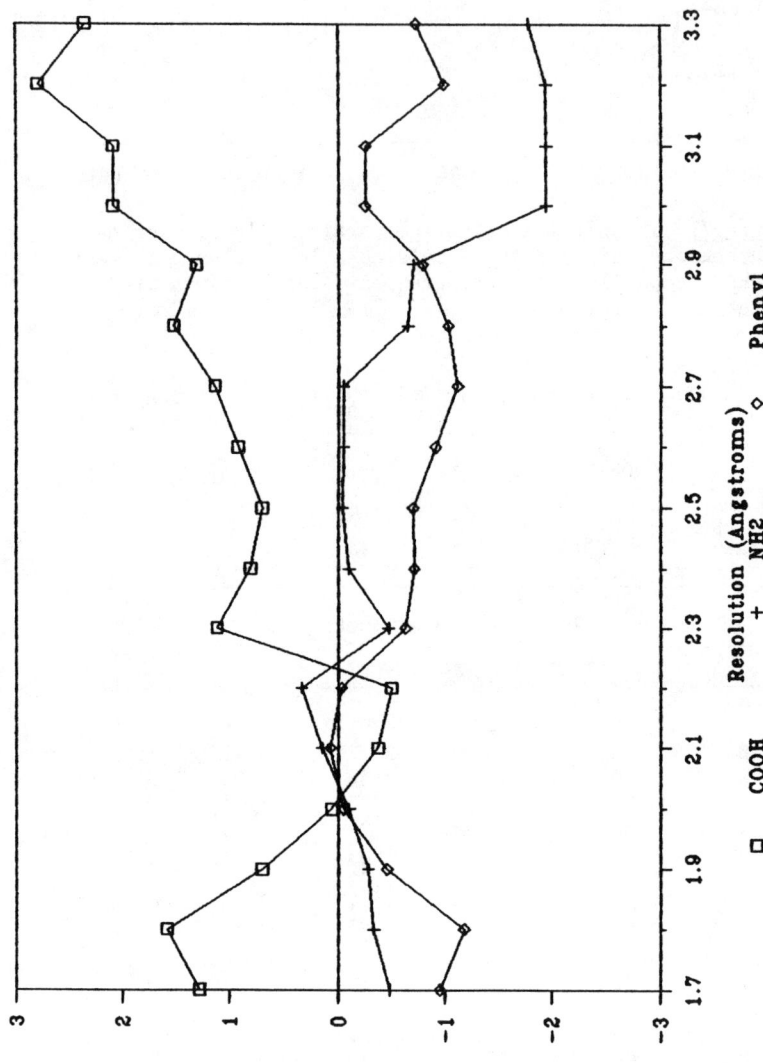

Figure 8. Error in the estimation of partial pKi values for COOH, NH2, and phenyl groups (Figure 7) as a function of resolution.

6. DOWEYKO Tool for SAR Study in Three Dimensions

Table IV. Dihydrofolate Reductase Inhibitor Set

#	R	pKi	#	R	pKi	#	R	pKi
D01*	H	6.18	D28	4-OCH2CH2OCH3	6.40	D55*	3-CF3	5.69
D02	3-F	6.23	D29	3-OH	6.47	D56*	3-F	5.85
D03	4-NH2	6.30	D30*	3-OCH2CH2OCH3	6.53	D57*	H	4.51
D04	4-F	6.35	D31	3-CH2O(CH2)3CH3	6.55	D58	3-Cl	5.87
D05	4-Cl	6.45	D32	3-OCH2CONH2	6.57	D59*	3-I	5.58
D06	3,4-(OH)2	6.46	D33	3-OCH2CH3	6.59	D60*	3-CN	5.51
D07*	4-CH3	6.48	D34	4-N(CH3)2	6.78	D61*	3-CH3	5.42
D08	3-Cl	6.65	D35	3-O(CH2)3CH3	6.82	D62	3-(CH2)5CH3	5.75
D09	3-CH3	6.70	D36*	3-O(CH2)5CH3	6.82	D63*	3-C(CH3)3	4.72
D10	4-Br	6.82	D37	4-O(CH2)3CH3	6.89	D64	3-O(CH2)3CH3	6.02
D11	4-OCH3	6.82	D38*	3-OCH2C6H5	6.99	D65*	3-OCH2C6H5	5.31
D12*	4-NHCOCH3	6.89	D39*	3,4-(OCH2CH2OCH3)2	7.22	D66*		8.36
D13	3-OCH3	6.93	D40*	3,5-(OCH3)2-4-O(CH2)2OCH3	8.35	D67*		7.17
D14	3-Br	6.96	D41*	3,5-(OCH3)2-4-Br	9.22	D68		6.55
D15	3-CF3	7.02	D42*	3,5-(OCH3)2-5-OCH2COOH	8.59	D69*	3,5-(OCH3)2	9.31
D16*	3-I	7.23	D43*	3-OCH3-4-Br-5-OCH2COOH	8.80	D70	4-OCH3	8.30
D17*	3-CF3	7.69	D44*	3,4-(OCH3)2-5-O(CH2)2COOH	9.23	D71*	3,5-(OCH3)2	5.95
D18*	3,4-(OCH3)2	7.72	D45*	3-OCH3-4-Br-5-O(CH2)2COOH	10.46	D72	4-OCH3	6.89
D19	3,5-(OCH3)2	8.38	D46*	3,4-(OCH3)2-5-O(CH2)3COOH	10.46			
D20*	3,4,5-(OCH3)3	8.87	D47	3-OCH3-4-Br-5-O(CH2)3COOH	10.49	Inhibition constants obtained from:		
D21*	3,4-(OH)2	3.04	D48*	3,4-(OCH3)2-5-O(CH2)4COOH	10.18	Ref. 18 (D01-D40), Ref. 19 (D41-D53),		
D22*	4-O(CH2)6CH3	5.60	D49*	3-OCH3-4-Br-5-O(CH2)4COOH	10.40	Ref. 20 (D54-D65), Ref. 21 (D68-D72).		
D23	4-O(CH2)5CH3	6.07	D50*	3,4-(OCH3)2-5-O(CH2)5COOH	10.62	Asterisk (*) denotes learning set member.		
D24	3-O(CH2)7CH3	6.25	D51	3-OCH3-4-Br-5-O(CH2)5COOH	10.92			
D25	3-CH2OH	6.28	D52	3,4-(OCH3)2-5-O(CH2)6COOH	10.30			
D26	3,5-(CH2OH)2	6.31	D53	3-OCH3-4-Br-5-O(CH2)6COOH	10.54			
D27*	3-O(CH2)6CH3	6.39	D54*	3-CONH2	3.48			

pKi distribution among the HASL points was solved to within a predictivity (actual pKi - predicted pKi) of 0.1 pKi units, with the exception of the 4.0 Å HASL, which was minimized to a predictivity of 1.14 pKi units.

Binding predictions were obtained for the entire inhibitor set at each resolution and the results plotted separately for learning and test set members in Figure 9. As expected, good predictivity was observed for the members of the learning set, with some perceptible scatter in evidence at 4 angstroms. For members of the test set, scatter was observed to increase with increasing resolution, a result essentially mirrored in the correlation coefficient (r). Since the learning set does not contain all the test set structural information, the observed scatter is expected. The results obtained at 2.8 angstroms are encouraging since a reasonable HASL was obtained (r = 0.753).

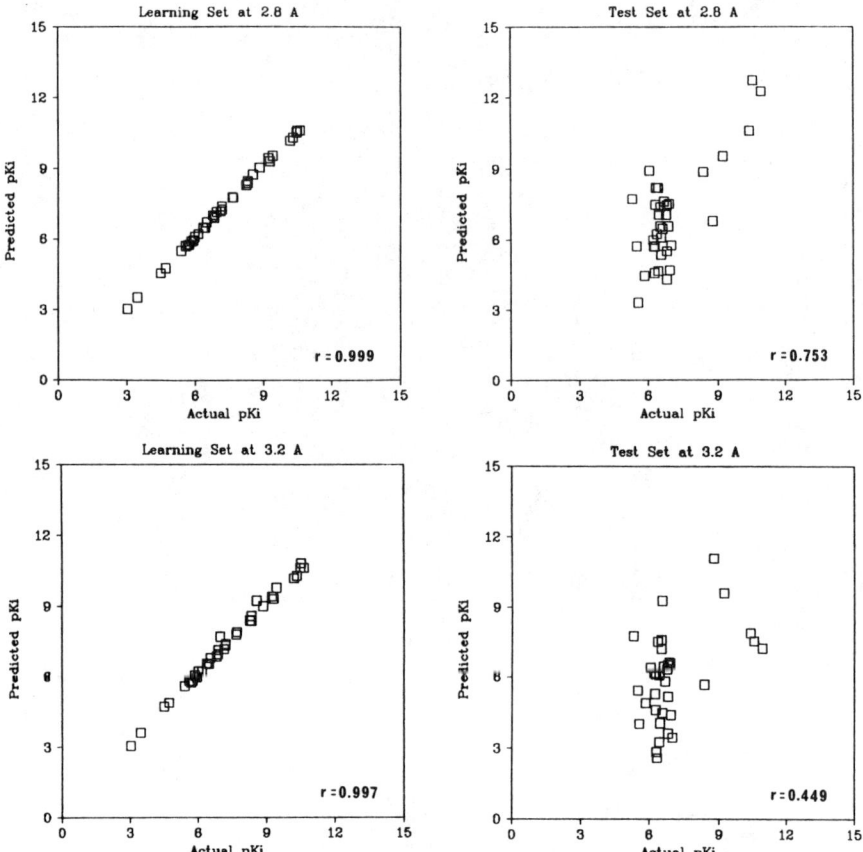

Figure 9. The effects of resolution choice on HASL predictivity using comparisons between learning set and test set DHFR inhibitor data. Plots compare predictivities for both sets at 2.8- and 3.2-angstrom resolutions. Correlation coefficients (r) are shown for each plot. (Continued on next page.)

The entire 72 compound set was used to construct a HASL at 2.8 angstroms in order to characterize the active site and make comparisons with crystal data. The resulting HASL was found to consist of 160 points. Although it is difficult to draw insight from the geometric representation of a HASL, an attempt is made in Figures 10A and 10B. The HASL points which represent H=-1 are depicted as small spheres in 3D space set apart in increments of 2.8 angstroms. The colors green, yellow, and red indicate strong binding (partial pKi > 1.0), weak binding (0.0 < partial pKi < 1.0), and poor binding (partial pKi < 0.0), respectively. To help orient the viewer, panel 10B includes inhibitor D66 docked to the HASL with its 2,4-diamino groups marked by arrows. In examining the relative orientations of other inhibitors after fitting, it was found that they too have their 2,4-diamino groups oriented in a similar manner.

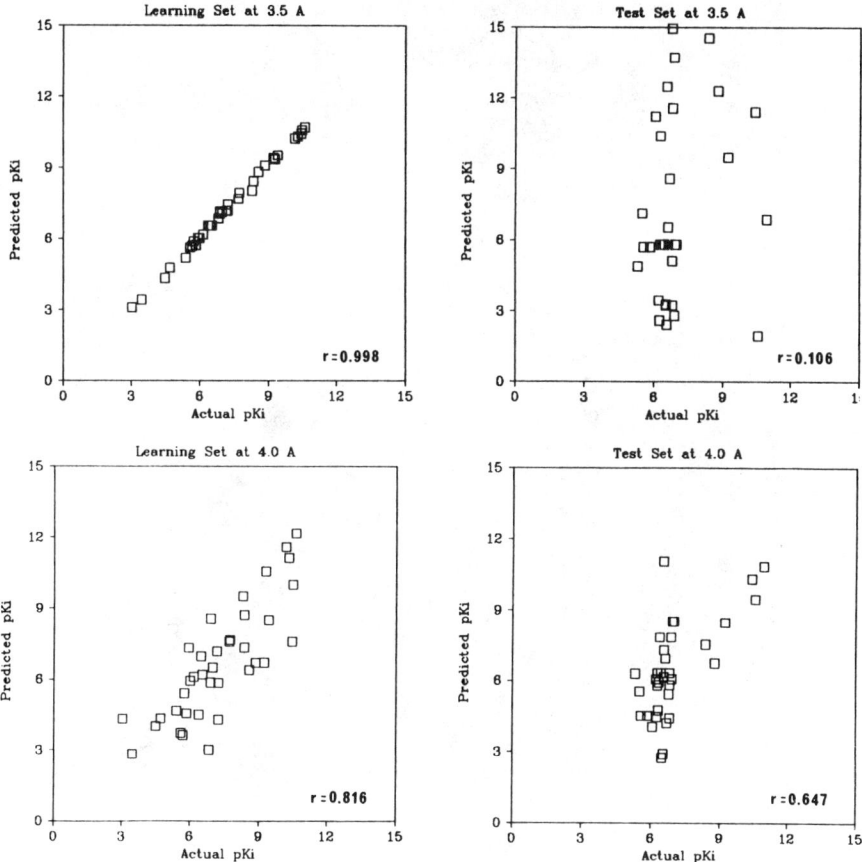

Figure 9. Continued. Plots compare predictivities for both sets at 3.5- and 4.0-angstrom resolutions. (Reproduced from ref. 13. Copyright 1980 American Chemical Society.)

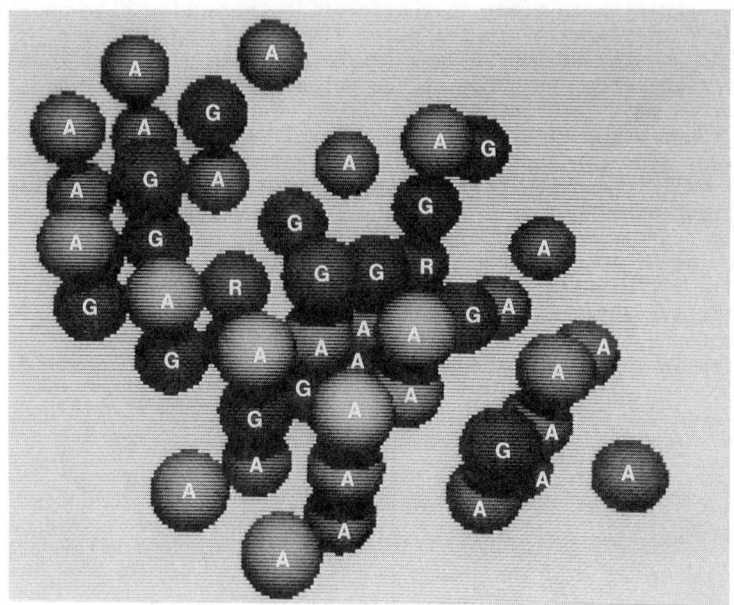

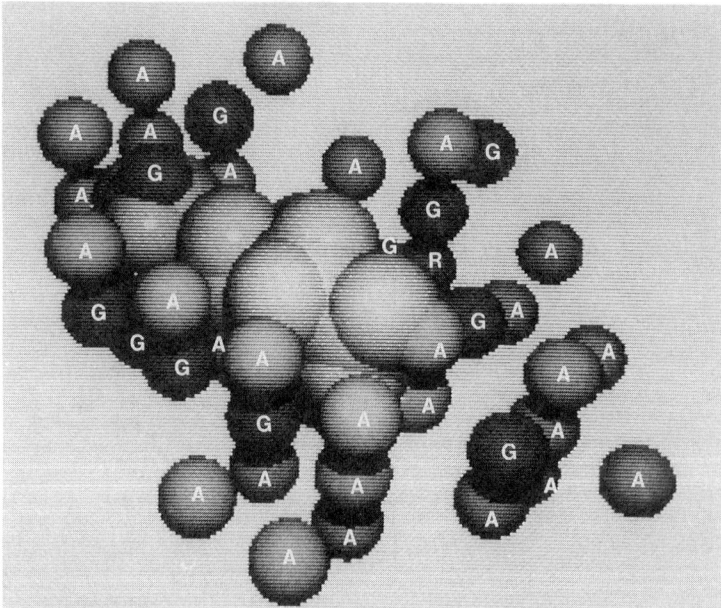

Figure 10. (Top) A partial representation of the DHFR HASL showing only those points with H = −1. The relative binding energy at each point is indicated by letter: G, pKi > 1.0; A, 0.0 < pKi < 1.0; R, pKi < 0.0. (Bottom) The same view with compound D66 fitted in the HASL.

In order to test how closely this 2.8 Å E. coli DHFR HASL comes to mimicking the actual active site, MTX, in its bound conformation, was used as a probe. The fitting of MTX resulted in 65% of its molecular lattice points coinciding with those of the HASL with a predicted pKi of 10.11 (actual pKi = 10.89). Considering that no MTX-like structures were used in HASL model, the predicted pKi is in quite good agreement with the experimental value. The orientation of methotrexate in the HASL was different from the other 2,4-diamino inhibitors. This prediction is consistent with published observations (25) that the 2,4-diamino portion of DHFR-bound MTX appears in a different orientation from that observed for the same moiety in DHFR-bound trimethoprim (D20). The relative orientations adopted after fitting to the DHFR HASL for both MTX and trimethoprim (D20) are illustrated in Figure 11. The positions of the 2,4-diamino groups indicate the relative molecular orientations.

The HASL Methodology

The logic flow chart in Figure 12 summarizes the key steps in the creation and use of a hypothetical active site lattice (HASL). The process begins with structural input, typically in the form of energy-minimized atomic Cartesian coordinates along with MM2 atom type designations (A). From this data a molecular lattice is created based on a resolution value selected by the user (B). This lattice can either be compared with other molecules, or considered as the initial HASL. As new molecules are brought into the system, each is put through the fitting routine (C). Predicted binding is immediately available (D). The data can then be merged into the existing HASL (E). Partial pKi distribution can be computed using HISTORY files containing all previously fitted molecular coordinates (F).

Conclusions

A new method for quantitative structural comparisons, creation of a hypothetical active site (HASL), and modelling potential inhibitor binding has been developed. The method accommodates a wide variety of structural types and makes no assumptions about relative orientations between them. The HASL can be used as a predictive tool to assess potentially useful structures and provides the means to create and test structures completely outside the learning set.
Further enhancements to HASL methodology are expected to include faster and more intelligent fitting algorithms, the incorporation of some structural flexibility options to free inhibitor molecules from the static conformation assumption, and algorithms to electronically synthesize potential inhibitors based on the information content of a HASL. All programs are written in FORTRAN and BASIC for use on IBM-PC compatible systems and are available from the author.

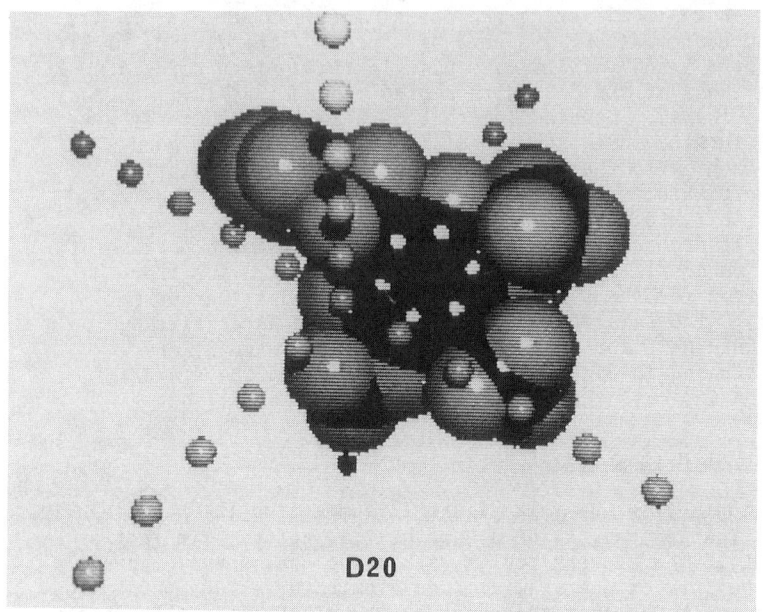

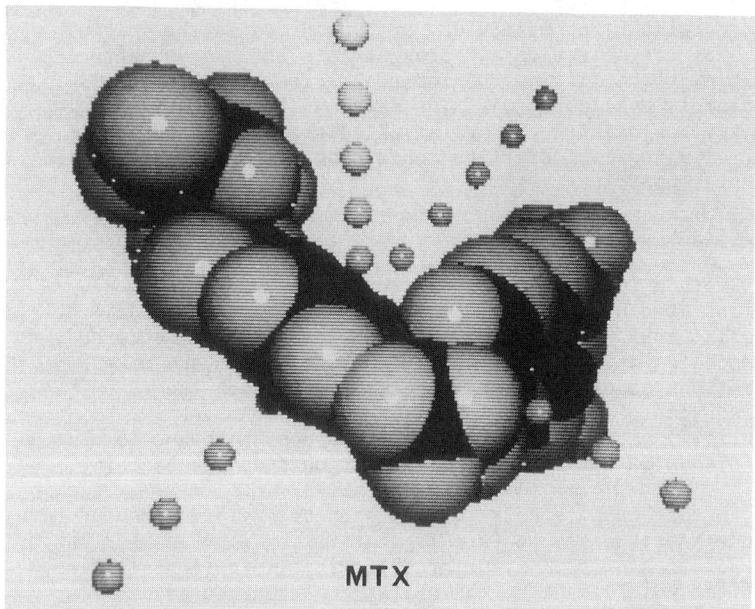

Figure 11. A comparison of the fitted orientations on the DHFR HASL for two inhibitors, trimethoprim (D20) and methotrexate (MTX). (Reproduced from Ref. 15. Copyright 1988 American Chemical Society.)

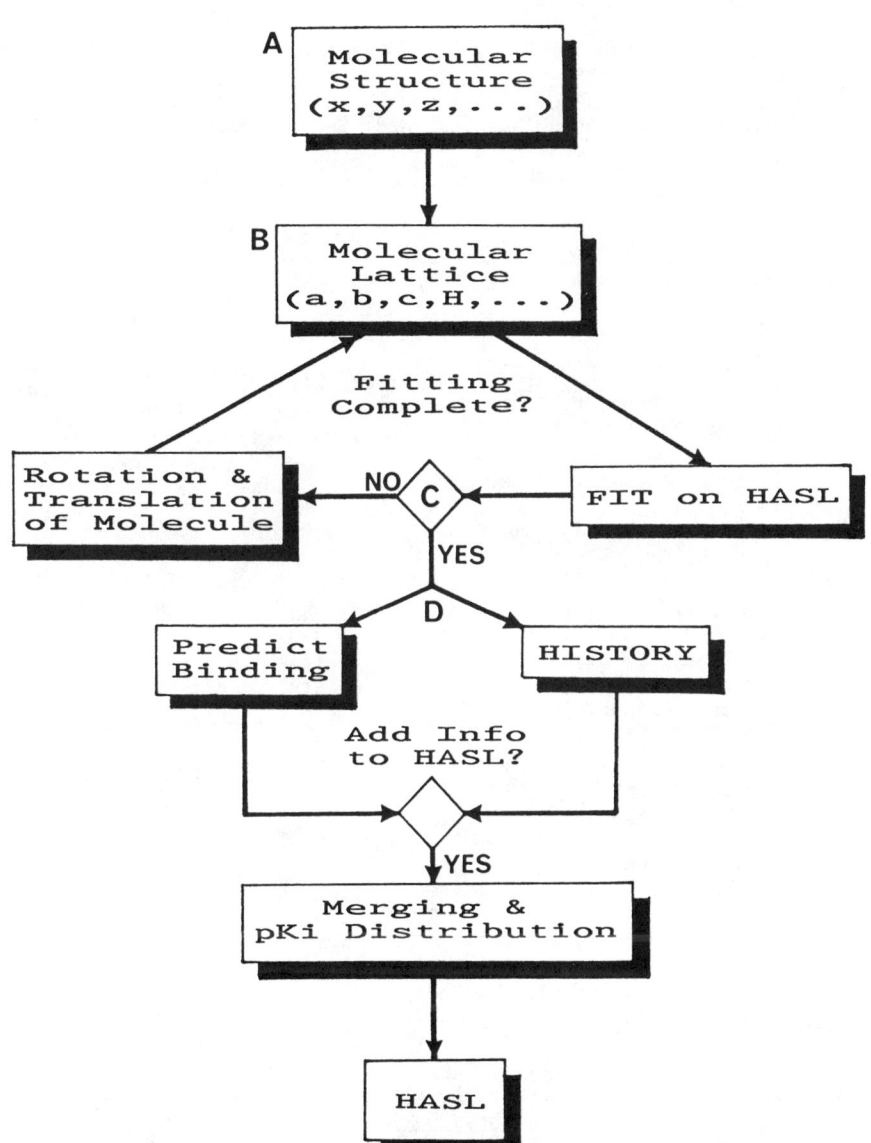

Figure 12. HASL logic flow chart. [A] input of Cartesian coordinates for structure, [B] generation of lattice containing spatial and physiochemical information, [C] fitting routine involving superposition of molecular lattice on HASL, [D] generation of results, which include binding prediction, record (HISTORY) files, and options to merge lattices and/or calculate partial pKi distribution.

Literature Cited

1. Silipo, C.; Hansch, C. J. Am. Chem. Soc. 1975, 97, 6849.
2. Jurs, P. C.; Isenhour, T. Chemical Applications of Pattern Recognition; Wiley-Interscience: New York, 1975.
3. Cammarata, A. Menon, G. K. J. Med. Chem. 1976, 19, 739.
4. Henry, D. R.; Block, J. H. J. Med. Chem. 1979, 22, 465.
5. Marshall, G. R. Comput.-Aided Mol. Des. (Proc. 2-Day Conf.), 1984, 1.
6. Jurs, P. C.; Stouch, T. R.; Czerwinski, M.; Narvaez, J. N. J. Chem. Inf. Comput. Sci. 1985, 25, 295.
7. Hopfinger, A. J. J. Med. Chem. 1985, 28(9), 1133.
8. Bowen-Jenkins, P. Laboratory Practice 1985, Dec., 10.
9. Brint, A. T.; Willett, P. J. Chem. Inf. Comput. Sci. 1987, 27, 152.
10. Unger, S. H. Drug Inf. Journal 1987, 21, 267.
11. Gund, T.; Gund, P. "Three Dimensional Molecular Modeling by Computer," VCH Publishers, Inc., NY, 1987, pp. 319-340.
12. Tollenaere, J. P.; Jansen, P. A. J. Med. Research Rev. 1988, 8(1), 1.
13. Hopfinger, A. J. J. Am. Chem. Soc. 1980, 102, 7196.
14. Crippen, G. M. J. Med. Chem. 1979, 22, 988.
15. Doweyko, A. M. J. Med. Chem. 1988, 31, 1396.
16. Ghose, A. K.; Crippen, G. M. J. Med. Chem. 1985, 28, 333.
17. Bernstein, F. C.; Koetzle, T. F.; Williams, G. J. B.; Meyer, E. F., Jr.; Brice, M. D.; Rodgers, J. R.; Kennard, O.; Shimanouchi, T.; Tasumi, M. J. Mol. Biol. 1977, 112, 535.
18. Hansch, C.; Li, R.; Blaney, J. M.; Langridge, R. J. Med. Chem. 1982, 25, 777.
19. Muller, K. Actual Chim. Ther. 1984, 11, 113.
20. Coats, E. A.; Genther, C. S.; Selassie, C. D.; Strong, C. D.; Hansch, C. J. Med. Chem. 1985, 28, 1910.
21. Burchall, J. J.; Hitchings, G. H. Mol. Pharmacol. 1965, 1, 126.
22. Maag, H.; Locher, R.; Daly, J. J.; Kompis, I. Helv. Chim. Acta 1986, 69, 887.
23. Wooton, R.; Cranfield, R; Sheppey, G. C.; Goodford, P. J. J. Med. Chem. 1975, 18, 607.
24. Doweyko, A. M.; Bell, A. R.; Minatelli, J. A.; Relyea, D. I. J. Med. Chem. 1983, 26, 475.
25. Champness, J. N.; Kuyper, L. F.; Beddell, C. R. In Molecular Graphics and Drug Design; Burger, A. S. V., Roberts, G. C. K., Tute, M. S., Eds.; Elsevier: New York, 1986.

RECEIVED March 21, 1989

Chapter 7

Finding Lead Structures from Amino Acid Sequence Similarities of Target Proteins

Takaaki Nishioka, Kazuo Sumi[1], and Jun'ichi Oda

Institute for Chemical Research, Kyoto University, Uji, Kyoto 611, Japan

>When a target enzyme for drug development is related to
>another enzyme by genetic evolution, these two
>homologous enzymes retain some similarities in the
>amino acid sequences and substrate specificities. The
>target enzyme shows some degree of affinity to the
>substrates, cofactors, and inhibitors of the other
>homologous enzyme. These compounds contain lead
>structures for new inhibitors of the target enzyme. To
>extend this approach to the cases in which two enzymes
>show only a weak and local similarity in their
>sequences, it must be decided whether the similarity
>found is due to chance or due to their functional
>similarity. A method called 'homology graphing' for
>sequence analysis was developed to detect sequence-
>function relationships in proteins. Its application to
>find lead structures is discussed.

The advancement of molecular biology techniques makes DNA sequencing of genes easy. GenBank Genetic Sequence Data Bank (IntelliGenetic Inc., Los Alamos National Laboratory) estimated that about five million DNA base pairs were sequenced in 1987. More than 90% of the known amino acid sequences have been deduced from translation of the DNA base sequences of their genes. By April 1988, 5,251 amino acid sequences had been registered in the NBRF Protein Sequence Database (NBRF; Protein Identification Resource at the National Biomedical Research Foundation). Sequence data have gradually increased for the proteins of interest in the development of medicines and agrochemicals; sequences are available in GenBank (June 1988) for acetylcholine, insulin, estrogen, GABA, and beta-adrenergic receptors, proteins of photosynthetic reaction centers and photosystems, and acetolactate synthase. These rapid increases in the sequence data stimulated us to develop a method for sequence-based drug design.

[1]Current address: Central Research Laboratory, Idemitsu Kosan and Company, Kimitsu, Chiba 299–02, Japan

In contrast to the sequence data, crystallographic data of proteins are still limited and are not rapidly increasing. The crystal structures available in the Protein Data Bank (Brookhaven National Laboratory) are 143 structures of only 75 different proteins on January 1987. Since this is far short of the number with known sequences, the need for accurate prediction of three-dimensional structure from sequence increases rapidly in the field of drug design.

Present practical methods to derive a three-dimensional model of a protein from its sequence, however, use a tertiary template such as crystal structures of other proteins closely related by sequence to the protein to be modeled (1, 2). When the crystallographic data are not available for related proteins, predictions have so far been too low in reliability to evaluate the energy of interactions between the protein and a drug molecule, and to optimize the chemical structure of a drug to fit into the active site in a satisfactory way. Even in the prediction of secondary structures, the match between the observed and the estimated is as low as 50-60% at most (3, 4). The low success score of structure prediction limits the availability of sequence information for drug design.

Another factor that retards the application of the data about structure and sequence of protein to drug design is the shortage of available knowledge about amino acid sequence-function relationships in proteins. Although there are several different kinds of protein functions, we here use the term 'function' to mean 'molecular recognition' such as the ability of enzymes and hormonal receptors to differentiate the chemical structures of substrates and hormones from those of other chemicals. In other words, we are actually interested in the relationship between a sequence and a chemical structure recognized by the sequence. Even when the three-dimensional structure of a target protein is known, without this functional knowledge it remains difficult to find a chemical structure of a lead compound that fits into the binding site of the protein with higher affinity .

In this chapter, we first review two examples of searching for enzyme inhibitors using amino acid sequence similarity between a target enzyme and another enzyme. There are two extreme cases of similarity; two sequences show (1) similarity along the entire span from the N-terminal to the C-terminal and (2) local similarity within a short region. In the first case, the two enzymes might have derived from a common ancestral protein and retain some similarities in their substrate specificities. The target enzyme recognizes the chemical structures similar to the substrates, cofactors, and inhibitors of the other homologous enzyme. Therefore it is expected that new lead structures could be found out of the structures of the ligands of the other homologous enzyme. In the second case, the sequence similarity found is by chance or due to sharing a piece of the ancestral protein. The two enzymes must retain similarity only in part of their function.

Next, we explain evolutional and structural bases for searching for sequence-function relationships in proteins. Then, we describe 'homology graphing', a method for sequence analysis to identify which regions in a sequence are those of functionally importance, and its applications to sequence-function relationships and lead-structure findings.

Target Protein is Inhibited by the Ligands of Other Proteins Related by Sequence Similarity.

Similarity of Proteins. When proteins are compared to find whether they are in the same group, their similarity can be tested in two aspects. One is protein function; proteins that are identical in their catalytic functions have the same enzymatic name and EC-number given by the International Union of Biochemistry (IUB). The other way of grouping is based on amino acid sequence. Proteins of similar sequences are classified into the same group.

Each protein is identified only by its amino acid sequence and related to the gene which codes the amino acid sequence. The proteins of a certain enzyme are identical in their functions, but they are slightly different, in some cases quite different, in their amino acid sequences from species to species which they are isolated from. For example, about twenty different proteins are called lysozyme based on their reaction type and substrate specificity, but they are classified into four different groups of proteins based on the similarity in their amino acid sequences. Therefore, one enzyme is usually related to several different amino acid sequences.

Owing to the development of computer programs for similarity ('homology') search between sequences (5-7), the sequence of a protein can be compared against all of the sequences registered in sequence databases. One protein will be related to other proteins through sequence similairty of various degrees from strong to weak and from global (between entire sequences) to local (between sequence segments).

Global Sequence Similarity: Family and Superfamily. NBRF organized proteins with the known sequences into hierarchical groups of families and superfamilies based on their global sequence similarities (8). Proteins are grouped into one family when their sequences differ from each other at fewer than half of their amino acid positions. Within a superfamily, sequences of any two proteins are similar at the level that the probability of finding the similarity by chance is less than 10^{-6}. NBRF ver.16 (April 1988) classified 5,251 sequences into 1,629 superfamilies.

Proteins that belong to the same family or superfamily are almost identical in the folding of their polypeptide chains and are very similar in their functions, because protein folding is uniquely determined by the amino acid sequence under physiological conditions. Alpha-chymotrypsin and elastase, for example, are identical at about 40% of their amino acid positions. Following Dayhoff's classification, these two hydrolytic enzymes belong to the same superfamily. In fact, difference between the three-dimensional structures of the two enzymes was only 1.8 Å when compared by the root mean square value of the differences in the positions of the corresponding alpha-carbon atoms (9). No exceptions in which the known three-dimensional structures are different from each other have been found among proteins in the same family or superfamily.

Proteins of interest for drug design, in general, are those found in pest insects, weeds, and pathogenic microorganisms, but not

widely found in animals and plants. There have been very few
target proteins that are in the same family or superfamily as other
proteins.

Application of Global Sequence Similarity to Find an Inhibitor of
Acetolactate Synthase. Acetolactate synthase (ALS) is the site of
action of sulfonylurea, imidazolinone, and triazolo pyrimidine
herbicides (10-14). Their mode of inhibition and binding sites on
ALS were ambiguous, because (1) these herbicides bear no obvious
similarity in their chemical structures to those of ALS substrates
(pyruvate and acetolactate), cofactors (thiamine pyrophosphate, FAD,
and Mg^{++}) and effectors (valine, isoleucine, and leucine) and (2)
they inhibit ALS in a mode too complex to be analyzed.

ALS genes have been cloned and sequenced from bacteria
(Escherichia (15-17)), yeast (Saccharomyces (18)), and higher plants
(Arabidopsis and Nicotiana (19)). These sequences are very simlar
to that of the pyruvate oxidase (POX) gene from E. coli (20). Since
ALS and POX are about 30% identical in the amino acid positions of
their sequences composed of about 540 residues, these two enzymes
are classified into the same superfamily. Some of their substrates
and cofactors (pyruvate, FAD, and thiamine pyrophosphate) are common
to the two enzymes. In addition, the chemical reaction catalyzed by
ALS proceeds very similarly to that of POX (Figure 1). In both of
the first reaction steps, thiamine and pyruvate react to form inter-
mediate $\underset{\sim}{1}$. FAD and ubiquinone-8 (Qox) oxidize the intermediate $\underset{\sim}{1}$ in
POX to form acetate (path $\underline{a}$ in Figure 1), while another pyruvate is
added to the intermediate $\underset{\sim}{1}$ in ALS to form acetolactate (path $\underline{b}$ in
Figure 1). Therefore, ALS and POX may have evolutionarily diverged
from a common ancestral protein.

These similarities between ALS and POX were noticed by Schloss
et al. at E.I.du Pont (21). They proposed that ALS has an almost
identical binding site to that of POX. It was expected that ALS
still retains a latent ubiquinone-binding site, because POX has the
binding site for ubiquinone-8 (Qox in Figure 1). Homologues of
ubiquinone-8 potently inhibited the ALS activity and also inhibited
the specific binding of sulfometuron methyl to ALS. These results
revealed that ALS shares a common evolutionary heritage with the
ubiquinone-binding site of POX and that this site in ALS is close to
or overlaps with the herbicide-binding site.

In terms of drug design, the du Pont chemists found a quinone
structure as a lead structure of new inhibitors of ALS based on the
sequence similarity.

Application of Local Sequence Similarity to Find Inhibitors of
Glutathione Synthetase. The other extreme case is one in which
local sequence similarity between sequence segments stretching 20 to
50 amino acid residues is found between a target enzyme and other
proteins which seem not to be mutually related by any biochemical
context.

During the protein engineering of E. coli B glutathione
synthetase (GSH) (22), we happened to find that a sequence segment
of the amino acid sequence of GSH, from Arg-55 to Ile-96, is similar
to those of mammalian and bacterial dihydrofolate reductases (DHFRs)
(23) (Figure 2). In this figure, similarities between the
sequence segment of GSH and those of DHFRs were quantitatively

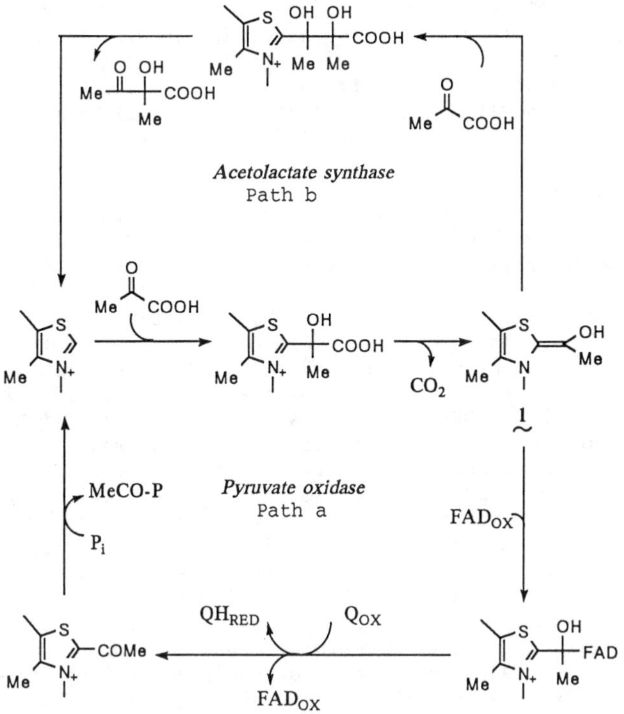

Figure 1. Catalytic reactions of pyruvate oxidase (Path a) and acetolactate synthase (Path b).

Residue No. of dihydrofolate reductase (Mouse)

```
47--+---- ----- +---------+---------+-------89
P   VIMGR HTWES I G   RPLPGRKNIILSSQPGTDDRVT WV
P   VIMGR HTWES I G   RPLPGRKNIILSSQPGTDDRVT WV
P   VIMGR KTWES LPVK  PLPGRRNIVISRQADYCAAGAETV
I   MVV GR RTYES FP K RPLPERTNVVLTHQEDYQAQGA VV
QNLVIMGK KTWFS IPEKNRPLKGRINLVLSRELKEPPQGAHFL
QNLVIMGK KTWFS IPEKNRPLKDRINIVLSRELKEPPKGAHFL
QNLVIMGR KTWFS IPEKNRPLKDRINIVLSRELKEPPQGAHFL
QNAVIMGK KTWFS IPEKNRPLKDRINIVLSRELKEAPKGAHYL
QNLVIMGR KTWFS IPEKNRPLKDRINIVLSRELKEPPQGAHFL
QNLVIMGR KTWFS IPEKNRPLKDRINIVLSRELKEPPRGAHFL
         ::: :  :: :       : : :::      :
RTLNVKQNYEEWFSFVGEQDLPLAD LDVILMR   KDPPFDTEFI
55----+---------+--------- +------  ---+----96
```

Source	Score
E. coli K12	58
E. coli B	58
N. gonorrheae	60
L. casei	58
Human	66
Bovine liver	69
Pig liver	70
Chicken	65
Chinese hamster	70
Mouse	71

Residue No. of glutathione synthetase (*E. coli* B)

Figure 2. Amino acid sequence similarities between glutathione synthetase from **E. coli** B and dihydrofolate reductases.

evaluated as 'Score' values that were calculated from the amino acid mutation data defined by Dayhoff (24). Degree of similarities increases with score values. As expected, because GSH and DHFR have different enzymatic names and belong to different superfamilies, they are different in both their reaction mechanisms (synthase and dehydrogenase) and ligand requirements; gamma-L-glutamylcysteine, glycine and ATP are the substrates of GSH, while folate and NADPH are those of DHFR. GSH did not seem to be related to DHFR by any functional similarity.

Effects of the substrates and inhibitors of DHFR on the activity of GSH were examined to test whether GSH is also related to DHFR by ligand specificity or not. GSH was potently inhibited by 7,8-dihydrofolate (46% inhibition at the concentration of 0.1 mM), methotrexate (64% inhibition at 0.1 mM), and trimethoprim (42% inhibition at 0.3 mM) (23). No compounds had been reported to inhibit GSH as potently as these DHFR ligands. That is, we found folate, methotrexate, and trimethoprim as lead-compounds of GSH inhibitors. GSH and DHFR are partially related not only by their sequences but also by their substrate specificities.

Stone and Morrison (25) reported that both methotrexate and trimethoprim were good inhibitors of E. coli DHFR (Ki=3.6 and 0.49 nM, respectively), while trimethoprim was much less potent on the chicken DHFR than methotrexate (Ki=3,530 and 1.3 nM, respectively). As expected from the sequence similarities between GSH and DHFRs in Figure 2 that indicated that GSH is similar to avian rather than bacterial DHFRs, trimethoprim was a less potent inhibitor of GSH than methotrexate by about five-fold. Thus the inhibition spectrum of GSH by the two DHFR inhibitors corresponded well to the degree of sequence similarities of GSH to avian and bacterial DHFRs.

Sequence Segment of GSH Similar to DHFR is Part of ATP-Binding Site.
Here arises a question whether the GSH inhibition by DHFR ligands is due to the sequence similarity found between GSH and DHFRs. We kinetically analyzed the mode of the GSH inhibition of methotrexate in detail and found that methotrexate competitively bound to the ATP-binding site of GSH with a Ki of 0.1 mM (23). Methotrexate inhibits DHFRs by binding to the dihydrofolate-binding site and/or the NADP-binding site (26). This suggests that the ATP-binding site of GSH is functionally similar to one of the two methotrexate-binding sites on DHFR.

The polypeptide portion of the Lactobacillus casei DHFR, a sequence segment from Ile-38 to Val-75 which shows a local similarity to the sequence segment from Arg-55 to Ile-96 of GSH, folds to construct a part of the NADP-binding site in the crystal structure (27-29). Inhibition of GSH by methotrexate can be rationalized if the ATP-binding site of GSH is similar in sequence and function to the NADP-binding site of DHFR.

Structural and functional similarity of nucleotide-binding site in proteins was first recognized by Rossmann (30). In kinases and dehydrogenases, nucleotides such as NAD(P) and ATP bind to a region called nucleotide-binding domain. A domain is a unit of structure and function composed of 100 to 150 amino acid residues. When Rossmann compared the three-dimensional structures of lactate dehydrogenase, malate dehydrogenase, alcohol dehydrogenase, and glyceraldehyde-3-phosphate dehydrogenase, he discovered that the nucleotide-binding domains of these enzymes were similar to each other in the topology of helix and sheet structures, although sequence similarities between the amino acid sequences forming these domains were not so strong. This topology called a Rossmann fold has been found commonly in the other enzymes requiring a nucleotide as a substrate or cofactor.

Recently we have crystallized GSH and obtained diffraction data enough to determine the positions of the polypeptide side chains (Katoh et al. J. Mol. Biol., submitted). In a few years, we hope to confirm that the sequence segment from Arg-55 to Ile-96 of GSH folds into a Rossmann fold.

Salicylic acid is an another example of the inhibitors that bind to both an ATP-binding site and an NAD(P)-binding site. Two groups of biochemists have reported separately that salicylic acid inhibited adenylate kinase by binding to the ATP-binding site (31) and alcohol dehydrogenase by binding to the NAD-binding site (32), although at that time there was little agreement that the two enzymes were related to each other in terms of functional and structural similarity in their nucleotide-binding domains.

Biological Bases of Local Similarity Found Between Sequences of Different Proteins.

When one searches for sequence similarity search of an amino acid sequence against a sequence database, he finds that the sequence shows local similarities at various regions to those of different kinds of proteins. For example, a similarity search of the sequence of GSH against the NBRF sequence database using IDEAS system (33) revealed that GSH was locally similar at 15 regions (subsequences stretching 20-50 residues) along the entire sequence to 62 different kinds of enzymes (Table I). The local similarity between GSH and DHFR we described in the above section is one of the similarities in this table. There are two biological reasons why local sequence similarities are found between the subsequences of proteins that are different families or superfamilies: molecular evolution of proteins and conservation of sequences at functionally important regions.

Sequence Similarity Due to Molecular Evolution of Proteins.
According to a theory of the molecular evolution of proteins, the gene of a new protein evolves not by random mutations of some other gene, but by 'exon-shuffling' (34-37). In the exon-shuffling theory, exons coding sequence segments composed of 30-50 amino acid residues are supposed to be the units of evolutionary rearrangements of genes. By gene duplications, exons are transferred and mixed with other exons to form a new gene that codes for a protein with a novel function. Then proteins the genes of which have inherited the same ancestral exon would show local sequence similarities with each other (38). Just after divergence into two separate genes the two duplicated exons from an ancestral exon are identical in sequence, but they begin to accumulate amino acid substitutions by random mutations. With time it becomes difficult to find the boundaries of the duplicated exons only from the analysis of sequence similarity.

Sequence Similarity Found in Functionally Important Regions. When amino acid sequences are compared between the enzymes that are

Table I. Local Amino Acid Sequence Similarities between E. coli B Glutathione Synthetase (GSH) and Other Enzymes

Sequence segments of GSH[a]	Enzymes showing sequence similarity with the sequence segments of GSH[b]
1- 30	Chymotrypsin, Pyruvate kinase
20- 50	Carbamoylphosphate synthetase, Flavodoxin, DNA polymerase, Dihydrofoalte reductase, Hygromycin B phosphotransferase, Glutamate dehydrogenase, Cytochrome oxidase 1, Ferredoxin, DNA ligase, ATPase B, p-Hydroxybanzoate hydroxylase
40- 70	Tyrosyl-tRNA syntetase, Asparagine synthetase, Inorganic pyrophosphatase, Endonuclease, Cytochrome b, Endodeoxyribonuclease 1, Cytochrome p-450, Ceruloplasmin, Chymotrypsin
60- 90	Dihydrofolate reductase, Cytochrome c oxidase, Ferredoxin, Cytochrome oxidase, ATPase, Ribonuclease, Dinitrogenase, Prothrombin, Phosphoglycerate mutase, RNA replicase, Acetolactate synthase
80- 110	Ferredoxin, Anthranilate synthetase, Tryptophan synthetase, Cytochrome p-450, Cytochrome c1, Phosphoglucomutase, Fructose-1,6-diphosphatase
100- 130	Cytochrome oxidase, NADH-cytochrome B5 reductase, Phosphoglycerate mutase
120- 150	Lactate dehydrogenase, Prothrombin, Tyrosinase, p-Aminobenzoate synthase, Diaminopimelate decarboxylase, Carboxyl/oxigenase, Cytochrome c2, Nitrogenase,
140- 170	Glycyl-tRNA synthetase A, Dinitrogenase B, Citrate synthase, Cytochrome c oxidase 3, Tyrosinase, Cytochrome oxidase, DNA polymerase 1
160- 190	Carbonic anhydrases c and 1, Cytochrome c6, ATPase B, ATP phosphoribosyl transferase, RNA polymerase B
180- 210	Tryptophan synthase, Tryptophan-tRNA synthetase, Methionyl-tRNA synthetase, Tryptophanase
200- 230	NADH-cytochrome b5 reductase, Thioredoxin
220- 250	Prothrombin, Cytochrome b5, Hygromycin B phosphotransferase
240- 270	DNA invertase, Cytochrome oxidase, RNA polymerase s, Protein kinase, Ceruloplasmin, Threonyl-tRNA synthetase, Lactate dehydrogenase
260- 290	Alcohol dehydrogenase, Cytochrome b, RNA polymerase B
280- 316	Cytochrome oxidase 2, Triosephosphate isomerase, Trypsin, Tyrosyl-tRNA synthase, Pyruvate dehydrogenase, D-Serine dehydrogenase, Alpha-glucan phsophorylase

a) Each pair of numbers represents the terminal residues of a sequence segment of GSH; for example, '1 - 30' is the segment from residue 1 to residue 30.
b) Enzymes the score values of which are greater than 30 are listed.

closely related to each other in terms of reaction mechanism or
substrate specificity, their sequences are locally similar only at
the regions where their polypeptide chains fold to construct
functionally important local structures. Sequence is usually
conserved at the regions of functional importance where amino acid
residues interact with substrate or cofactor molecules. Amino acid
substitutions at the regions of functional importance by random
mutations cause the fatal loss of the physiological function of the
protein and this mutation is not inherited. In other words,
sequence regions that show local similarity with other sequences are
the regions of functional importance.

Sequence-Function Relationships in Proteins. Based on the above two
aspects of sequence similarity, regions the sequences of which show
local similarity with those of other proteins are the regions of
evolutional and functional importance. This is the biological basis
for analyzing sequence-function relationships in proteins using
sequence similarity searches.

If we could find a sequence segment commonly present among the
sequences of the functionally related proteins that can recognize
the molecules containing a common chemical structure (or substructure), the sequence segment found is characteristic of the
chemical structure (or substructure).

Several examples of sequence segment-chemical structure
relationships have been found as 'consensus sequences' (sometimes
called 'functional motifs' or 'functional fingerprints' (39). The
sequence Gly-X-Gly-X-X-Gly (X means any amino acid) is typically
found as a segment in the sequences of nucleotide-binding proteins
such as adenylate kinase and ATPase (40, 41). This glycine-rich
segment called the consensus sequence of the nucleotide-binding
proteins interacts with the 5'-pyrophosphate moiety of a bound
nucleotide (42). The sequence Gly-X-Gly-X-X-Gly is actually a
sequence-5'-pyrophosphate moiety relationship rather than a
sequence-nucleotides relationship. Other examples are a EF-hand
structure for Ca-ion-binding proteins (43) and a Cys-rich sequence
(Zn-binding finger) for Zn-binding proteins (44).

These relationships are very useful to design chemical
structures for drugs based on the amino acid sequece of a target
protein, because the reverse of these relationships is also true;
that is, if we found a Zn-binding finger or a sequence segment
similar to a Zn-binding finger in the sequence, the target protein
is expected to have a Zn-binding site or to show some affinity to
Zn, so there is a better possibility of finding new inhibitors of
the target protein among the chemical structures of the inhibitors
of Zn-binding proteins. Before drug design using sequence-chemical
structure relationships, we have to compile the relationships
systematically for various types of chemical structures.

Homology Graph.

There are several difficulties in searching for sequence-chemical
structure relationships. There is no available computer algorithm
for detecting a sequence similarity among three sequences or more
(5, 45). We can search for sequence similarity only between a pair
of sequences. Most sequence similarities found between the

sequences of distantly related proteins usually have sequence
segments 20-30 amino acid residues in length and 20-30% identical.
In such cases, it is difficult to decide whether the two sequence
segments are functionally related ones or are similar by chance. It
is also difficult to determine how much similarity is necessary to
identify regions of evolutionary and functional importance. When
the threshold for detecting similarity is too low, subsequences that
are not of functional importance are also detected by chance. On
the other hand, when the threshold is too high, the sequence regions
detected are scarce and not in common among the set of sequences
searched.

Homology graphing has been developed to detect sequence segments
of functional importance by calculating and showing segment
similarities of a given sequence with a set of other sequences, not
with a single other sequence (Sumi, Nishioka, and Oda, manuscript in
preparation).

<u>Definition of Homology Graphing</u>. When a window of fixed length
moves along the amino acid sequence of a target protein from the
N-terminal to the C-terminal, the window cuts the sequence into
segments; for example, the segment from residue 1 to residue 30, the
segment from residue 6 to residue 35, the segment from residue 11 to
residue 40, and so on (Figure 3). In this example, each segment is
30 residues long. Since the i-th segment and the (i+1)-th segment
overlap by 25 residues, the increment between two successive
segments is five residues. Thus, a set of segments is defined by
the length of the window and the increment.

A set of amino acid sequences called 'reference sequences' is
collected from the NBRF database for the proteins that have the same
function. For the i-th segment of the target protein, similarity
search is done against one of the reference sequences. When the
similarity score of the best local alignment is above a certain
level (maxd score), the similarity score is saved. This step is
repeated against the other one of the reference sequences until all
the reference sequences are searched in pairs by the i-th segment.
The total value of similarity scores saved is defined as the
'homology value of the i-th segment' (Equation 1).

$$\text{Homology value of the i-th segment} = \sum_{\text{Reference sequences} = 1}^{n} \text{score \{if score > maxd score\}} \quad (1)$$

This procedure is repeated for the next (i+1)-th segment of the
target protein. Thus, the homology value for each of the all
segments is calculated. To show this graphically, the homology
value of the i-th segment is plotted at the middle residue of the
i-th segment. This graph is called a 'homology graph'.

The homology value increases not only with the degree of
similarity between sequence segments of a target protein and
reference sequences but also with the number of reference sequences
which show similarity above the maxd score. Homology value is
therefore an index of both the degree and the frequency of local
similarities of a segment with a set of reference sequences. By

comparing homology values, it is found which regions of the target sequence are more similar to the set of reference sequences.

In the following example, we will show that homology graphing is useful for analyzing functionally important regions along the sequence of a target protein. A similarity search between the i-th segment and the reference sequences was done using SEQHP program in the IDEAS system (33) installed on a FACOM M-380 computer in the Institute for Chemical Research, Kyoto University. In the IDEAS system, similarity scores are evaluated with the mutability values defined by Dayhoff (24).

Homology Graph of Glutathione Reductase. Glutathione reductase (GR) catalyzes the reduction of oxidized glutathione (glutathione disulfide) at the expense of NADPH. GR is a flavoprotein composed with two identical subunits. Since the three-dimensional structure of the FAD-complex of human GR have been analyzed at the resolution of 1.54 Å (46, 47), this enzyme is suitable to test whether homology graphing is valid and useful to analyze an amino acid sequence.

The target sequence is the sequence of human GR (478 residues; NBRF entry name = RDHUU). Two sets of reference sequences were prepared. One is a set of the sequences of NAD(P)/FAD-related enzymes whose substrate or cofactor is NAD(P) or/and FAD. In the version 16 NBRF database, there were registered 70 sequences of 45 NAD(P)/FAD-related enzymes including human GR. The sequence of human GR was not included in the reference sequences.

The other reference sequence set is composed of the 70 sequences of 70 nucleotide-nonrelated enzymes whose substrates or cofctors are neither NAD(P), FAD, ATP, nor CoA. Seventy enzymes were randomly selected from the nucleotide-nonrelated enzymes listed in the database. One sequence from each of the 70 enzymes was collected for the reference sequences.

The first set of reference sequences is to analyze the sequence segments involved in the recognition of FAD and NADPH by GR. The second set is for the control experiment to analyze similarities found by chance.

Figure 4 shows two homology graphs of the amino acid sequence of human GR obtained against the two different sets of reference sequences. One (- - -) is a graph the reference sequences of which are the sequences of NAD(P)/FAD-related enzymes. In the other graph (———), the reference sequences are those of nucleotide-nonrelated enzymes. Analytical conditions are window = 50 residues, increment = 5 residues, and maxd score = -45.

Peaks in these two graphs show several sequence regions whose homology values are higher than the other regions. The peaks at five regions (25-157, 157-245, 245-340, 340-405, and 405-450) in the graph (- - -) are significantly higher when compared with the peaks shown in the graph (———). These five regions in the amino acid sequence of human GR are expected to contribute to the binding of NADPH and FAD or to the reduction of oxidized glutathione.

It is interesting to examine in the crystal structure what are the functional roles of these five regions in human GR. The first region from residue 25 to residue 157 in the graph corresponds to the FAD domain from residue 18 to residue 157 in the crystal structure. The second and a half of the third regions in the graph constructs the NADPH doamin from residue 158 to residue 293 in the

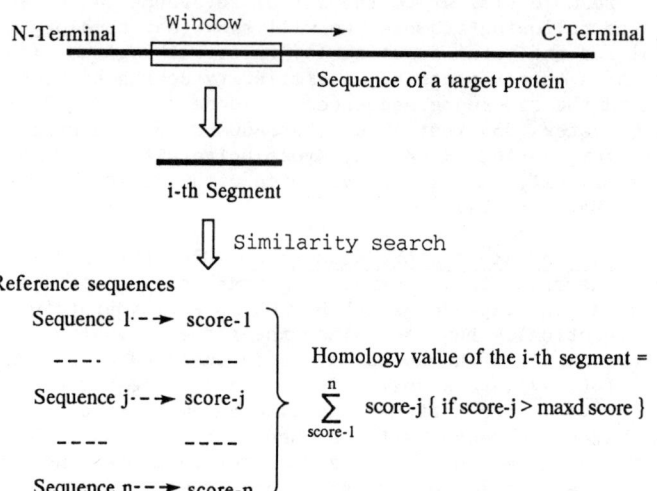

Figure 3. Procedure to calculate homology values.

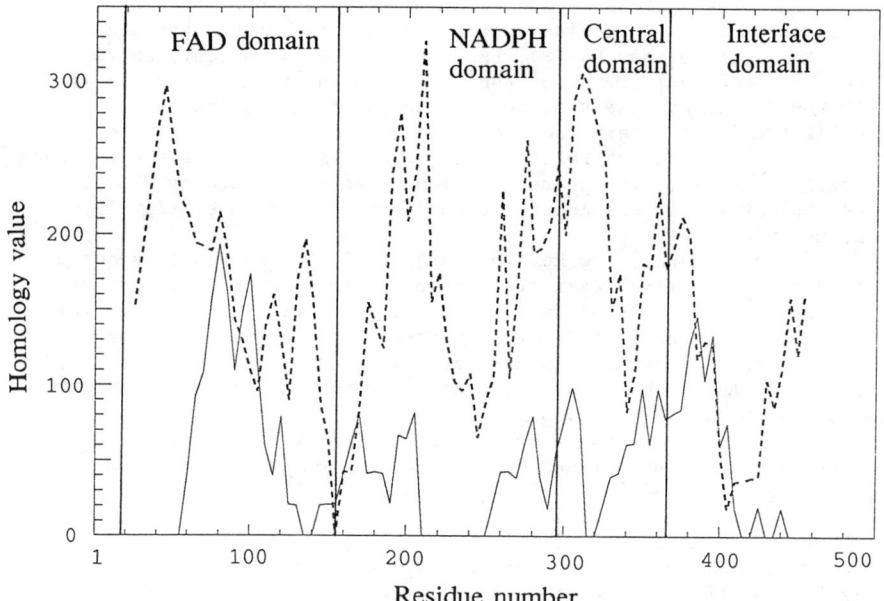

Figure 4. Homology graphs of human glutathione reductase. Reference sequences are NAD(P)/FAD-related enzymes (- - -) and nucleotide-nonrelated enzymes (———).

crystal structure. The second region shows the highest peak in the graph from residue 176 to residue 240 where the polypeptide chain forms a Rossmann fold to interact with a bound NADPH molecule.

In the crystal structure of a human GR-FAD complex, Cys-58, Cys-63, Lys-46, Tyr-114, Tyr-197, Glu-201, Arg-271, Val-370, Asp-331, His-467, and Glu-472 are assigned as the residues that are in contact with a bound FAD molecule or that catalyze the reaction of human GR (47). Except for Asp-331, all of these important residues are around the peak maxima in the homology graph.

Sequence-Chemical Structure Relationships Obtainable from Homology Graph   Without any information about three-dimensional structure, homology graphing identified the FAD- and NADPH-binding regions in the sequence of human GR. We are now accumulating other examples of the applications of homology graphing to analyze the sequences of several other proteins with a known crystal structure. Up to now, it is safe to say that with a proper set of reference sequences, homology graphing can identify the sequence regions used in the recognition of the chemical structures of substrate and cofactors. Further studies on a method of homology graphing are in progress to provide more details about the amino acid residues engaged in molecular recognition of proteins.

In the above analysis of the sequence of human GR by homology graphing, four sequence segments are assigned as those which have affinity with or make recognition of the chemical structures (or substructures) of FAD and NADPH.

(1) Two subsequeces from residue 20 to residue 75 and from 95 to 155 are those recognizing FAD. Especially the subsequence 20 - 75 contains the sequence Gly27--Gly29---Gly32 that is matched to the consensus sequence of nucleotide-binding proteins, Gly-X-Gly-X-X-Gly. In the crystal structure of a human GR-FAD complex, the subsequence from Gly27 to Gly32, in fact, interacted with the 5'-pyrophosphate group of the bound FAD (48).

(2) The other two subsequences from residue 185 to 225 and from 240 to 270 are those recognizing NADPH. Gly194-X-Gly196 is believed to be a part of the consensus sequence of nucleotide-binding proteins and a part of a Rossmann fold (48).

Peaks in the homology graph (Figure 4) are much higher around the consensus sequence of nucleotide-binding proteins. Thus, the homology graph covers a wider sequence segment than the corresponding consensus sequence. Sequence-function relationships are slightly different between the consensus sequence and the homology graph. The amino acid 'residues matched' commonly among a set of sequences are defined as a sequence-function relationship in a consensus sequence, while the amino acid 'sequence segment similar' to each other among a set of sequences is identified in the homology graph.

Since the number of amino acid residues defined in consensus sequences are usually less than 10 residues, consensus sequences are too small as a unit of function and molecular evolution of proteins. Therefore sequence segments found in homology graphs are favored for sequence-function relationships over consensus sequences.

Application of Homology Graphing to Finding Lead Structures.

When a database collecting sequence-chemical structure relationships becomes available for the chemical structures appearing in the substrates, cofactors, regulators, inhibitors, and hormones of proteins, the collected sequence segments would be useful as a functional template to search and assign the all possible chemical structures recognized by a given target protein. But sequence-chemical relationships have to be collected systematically and this may take several years before we can apply them to drug design.

In this chapter we will describe the application of homology graphing to find lead chemical structures from the amino acid sequence of a target protein. This application does not require any functional templates. Sequence segments with higher peaks in a homology graph are those which are responsible for the same molecular recognition as that of the reference proteins. This is the basis of our approach to find lead structures using homology graphing.

Procedure to Find Lead Structures. The procedure to find lead structures of a target protein for drug development is summarized in Figure 5. The amino acid sequence of a target protein is analyzed by homology graphing in which the reference sequences are all the sequences registered in the NBRF database. Regions with higher peaks in the homology graph are identified as sequence segments responsible for chemical recognition. Then, the following steps are repeated for each sequence segment identified.

Step 1. With the i-th segment, for example with a segment from residue 250 to 300 in Figure 4, sequence similarity is searched for against all the sequences registered in the NBRF database. All the enzymes whose sequences contain a local sequence similar to the i-th segment are listed by their entry code in the NBRF database.

Step 2 amd Step 3. Chemical structures of the substrates, cofactors, and inhibitors of the enzymes listed in Step 1 are searched for in the Enzyme-Reaction database (see the next section) by the entry codes of these enzymes.

Step 4. Some of the chemical structures searched for in Step 3 must contain the lead structures of agonists and antagonists for the target protein. Bioassay of the compounds will reveal the structural features necessary for the lead compounds.

Enzyme-Reaction Database. We have built a database called the Enzyme-Reaction database for drug design based on amino acid sequence (Nishioka and Oda, unpublished). This database contains the following items for each enzyme; enzyme name including common names, EC-number and reaction type classified by IUB, names of substrates, cofactors, inhibitors, and products, and entry codes in the NBRF sequence database and the Brookhaven Protein Databank.

The entry code for the amino acid sequences of all the enzymes registered in the NBRF database are collected in the Enzyme-Reaction database. The number of NBRF entry codes collected in this database (July 1988) is 1,497 for 430 enzymes. Since IUB gave a name for 2,477 enzymes in 1984 (49), the number of enzymes with a known sequence is about 17% of the known enzymes. The Enzyme-Reaction database is updated with the version of the NBRF database.

7. NISHIOKA ET AL. *Lead Structures from Amino Acid Sequences* 119

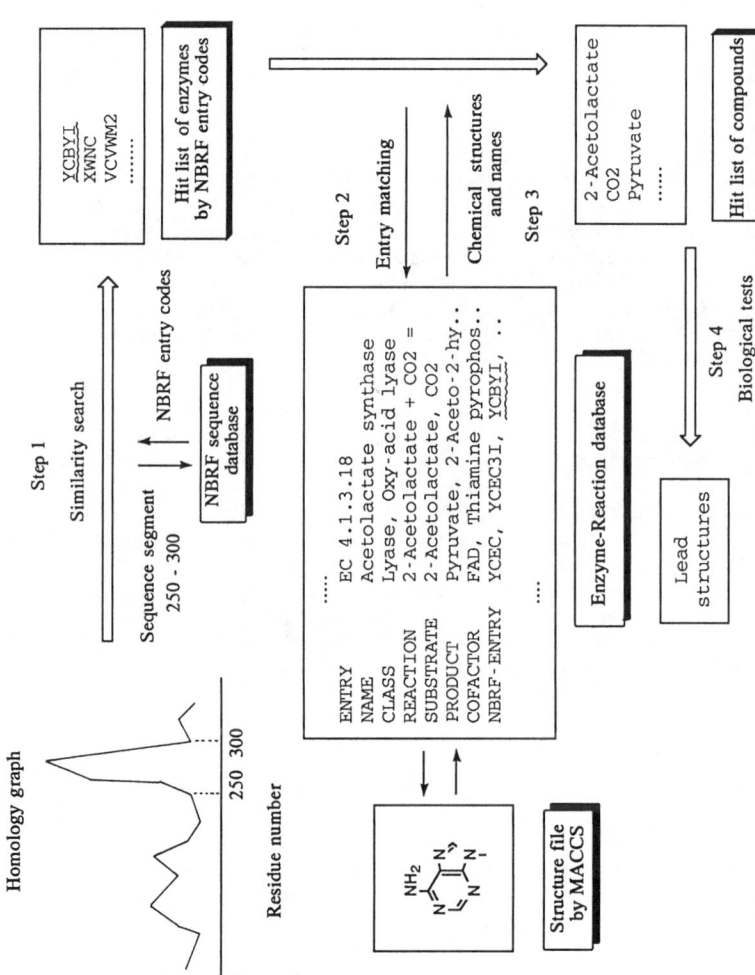

Figure 5. Flow chart of finding lead structures using homology graph of a target protein.

The total number of chemical compounds registered as their substrates, cofactors, and products are 842 on July 1988 and increase with each update of the database. Chemical structures of these chemical compounds are stored with the related enzyme names on a MACCS system (Molecular Design Ltd) installed on a FACOM-380 in the Institute for Chemical Research, Kyoto University. Few inhibitors are registered now, because we have just started collecting the chemical structures of the inhibitors of the enzymes in this database.

Two copies of the Enzyme-Reaction database are prepared; one copy is accessible through the FORTRAN 77 programs for lead-structure findings discussed in the above section and the other is on a FACOM relational-database system.

Future Prospects. When the above procedure from Step 1 to Step 3 was done for the sequence segment from residue 150 to residue 210 of the sequence of tobacco ALS, our system gave us a list of 33 compounds as test compounds (Table II). The compounds listed are substrates, cofactors, and products, but only one inhibitor (diisopropyl fluorophosphate), because only a few inhibitors have been registered in the Enzyme-Reacton database. When the number of inhibitors increases in the database, this system will become a practical one for drug design.

Table II. List of Compounds Hit by the Steps 1-3 for the Sequence Segment from Residue 150 to 210 of Tobacco ALS.

NADPH, NADH, Oxidized glutathione, FAD, Starch, Glycogen, 3-Hydroxy-4-methyl-3-carboxypentanoate, Coenzyme A, Reduced ferredoxin, NADP+, (S)-Malate, Protein(Glu-, Asp-), Peptide, RNA, ATP, deoxyATP, Pyruvate, Acetylcholine, Acetic ester, Diisopropyl fluorophosphate, Stearoyl-CoA, Carbonate, Zinc, Copper, Manganese, Iron, L-Tyrosine, L-Dopa, 1,2-Benzenediol, Acetyl-CoA, beta-D-Galactoside, Thiogalactoside, Phenylgalactoside

---

Compounds are ordered by score values in Step 1.

In addition, we have to develop a computer algorithm to extract the chemical structure common to or most similar to the structures listed by the system. This algorithm should be inserted as an additional step between Steps 3 and 4 in the above procedure. Insertion of this step is expected to increase the probability of a hit-list in biological assays.

Acknowledgments

The authors thank Dr. Minoru Kanehisa, Institute for Chemical Research, Kyoto University, for his valuable discussions for searching for and interpretation of sequence similarity by IDEAS system. This work was partly supported by a research grant from the Ministry of Education, Japan.

Literature Cited

1. Blundell, T. L.; Sternberg, M. J. E. Trends Biotech. 1985, 3, 228-235.
2. Blundell, T. L.; Sibanda, B. L.; Sternberg, M. J. E.; Thornton, J. M. Nature 1987, 326, 347-352.
3. Kabsch, W.; Sander, C. FEBS Lett. 1983, 155, 179-182.
4. Nishikawa, K.; Ooi, T. Biochim. Biophys. Acta 1986, 871, 45-54.
5. von Heijine, G. Sequence Analysis in Molecular Biology; Academic Press: New York, 1987.
6. Doolittle, R. F. Of Urfs and Orfs - A Primer on How to Analyze Derived Amino Acid Sequence; Oxford University Press: London, 1987.
7. Nucleic Acid and Protein Sequence Analysis, A Practical Approach; Bishop, M. J.; Rawlings, C. J.; Ed.; IRL Press: Oxford, 1987.
8. Dayhoff, M. O.; Barker, W. C.; Hunt, L. T. Methods Enzymol. 1983, 91, 524-545.
9. Tsukihara, T.; Kobayashi, M.; Nakamura, M.; Katsube, Y.; Fukuyama, K.; Hase, T.; Wada, K.; Matsubara, H.; Biosystems 1982, 15, 243-257.
10. LaRossa, R. A.; Schloss, J. V. J. Biol.Chem. 1984, 259, 8753-8757.
11. Chaleff, R. S.; Mauvais, C. J. Science 1984, 224, 1443-1445.
12. Ray, T. B. Plant Physiol. 1984, 75, 827-831.
13. Shaner, D. L.; Anderson, P. C.; Stidham, M. A. Plant Physiol. 1984, 76, 545-546.
14. Muhitch, M. J.; Shaner, D. L.; Stidham, M. A. Plant Physiol. 1987, 83, 451-456.
15. Wek, R. C.; Hauser, C. A.; Hatfield, G. W. Nucl. Acids Res. 1985, 13, 3995-4010.
16. Lawther, R. P.; Calhoun, D. H.; Adams, C. W.; Hauser, C. A.; Gray, J.; Hatfield, G. W. Proc. Nat. Acad. Sci. USA 1981, 78, 922-925.
17. Squires, C. H.; DeFelice, M.; Devereux, J.; Calvo, J. M. Nucl. Acids Res. 1983, 11, 5299-5313.
18. Falco, S. C.; Dumas, K. S.; Livak, K. J. Nucl. Acids Res. 1985, 13, 4011-4027.
19. Mazur, B. J.; Chui, C. -F.; Smith, J. K. Plant Physiol. 1987, 85, 1110-1117.
20. Grabau, C.; Cronan, J. E. Nucl. Acids Res. 1986, 14, 5449-5460.
21. Schloss, J. V.; Ciskanik, L. M.; Van Dyk, D. E. Nature 1988, 331, 360-362.
22. Kato, H.; Tanaka, T.; Nishioka, T.; Kimura, A.; Oda, J. J. Biol. Chem. 1988, 263, 11646-11651.
23. Kato, H.; Chihara, M.; Nishioka, T.; Murata, K.; Kimura, A.; Oda, J. J. Biochem. 1987, 101, 207-215.
24. Dayhoff, M. O.; Schwartz, R. M.; Orcutt, B. C. In Atlas of Protein Sequence and Structure; National Biomedical Research Foundation; Silver Spring, MD.; 1978; Vol.5, suppliment 3, pp 345-352.
25. Stone, S. R.; Morrison, J. F. Biochim. Biophys. Acta 1986, 869, 275-285.
26. Stone, S. R.; Morrison, J. F. Biochemistry 1982, 21, 3757-3765.

27. Bolin, J. T.; Filman, D. J.; Matthews, D. A.; Hamlin, R. C.; Kraut, J. J. Biol. Chem. 1982, 257, 13650-13662.
28. Filman, D. J.; Bolin, J. T.; Matthews, D. A.; Kraut, J. J. Biol. Chem. 1982, 257, 13663-13672.
29. Stammers, D. K.; Champness, J. N.; Beddell, C. R.; Dann, J. G.; Eliopoulos, E.; Geddes, A. J.; Ogg, D.; North, A. C. T. FEBS Lett. 1987, 218, 178-184.
30. Rossmann, M. G.; Liljas, A.; Branden, C. -I.; Banaszak, L. J. In The Enzymes; Boyer, P. D., Ed.; Academic Press: New York, 1975; pp 61-102.
31. Pai, E. F.; Sachsenheimer, W.; Schirmer, R. H.; Schulz, G. E. J. Mol. Biol. 1977, 114, 37-45.
32. Einarsson, R.; Eklund, H.; Zeppezauer, E.; Boiwe, T,; Branden, C.-I. Eur. J. Biochem. 1974, 49, 41-47.
33. Goad, W. B.; Kanehisa, M. Nucl. Acids Res. 1982, 10, 247-263.
34. Blake, C. C. F. Nature 1978, 273, 267.
35. Rogers, J. Nature 1984, 315, 458-459.
36. Cornish-Bowden, A. Nature 1985, 313, 434-435.
37. Marchionni, M.; Gilbert, W. Cell 1986, 46, 133-141.
38. Lonberg, N.; Gilbert, W. Cell 1985, 40, 81-90.
39. Hodgman, T. C.; CABIOS Rev. 1986, 2, 181-187.
40. McCormick, R.; Clark, B. F. C.; La Cour, T. F. M.; Kjeldgaard, M.; Norskov-Lauritsen, L.; Nyborg, J. Science 1985, 230, 78-82.
41. Dever, T. E.; Glynias, M. J.; Merrick, W. C. Proc. Natl. Acad. Sci. USA 1987, 84, 1814-1818.
42. Fry, D. C.; Kuby, S. A.; Mildvan, A. S. Proc. Natl. Acad. Sci. USA 1986, 83, 907-911.
43. Van Eldik, L. J.; Zendegui, J. G.; Marshak, D. R.; Watterson, D. M. Int. Rev. Cytol. 1982, 77, 1-61.
44. Berg, J. M. Science 1986, 232, 485-487.
45. Murata, M.; Richardson, J. S.; Sussman, J. L. Proc. Natl. Acad. Sci. USA 1985, 82, 3073-3077.
46. Thieme, R.; Pai, E. F.; Schirmer, R. H.; Schulz, G. E. J. Mol. Biol. 1981, 152, 763-782.
47. Karplus, P. A.; Schulz, G. E. J. Mol. Biol. 1987, 195, 701-729.
48. Schulz, G. E.; Schirmer, R. H.; Sachsenheimer, W.; Pai, E. F. Nature 1978, 273, 120-124.
49. Enzyme Nomenclature; International Union of Biochemistry. Nomenclature Committee; Academic Press; Orlando, FL.; 1984.

RECEIVED March 30, 1989

Chapter 8

Application of Micellar Liquid Chromatography to Modeling of Organic Compounds by Quantitative Structure–Activity Relationships

Barry K. Lavine, Anthony J. I. Ward, Jian Hwa Han, and Orla Donoghue

Department of Chemistry, Clarkson University, Potsdam, NY 13676

> In qualitative and quantitative structure activity relationship studies, the lipophilic character of a compound is usually modelled by the logarithm of the octanol water partition coefficient, i.e., log P. However, experimental values for log P are sometimes difficult to obtain. Recently, we have investigated the utility of micellar liquid chromatography for assessing the lipophilic character of organic molecules in biological media. In our study we used a liquid crystalline stationary phase to generate retention data for a set of 22 mono-, di-, and tri-substituted benzenes. The retention factor (i.e., log k') of these compounds was found to be correlated to the log of their octanol/water partition coefficient. Because retention times (hence log k' values) can be accurately measured for even impure compounds, we conclude that log k' values obtained in MLC experiments can be substituted for log P values as a convenient hydrophobic parameter for many different types of organic compounds.

In QSAR studies the lipophilic character of a compound is usually expressed in terms of the logarithm of the octanol/water partition coefficient (i.e., log P). Experimentally, the octanol/water partition coefficient is determined by the shake-flask method. Unfortunately, it is often difficult to obtain precise values for the octanol/water partition coefficient because solute impurities will affect the measured distribution coefficient. For example, reported log P values for benzene (1-5) vary from 1.56 to as high as 2.34. (In an experienced laboratory, log P values in the range of -2 to 4 are usually measured with a reproducibility of 0.1 log units.) Furthermore, reliable shaker-flask measurements are time consuming and are often difficult to make for very hydrophobic compounds.

Recent studies have shown that reversed phase liquid chromatography (RPLC) can be used to estimate the octanol/water partition coefficient for a large variety of organic molecules (6,7). In fact, some workers have even proposed that RPLC is a better technique than the classical shake-flask technique for estimating the octanol/water partition coefficient of compounds which are of biological interest (8,9). However, there is a fundamental problem limiting the applicability of RPLC to problems in this rapidly growing field - the inability of the packed liquid chromatographic (LC) column to mimic the biological environment of the receptor site.

To overcome this problem, our laboratories have initiated a program of study in the area of micellar liquid chromatography (MLC). The mobile phase in a MLC experiment consists of a surfactant that is at a concentration above the critical micellization concentration (cmc). We have learned that the addition of a co-surfactant to a micellar mobile phase will result in the formation of lamellar liquid crystals at the surface of the reversed phase (i.e., C_{18}) material (10). Since lamellar liquid crystals (11) have a structure similar to that of a cell membrane (see Figure 1), we have, therefore, created an environment in the LC column which may serve as a model for the transport of organic molecules in a biological system.

The utility of the chromatographic system to model the hydrophobicity of organic compounds in a biological system has been evaluated using a set of 22 mono-, di-, and tri-substituted benzenes as retention probes. The retention data for these known environmental pollutants was found to be highly correlated to the log of the octanol/water partition coefficient. Although previous workers (9,12) have shown that log k_w' is a useful descriptor for hydrophobicity, the results from this study suggest that the log of the capacity factor measured on a reversed phase column possessing a lamellar liquid crystalline stationary phase is an even better predictor for log P.

Experimental

All HPLC measurements were made with a Perkin-Elmer Tridet HPLC system. The dead volume of the system was calculated to be less than 60 micro liters. Two columns were used in the study: (1) a Rainin Microsorb 3 micron ODS (Octyldecyisiline) column (4.6 x 50 mm), and (2) a Rainin Microsorb 5 micro C_{18} column (4.6 x 150 mm) which was water jacketed and temperature controlled. A silica precolumn was placed before the injector to ensure saturation of the mobile phase with silicates.

The mono-, di-, and tri-substituted benzenes were obtained from Aldrich and were used as received (see Table 1). Sodium dodecyl sulfate (SDS) was obtained from BDH Chemicals Ltd. and unless otherwise noted was recrystallized in ethanol and dried in an oven at about 65 degrees Centigrade. $C_{12}EO_6$ was obtained from Nikko Ltd. (Japan) and was 100% homogeneous in the alkyl chain and greater than 98% in the EO-chain length. HPLC grade ethyl acetate and HPLC grade water was purchased from J.T. Baker. Dodecanol was purchased from Aldrich and was used as received.

The appropriate weight of surfactant and cosurfactant was dissolved in HPLC grade water (J.T. Baker), and ethyl acetate was then added (2% by volume). Particulate matter was removed from the solu-

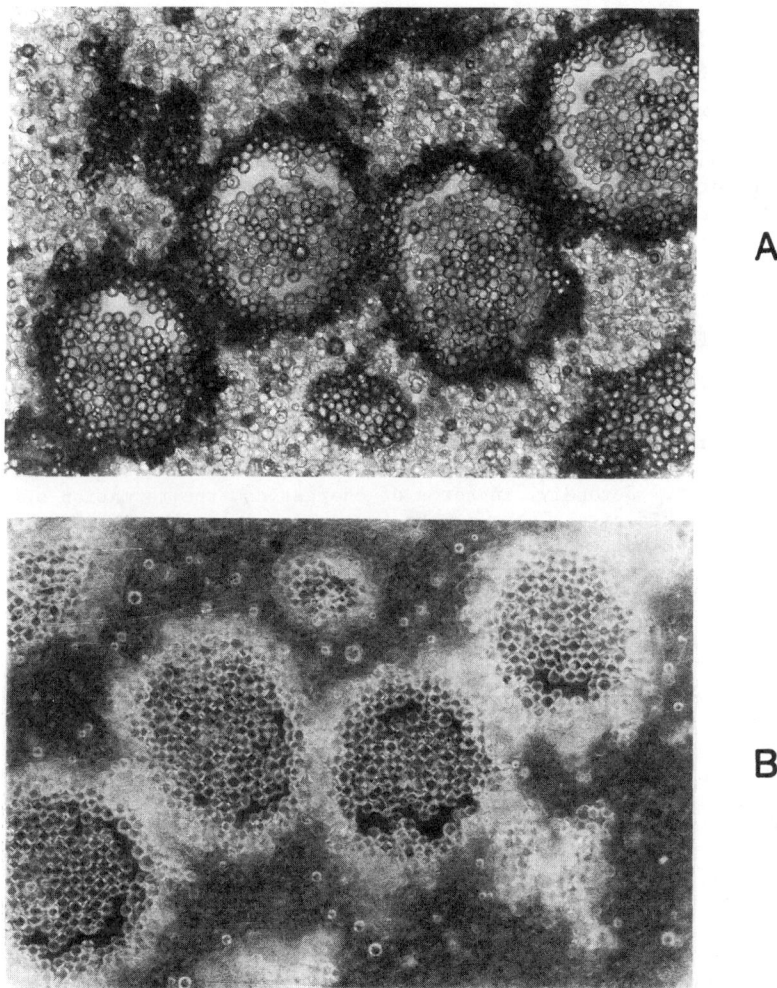

Figure 1. Microphotographs of the C_{18} stationary phase. In (A) the ODS particles are in a solution of SDS with ethyl acetate, and (B) a solution of SDS with dodecanol and ethyl acetate. In the photograph of the system containing dodecanol, the white region surrounding the aggregates of silica gel particles is indicative of a liquid crystalline phase. The same region in the photograph of the system without dodecanol is dark, indicating an isotropic solution.

tion by filtering it through a 0.45 micron Nylon-66 membrane filter (Rainin Instruments, Woburn, MA). Capacity factor values determined in this study were averages of at least triplicate determinations.

Results and Discussion

In a previous study (10) Lavine and co-workers observed that liquid crystals were formed *in situ* on the surface of ODS particles when a solution of SDS, dodecanol and ethylacetate was percolated through a C_{18} column (see Figure 1). Interesting enough, the efficiency of the column for some compounds (e.g., acetophenone) was actually superior to conventional MLC and comparable to RPLC with hydro-organic mobile phases (see Table 2). This efficiency enhancement was attributed to a change in the retention mechanism. Instead of adsorption, the authors believed that partitioning was occurring. Since partitioning is a more efficient retention mechanism, it is reasonable to expect better column efficiency and improved peak shape.

With respect to the structure of the anisotropic phase, the authors stated the liquid crystals were lamellar. They presented two pieces of evidence to support this contention. First, on the basis of the phase diagram for the dodecanol/SDS/water system, only two liquid crystalline stationary phases will exist: (1) hexagonal, or (2) lamellar. Secondly, in terms of energetics, the formation of lamellar liquid crystals is a more favorable process. Details regarding the energetics for these types of systems are given elsewhere (13). Therefore, the authors concluded that liquid crystals formed on the surface of ODS particles were, in all probability, lamellar. Presently, low angle X-ray scattering experiments are underway to confirm the structure of these crystals.

Since lamellar liquid crystals have a structure similar to that of a cell membrane, these phases can be used to investigate the transport properties of organic molecules in biological systems. Therefore, we have generated retention data using these lamellar phases for a set of aromatic compounds. In Figure 2, the log P value is plotted against the log k' for the set of 22 mono-, di-, and trisubstituted benzenes. These data were generated on a 3 micron C_{18} column using a mobile phase of 0.05 M SDS, 0.001 M dodecanol, 0.001 M H_2SO_4, and 2% ethyl acetate (v/v). An examination of this plot reveals a very interesting result. The compounds are divided into two sets. The first set of compounds consists entirely of phenols, whereas the second set is mainly monosubstituted benzenes. When the correlation coefficient was computed for each set of compounds, the r^2 value obtained for the first set of compounds was 0.97 and the r^2 value obtained for the second set was 0.98. The r^2 value obtained for the entire data set (i.e., all 22 compounds) was 0.85. The results of this experiment are significant for two reasons. First, the log k' values that were obtained using a column with a lamellar phase are well correlated to the log P values of the compounds (i.e., the r^2 value was 0.85 for the entire data set). Second, the dichotomy in the data set suggests that log k' values may convey more specific information about the interaction of organic molecules in biological media than log P values.

A second study was undertaken to assess the influence of surfactant type on the degree of correlation between log k' and log P. A

Table I. 22 Mono-, Di-, and Tri-Substituted Benzenes

1.	Benzene	12.	p-Nitrophenol
2.	Benzaldehyde	13.	o-Chlorophenol
3.	Benzonitrile	14.	o-Bromophenol
4.	Acetophenone	15.	2,4 Dichlorophenol
5.	Nitrobenzene	16.	Ethylbenzene
6.	Methylbenzoate	17.	2,4 Dinitrophenol
7.	Anisole	18.	Benzyl Alcohol
8.	Chlorobenzene	19.	Toluene
9.	Bromobenzene	20.	p-Nitroanisole
10.	Resorcinol	21.	o-Nitrophenol
11.	Phenol	22.	Catechol

Table II. Comparison of Different Reversed Phase HPLC Modes

Mobile Phase	Flow Rate (ml/min)	k'	Plates[1,2]	B/A
59% methanol/ 41% water	1.0	4.3	2800	1.4
0.05 M SDS[3] with 2% ethyl acetate	1.0	9.2	3100	1.4
0.05 M SDS, 0.01 M dodecanol, 0.001 M H_2SO_4, 2% ethyl acetate	1.0	11.2	4000	1.0

[1]The Foley-Dorsey method was used to computer the number of theoretical plates. All plate counts were averages of at least four determinations.

[2]The compound used to assess column efficiency was acetophenone. A 3 micron C_{18} column was used in this experiment.

[3]The cmc of SDS in a 2% ethyl acetate solution is 4 mM.

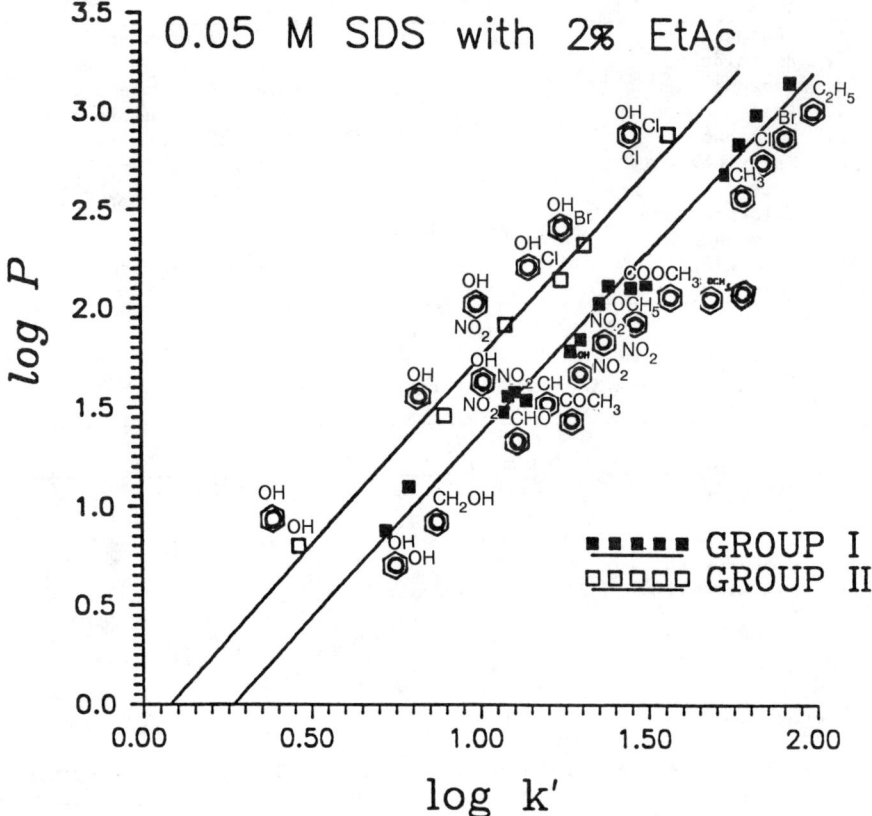

Figure 2. A plot of Log P vs Log k' for the 22 mono-, di-, and tri-substituted benzenes. An aqueous mobile phase consisting of 0.05 M SDS, 0.001 M dodecanol, 0.001 M H_2SO_4, and 2% ethyl acetate (v/v) was used to generate the retention data.

mobile phase consisting of 0.01 M $C_{12}EO_6$, 0.001 M H_2SO_4, and 2% ethyl acetate was percolated through a 3 micron C_{18} column. On the basis of previous experiments using polarized light microscopy, lyotropic liquid crystals were observed to form on the surface of the ODS particles under these conditions. In Figure 3 the log P values are plotted against the log k' values. (The r^2 value was 0.91.) Although the r^2 value for both surfactant systems was similar, no dichotomy was observed for the nonionic surfactant data set (i.e., the 22 compounds are not divisible into two distinct groups as was the previous case). This difference is attributed to the fact that in this case the surfactant aggregates do not have a charged surface - meaning, there is a more continuous distribution of the solute molecules within the aggregated surfactant structure and consequently the surfactant aggregate will exhibit less selectivity toward the organic solutes.

The significance of these results can be understood by recalling the range of different lipid components that are found in cell membranes. For example, in erythrocytes and in myeline the lipids are basically phospholipids (i.e., the headgroup is charged), whereas in the lipid fractions constituting the stratum corneum the lipids are polar lipids, i.e., they have uncharged headgroups. Thus, an organic molecule in biological media is exposed to differing environments. However, different environments can be simulated by properly selecting the appropriate surfactant system, and in principle more information can be obtained from a chromatographic experiment using surfactant based mobile phases than from simple shake-flask measurements.

The results from our studies were compared to some previous studies (14-16) where retention data was generated using a C_{18} column for two different hydro-organic mobile phases - acetonitrile/water and methanol/water. For those compounds in our study for which comparable data existed in the literature, we computed log k_w values. (The log k_w is obtained by extrapolating retention data from binary solvent systems to an aqueous eluent. It is of interest because it has been reported by other investigators that log k_w values may be substituted for log P values (17,18).) In Figure 4, the log P values are plotted against log k' or the log k_w for a set of eight compounds (e.g., benzonitrile, nitrobenzene, acetephenone, benzaldeyde, anisole, chlorobenzene, bromobenzene, and methylbenzoate). It is evident from the figure that a better correlation exists for log k' than log k_w. This result suggests that log k' values generated on a lamellar phase may be a better predictor of log P than log k_w.

Conclusion

These studies suggest that micellar liquid chromatography with lamellar stationary phases may yield a useful hydrophobic parameter. In view of the fact that retention times (and hence log k' values) can be determined with high precision for impure, unstable, and even volatile compounds, we conclude that log k' values obtained with lamellar phases may provide useful estimates for log P.

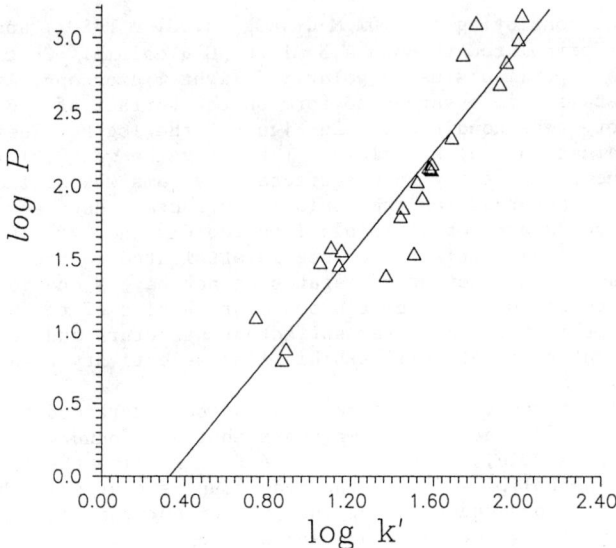

Figure 3. A plot of Log P vs Log k' for the 22 mono-, di-, and tri-substituted benzenes. An aqueous mobile phase consisting of 0.01 M of $C_{12}EO_6$ (nonionic surfactant), 0.001 M H_2SO_4, and 2% ethyl acetate (v/v) was used to generate the retention data.

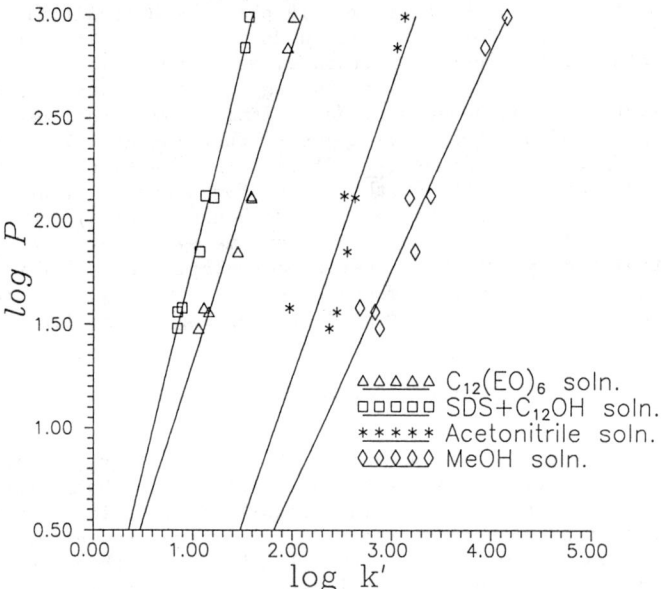

Figure 4. A plot of Log P vs Log k' or log k_w for a set of eight organic compounds. It is evident that a better correlation exists for log k' than log k_w.

Acknowledgment

The authors wish to thank Professor Stig Friberg for many valuable discussions. BKL acknowledges the financial support of the R.J. Reynolds Tobacco Company.

Literature Cited

1. Harnisch, M.; Mockel, M.J.; Schulze, G. J. Chromatogr. 1983, 282, 315.
2. Hansch, C.; Leo, A. Substituent Constants for Correlation Analysis in Chemistry and Biology; Wiley, New York, 1979.
3. Smith, R.M. J. Chromatogr. 1982, 236, 313.
4. Sarna, L.P.; Hodge, P.E.; Webster, G.R.B. Chemosphere. 1984, 13, 975.
5. Log P and Parameter Database: A Tool for the Quantitative Prediction of Bio-Activity; Pomona College Medicinal Chemistry Project: Claremont, CA, 1988.
6. Unger, S.; Feuerman, T. J. Chromatogr. 1979, 176, 426.
7. Mirrlees, R.; Taylor, P. J. Med. Chem. 1976, 19, 615.
8. Kaliszan, R. J. Chromatogr. 1981, 220, 71.
9. Braumann, Th. J. Chromatogr. 1986, 373, 191.
10. Bonanno, A.; McMillan, S.; Donoghue, O.; Ward, A.J.I.; Lavine, B.K. Microchem. Jour., submitted.
11. Chapman, D. Quart. Rev. Biophysics. 1975, 8, 185.
12. Braumann, Th.; Genieser, H.G.; Lullman, C.; Jastorff, B. Chromatographia. 1987, 24, 777.
13. Roberts, B.L.; Scamehorn, J.F.; Harwell, J.H. In Phenomena in Mixed Micellar Systems; Scamehorn, J.F., Ed.; ACS Symposium Series No. 311: Washington, DC, 1986; p 211.
14. Hanai, T.; Hubert, J. Jour. High Resol. Chromat. & Chromatog. Commun. 1983, 6, 20.
15. Jandera, P. Chromatographia. 1984, 19, 101.
16. Hanai, T.; Hubert, J. Jour. Liq. Chromatog. 1985, 8, 2463.
17. Hafkenscheid, T.L.; Tomlinson, E. J. Chromatography. 1981, 218, 409.
18. Tayar, N.El.; van de Waterbeemd, H.; Testa, B. Quant. Struct. Act. Relat. 1985, 4, 69.

RECEIVED June 6, 1989

AGROCHEMICAL MECHANISMS

Agrochemical Mechanisms: Introduction

Modern agrochem and medicinal QSAR began on even terms in 1963 with the classic publications of Corwin Hansch and Toshio Fujita. One of the early studies treated in their papers was the auxin effect of substituted phenoxyacetic acids on Avena seedlings (oats).

Since then, SAR studies in medicine have far outstripped those in agrochemistry for a number of reasons. While enormous sums of money have been pumped into medical research to solve crisis problems in health care, agriculture in America has done a superb job of increasing food production in step with a growing population. Actual crises have been rare and the corresponding funding of agrochemical research has been minimal. Hundreds of medically important enzymes and enzyme-substrate complexes are known in full 3-dimensional detail, while not a single significant agrochemical enzyme has been purified or sequenced. This trend will likely continue until food production becomes as important as cancer and AIDS. As a direct result of underfunding, certain forms of molecular modelling approaches, in particular, computer-assisted docking of experimental compounds to active-site models, remain the exclusive domain of medical SAR research. The game is more even in the quality of statistical research where identical techniques are used with equal success.

At the fundamental levels of biomechanism, there is common molecular ground in cuticle penetration, passive and active transport, metabolism, active-site binding and irreversible inhibition. These microscopic physicochemical processes are described in each field with the same transport, electronic, steric and substructure descriptors. Thus, at the microscopic level, the biochemistry of medicine and agriculture share more common features than their macroscopic diversity would suggest.

This section explores selected topics that reveal the sophistication of modern agrochemical research and at least suggest the enormous scope of the field. In an opening paper by Mitsubishi Kasei chemists and contract researcher Magee, the subtleties of enzyme variations in susceptible and resistant insects are explored by SAR analysis. The genetic changes are clearly revealed in the differing steric and electronic responses, providing a new technique for mutational changes in mechanism. In the second paper, Magee presents a new proposal based on transition-state models to explain differences in response between aryl phosphate and carbamate AChE inhibitors. The interpretation clarifies thirty years of conflicting results between these mechanistically similar classes. Plummer and associates then explore two areas of insecticide chemistry where the details of mechanism are poorly understood, pyrethroids and benzoylurea insect development disrupters. Their study centers on active-site requirements for these compounds. Hammock and a university group of Hungarian associates develope quantitative relations between the molecular structures of aromatic trifluoromethyl ketones as inhibitors of insect juvenile hormone esterase. This critical enzyme is of key importance in the metamorphosis of insects, controlling the level of juvenile hormone. In another complex area, a team of Sumitomo chemists employ molecular orbital methods to explore the active conformations of fenvalerate and a new type of ether pyrethroid. Conformation appears to be of critical importance in designing for the highest pyrethroid activity. Sisler and Ragsdale then lead us through the intricacies of fungicide biochemistry, with particular emphasis on ergosterol synthesis inhibitors, tubulin polymerization and succinate oxidation. These mechanisms of inhibition have led to some of agrochem's most important crop fungicides. In a final paper, Draber and coworkers at Bayer AG examine the complexities of the herbicide binding protein where photosystem II is interrupted. A variety of statistical and quantum chemical methods are employed to study the action of quinolones, chromones and naphthoquinones on thylakoid membrane preparations (spinach).

These studies do not cover the massive scope of modern agrochemical SAR research, but provide incisive views into critical areas that will shape future agricultural practice. While the total funds and research effort are dwarfed by those of medicinal chemistry, there is little doubt that the problems faced in agrochemistry are equally enormous and must be addressed with equal energy in the near future.

PHILIP S. MAGEE
BIOSAR Research Project
Vallejo, CA 94591 and
School of Medicine
University of California
San Francisco, CA 94143

Chapter 9

Inhibition of Susceptible and Resistant Green Rice Leafhopper Acetylcholinesterase by N-Methylcarbamate and Oxadiazolone Insecticides

Hiroki Ohta[1], Noburo Kyomura[1], Yoji Takahashi[1], and Philip S. Magee[2]

[1]Mitsubishi Kasei Corporation, Yokohama, Japan
[2]BIOSAR Research Project, Vallejo, CA 94591 and School of Medicine, University of California, San Francisco, CA 94143

> Inhibition of AChE preparations from S- and R-strains of green rice leafhopper by aryl N-methylcarbamates (n = 20) and aryl oxadiazolones (n = 81) were evaluated with pI50 as the end-point. Regression against binding and reactivity descriptors was used to explore differences in mechanistic behavior.
> Ortho-substituted carbamates showed similar behavior against S- and R-AChE, but with different loadings and electronic effects that suggest a shift in mechanism. Oxadiazolones show a similar dependance on binding and steric factors with non-equivalent loadings for S- and R-AChE. Steric effects are much larger for ortho-substituted oxadiazolones than the related carbamates.
> These studies show that carbamates and oxadiazolones bind differently at the active sites of AChE. The basic mechanism of resistant AChE inhibition remains the same, but occurs in a physically modified active site.

One of the major pest insects in rice fields is the green rice leafhopper (Nephotettix cincticeps). Recently acquired resistance to carbamate insecticides has greatly complicated control of this insect. Resistance can develop by a number of different mechanisms such as cuticle thickening to impede transport, enhanced metabolic degradation or molecular changes within the target enzyme (AChE) (1). In the case of carbamate insecticides, resistance occurs mainly at the target enzyme which expresses reduced sensitivity to the inhibitors (2).

The mechanistic details of carbamate resistance have been studied by reaction kinetics to show that decreased formation of the receptor complex (ECX) is the most important factor in the process (3).

$$E + CX \underset{k_{-1}}{\overset{k_1}{\rightleftharpoons}} ECX \underset{-X}{\overset{k_2}{\rightleftharpoons}} EC$$

E = AChE   CX = Carbamate Inhibitor   EC = Carbamoylated AChE

ECX = Receptor Complex   X = Phenolate Leaving Group

Early studies of the relation of aryl N-methylcarbamate structure with activity and enzyme inhibition were first reported in 1966 and showed rather simple dependance on substituent effects for limited sets of compounds (4, 5). The apparent simplicity was partly due to the fact that QSAR techniques were still in the first decade of development. The true complexity of carbamate inhibition is revealed in later studies by Fujita and co-workers on brown planthopper AChE (6, 7), and by Hansch and co-workers on housefly head AChE (8). The latter study shows that 12 significant factors are required to correlate the pI50 data for a set of 269 carbamates. The details of these studies have been reviewed by Magee (9, 10).

Recently, another important class of chemicals based on the oxadiazolone ring have shown activity similar to the carbamates, but with surprising activity on carbamate-resistant green rice leafhopper (Structure, Table 2) (11-13). To investigate the resistance mechanism in green rice leafhopper, we decided to explore the structure-activity relations of both insecticide classes against AChE preparations from susceptible and resistant insects. The difference in the substituent effects was expected to provide critical information at the molecular level of the receptor complex.

Materials and Methods

Compounds Studied. Twenty Ortho-substituted phenyl N-methylcarbamates were prepared in the conventional manner by reacting the selected phenols with methyl isocyanate using a catalytic amount of triethylamine (14). The structures are listed in Table 1.

Over eighty N-aryl oxadiazolones were synthesized by a procedure described by Boesch in a German patent (15). This consists of reacting aryl hydrazines with a chloroformic ester followed by ring closure with phosgene. Representative structures are listed in Table 2.

AChE Preparation and Assay. Whole body enzyme preparations of susceptible(S) and resistant(R) strains of green rice leafhopper were prepared according to a standard method used for fly head AChE (16). The crude enzyme preparation was inhibited by serial dilutions of carbamate and oxadiazolone insecticides at 25°C for 15 minutes. Assays of residual AChE were made by the Ellman spectrophotometric procedure based on acetylthiocholine and the cleavable indicator, 5,5'-dithiobis-2-nitrobenzoic acid (17). The negative logarithm of the molar concentration causing 50% inhibition, pI50, is derived from the inhibition-concentration plot.

Statistical Data Analysis. The pI50 data were analyzed by standard multiple regression analysis using the Hansch program (18) at Mitsubishi and related programs at BIOSAR. All programs were cross-validated using standard data sets. Values of the hydrophobic constant(π) based on logP (octanol/water), the Hammett sigma constant(σ_p) and sigma inductive (σ_I) constant were taken from the Pomona College Medicinal Chemistry Project database (19). Sigma resonance (σ_R) values were derived from $\sigma_p - \sigma_I$. Steric descriptors scaled to van der Waals radii (Charton's υ) were selected from the literature (20, 21).

Results and Discussion

Ortho Substituted Phenyl N-Methylcarbamates.

Table 1 shows the enzyme inhibition data for all of the tested carbamates. With the single exception of the 2-phenyl compound, all of the carbamates were stronger inhibitors (higher pI50) of AChE from S-strain of the leafhopper. This clearly confirms the supposition that resistance to carbamates is due mainly to changes in enzyme inhibition. It is also consistent with a study by Hama and Iwata showing a correlation between the degree of resistance and both insecticidal activity and AChE inhibition for several carbamate insecticides (22).

Equations (2) and (3) are based on multiple regression analysis of the data in Table 1. In the equations, π is the hydrophobic constant, υ_o is Charton's steric constant, σ_I and σ_R are the inductive and resonance components of Hammett electronic constants. Values in parentheses are the 95% confidence intervals of the regression coefficients. The other values are n (number of samples), r (correlation coefficient) and s (standard error of estimate).

$$pI50(S) = 0.34\ \pi\ (0.35) + 1.65\ \upsilon_o\ (0.87) - 0.74\ \sigma_R\ (0.51) + 3.74\ (0.45) \quad (2)$$

$$n = 19 \quad r = 0.900 \quad s = 0.320 \quad F = 21.62$$

$$pI50(R) = 0.56\ \pi\ (0.23) + 0.99\ \upsilon_o\ (0.59) + 1.63\ \sigma_I\ (0.68) + 2.57\ (0.36) \quad (3)$$

$$n = 20 \quad r = 0.916 \quad s = 0.233 \quad F = 27.66$$

Both equations show similar but different responses to hydrophobic character and steric size. The steric sensitivity of the S-strain is substantially more accomodating than the R-strain, though the binding effect is positive for both enzymes. The most striking difference is observed in the electronic behavior with resonance important in the S-strain and induction in the R-strain. This strongly suggests a difference in the rate-determining step for carbamoylation of S and R AChE.

In studies on house fly head AChe, Kamoshita et al developed equation (4) for 16 ortho-substituted phenyl N-methylcarbamates (6). Log $1/K_d$ is reasonably collinear with pI50, but differently scaled, so we cannot compare the magnitudes of the regression coefficients in equations (2-4). The longer incubation times used in this study may reduce the collinearity and make comparison less precise. However, they observe a positive dependence on hydrophobicity as we do. As σ^o is largely inductive, their dependence is similar to that found in equation (3). The HB descriptor (HB = 1.0 for H-bonding Ortho group) in equation (4) was insignificant in our studies despite

TABLE 1. Anti-AChe activity of aryl N-methylcarbamates on susceptible(S) and resistant(R) Nephotettix cincticeps

X	$pI_{50}(S)$	$pI_{50}(R)$
H	3.89	2.50
Me	4.43	3.20
Et	4.89	3.74
i-Pr	5.70	4.21
sec-Bu	6.04	4.60
$CH_2CH=CH_2$	5.07	3.96
OMe	4.77	3.26
OEt	4.60	3.52
O-i-Pr	6.30	3.82
O-sec-Bu	5.68	4.17
$O-CH_2CH=CH_2$	5.15	3.96
$O-CH_2C\equiv CH$	5.09	4.10
Cl	5.52	4.26
Br	5.66	4.64
NH-i-Pr	5.64	3.77
NMe_2	5.02	3.89
$N(Allyl)_2$	6.14	4.80
CN	4.15	3.64
Phenyl	3.74	4.10
CH=NOMe	4.77	3.74

the presence of many H-bonding groups in our data set (Table 1). The other significant difference is the lack of a positive steric effect in their study. As n, r and s are comparable in equations (2-4), the differences point to real variations in the binding mechanisms of each AChE.

$$\text{Log } 1/K_d = 1.558 \, \pi \, (0.354) + 1.009 \, \sigma^o \, (0.508) + 1.26 \text{ HB } (0.313) +$$
$$3.998 \, (0.294) \qquad (4)$$

n = 16   r = 0.953   s = 0.255

<u>N-Aryl Oxadiazolones</u>. Compounds studied (n = 81) are shown in Table 2 with the associated pI50 values for R and S leafhopper AChE. As a class, these compounds are substantially stronger inhibitors than the aryl N-methylcarbamates (compare Tables 1 and 2). Moreover, there is no cross-resistance with carbamates as all are stronger inhibitors of the R AChE than S AChE. In addition, Ambrosi <u>et al</u> report the 2-methoxy analog (RP-32,861) to show high insecticidal activity against resistant green rice leafhopper (<u>11</u>), while Bakry <u>et al</u> report potent inhibition of housefly and electric eel AChE's for the same analog (<u>13</u>). There is little doubt that the oxadiazolones are broad, general AChE inhibitors.

Huang <u>et al</u> have recently studied the chemical reactivity of the oxadiazolone ring and propose nucleophilic attack of serine-oxyl anion on carbonyl as the probable inhibition mechanism (<u>23</u>). Our own experiments on oxadiazolone ring-opening with methoxide ion support their work (5). The driving force for this reaction is substantial as a delocalized carbamate group is formed from the iminoester leaving group. We believe the same ring-opening to be responsible for the observed lability of these compounds in both insect testing and pI50 measurement.

$$CH_3O^- + \underset{\text{oxadiazolone (Ar, OCH}_3\text{)}}{\text{[ring structure]}} \xrightarrow{CH_3OH} CH_3O\overset{O}{\underset{\|}{C}}\underset{Ar}{N}\text{-}NH\overset{O}{\underset{\|}{C}}OCH_3 \qquad (5)$$

There are both similarities and differences in the AChE inhibition mechanisms of carbamates and oxadiazolones. Both proceed by reversible binding and irreversible carbonyl attack by activated serine-oxyl ion. However, the oxadiazolones react by ring-opening leaving a large, complex structure at the site of inhibition. By contrast, the carbamates react by phenolate displacement leaving a simple carbamoylated enzyme. When the pI50 data are examined by regression analysis, it becomes clear that the mode of binding is quite different from the aryl N-methylcarbamates.

Equations (6) and (7) are statistically significant but do not correlate with the sharpness of the carbamate data, a fact we attribute to ongoing degradation during the pI50 assay. Unlike the carbamate study, the correlating factors for R and S AChE are identical but with different responses. The ortho steric effect is negative for the oxadiazolones and nearly twice as strong for the S AChE, accounting for much of the reverse selection [pI50(R) > pI50(S)].

Table 2. Anti-AChE activity of N-aryl oxadiazolones on susceptible(S) and resistant(R) Nephotettix cincticeps

X	Y	Z	R	$pI_{50}(S)$	$pI_{50}(R)$
H	H	H	Me	4.82	4.82
Me	H	H	Me	5.92	6.82
Cl	H	H	Me	5.92	6.82
Br	H	H	Me	6.43	7.39
OMe	H	H	Me	5.64	6.25
O-i-Pr	H	H	Me	5.66	6.20
OMe	H	H	Et	5.22	6.66
O-i-Pr	H	H	Et	4.80	6.49
H	H	3-Me	Me	5.82	5.89
H	H	3-i-Pr	Me	6.85	6.43
H	H	3-t-Bu	Me	6.59	6.85
H	H	3-Cl	Me	5.14	5.16
H	H	3-NO_2	Me	4.80	5.44
H	H	3-t-Bu	Et	5.92	6.33
H	Me	H	Me	5.24	5.41
H	F	H	Me	4.20	5.06
H	Cl	H	Me	4.64	4.15
i-Pr	Me	H	Me	4.77	5.70
OMe	Me	H	Me	6.22	7.00
OMe	F	H	Me	5.41	5.96
OMe	NO_2	H	Me	6.32	6.33
OEt	Me	H	Me	6.30	6.96
OEt	Et	H	Me	6.59	7.57
OEt	F	H	Me	5.82	6.22
OEt	Cl	H	Me	6.05	6.48

Continued on next page

Table 2. Continued

X	Y	Z	R	$pI_{50}(S)$	$pI_{50}(R)$
O-i-Pr	Et	H	Me	5.03	5.37
O-i-Pr	Cl	H	Me	4.82	5.82
O-i-Pr	F	H	Me	4.82	5.72
O-i-Pr	CF_3	H	Me	4.72	5.59
O-n-Pr	Me	H	Me	6.17	6.14
Cl	Me	H	Me	6.33	6.96
Cl	CF_3	H	Me	5.89	6.17
OMe	Cl	H	Et	5.07	6.85
OEt	Me	H	Et	5.41	7.20
OEt	F	H	Et	4.70	6.85
OEt	Cl	H	Et	5.05	6.96
O-n-Pr	Me	H	Et	4.82	7.49
O-i-Pr	F	H	Et	4.30	6.19
Cl	H	5-Me	Me	6.77	7.39
OMe	H	5-Me	Me	5.47	6.36
OMe	H	5-Et	Me	5.57	6.36
OMe	H	5-i-Pr	Me	5.74	6.92
OMe	H	5-s-Bu	Me	6.38	6.74
OMe	H	5-t-Bu	Me	5.35	6.35
OMe	H	5-F	Me	5.46	6.07
OMe	H	5-Cl	Me	5.08	6.17
OMe	H	5-Cl	Me	5.44	6.10
OMe	H	5-OMe	Me	4.66	5.55
OMe	H	5-NO_2	Me	5.28	5.59
OMe	H	5-CF_3	Me	5.03	5.89
OEt	H	5-Me	Me	5.05	6.21
OEt	H	5-Et	Me	5.00	6.52
OEt	H	5-i-Pr	Me	4.66	6.96
OEt	H	5-F	Me	5.59	6.06
OEt	H	5-Cl	Me	5.68	7.16
O-n-Pr	H	5-Me	Me	5.02	6.28
O-i-Pr	H	5-Me	Me	4.46	5.82
O-i-Pr	H	5-Cl	Me	4.59	5.80

Continued on next page

Table 2. Continued

X	Y	Z	R	$pI_{50}(S)$	$pI_{50}(R)$
O-i-Pr	H	5-F	Me	4.96	5.80
O-s-Bu	H	5-Cl	Me	4.08	5.66
O-s-Bu	H	5-F	Me	4.66	5.31
OMe	H	5-s-Bu	Et	5.62	7.12
OEt	H	5-Et	Et	4.74	7.00
OEt	H	5-Cl	Et	5.24	7.74
O-n-Pr	H	5-Me	Et	4.92	6.80
O-i-Pr	H	5-Me	Et	4.60	6.46
O-i-Pr	H	5-Cl	Et	5.68	6.46
OMe	H	5-t-Bu	Et	4.80	6.59
OMe	H	3-Me	Me	5.80	6.70
OMe	H	3-i-Pr	Me	5.24	5.92
OMe	H	3-Cl	Me	5.70	6.41
OMe	H	3-OMe	Me	5.89	6.72
OEt	H	3-Me	Me	5.72	6.24
H	H	3,5-diCl	Me	5.52	5.51
OMe	Me	5-Me	Me	5.96	6.96
OMe	Me	5-Cl	Me	5.89	5.80
OMe	Cl	5-Me	Me	5.82	6.89
OMe	Me	5-SMe	Me	5.49	6.89
O-i-Pr	Me	5-Cl	Me	4.40	5.82
O-i-Pr	Me	3-Cl	Me	4.54	6.14
O-i-Pr	Me	3-Cl	Et	4.74	6.51

The electronic effect ($\Sigma\sigma$) is opposite to that of equations (3) and (4), and several special factors are important. These are $IX_o = 1.0$ for special enhancement by ortho-halogen groups, IOR = 1.0 for EtO on the hetero-ring (vs. 0.0 for MeO) and HB = 1.0 for ortho-OR groups (H-bonding). The effect of ortho-halogen groups is large, similar for both R and S AChE and of unknown mechanistic origin. No other descriptors were found to accommodate the mix of alkyl, halo and OR groups at the ortho position. The effect operates independently of the group steric effect and the H-bonding effect of the OR groups. The H-bonding effect of the OR groups is similar to that observed in equation (4), but much weaker. The effect of the hetero-ring OR (R = Me, Et) is interesting as R AChE (MeO > EtO) and S AChE (EtO > MeO) are oppositely selective.

pI50(S) = -1.50 υ_o (0.69) - 0.55 $\Sigma\sigma$ (0.49) + 1.66 IX_o (0.65) -

0.32 IOR (0.29) + 0.41 HB (0.43) + 5.62 \hfill (6)

n = 81   r = 0.608   s = 0.627   F = 8.80

pI50(R) = -0.83 υ_o (0.64) - 0.82 $\Sigma\sigma$ (0.46) + 1.89 IX_o (0.61) +

0.56 IOR (0.28) + 0.77 HB (0.41) + 5.77 \hfill (7)

n = 81   r = 0.694   s = 0.494   F = 13.94

The low explained variance of these equations ($100r^2$ = 37 - 48) is not attributable to experimental error in the pI50 measurement as the same technique was used for the carbamates ($100r^2$ = 81 - 84). The probable loss in precision is due to variable degrees of hydrolysis during the measurement time span. This was tested by looking at the difference between pI50(R) and pI50(S), a procedure that should remove much of the variance due to selective hydrolysis of each compound. We find the difference to correlate significantly better than either pI50 ($100r^2$ = 58), supporting differential hydrolysis as a wild variable. Note that the electronic effect ($\Sigma\sigma$) and variable IX_o were not sufficiently different to show up statistically, although r, s and F are markedly improved. The key factors supporting greater inhibition of the R AChE are seen to be a less unfavorable steric effect and stronger H-bonding of the ortho-OR groups. There may, of course, be a missing factor that would bring r to 0.85-0.90, but it cannot be one normally explored in SAR studies.

pI50(R) - pI50(S) = 0.79 υ_o (0.50) + 0.90 IOR (0.25) + 0.34 HB (0.28) +

\hfill 0.14   (8)

n = 81   r = 0.760   s = 0.443   F = 34.99

<u>Conclusions</u>

It is clear from the SAR analysis of the pI50 response that carbamates and oxadiazolones do not inhibit either susceptible- or resistant-strain GR leafhopper AChE by analogous mechanisms. This is somewhat surprising as irreversible inhibition proceeds by analogous attack of serine-oxyl ion on positionally related carbonyl groups of both classes. The dramatic differences observed must be related to the mode of binding, in particular, to non-analogous binding regions. While both classes must bind to favorably position the carbonyl groups for serine attack, the aromatic rings are bound in distinctly different sites. This is revealed by several factors. The carbamates show a dependence on lipophilicity, an accommodation of ortho steric effects and no dependence on H-bonding potential. Conversely, the oxadiazolones show no dependence on lipophilicity, negative ortho steric effects and substantial pI50 enhancement for H-bonding groups. These effects are inconsistent with aryl ring binding at the same site location. In addition, carbamates are more active on the susceptible strain while oxadiazolones show greater activity on the resistant strain. While not tested in detail, this would suggest the probable absence of cross-resistance.

There appears to be a shift in electronic mechanism in the carbamate inhibition of the S- and R-strains of AChE. The dependence of ortho groups on lipophilicity and an accommodating steric effect is similar in both cases though different in magnitude. The much lower steric accommodation of the R-strain is probably the major factor in weakening its response to carbamate inhibition. The differing electronic effects suggest a shift in the rate-determining transition state for carbamoylation of the serine hydroxyl group.

Inhibition of S- and R-strains of AChE by oxadiazolones are found to correlate with identical factors. The difference equation [pI50(R) - pI50(S)] shows major variances in response to orthosteric effects, to OMe vs. OEt in the hetero-ring and to the H-bonding strength of some groups. In brief, higher pI50 values for the R-strain are supported by lower steric repulsion, stronger H-bonding, and by OEt substitution of the hetero-ring (OMe favored by S-strain).

These studies clearly identify the major resistance mechanism in the green rice leafhopper as involving a modified AChE target site. This could arise at the genetic level or more probably by selective concentration of an existing isozyme in the S-strain. In either case, it is important to note that the basic inhibition mechanism remains the same for both carbamates and oxadiazolones vs. S- and R-strain AChE. The differences detected by the pI50 response analysis are consistent with isozyme selection where the active sites differ by one or two structural amino acids not directly involved in the basic AChE mechanism. In brief, a simple change of cavity shape could easily account for the observed differences. Of greatest importance is the ease with which these SAR analyses were able to pin-point the mechanistic differences both between the two classes and within each class.

Literature Cited

1. Takahashi, Y. Pesticide Design-Strategy and Tactics; Eto, M. Ed; Soft Science, Inc. 1979, 674.
2. Iwata, T.; Hama, H. J. Econ. Ent. 1971, 65, 643.
3. Yamamoto, I.; Kyomura, N.; Takahashi, Y. J. Pestic. Sci. 1977, 2, 463.
4. Metcalf, R. L. Bull. W. H. O. 1971, 44, 43.

5. Hansch, C.; Deutsch, E. W. Biochem, Biophys. Acta 1966, 126, 177.
6. Kamoshita, K.; Ohno, I.; Fujita, T.; Nishioka, T.; Nakajima, M. Pestic. Biochem. Physiol. 1979, 11, 83.
7. Kamoshita, K.; Ohno, I.; Kasamatsu, K.; Fujita, T.; Nakajima, M. Pestic. Biochem. Physiol. 1979, 11, 104.
8. Goldblum, A.; Yoshimoto, M.; Hansch, C. J. Agric. Food Chem. 1981, 29, 277.
9. Magee, P. S. In Quantitative Structure-Activity Relationships of Drugs; Topliss, J. G., Ed.; Academic Press, New York 1983, 393-436.
10. Magee, P. S. In Insecticide Mode of Action; Coats, J. R. Ed.; Academic Press, New York 1982, 101-161.
11. Ambrosi, D.; Bic, G.; Desmoras, J.; Gallinelli, G.; Roussel, G. Proc. Br. Crop Prot. Conf. Pests. Dis. 1979, 533.
12. Ambrosi, D.; Boesch, R.; Desmoras, J. J. Phytiatrie-Phytopharmacie 1980, 199.
13. Bakry, N. M.; Sherby, S.; Elderfrawi, A. T.; Elderfrawi, M. E. Neurotoxicology 1986, 7, 1.
14. Kolbezen, M. J.; Metcalf, R. L.; Fukuto, T. R. J. Agr. Food Chem 1954, 2, 864.
15. Boesch, R. Ger. Offen. 1976, 2,603,877.
16. Reed, W. D.; Fukuto, T. R. Pest. Biochem. Physiol. 1973, 3, 120.
17. Ellman, G. L.; Courtney, K. D.; Andres, V. Jr.; Featherstone, R. M. Biochem. Pharmacol. 1961, 7, 88.
18. Hansch, C.; Fujita, T. J. Am. Chem. Soc. 1964, 86, 1616.
19. Hansch, C.; Leo, A. Substituent Constants for Correlation Analysis In Chemistry and Biology, Wiley, New York 1979.
20. Charton, M. J. Am. Chem. Soc. 1975, 97, 1552.
21. Charton, M. J. Org. Chem. 1976, 41, 2217.
22. Hama, H.; Iwata, T. Appl. Ent. Zool. 1971, 6, 183.
23. Huang, J.; Bushey, D. F. J. Agr. Food Chem. 1987, 35, 368.

RECEIVED June 19, 1989

Chapter 10

Critical Differences in the Binding of Aryl Phosphate and Carbamate Inhibitors of Acetylcholinesterases

Philip S. Magee

BIOSAR Research Project, Vallejo, CA 94591 and School of Medicine, University of California, San Francisco, CA 94143

> A study of bulk tolerance in the ring substituents of commercial aryl carbamate and phosphate acetylcholinesterase inhibitors strongly suggests that active site binding must be different for these related classes. This is confirmed by transition state modelling of the serine hydroxyl ion raction with the N-methylcarbamoyl and dimethyl phosphoryl derivatives of 3,4-dimethyl-phenol. Distance measurements from the esteratic site (serine oxygen) to the meta- and para-methyl groups show that binding must be different in both spacing and direction. Meta-alkyl groups of aryl carbamates bind in the lipophilic region adjacent to the anionic site. The compounds are efficiently held for reaction with the serine hydroxyl ion. To react with similar efficiency, the aryl ring of a phosphate must bind about 1.0 Å further from the esteratic site, placing the meta position beyond the lipophilic site used by the aryl carbamates. Many differences between aryl carbamate and phosphate inhibitors are clarified by this new binding model.

Inhibition of acetylcholinesterases by organophosphate (OP) and organocarbamate (OC) inhibitors proceeds by reversible binding followed by substantially irreversible blocking of the active site serine hydroxyl (1,2). This is irrefutable, and whether inhibited for hours by carbamoylation or days by phosphorylation, the neural response mechanism depending on microsecond clearance of acetylcholine is effectively blocked. In the case of aryl phosphates and carbamates, there is also no question that the phenolate anion is an electronegative leaving group (3,4). Cross-resistance of OC's to phosphate resistant house flies provides additional mechanistic overlap (5). The general principles involved have led to design of many commercial OC's and many more OP's for crop protection, animal health and human disease vector control (6,7). With so many obvious similarities in mechanism and general structural features, one would expect many more parallels when examined in detail. This is not the case for either symptomology or local molecular structure.

Studies by Miller and co-workers show clear differences in the temporal behavior of flies poisoned by OC's and OP's (8,9). This occurs despite similar rates of entry and appears to be a fundamental difference in mechanism "not wholly explainable by cholinesterase inhibition" (authors' quote)(9). It might be explainable, however, by cholinesterase selectivity. The housefly is known to contain at least nine AChE isozymes that respond quite differently to inhibition by standard OP's (10). If these isozymes are assigned specifically (one on one) to different neural functions, rather than randomly distributed, then differential shutdown by OP's and OC's is understandable. All that would be required for a different symptomology would be a different sequence of inhibition of the AChE isozymes. This is clearly achievable by the selectivity built into two parallel, but not identical, inhibition mechanisms.

In terms of molecular structure, some differences between OC's and OP's are very clear. The earliest structure-activity studies of diethyl aryl phosphates unequivocally define P-O bond breaking with a phenolate leaving group as the phosphorylation step (3). This follows from the dominant dependence of pI50 (HF head AChE) on Hammett's sigma with positive rho (3). Later studies have modified sigma to sigma minus and revealed the presence of a steric effect without altering the basic concept (11). The mechanistic simplicity of OP inhibition (12), was reflected in early structure activity studies on OC inhibition (13, 14, 15). Electronic effects in OC inhibition were much weaker and oppositely directed (negative rho). Ortho substituted carbamates, however, displayed a strong positive rho similar to the phosphates (15). Later QSAR studies revealed exceptionally complex relations for both insects (16) and isolated insect AChE (16, 17). In both studies, electroneutral rather than electronegative substituents are favored for maximum activity. This is due partly to the sensitivity of aromatic OC's to degradation by simple hydrolysis, a factor less important in related OP's and far less important for the pro-insecticidal thionophosphates. Despite the lack of a clear-cut electronic effect to support the mechanism, there is no question that carbamoylation occurs with a phenolate leaving group. The two inhibition mechanisms are identical in this respect.

Much greater differences are observed when bulk tolerances are considered in the binding step prior to irreversible inhibition. Though not the subject of this paper, differences in bulk tolerance at the esteratic site between OC's and OP's are simply immense in molecular terms. The carbamate N-alkyl group is limited in size to methyl for commercial activity while phosphates, phosphoramidates, and phosphonates typically accommodate isopropyl and phenyl groups. The variation is extensive, however, with some esteratic sites (OP resistant mites) (18), unable to accept an O,O-dimethylphosphoryl group while others (electric eel AChE) are able to bind a diphenylphosphinyl group (19).

Bulk tolerance of the ring substituents is a direct concern of this study. Tables 1 and 2 list the aromatic substituents on commercial OP's and OC's having phenol leaving groups (6). The largest groups accommodated in the various positions are summarized at the

Table 1. Aromatic Substituents on Commercial Organophosphate Insecticides

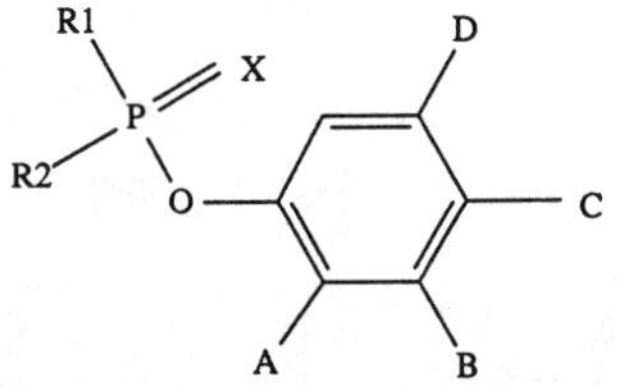

X = O, S

R_1, R_2 = OR, SR, Et, Ph

R = C_1 - C_3 Alkyl

Common Name	A	B	C	D
bromophos	Cl	H	Br	Cl
chlorthiophos	Cl	H	SCH$_3$	Cl
cyanofenphos	H	H	CN	H
cyanophos	H	H	CN	H
dicapthon	Cl	H	NO$_2$	H
dichlofenthion	Cl	H	Cl	H
EPN	H	H	NO$_2$	H
fenitrothion	H	CH$_3$	NO$_2$	H
fensulfothion	H	H	S(O)CH$_3$	H
fenthion	H	CH$_3$	SCH$_3$	H
iodofenphos	Cl	H	I	Cl
leptophos	Cl	H	Br	Cl
parathion	H	H	NO$_2$	H
profenofos	Cl	H	Br	H
prothiofos	Cl	H	Cl	H
ronnel	Cl	H	Cl	Cl
sulprofos	H	H	SCH$_3$	H
trichloronate	Cl	H	Cl	Cl
Largest Group	Cl	CH$_3$	S(O)CH$_3$	Cl

Table 2. Aromatic Substituents on Commercial Organocarbamate Insecticides

Common Name	A	B	C	D
aminocarb	H	CH_3	$N(CH_3)_2$	H
bendiocarb	$-O-C(CH_3)_2-O-$		H	H
BMPC	$s-C_4H_9$	H	H	H
bufencarb	H	$s-C_5H_{11}$	H	H
butacarb	H	$t-C_4H_9$	H	$t-C_4H_9$
carbaryl	$-CH=CH-CH=CH-$		H	H
carbofuran	$-O-C(CH_3)_2CH_2-$		H	H
CPMC	Cl	H	H	H
fenethacarb	H	C_2H_5	G	C_2H_5
methiocarb	H	CH_3	SCH_3	CH_3
promecarb	H	$i-C_3H_7$	H	CH_3
propoxur	$-O-i-C_3H_7$	H	H	H
matacil	H	CH_3	$N(CH_3)_2$	H
zectran	H	CH_3	$N(CH_2CH=CH_2)_2$	CH_3
Largest Group	$s-C_4H_9$	$s-C_5H_{11}$	$N(CH_2CH=CH_2)_2$	$t-C_4H_9$

end of each table. For OP's with heterocyclic leaving groups, substituents can be larger than those listed in Table 1. A good example is pyridaphenthion (6) which has a phenyl group in the meta-equivalent position. Ring nitrogens, however, increase the likelihood of specific binding that differs from the simple aromatics and heterocyclic examples are not considered in this study.

Physical Nature of the AChE Binding Site. The often quoted distance from the esteratic site (serine hydroxyl) to the anionic site (carboxylate group) is relevant only for acetylcholine (ACh) and mimics with charged amino-residues. None of these are important commercial inhibitors as cationic structures do not transport well through phospholipid membranes in living targets. The natural process with ACh is a diffusion controlled ion-pairing reaction of very high velocity, a necessary requirement for a cyclical microsecond response (1). The fact that aldicarb (Temik) (6) and ACh have nearly identical carbonyl to tertiary center distances (extended) is irrelevant as aldibarb cannot bind to a carboxylate site. Nevertheless, the distance analogy has been valuable in both OP and OC design.

$$CH_3\overset{O}{\overset{\|}{C}}OCH_2CH_2\overset{+}{\underset{CH_3}{\overset{CH_3}{\underset{|}{N}}}}-CH_3 \qquad\qquad CH_3NHCON=CH-\underset{CH_3}{\overset{O}{\overset{\|}{C}}}-SCH_3$$

$$\overset{*}{C}=O \text{ to } N^+ = 5.05 \text{ Å} \qquad\qquad \overset{*}{C}=O \text{ to } t\text{-}C = 5.04 \text{ Å}$$

Table 3 gives some examples that we have modelled. Binding of these tertiary centers is critical for OC activity and must occur in a lipophilic region adjacent to the anionic site.

Mapping of the lipophilic regions near the anionic site has been carried out by several investigators. The earliest work by Kabachnik et al. is the most extensive and clearly indicates two binding regions, one surrounding the anionic site and one beyond it that can accommodate an 8-carbon chain (20). Their work was done entirely with alkyl-substituted phosphates and phosphonates having total molecular flexibility. Thus, the position of the region "beyond" the anionic site is not defined. Moreover, their work with bovine erythrocyte AChE may not translate in detail to insect AChE's. Later, Steinberg and co-workers used rigid, reversible inhibitors to probe an area adjacent to the anionic site described as "a conformationally flexible, hydrophobic (lipophilic) area which tends readily to assume a near planar form" (21). This is clearly a region that could accommodate an OC or OP aryloxy-group. Studies using spin labelled ACh analogs led Abou-Donia and co-workers to describe a planar, lipophilic binding site of large radius of curvature (>10 Å) in general agreement with Steinberg (22).

Thus, it is clear that regions suitable for the binding of OC and OP aryloxy-groups exist near the anionic site. The purpose of this study is to decide if this area is used identically by both classes of inhibitors.

Experimental Section

Modelling. Two-dimensional models of acetylcholine, aldicarb (Temik), bufencarb (Bux) and fenitrothion (Sumithion) were created in the draw mode of a MACCS database and transferred to the PRXBLD modelling program for approximate energy minimization. Identical operations were carried out on the N-methylcarbamoyl and dimethyl phosphoryl derivatives of 3,4-dimethylphenol. All software programs were accessed on a Prime 9950 residing at Molecular Design Ltd. (MDL) in San Leandro, California, through an Envision 230 graphics terminal. While the PRXBLD program is much less precise than MM2, it has the advantage of handling unusual groups such as phosphoryl. Moreover, all of the structures modelled are of sufficient simplicity that further refinement is unlikely to yield new information. For physical comparison, all structures were modelled by seeding PRXBLD in the conformations indicated in Table 3. This is extended for the aliphatics and syn-planar for the aromatics. After the modelling process, minor adjustments were made by simple bond rotations in the MDL DISP program. While these may not be the precise conformations during bioactivity, it provides standard conformations for critical distance comparisons. Distances to the tertiary center or alpha-carbon center from the carbonyl or phosphoryl group were measured by the LOOK program in DISP. The modelling process is subject to small positional errors, but the distance measurements are precise. Results are shown in Table 3.

Table 3. Distance from Carbonyl or Phosphoryl to Possible Binding Center

Compound	Center	Distance, Å[a]
acetylcholine[b]	$\overset{+}{N}(CH_3)_3$	5.05
aldicarb[b]	$C(CH_3)_2SCH_3$	5.04
bufencarb[c]	m-$CH(CH_3)(C_3H_7)$	5.41
fenitrothion[c]	m-CH_3	5.24
3,4-dimethylphenyl N-methyl carbamate[c]	m-CH_3	5.50
	p-CH_3	6.89
3,4-dimethylphenyl dimethyl phosphate[c]	m-CH_3	5.48
	p-CH_3	6.88

[a] Measured from the carbonyl C or phosphoryl P atoms.
[b] Extended conformation.
[c] Syn conformation. Carbonyl or phosphoryl planar with ring.

Transition state models of serine hydroxyl anion reacting with the 3,4-dimethylphenyl carbamate and phosphate were created from the 3-D structures with the following assumptions. The carbamate intermediate is assumed to be tetrahedral with normal C-O bond lengths (1.43 Å). The phosphoryl intermediate is assumed to be bipyramidal (linear displacement) with normal P-O bonds (1.57 Å) (23). Figure 1 shows the construction of these models. The tetrahedral carbamate structure was modelled directly by PRXBLD, then rotated in DISP to bring the C-O bond coplanar with the ring. The bipyramidal phosphate intermediate required mapping on graph paper as the pentacovalent P atom was not acceptable in the modelling program. Using the serine oxygen atom as a fixed site, distances were measured to the meta and para-methyl groups as shown in Figure 1. A graphical solution by triangulation was used to measure the phosphate intermediate distances. These values have a somewhat larger error than the OC measurements.

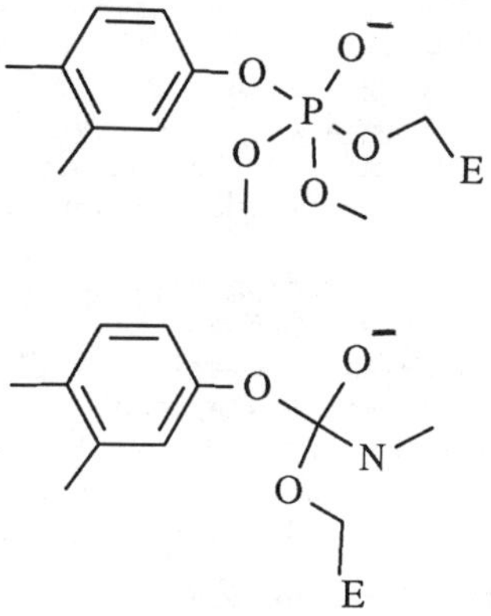

Serine Oxyl to Methyl Distance, Angstroms

Meta	5.48
Para	7.03

Meta	6.20
Para	8.10

Figure 1. Transition State Models of Serine Hydroxyl Displacement of 3,4-Dimethylphenolate Ion from the N-Methylcarbamate and Dimethylphosphate

Discussion

If binding were the only issue, the results of Table 3 would support similar binding for aryl carbamates and phosphates. Models of the analogs are superimposible. Aldicarb is a nearly perfect model for acetyl choline in the extended form, and it is easy to visualize a lipophilic region adjacent to the anionic site with structures favorable for binding tertiary centers. Moreover, the aryl carbamates in the syn-planar configuration are close enough in carbonyl-meta-α-carbon distance to reasonably bind to the same site. In support of this hypothesis, there is no significant movement required of the bound carbamate during the tetrahedral addition of the serine hydroxyl ion (Figure 1). The meta-α-carbon remains within 5.5 Å of the esteratic site. Hence, the general lore of favorable secondary and tertiary-alkyls in the meta-position of aryl carbamates is supported by a favorable sequence of binding and reactivity. Table 2 shows the high frequency of meta-alkyl groups in commercial aryl OC's. Good inhibitors with meta groups as large as hexyl and heptyl are known (24).

The situation with aryl OP's is quite different. Table 1 reveals that meta-groups of any type are uncommon and tend to modify rather than promote bioactivity. Bufencarb (Bux) has no activity without the meta-alkyl group while fenitrothion (Sumithion) simply reverts to methyl parathion, another commercial OP. As the function of the meta-alkyl group is clearly different for OP's and OC's, there is no necessary condition for identical binding. Another major factor is bulk tolerance at the meta-postion in aryl OC's and OP's. A study of Tables 1 and 2 reveals some startling differences in the leaving groups that have little to do with electronic effects. Meta-groups in commercial aryl OC's are both common and large, a consequence of favorable binding. Meta-groups in commercial aryl OP's are both rare and small, having little to do with enhancement of activity. The general absence of larger meta-groups from the aryl OP literature for the last forty years is a clear message that these groups do not enhance binding and may, in fact, be non-binding. This conclusion is completely unreasonable if the OC's and OP's occupy exactly the same binding site.

A final argument for different aryl binding sites depends on the reaction transition state. The versatility of phosphorus d-orbitals allows the potential of a non-linear displacement by serine hydroxyl anion. However, the normal model of displacement is by inversion, presumably through an intermediate of the PCl_5 structure (25-28). If we assume aryl OP to bind at the same location as a related OC, then one of two things must occur during reaction. Either the serine hydroxyl ion must undergo nearly 1 Å of distortion or the aryl ring must desorb and move a comparable distance away from the initial binding site (Figure 1). Positioning is also important as the models show a difference between the meta-position (OP - OC = 0.72 Å) and the para-position (OP - OC = 1.07 Å). This indicates the most favorable binding location for the OP would be both further from the esteratic site and in a sharply different direction, i.e., into a totally different region. At a distance of 6.20 Å from the esteratic site, it is no longer surprising that m-alkyl groups fail to support binding in a region 5.0-5.5 Å away. It is also clear that a binding position further away from the esteratic site is consistent with the greater bulk tolerance for substituents on phosphorus, a previously unexplainable fact.

Further discussion is speculative and unsupported by evidence. However, we can ask what factors might cause the aryl ring of an OP to select a binding region that differs in distance and direction from that selected by an aryl OC. As mapped by the studies of Kabachnik (20) and others (21,22) the lipophilic regions near the anionic site can accommodate an 8-carbon chain. Abou-Donia and co-workers suggest a large radius of curvature (>10 Å) for this region (22). These areas were mapped by flexible aliphatic chains and could describe elliptical regions with dissimilar axes. As seen by a study of Tables 1 and 2, the different leaving group requirements separate the OC and OP aryls into two distinct classes. Carbamate aryls are lipophilic and of low dipolarity, ideal for binding in the described region. By contrast, the phosphate aryls are much less lipophilic and highly dipolar. We can speculate then that the energetics of aryl phosphate binding are enhanced when the dipolar ring stretches across a long but narrow lipophilic region toward a more compatible dipolar area. Firm evidence for this speculation will require more extensive mapping studies or better, sequencing and modelling of a purified AChE.

Acknowledgment

The author gratefully thanks Molecular Design Ltd. (San Leandro, California) for access to their modelling programs and for a generous grant-in-aid to support this study.

Literature Cited

1. Engelhard, N.; Prchal, K.; Nenner, M. Angew. Chem. Internat. Edit.; 1967, 6, 615.
2. O'Brien, R. D. In Insecticide Biochemistry and Physiology; Wilkinson, C. F., Ed.; Plenum Press, New York and London, 1976, Chapter 7.
3. Fukuto, T. R.; Metcalf, R. L. J. Agr. Food Chem., 1956, 4, 930.
4. Hastings, F. L.; Main, A. R.; Iverson, F. J. Agr. Food Chem., 1970, 18, 497, and references cited therein.
5. Tripathi, R. K. Pest. Biochem. Physiol., 1976, 6, 30.
6. The Agrochemicals Handbook, Royal Society of Chemistry, Unwin Brothers Ltd.; Hartley, D.; Kidd, H. Eds.; Old Woking, Surrey, England, 1983.
7. The Merck Index; Windholz, M.; Budavari, S.; Blumetti, R. F.; Otterbein, E. S., Eds.; Merck & Co., Inc., Rahway, New Jersey, Tenth Edition, 1983.
8. Miller, T.; Kennedy, J. M.; Collins, C.; Fukuto, T. R. Pest. Biochem. Physiol., 1973, 3, 447.
9. Miller, T. Pest. Biochem. Physiol., 1976, 6, 307.
10. Tripathi, R. K.; O'Brien, R. D. Pest. Biochem. Physiol., 1973, 2, 418.
11. Hansch, C. J. Org. Chem., 1970, 35, 620.
12. Fukuto, T. R. Bull. Wld. Hlth. Org., 1971, 44, 31.
13. Metcalf, R. L.; Fukuto, T. R.; Frederickson, M. J. Agr. Food Chem., 1964, 12, 231.
14. Metcalf, R. L.; Fukuto, T. R. J. Agr. Food Chem., 1965, 13, 220.
15. Hansch, C.; Deutsch, E. W. Biochem. Biophys. Acta, 1966, 126, 117.
16. Kamoshita, K.; Ohno, I.; Kasamatsu, K.; Fujita, T.; Nakajima, M. Pest. Biochem. Physiol., 1979, 11, 104.
17. Goldblum, A.; Yoshimoto, M.; Hansch, C. J. Agr. Food Chem., 1981, 29, 277.
18. Zahavi, N.; Tahori, A. S.; Klimer, F. Mol. Pharmacol., 1971, 7, 611.

19. Lieske, C. N.; Clark, J. H.; Meyer, N. G.; Lowe, J. R. Pest. Biochem. Physiol., 1980, 13, 205.
20. Kabachnik, M. I.; Brestkin, A. P.; Godovikov, N. N.; Michelson, M. J.; Rozengart, E. V., Rozengart, V. I. Pharmacol. Rev., 1970, 22, 355.
21. Steinberg, G. M.; Mednick, M. L.; Maddox, J.; Rice, R. J. Med. Chem., 1975, 18, 1056.
22. Abou-Donia, M. B.; Rosen, G. M.; Paxton, J. Int. J. Biochem., 1976, 7, 371.
23. Baughman, R. G.; Jacobsen, R. A. J. Ag. Food Chem., 1976, 24, 1036. Average P-O distance in bromophos = 1.57 A.
24. Kohn, G. K.; Ospenson, J. N.; Moore, J. E. J. Ag. Food Chem., 1965, 13, 232.
25. Aaron, H. S.; Uyeda, R. T.; Frack, H. F.; Miller, J. I. J. Am. Chem Soc., 1962, 84, 617.
26. Green, M.; Hudson, R. F. J. Chem. Soc., 1963, 3883.
27. Michalski, J.; Mikolajczyk, M.; Mlotkowska, B.; Omelanczuk, J. Tetrahedron, 1969, 25, 1743.
28. Wadsworth, W. S. Jr.; Tsay, Y.-G. J. Org. Chem., 1974, 39, 984.

RECEIVED June 14, 1989

Chapter 11

Contribution of Quantitative Agrochemical Design Strategies to Mechanism-of-Action Studies

E. L. Plummer, J. A. Dixson, and R. M. Kral

Agricultural Chemical Group, FMC Corporation, P.O. Box 8, Princeton, NJ 08543

The context provided by quantitative design strategies offers many benefits towards improving the efficiency of agrochemical discovery programs. Amongst these are an understanding of mechanism of action and the factors, other than active-site interactions, that govern the efficacy of agrochemicals. Two areas of insecticide chemistry, where the mechanism of action is poorly understood, are pyrethroids and benzoylurea insect development disrupters (IDDs). The QSAR approach has been used to provide an insight into the active-site requirements for these compounds. From a knowledge of the active-site structure a greater understanding of the exact mechanism of action can be obtained. In addition, once the mechanism is better understood, mechanism based intrinsic assays can be designed to separate active-site interactions from factors of transport and metabolism. The value of such mechanism based intrinsic assays will also be discussed particularly with regard to the IDDs.

The application of the strategies that have evolved from the QSAR paradigm offer many benefits to their practitioners which in total lead to higher efficiency in the design process. To many, the primary goal of QSAR strategies is to develop rules that will efficiently lead to the most active compound in a series. Clearly this is not the only benefit. In providing a context within which one can understand chemical structure-activity relationships these strategies also allow one to recognize compounds that fail to fit the rules, i.e. outliers, and thus form the basis for new leads. Molecular modeling programs are strengthened by the selection of compounds to effectively represent the factors important to activity, by previous development of quantitative rules and by the use of standard QSAR strategies for validation of their often subjective results. Together these tools might be used to generate new lead molecules having significantly different connectivity but

0097-6156/89/0413-0157$06.00/0
© 1989 American Chemical Society

maintaining the biological activity of the compound used to develop the model.

Today's most challenging goal for pesticide design projects is the development of selectivity: compounds to control harmful insects while sparing man and other non-target organisms or herbicides that control weeds without damaging crops. Art based approaches to that goal have been only moderately successful. An alternative is to develop quantitative models for target and non-target and then to seek differences in these models which can be exploited for selectivity.

As a planning tool QSAR can improve the overall management of design projects. Perhaps the single most important outcome of the strategy is the ability to know when a project is complete, to understand when one more compound will not lead to the elusive commercial compound. Often the best compound is among the first made in a project. Thousands of additional compounds provide no improvement. If a project is well designed and carried out significant savings in resources can be realized

Finally the context of structure activity relationships that exists as the result of applying good experimental design strategies can also provide information vital to understanding the mode of action of a compound. At the very least it can provide clues to understanding factors that influence the compound's biological fate and can be separated from specific interaction with the active site. It is this latter aspect of pesticide design that this addresses with two examples, one from an older project on pyrethroid insecticides and the other from a more recent project on Insect Development Disrupters (IDDs).

A Pyrethroid Discovery Example

There is no more important step in pesticide design than the original selection of substituents. Many applications of QSAR appear to fail because the original derivatives of the lead covered only a small portion of the available physical chemical space or contained substituents whose physicochemical parameters where highly cross correlated. The cluster analysis method of Hansch, et al. [Hansch, et al., 1973] was one of the first attempts to address this issue. When we first investigated pyrethroid insecticides we realized, as had others before us, that the most active derivatives of pyrethroid esters based on benzyl alcohols bore a substituent in the meta position of the benzyl ring. New leads were sought by preparing a set of meta substituted benzyl esters of cis,trans-dichlorovinyl-2,2-dimethylcyclopropane carboxylic acid (DVA) in which the substituents were chosen by reference to the cluster sets suggested by Hansch. The original set was made up of 18 substituents. From these and subsequent sets prepared to test the early structure-activity relationships that were found, the equivalent α-cyano esters of a total of 15 of the selected substituents were prepared. These compounds (Figure 1) represent a wide range of physicochemical properties.

The compounds were tested in a topical assay. Each set of assays include the standard permethrin to help account for intertest variation. This was accomplished by calculating the individual LD_{50} of the standard and the test sample and then calculating the potency of the test compound relative to the standard. The biological response used for regression analysis was the log of the relative potency (RP). The use of a topical test was a compromise required by the fact that a specific active site assay was not available. It has advantages over foliar spray tests, in that the environmental factors of light, leaf penetration and plant metabolism are removed and the dose that the insect receives is better controlled. The resultant biological response is clearly a composite of active site interaction, penetration, metabolism and sequestering but does better reflect the relationship of chemical structure to mode of action than a foliar assay.

When the biological data for the esters and the α-cyano esters was analyzed the following models were developed:

$$\log RP_{(X = CN)} = -1.02 \, \pi \, (\text{meta}) + 2.6$$
$$(t = 5.93)$$

$$n = 10 \quad r = 0.903 \quad s = 0.33 \quad F_{1,8} = 35.2$$
$$(p = 0.0003)$$

$$\log RP_{(X = H)} = -1.09 \, \pi \, (\text{meta}) + 2.7$$
$$(t = 4.20)$$

$$n = 11 \quad r = 0.814 \quad s = 0.56 \quad F_{1,9} = 17.7$$
$$(p = 0.002)$$

Although π, F and R, the field and resonance electronic affects, and the STERIMOL parameters were studied, one parameter, the hydrophobic substituent constant π, explained the majority of biological variance for each set. The localized hydrophobic pocket which has been suggested [Plummer 1984] as present at the active site apparently binds in a similar manner for the two sets of compounds, that is, as the lipophilicity of the meta substituent increases, so does the activity. However, the α-cyano esters (Table I) are generally 1.5 to 2.0 fold more active than their respective esters. The exception is the biphenyl and heteroaromatic benzyl esters which were essentially inactive and the benzoyl ester which was also significantly less active than predicted. It is important to note, however, that the phenylethyl and the phenylethenyl substituent, which because of their length were outliers in the ester set, have doubled activity like other members of the set. The fact that the biphenyls, which fit so well in the ester series, are now outliers might be explained by a change in the way the α-cyano esters bind at the active site or by a change to a different active site. This result probably reflects the latter, a change in active site that has since been defined as a Type II pyrethroid site as opposed to the Type I site where pyrethroid esters, including the biphenyl ester leads generated from this study, are active [Gammon, 1981].

However, at the time of the original study the alternative cause was investigated. If one assumes that the second aromatic ring of the biphenyl alcohol and the second aromatic ring of the

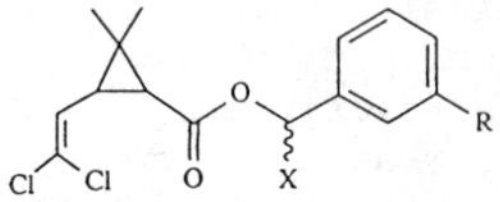

R
OC$_6$H$_5$
C$_6$H$_5$
C=OC$_6$H$_5$
thien-2-yl
furan-2-yl
I
OCH$_2$C$_6$H$_5$
*CH$_2$CH$_2$C$_6$H$_5$
*CH=CH$_2$C$_6$H$_5$
CF$_3$
Br
Cl
OCH$_3$
F
NHC=OCH$_3$

*Outliers in the Ester Study

Figure 1. α-Cyanobenzyl Esters of DVA

Table I. α-Cyanobenzyl Esters of DVA

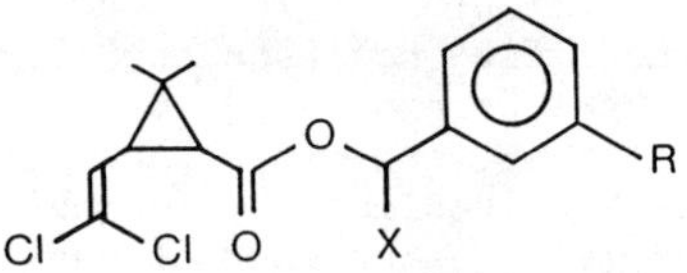

		Relative Potency		
R	π	X = H	X = CN	CN/H
OC$_6$H$_5$	2.08	1.00	1.20	1.20
C$_6$H$_5$	1.96	0.60	0.001	0.002
C=OC$_6$H$_5$	1.05	0.40	0.06	0.15
Thien-2-yl*	1.61	0.33	Inactive	–
Furan-2-yl*	1.36	0.05	Inactive	–
I	1.12	0.03	0.03	1.00
OCH$_2$C$_6$H$_5$	1.66	0.02	0.05	2.50
CH$_2$CH$_2$C$_6$H$_5$	2.66	0.01	0.03	2.50
CH=CHC$_6$H$_5$	2.68	0.01	0.02	2.50
CF$_3$	0.88	0.01	0.01	1.00
Br	0.86	0.007	0.01	1.43
Cl	0.71	0.006	0.01	2.00
CH$_3$	0.56	0.005	0.01	2.00
OCH$_3$	–0.02	0.005	0.005	1.00
F	0.14	0.004	0.005	1.25
NHC=OCH$_3$	–0.98	Inactive	Inactive	–

*cis isomer – all other 60/40 cis/trans.

phenoxy benzyl alcohol interact with the active site in exactly the same orientation and that the gem-dimethyl groups and vinyl groups are themselves totally coincident, then an examination of Dreiding models shows that for the two α-cyano esters the cyano groups point in rather different directions. However, if one replaces the biphenyl benzyl ester with a 2-(biphenyl-3-yl)-1-cyanoethyl ester the groups are coincident. This compound was prepared by the method in Figure 2. Both the ester and α-cyanoester were totally devoid of activity, reinforcing the conclusion that the site of action or mechanism of action was not the same.

An Example from Insect Development Disrupter (IDD) Discovery

The experimental design strategy proposed by Hansch has been proven to have faults with its many virtues. Although it goes a long way, the method assures neither good spread of substituents in parameter space nor a lack of collinearity. This has to be done on a trial and error basis using factor analysis. Perhaps the major objection that one might have to this or a more precise factorial design is that to represent even three or four parameters requires a relatively large substituent set (e.g. 2^n compounds). If more than one position is investigated the design becomes difficult to handle or at least very consumptive of resources. An alternative approach is sequential simplex optimization (SSO), first introduced to drug design in 1974 by Darvas [Darvas, 1974]. This method has been applied to many projects in our laboratory and has been found to be both economical and effective. Although it is primarily an optimization scheme, it can also provide the necessary context within which other elements of pesticide design, such as mechanism of action studies, can be pursued. Our approach has been to chose the original simplex set to maximize coverage of parameter space rather than selecting an original simplex that is centered in the parameter space. The latter approach generally assures that the optimal compound will be outside the simplex. Although this is likely to be more efficient, our large simplex approach, although suffering from some initial wandering, will provide both parameter space coverage and rapid optimization. The SSO experimental design has been used extensively in the study of benzoylurea insect development disrupters (IDDs).

Since the introduction of diflubenzuron by Phillips-Duphar in the mid 1970's there has been considerable interest in the benzoylureas. Current information suggests that they exert their effect on insects by interfering with the formation of chitin, the primary structural component of the arthropod exoskeleton. The biochemical process that is responsible for chitin formation is shown in Figure 3.

The most likely steps where benzoylureas may be active are the last two steps: the assembly of Uridine diphosphate N-acetylglucosamine from UTP and N-acetylglucosamine-1-phosphate and the polymerization of N-acetylglucosamine from UDP-N-acetylglucosamine under the mediation of chitin synthetase. Since studies [Gijswijt, et al., 1979] have indicated that benzoylurea intoxication is accompanied by the buildup of UDP-N-acetylglucosamine, the latter

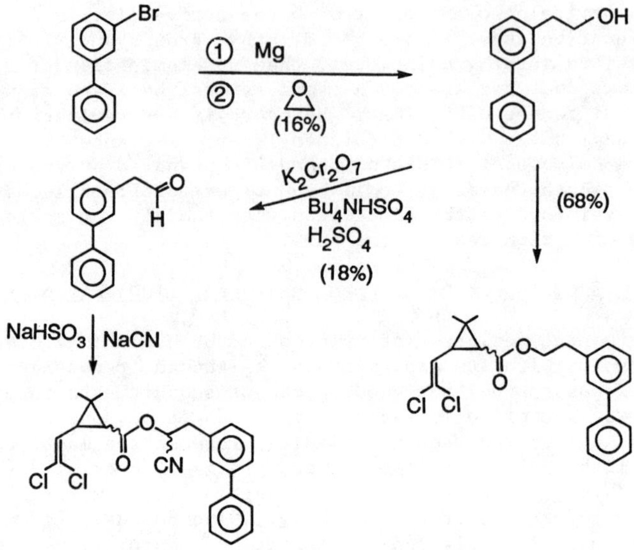

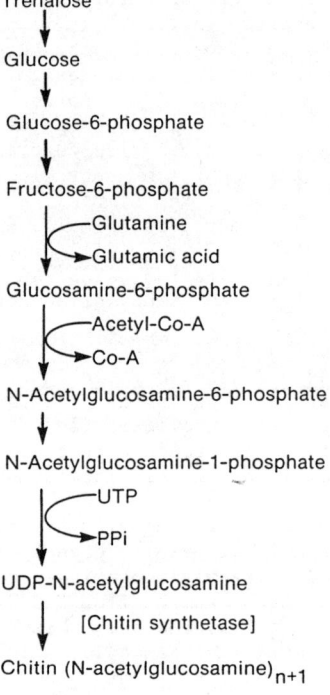

Figure 2.   Synthesis of 2-Biphenylethanol Analogs

Trehalose
↓
Glucose
↓
Glucose-6-phosphate
↓
Fructose-6-phosphate
↓ ⟵ Glutamine
  ⟶ Glutamic acid
Glucosamine-6-phosphate
↓ ⟵ Acetyl-Co-A
  ⟶ Co-A
N-Acetylglucosamine-6-phosphate
↓
N-Acetylglucosamine-1-phosphate
↓ ⟵ UTP
  ⟶ PPi
UDP-N-acetylglucosamine
↓ [Chitin synthetase]
Chitin (N-acetylglucosamine)$_{n+1}$

Figure 3.   The Biochemical Pathway Affected by Benzoylphenylureas

step has been suggested as the fatal lesion. The most likely mechanism would be the inhibition of chitin synthetase, but this does not seem to be the case [Reynolds, 1987]. Other workers [Mitsui, et al., 1985] have indicated that benzoylureas inhibit the transport of UDP-N-acetylglucosamine from the internal to the external surface of the cell membrane, where the chitin polymer is assembled by chitin synthetase.

Modeling studies in our laboratory using the facilities of MDL's CHEMLAB suggest that the benzoylureas could mimic the uridine phosphates or UDP-N-acetylglucosamine at a binding site. The crystal structure of diflubenzuron was published in 1978 by Cruse [Cruse, 1978]. Figure 4 is an ORTEP representation of that crystal structure.

The urea forms a six membered ring by virtue of a strong hydrogen bond between the aniline hydrogen and the benzoyl group carbonyl. There are many points of similarity between the benzoylurea structure and the structure of uridine. CNDO2 calculation of electron density for the urea six membered ring and uridine are shown in Figure 5.

The three points of similarity shown, as well as several others, were used to form the working hypothesis that the benzoylureas mimic the uridine moiety at the active site. A model of UDP-N-acetylglucosamine was assembled from published crystal structures of Uridine and N-acetylglucosamine. An ORTEP representation of that molecule in a linear form is shown in Figure 6.

If indeed the benzoylureas mimic this molecule at a catalytic or binding site, it is apparent that there is considerable room at the site for the inhibitor molecule. This is a contradiction to the original QSAR study done by the Duphar group in which they concluded that a short, thick substituent in the para position enhances activity [Verloop, et al., 1976]. However, benzoylureas introduced since that time have tended to have longer, bulkier groups in the para position, supporting this qualitative picture.

Some of the products of our own work in this area appear in Figure 7. The QSAR models developed in the studies are included along with the general structures .

One overall conclusion drawn from these compounds was that the length could in fact be extended in the direction of the para position of aniline ring. The series that best exemplified this was the 2-phenyl-1,1,2,2-tetrafluoroethoxy analogs. The approach to this series involved not only basic QSAR design but also the use of intrinsic assays. These were selected to contrast with in vivo testing to take full advantage of the context provided by a QSAR design to understand environmental factors as well as probing the mechanism of action. The initial simplex set was prepared to cover π, F, R, and the STERIMOL parameters L and B_1 representing the essential factors lipophilicity, electronics and size/shape. The STERIMOL parameter B_4 was not included because of its high collinearity to π. Since five parameters (n) were to be evaluated, the six substituents (n + 1) in Figure 8, chosen by cluster analysis, were prepared.

Several test methods, Table II, were used to evaluate the compounds. The chitin synthesis assay was designed to minimize

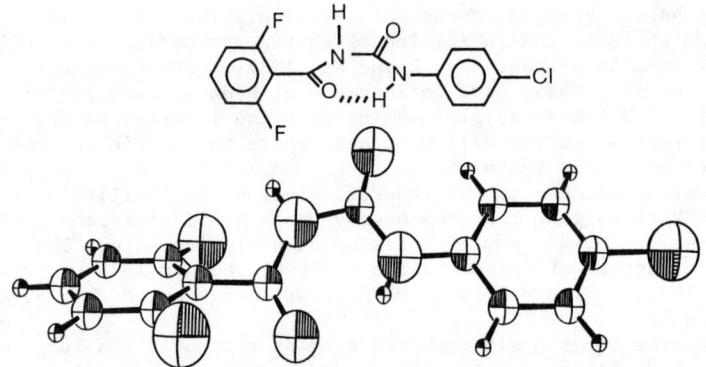

Figure 4.   Diflubenzuron Crystal Structure (Reprinted with permission from ref. 1. Copyright 1978 Elsevier.)

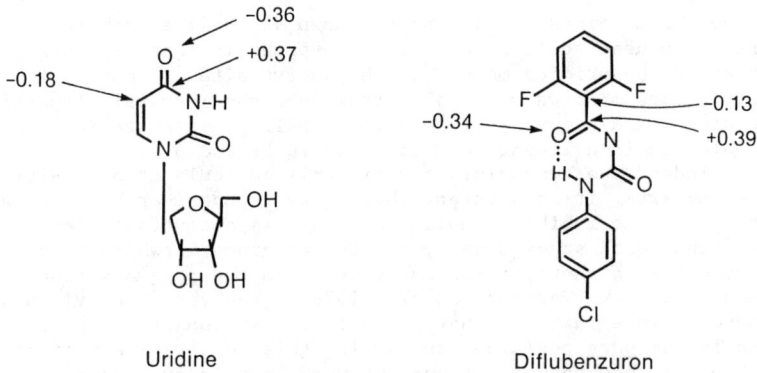

Figure 5.   Partial CNDO2 Electron Density Calculations

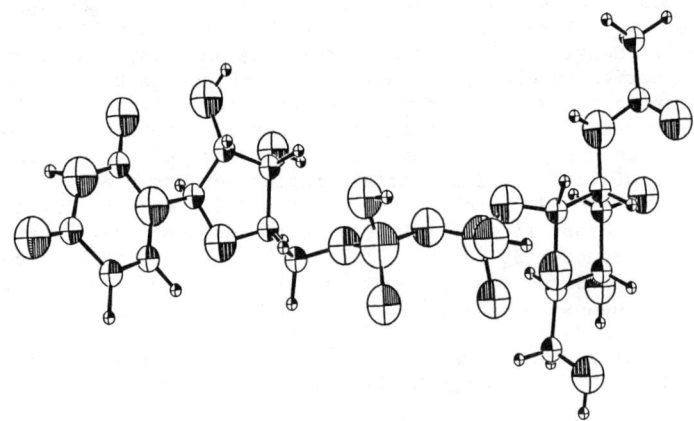

Figure 6.   UDP-N-Acetylglucosamine

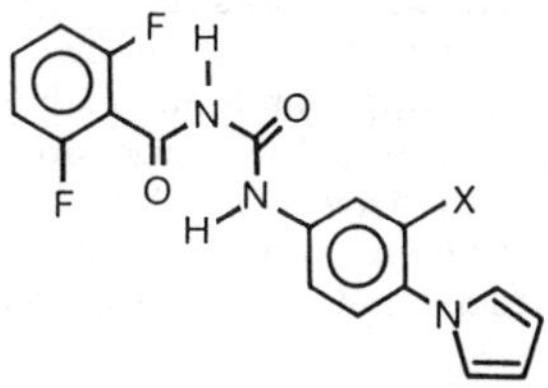

$\log(1/LC_{50}) = -9.9\,R + 2.4\,B_1 - 6.4$
$n = 8 \quad r^2 = 0.85 \quad s = 0.444 \quad F = 14.4$

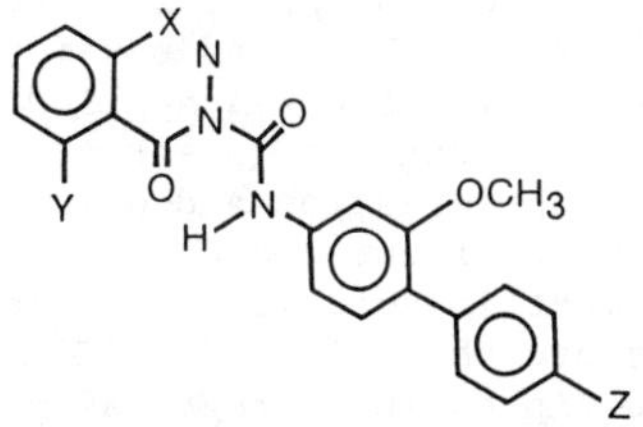

$\log(1/LC_{50}) = 3.6\,R + 2.6\,\pi - 1.6\,\pi^2 + 1.5\,I_{2-6} - 0.2$
$n = 22 \quad r^2 = 0.75 \quad s = 0.999 \quad F = 12.4 \quad \pi_0 = 0.8$

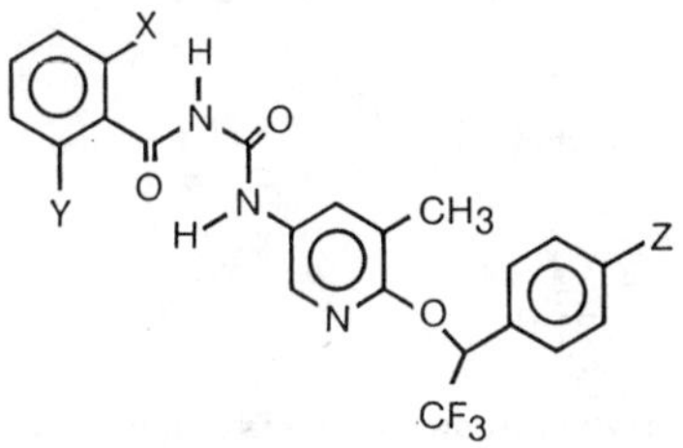

$\log(1/LC_{50}) = 0.3\,(\pm 0.1)\,\pi - 0.4\,L\,(\pm 0.1) + 1.7$
$n = 20 \quad r^2 = 0.60 \quad s = 0.50 \quad F_{2,17} = 12.5$

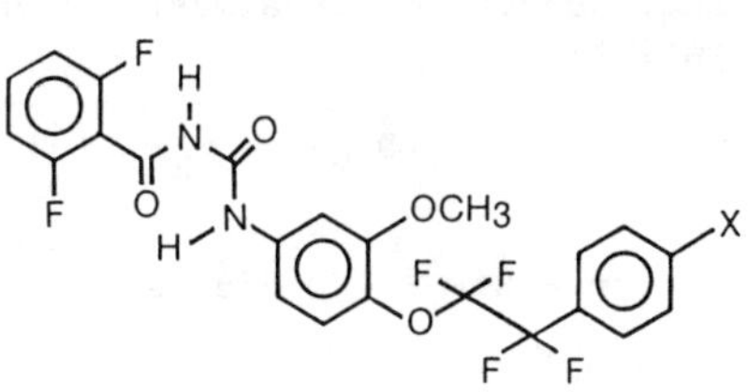

Figure 7. Benzoylurea Insecticides - QSAR

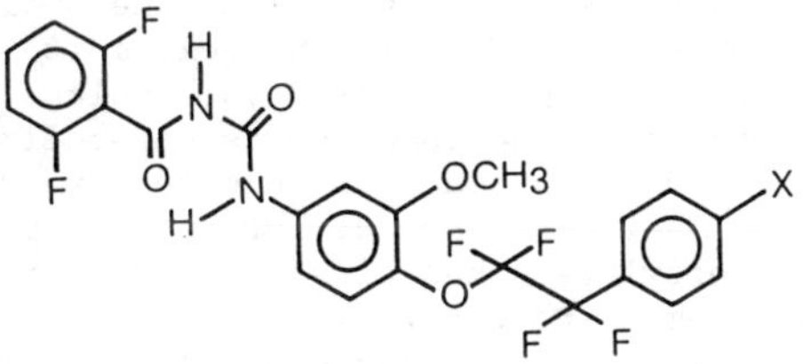

X
H
Cl
N(CH$_3$)$_2$
C=OC$_6$H$_5$
OC$_5$H$_{11}$
OCH(CH$_3$)$_2$

Figure 8. Initial SSO Design Set

environmental influences on the structure-activity analysis while the topical, diet and foliar assays were meant to bring the compounds into the real world by stepwise addition of environmental factors. The test primarily used for the optimization was the topical assay. This was particularly desirable for this purpose since a single dose was applied rather than a dose dependent on how much or how long the insect ate the medium on which it is exposed.

The optimization and original probe strategy involved the preparation of three additional compounds and was abandoned when no improvement in activity was noted beyond fluorine and chlorine. The results are shown in Table III.

Multiple linear regression analysis of topical data for this short series was conducted. Although no single parameter in the original set, π, F, R, L, B_1, showed significant correlation, it was found that the substitution of sigma p for F and R gave a model that accounted for almost seventy percent of the biological variation:

$$\log(1/LD_{50}) = 3.0 \text{ Sigma P} + 0.6$$
$$(t = 3.74)$$

$$n = 8 \quad r = 0.836 \quad s = 0.70 \quad F_{1,6} = 13.96$$
$$(p = 0.01)$$

That is, for the series tested electron withdrawing substituents in the para position increased activity.

It is particularly interesting that the unsubstituted compound and the 4-Cl analog have equal activity in the in vitro assay but dramatically different activity in the topical, diet and foliar assay. It is very likely that this difference is due to a difference in susceptibility to metabolism. Specifically we suggest that aryl hydroxylation is the metabolic step affected. The importance of metabolism is not too surprising. Since the biological response to IDDs is often not expressed until the next molt, the effect of the compound must persist for several days. Unless the fatal lesion persists long after the compound is metabolized and eliminated, then the compound must remain in an effective dose for most of the premolt period. The importance of metabolism to the differential activity of benzoylureas has also been noted recently by Neuman and Guyer [Neuman and Guyer, 1987]. In contrast to our studies, which demonstrate the absence of metabolism in the intrinsic system, they followed ^{14}C-labelled glucose incorporation in chitin in a living insect and thus observed the loss of benzoylurea with time. Our data was also generated on a congeneric series allowing direct comparison of the substituent presents. Together these studies emphasize the importance of accounting for metabolic differences of substituents when comparing benzoylureas.

It is also of significance that the compounds in this series have activity in excess of diflubenzuron and comparable to treflubenzuron, even though they are clearly significantly longer. Once again the concept that the benzoylureas act by mimicking UDP-N-acetylglucosamine cannot be dismissed.

Table II. Biological Evaluation Methods

- Chitin Synthesis Inhibition
  Southern armyworm pupae wings
  Incubate 24 hours with ^{14}C-N-acetylglucosamine
  Incubate 24 hours with chitinase
  Solubilize and count radioactivity
- Topical Assay
  Acetone solution placed mid-dorsal on Southern armyworm [3rd instar]
  Hold on casein diet
  96 Hour mortality counts
- Diet Assay
  Southern armyworm raised on artificial diet
  Compound formulated on clay and dispersed
  96 Hour mortality counts
- Foliar Assay
  Spray Pinto beans
  Infest with first instar larvae
  96 Hour mortality counts

Table III. Comparative Biological Response of 4'-(1,1,2,2-Tetrafluoro-2-phenylethoxy)benzoylphenylureas

X	SAW Topical LD_{50} (nmole/insect)	Chitin Synthesis Inhibition (pI90)	SAW Diet LC_{50} (ppm)	SAW Foliar LC_{50} (ppm)
H	0.24	6.6	2.5	67.0
Cl	0.03	6.9	0.4	1.8
OC_5H_{11}	0.42	5.9	19.0	–
$C=OC_6H_5$	> 80.0	5.8	> 250	–
$N(CH_2)_2$	73.0	–	700	–
$OCH(CH_3)_2$	2.0	–	75	–
OC_3H_7	35.0	–	117	–
CH_3	1.13	–	–	–
F	0.08	–	–	2.4
Diflubenzuron	0.99	7.3	1.4	48.2
Teflubenzuron	0.03	7.1	–	–

Certainly the early work of the Duphar group and later developments in benzoylurea chemistry support the observation that the para substituent on the aniline ring can be quite large. Although electron withdrawal at that position still seems to favor activity, groups as long as phenoxy, phenyl and the phenylethoxys just discussed, are still quite active. It appears that this position can be described as a ballast position; one that can be substituted freely without dramatically changing the activity in vitro. The identification of such a position has considerable advantage for design since such positions can be used to build in properties that favor in vivo activity. As indicated earlier, it also has some implications with regard to the mechanism of action of these compounds.

Early studies suggest that diflubenzuron and its analogs are stomach poisons, lacking significant activity if they are not ingested. However, when one extends the ballast groups to substituents with high lipophilicity, it becomes apparent that this is only a physical barrier. Penetration of the insect cuticle is quite effective when the lipophilicity is sufficiently high.

In conclusion, we wish to once again emphasize the value of the context that surrounds the application of QSAR strategies in synthesis planning and in analysis of biological data including insights relative to mode of action studies. If the first steps of experimental design are done properly the benefits can include activity optimization as well as effective overall project execution.

Literature Cited

1. Cruse, W. B. T. Acta. Cryst. 1978, B34, 2904-2906.

2. Darvas, F. J. Med. Chem. 1974, 17, 799-804.

3. Gammon, D. W.; Brown, M. A.; Casida, J. E. Pestic. Biochem. Physiol. 1981, 15, 181-191.

4. Gijswijt, M. J.; Deul, D. H.; Dejong, B. J. Pestic. Biochem. Physiol. 1979, 12, 87-94.

5. Hansch, C.; Unger, S.; Forsythe, A. B. J. Med. Chem. 1973, 16, 1217.

6. Neuman, R.; Guyer, W. Pestic. Sci. 1987, 20, 147-156.

7. Plummer, E. L. ACS Symposium Series 255, P.S. Magee, G. K. Kohn and J. J. Menn, Editors, 1984, 297-320.

8. Reynolds, S. E. Pestic. Sci. 1987, 20, 131-146 and references therein cited.

9. Verloop, A.; Hoogenstraaten, W.; Tipker J. Drug Design, 1976, 7, 165-206.

RECEIVED August 2, 1989

Chapter 12

Quantitative Structure–Activity Relationship Study of Aromatic Trifluoromethyl Ketones

In Vitro Inhibitors of Insect Juvenile Hormone Esterase

András Székács[1,2], Barna Bordás[2], György Matolcsy[2], and Bruce D. Hammock[1]

[1]Department of Entomology and Department of Environmental Toxicology, University of California, Davis, CA 95616
[2]Plant Protection Institute of the Hungarian Academy of Sciences, Budapest, Post Office Box 102, 1525 Hungary

> The in vitro inhibitory activity of 41 aryl substituted 3-phenyl-1,1,1-trifluoro-2-propanones against insect juvenile hormone esterase has been related to various electronic, hydrophobic and steric parameters using linear stepwise regression analysis. The pI_{50} values were found to be significantly correlated to the total lipophilicity of the molecule and the corrected molar volume (partial molar refractivity) of the substituents at the different substituent positions.

Trifluoromethyl ketones (TFKs) have been found to inhibit various hydrolytic enzymes (1-6). Series of aliphatic and aromatic trifluoromethyl ketone sulfides (7-10) proved to be exceptionally powerful inhibitors of insect juvenile hormone esterase (JHE), an enzyme of key importance in insect metamorphosis. The trifluoroketone moiety is believed to behave as a transition state mimic (11,12) of juvenile hormones (JHs), substrates of the enzyme. The β sulfur atom is anticipated to mimic the α-β double bond present in all natural JH substrates. In earlier structure-activity relationship (SAR) studies (7,11) clear correlation was found among the molar I_{50} values of these compounds against JHE and the calculated molar refractivity of the inhibitors. Examining 18 substituted 3-phenylthio-1,1,1-trifluoro-2-propanones, regression equations were obtained between the inhibitory activities and the Hammett (σ), Taft (E_s) steric and Hansch (π) hydrophobicity constants (11). In the hope of increasing the significance of these equations and to better distinguish between the importance of various substituent positions, several new compounds of the related structure were synthesized, a much larger set of substituent parameters was applied, and instead of the arbitrary choice of these values, the variables were selected into the equations by a more sophisticated tool, linear stepwise regression analysis.

Stepwise Regression Analysis

Considering the data set of the biological activity as dependent variable (Y_i) and different substituent parameters as independent variables ($X_{j,i}$), their correlation is approached by a linear equation (Equation 1):

$$Y_i = \beta_0 + \beta_1 X_{1,i} + \beta_2 X_{2,i} + \ldots + \epsilon_i \quad (1)$$

where β_0 is the constant of the equation, β_is are the various coefficients of the independent variables and ϵ_i is the error factor, a random deviation. Several multiple regression analysis techniques are known to fit the equation to the actual data set ($X_{j,i}, Y_i$), differing from each other in the mode or sequence of how to consider the different independent variables one by one. The stepwise regression analysis (SRA) procedure is known to be one of the best of the variable selection techniques and is very useful when there is a great number of independent variables to consider. The process is discussed thoroughly elsewhere (13, for a computational method see 14,15), hereby we mention only its brief summary.

The variables (X_j) are included in the equation one at a time, entering always the one which, at that point, is the most correlated to Y, and the significance of the new linear equation, obtained by selecting the new parameter in, is checked at each step. To enter a new parameter the partial correlation coefficient (F) has to exceed a critical value, set at the beginning of the process. Also, any other X_j variable selected in the equation at a previous stage is discarded if its partial F value decreases below a partial F criterion due to entering the new independent variable. If no more variables meet the partial F criterion, the procedure stops resulting in Equation 1. The system also terminates if no significant variables can be selected at all, if all independent variables are selected, if the variable to be added to the model is the one that just was deleted from it, and - in some cases - if the regression steps exceed a certain number. In order to "protect" the variables already selected in the equation, the partial F criterion can be set to be higher for selecting a new X_j in (entry F) than for discarding one out (exit F), to avoid overflow, however, it is advisable to set these two criteria to the same value, i.e. to 95% confidence level (α=0.05).

Generally, some difference (ϵ_i) exists between the Y_i values observed and calculated, the error in prediction might be caused by errors in measuring Y or by not considering all responsible parameters in the initial model. As it has been stated by Pope (16), SRA does not necessarily result in an equation with direct physical meaning since imitative variables may correlate well with the response. Therefore high correlations obtained by this method do not necessarily indicate primary physical connection and *vice versa*, no correlation does not prove a lack of physical relationship. For predictive purposes, however, imitative models are suitable as well. Stowe and Mayer (17) also warns of the possible limitations of various statistical methods for the selection of significant variables. Using ANOVA, t-test and SRA methods, the authors found significant correlation between the result of drawing tokens form a sack and various *ad hoc* variables. The error estimation, the validation test at each step and the consciousness at the conclusions are important to avoid misinterpretations (16,17) as well as to the correct evaluation of correlation coefficients (18). SRA has been successfully applied for QSAR purposes of pesticides (19-27) as well as in other areas of chemistry.

Synthesis and Enzyme Inhibition Assay of Trifluoromethyl Ketones

The title compounds, aryl substituted 3-phenyl-1,1,1-trifluoro-2-propanones of Structure A, were synthesized from the appropriate thiophenol using 3-bromo-1,1,1-trifluoro-2-propanone (BTFA) and triethylamine according to the previously published procedure (7,9). In general, the products were formed at good yields (80-95%), however, several compounds containing nitro and amino substituents on the aromatic ring were difficult to synthesize. Under the above reaction conditions, nitrothiophenols are oxidized to disulfides. Aminothiophenols, meanwhile, appear to react with BTFA on the carbonyl group, probably forming hemiaminals. Thus, the reaction of 4-nitro-thiophenol, 2-nitro-4-trifluoromethyl-thiophenol and 4-amino-thiophenol with BTFA was unsuccessful to give trifluoromethyl ketones of Structure A. Other amide, amine and imine derivatives (compounds 37-41, see Table I) were, although at a lower yield, successfully synthesized. Compounds 35 and 36 were gifts from Drs. Tim Schierling and Ken Musker (Chemistry Department, University of California, Davis).

The radiopartition assay (28,29) was applied to monitor the *in vitro* inhibitory activity of the title compounds against JHE. The enzyme activity is monitored by measuring the hydrolysis rate of the substrate, juvenile hormone. Diluted hemolymph from L_5D_2 larvae of *Trichoplusia ni* (cabbage looper) was the source of the enzyme, a mixture of ^3H-labeled and unlabeled JH-III was used as a substrate. Ethanol or acetone were used as solvents for the TFK inhibitors in the assay. The chemical structures and inhibitory potencies of the aromatic trifluoromethyl ketones involved in this study are summarized in Table I.

Stepwise Regression Analysis on Trifluoromethyl Ketones

In a previous study of 18 aromatic trifluoroketones, their inhibitory activity against JHE was correlated to the structure using four substituent parameters (11). The descriptors were chosen more or less arbitrarily and the effect of the different substituent positions could be determined semi-empirically, mainly by examining how the various data-points fit the equation. Important conclusions could, however, be drawn based on the empirical and intuitive method: the authors found that *meta*- and *para*-substituted compounds fit the regression line better and offered better selectivity against JHE.

A broad range of different electronic, hydrophobic and steric substituent parameters are available in the literature (30). In the present study 11 physicochemical parameters were compiled from standard tables (31) and applied to stepwise regression analysis. The parameters involved in the study are Hansch-Fujita's lipophilicity parameter (π) (32), Hammett constants in the meta and para positions (σ_m, σ_p) (33), molar refractivity (MR) (31), Swain and Lupton's electronic parameters characterizing the field and resonance effects (F,R) (34), Taft's steric parameter (E_s) (35), STERIMOL steric parameters representing the smallest and largest width of the substituent perpendicular to the bond connecting its α-atom to the rest of the compound (B1,B4) (36) and two indicator variables describing proton acceptor-donor properties (H-AC, H-DO: the parameters are 1 if the appropriate property applies to the substituent, otherwise their value is 0).

Table I. Inhibition of insect juvenile hormone esterase by arylthio-trifluoropropanones of structure A

compound number	substituent	Molar I_{50} [M][a]	compound number	substituent	Molar I_{50} [M][a]
1[b]	-	8.2×10^{-6}	21	3,4-Me$_2$	1.1×10^{-8}
2	2-Br	6.8×10^{-7}	22	3,5-Me$_2$	1.8×10^{-6}
3	3-Br	8.8×10^{-7}	23	2-Et	2.9×10^{-6}
4[b]	4-Br	4.1×10^{-7}	24[b]	2-iPr	3.4×10^{-7}
5[b]	2-Cl	2.1×10^{-6}	25[b]	4-tBu	7.5×10^{-9}
6[b]	3-Cl	4.0×10^{-7}	26	2-Me-4-tBu	3.0×10^{-9}
7[b]	4-Cl	1.3×10^{-6}	27	3-Br-4-Me	1.3×10^{-8}
8[b]	4-F	8.4×10^{-6}	28	2-MeO	2.0×10^{-6}
9[b]	2,5-Cl$_2$	2.3×10^{-8}	29[b]	3-MeO	1.3×10^{-6}
10	2,6-Cl$_2$	7.8×10^{-8}	30[b]	4-MeO	1.3×10^{-6}
11[b]	3,4-Cl$_2$	2.0×10^{-8}	31[b]	3-CF$_3$	2.5×10^{-6}
12[b]	2,4,5-Cl$_3$	3.2×10^{-8}	32[b]	4-OH	6.1×10^{-6}
13	2,3,5,6-F$_4$	1.2×10^{-6}	33	2-COOH	$> 10^{-4}$
14	F$_5$	1.1×10^{-5}	34	2-CH2OH	7.0×10^{-6}
15[b]	2-Me	4.3×10^{-6}	35	4-Me$_2$N	9.0×10^{-8}
16[b]	3-Me	1.3×10^{-7}	36	4-Me$_3$N$^+$	$\gg 10^{-4}$
17[b]	4-Me	1.1×10^{-7}	37	4-MeCONH	$> 10^{-4}$
18	2,4-Me$_2$	9.1×10^{-7}	38	2-PhNH	3.0×10^{-8}
19	2,5-Me$_2$	1.7×10^{-7}	39	4-PrNHCOO	$> 10^{-4}$
20	2,6-Me$_2$	1.2×10^{-7}	40	2(2-HOPh)CH=N	$> 10^{-4}$
			41	2(2-thienyl)CH=N	$> 10^{-4}$

[a] JH III was used as a substrate for JHE from *Trichoplusia ni* (cabbage looper).
[b] Data from Hammock et al. (7).

As it has been pointed out by Topliss and Edwards (37), the higher the number of the possible independent variables to consider in a QSAR study, the more probable the occurrence of chance correlations. Therefore, in order to enter a large number of parameters into the SRA, the number of compounds needed to be increased. Thus, 23 more compounds of Structure A were synthesized.

Thirty-five compounds provided exact numerical I_{50} values which could be used in the computations. There are 15 various substituents appearing in five possible substituent positions (numbered I-V in Structure A) in these compounds. In a program, written in BASIC, a descriptor matrix was created by copying the appropriate substituent parameters into the corresponding substitution position-compartments of the descriptor matrix if the actual substituent is present in the molecule in the actual position. From the Hammett constants, the equation $\sigma_o = \sigma_p$ was assumed (31,38,39), thus σ_p was applied for positions I, III and V, and σ_m for positions II and IV. Therefore, 10 parameters were used to enter the descriptor matrix for each substituent position. Two additional parameters, the total lipophilicity $[\sum\pi]$ and its square value $[(\sum\pi)^2]$ were also calculated for each compound and were added to the data-set.

Five parameters in the data-set were found to be unchanged for all 35 compounds and removed from the matrix. These parameters are H-DO for positions II, IV and V and H-AC for positions IV and V. After the redundant elements had been removed, the resulting [35x47] matrix was correlated to the vector of the biological activity. To perform the linear stepwise regression analysis, the STEPWISE procedure of the SAS statistical package (40) and BASIC programs were used.

Running SRA on the full sample gave the following equation:

$$-pI_{50} = 0.910(\pm 0.390) \text{ H-DO}_I + 0.153(\pm 0.046) \text{ MR}_{II}$$
$$- 1.653(\pm 0.720) \sigma_{mII} + 0.088(\pm 0.020) \text{ MR}_{III}$$
$$+ 1.561(\pm 0.900) \sigma_{mIV} + 0.169(\pm 0.084) \text{ MR}_V$$
$$+ 0.712(\pm 0.143) \sum\pi + 4.885 \qquad (2)$$

n = 35   r = 0.888   s = 0.51   F = 14.41

(In the equations, n represents the number of datum points used to derive the equation, r is the multiple correlation coefficient, s is the standard deviation from regression, F is the F statistic for variance of each additional variable, the values in parentheses after the equation coefficients are for construction of confidence intervals and the Roman numbers in subscript refer to the substituent position.)

The partial r value is the highest for $\sum\pi$ and MR_{III} (0.739 and 0.325, respectively) and does not exceed 0.183 for any of the other variables in the equation. This means that the total lipophilicity alone explains over 54% of the total variance (Figure 1). The predicted pI_{50} values and the standard error of the prediction are listed in Table II. and shown in Figure 2. The fact that σ_m has opposite signs in positions II and IV is seemingly contradictory, but the weight of σ_{mIV} and MR_V is very low in the equation, therefore they can be omitted resulting in Equation 3.

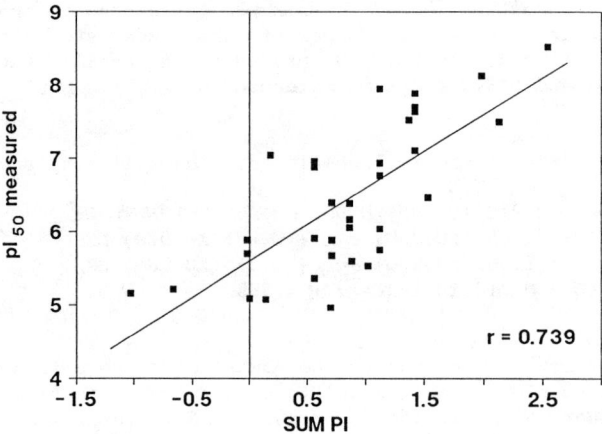

Figure 1. Correlation between the total lipophilicity and the inhibitory potency.

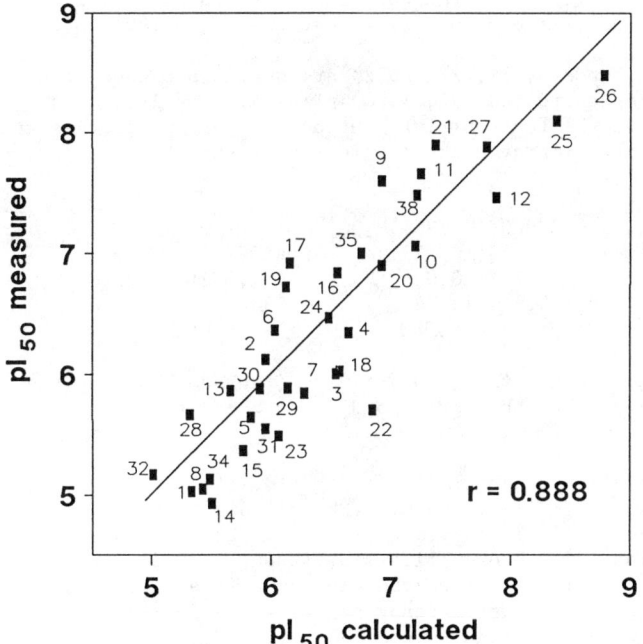

Figure 2. Correlation between the calculated and measured pI$_{50}$ values.

$$-pI_{50} = 0.796(\pm 0.412)\ \text{H-DO}_I + 0.108(\pm 0.046)\ \text{MR}_{II}$$
$$- 1.292(\pm 0.702)\ \sigma_{mII} + 0.071(\pm 0.020)\ \text{MR}_{III}$$
$$+ 0.842(\pm 0.140)\ \sum\pi + 5.220 \tag{3}$$

$n = 35 \qquad r = 0.862 \qquad s = 0.54 \qquad F = 16.79$

Requirements for good activity (ranked according the absolute value of the corresponding path coefficients given in brackets):

Large overall hydrophobicity ($\sum\pi$) [0.626];
in position III: large volume (MR) [0.368];
in position II: large volume (MR) [0.263];
in position II: electron-donating substituent [-0.194];
in position I: proton-donor substituent [0.189].

Compounds 9, 17, 22 and 23 are outliers because the residual of their measured and calculated pI_{50} value deviate most. Omission of the four outliers introduced more parameters and significantly improved the statistics of the new equation, Equation 4.

$$-pI_{50} = 0.960(\pm 0.296)\ \text{H-DO}_I + 0.165(\pm 0.034)\ \text{MR}_{II}$$
$$- 1.633(\pm 0.529)\ \sigma_{mII} - 0.608(\pm 0.300)\ \pi_{III}$$
$$+ 0.117(\pm 0.026)\ \text{MR}_{III} + 0.115(\pm 0.064)\text{MR}_V$$
$$+ 0.984(\pm 0.141)\ \sum\pi + 4.803 \tag{4}$$

$n = 31 \qquad r = 0.943 \qquad s = 0.38 \qquad F = 26.42$

Requirements for good activity (ranked according the absolute value of the corresponding path coefficients given in brackets):

Large overall hydrophobicity ($\sum\pi$) [0.754];
in position III: large volume (MR) [0.627];
in position II: large volume (MR) [0.403];
in position III: hydrophilic substituent [-0.341];
in position II: electron-donating substituent [-0.252];
in position I: proton-donor substituent [0.189];
in position V: large volume (MR) [0.138].

The partial r value is still the highest for $\sum\pi$ and MR_{III}, the total lipophilicity alone explains over 60% of the total variance and it is also visible that it has a particular importance in the *para* position. Since σ_m has a negative coefficient, electron donating substituents are preferred in the *meta* position. MR has almost the same coefficients for positions II, III and V, the latest, however, has the lowest absolute path coefficient [0.138]. Therefore, compounds 3,4-disubstituted with lipophilic substituents are expected to be highly active.

Test of Significance

A way to avoid chance correlations is to increase the size of the sample. The partial F value, however, is also informative, in

Table II. Observed and predicted pI_{50} values and statistics of prediction

compd. number	pI_{50} obs.	pI_{50}^a predicted	pI_{50}^b pred.	compd. number	pI_{50} obs.	pI_{50}^a predicted	pI_{50}^b pred.
1	5.09	5.31 ± 0.16	5.33	19	6.77	6.10 ± 0.21	4.40[c]
2	6.17	5.92 ± 0.27	5.90	20	6.94	6.89 ± 0.35	6.06[c]
3	6.06	6.48 ± 0.27	8.39[c]	21	7.95	7.34 ± 0.21	6.70
4	6.39	6.61 ± 0.13	6.63	22[d]	5.75	6.82 ± 0.22	8.20[c]
5	5.68	5.81 ± 0.15	6.21[c]	23[d]	5.54	6.03 ± 0.16	6.31
6	6.40	6.00 ± 0.22	5.87	24	6.47	6.40 ± 0.21	6.38
7	5.89	6.25 ± 0.12	6.28	25	8.13	8.36 ± 0.29	8.47
8	5.08	5.40 ± 0.15	5.43	26	8.52	8.76 ± 0.31	8.89
9[d]	7.64	6.90 ± 0.30	7.09	27	7.89	7.83 ± 0.23	7.82
10	7.11	7.17 ± 0.38	6.54	28	5.70	5.29 ± 0.16	5.20
11	7.70	7.21 ± 0.24	6.77[c]	29	5.89	6.14 ± 0.26	6.23
12	7.50	7.84 ± 0.31	8.58[c]	30	5.89	5.90 ± 0.18	5.90
13	5.91	5.64 ± 0.28	5.39[c]	31	5.60	5.83 ± 0.26	6.41
14	4.97	5.53 ± 0.28	4.79	32	5.22	4.99 ± 0.22	5.90[c]
15	5.37	5.71 ± 0.14	5.74	34	5.16	5.48 ± 0.40	3.46[c]
16	6.89	6.53 ± 0.22	6.09	35	7.05	6.72 ± 0.28	5.42
17[d]	6.96	6.11 ± 0.12	5.91	38	7.52	7.19 ± 0.40	6.19
18	6.04	6.51 ± 0.13	6.89[c]				

[a] Prediction by Equation 2.
[b] Prediction by the "leave one out" method.
[c] Parameters, not involved in Equation 2, were introduced into the "leave one out" equation at r=0.900 total confidence level.
[d] Compounds have been omitted from Equation 4.

this respect it is the most important statistical parameter. It takes automatically into account the number of observations and variables (degrees of freedom). According to this parameter, the significances of Equation 3 and Equation 4 are very high ($p<0.001$), at the level of 0.1%. The high F values provided by the F-statistics for significance show that the equations are descriptive, also informative is the fact that the F value of each equation is gradually increasing when a new parameter is considered, which also suggests non-accidental correlation. Additional indication of significance is the fact if the residues of the regression follow normal distribution, it demonstrates that there is no further correlation in the sample.

On the other hand, theoretically there is no way to tell the difference between chance correlation and true correlation. Only

the development and exploration of a plausible hypothesis that
corresponds to the meaning of the equation can test its relevance.
In our case, the use of well established chemical descriptors
which were found to be relevant with biological activities by many
authors is certain guarantee for a real QSAR. Further guarantee
is the test of the predictive power of the equation.

For this purpose the "leave one out" approach (also known as
the jack-knife method (13,20)) was used: the activity of each
compound was predicted by the equation obtained from linear
regression analysis on the sample leaving out the compound in
question. The calculated standard deviation of the difference
between the calculated and measured values is, of course, larger
than that obtained the usual way. Its increase, however, brings
plausible information on the predictive power of the regression
equation, as well as which compound's removal would increase the
significance of the equation most. Also informative, which
parameters and how frequently are considered in the equations
obtained in the "leave one out" approach: their continuous and
systematic appearance also suggests a "real" correlation.

Leaving the compounds out one by one 35 regression equations
were obtained, each on different 34 membered (n=34) samples. $\sum\pi$
and MR_{III} appeared in all 35 equations, meanwhile MR_{II}, $H-DO_I$, σ_{mII},
MR_V and σ_{mIV} appeared only in 85.7%, 82.9%, 82.9%, 74.3% and 62.9%
of the cases, respectively. In addition, occasionally additional
parameters were introduced, i.e. $\sum\pi^2$, $H-AC_I$, π_{III} and R_{III}. The r
values of the regression equations were between 0.865 and 0.950.
The pI_{50} values predicted by Equation 2 and the "leave one out"
method are listed in Table II.

Interpretation of the Regression Equations

It is interesting that in the individual substituent positions
molar volume (MR) was found to be the relevant parameter rather
than π, E_s or the STERIMOL descriptors. This fact and the
positive coefficients for MR suggest that the enzyme-inhibitor
interaction proceeds via London dispersion forces (31,41) and the
binding to the enzyme is favored if the bulky substituents are in
meta or para positions, which is in accordance with the earlier
results (11). The parameter H-DO in position I seems to be
important in the regression, it systematically appears in each
equation. Its path coefficient and partial r value, however, is
relatively low compared to those of $\sum\pi$ and MR_{III}. In addition, the
$H-DO_I$ variable is not very useful because these indicator
parameters are the most poorly defined and the number of proton
donor substituents (H-DO=1) is rather small.

The simultaneous presence of $\sum\pi$ and MR could be somewhat
contradictious, since π commonly relates to a desolvation step in
binding to the enzyme, meanwhile correlation with MR does not
appear to be associated with desolvation (31). However, it is
quite commonly observed that π and MR parameters appear together
in descriptive equations (42-45) and it indicates that the rate
limiting step in the interaction with the enzyme is the binding
step to it, rather than transport or the inhibition reaction
itself (46,47).

The function of π is obvious: the higher the lipophilicity of the compound, the better binding is possible to the enzyme. The good correlation of the pI_{50} values with $\sum\pi$ shows the presence of a hydrophobic pocket in JHE close to the active site. Such a hydrophobic pocket has been recently evidenced by non-QSAR methods as well (10). In addition, since trifluoromethyl ketones, in general, predominantly exist in their hydrated form (geminal diol), a dehydration step and/or condensation with the enzyme active site is involved, as it has been proposed previously (4,9).

The inhibitory potencies, suggested by the present equations, appear to be significant if found within the range of those represented in the sample ($I_{50} = 10^{-4} - 10^{-10}$ M). Extrapolations to lower ranges of I_{50}s are, however, not necessarily valid. I_{50} values for several 3,4-disubstituted 3-phenylthio-1,1,1-trifluoro-2-propanones, such as the 3,4-di-tert.butyl and 3,4-dicyclohexyl derivatives have been predicted even at the low range of $10^{-13} - 10^{-15}$ M. This prediction is against the observation that below $10^{-9} - 10^{-10}$ M range the inhibitory activity seems not to increase with the lipophilicity. In the present study, no square terms, such as $(\sum\pi)^2$, $(\sum\sigma)^2$, etc. showed significant correlation with the inhibitory potencies, therefore these correlations appear "truly" linear and not parabolic with these parameters. The seeming controversy can be explained by two reasons. First, the calculated molarity of the enzyme in diluted hemolymph is in the same range, $10^{-9} - 10^{-10}$ M, mentioned before. In other words, the most potent inhibitors show almost stoichiometric inhibition of the enzyme. I_{50} values below this level cannot be observed, even if the compounds are truly of higher inhibitory potency. I_{50}, which can be related to K_i under certain conditions, works well for moderately active compounds. For highly potent compounds one will observe smaller and smaller differences in I_{50} as the compounds approach stoichiometric inhibition. Second, the compounds in the present series might appear in a *quasi*-linear portion of a parabolic curve before reaching the $\sum\pi$ (or $\sum MR$) optimum.

Other predictions for the molar I_{50} values appear to be more realistic, i.e. that of 3-(4-phenoxy-phenyl)thio-1,1,1-trifluoro-2-propanone. (Some of the *para*-substituted derivatives of high predicted inhibitory potency are listed in Table III.) This also illustrates the importance of the professional interpretation: the selection of the plausible equation. The equations never guarantee strict cause-effect relationships, they only show the mathematical possibility of a model (while rejecting others) and give a useful guide for finding out the "real" explanation.

Although steric parameters in the *ortho* position have not been selected into the equation, the fact that *ortho* substituents often seem to decrease the inhibitory activity, relative to other compounds of the same lipophilicity, shows possible steric hindrance. It has been suggested (12) that the high specificity of JHE to the natural substrates (JHs) is due to a "recognition" of the α-β double bond in the substrate, and that the β sulfur in the title compounds is bioisosteric with the β sp^2 carbon of this double bond. The *ortho* substituents of the phenyl ring possibly

Table III. *Para*-substituted 3-phenylthio-1,1,1-trifluoro-2-propanones (Structure A) of high predicted inhibitory activity

substituent	I_{50} [M] predicted	substituent	I_{50} [M] predicted
4-Ph	1.43×10^{-9}	4-PhCH=N	9.12×10^{-10}
4-PhO	7.36×10^{-10}	4-PhN=CH	9.12×10^{-10}
4-PhCH$_2$	5.14×10^{-10}	4-PhN=N	6.68×10^{-10}
4-PhOCH$_2$	6.39×10^{-10}	4-PhCONH	2.43×10^{-9}
4-PhCO	2.33×10^{-9}	4-PhSO$_2$NH	1.35×10^{-9}
4-PhOCO	1.23×10^{-9}	4-PhCOO	7.92×10^{-10}
4-PhC≡C	9.42×10^{-11}	4-PhSO$_2$O	7.79×10^{-10}
4-PhCH=CH	7.38×10^{-11}	4-PhSO$_2$	4.67×10^{-9}
4-PhCOCH=CH	3.67×10^{-10}	4-PrCOCH=CH	2.41×10^{-9}
4-(Ph)$_2$N	2.38×10^{-13}	4-PrOCOCH=CH	9.61×10^{-10}

hinder the sulfur atom and therefore render its binding to the enzyme binding pocket, which is reflected in the relatively higher I_{50} values.

Acknowledgments

The authors wish to express their sincere thanks to Dr. Phil Magee for the invitation to present this paper and for his valuable advices and consultations. We are grateful to Dr. Tim Schierling and Prof. Ken Musker (Chemistry Department, University of California, Davis) for the synthesis of compounds 35 and 36 and to Dr. Mike Pitcairn (Department of Entomology, University of California, Davis) for his technical help in using the SAS statistical software package. This work was supported by grants ES02710-07, DCB-8518697, and 85-CRCR-1-1715 from NIEHS, NSF, and USDA, respectively. A.S. is a Fulbright Scholar (Fulbright Program #33917, Institute of International Education). B.D.H. is a Burroughs Wellcome Scholar in Toxicology.

Literature Cited

1. Hammock, B.D.; Wing, K.D.; McLaughlin, J.; Lowell, V.M.; Sparks, T.C. Pestic. Biochem. Physiol. 1982, 17, 76-88.
2. Brodbeck, U.; Schweikert, K.; Gentinetta, R.; Rottenberg, M. Biochim. Biophys. Acta 1979, 567, 357-69.
3. Gelb, M.H.; Svaren, J.P.; Abeles, R.H. Biochemistry 1985, 24, 1813-7.
4. Imperiali, B.; Abeles, R.H. Biochemistry 1986, 25, 3760-7.
5. Stein, R.L.; Strimpler, A.M.; Edwards, P.D.; Lewis, J.J.; Mauger, R.C.; Schwartz, J.A.; Stein, M.M.; Trainor, D.A.; Wildonger, R.A.; Zottola, M.A. Biochemistry 1987, 26, 2682-9.

6. Ashour, M.B-A.; Hammock, B.D. Biochem. Pharm. 1987, 36, 1869-79.
7. Hammock, B.D.; Abdel-Aal, Y.A.I.; Mullin, C.A.; Hanzlik, T.N.; Roe, R.M. Pest. Biochem. Physiol. 1984, 22, 209-23.
8. Prestwich, G.D.; Eng, W-S.; Roe, R.M.; Hammock, B.D. Arch. Biochem. Biophys. 1984, 228, 639-45.
9. Székács, A.; Hammock, B.D.; Abdel-Aal, Y.A.I.; Halarnkar, P.P.; Philpott, M.; Matolcsy, Gy. Pest. Biochem. Physiol. 1989, 33, in press.
10. Linderman, R.L.; Leazer, J.; Venkatesh, K.; Roe, R.M. Pest. Biochem. Physiol. 1987, 29, 266-77.
11. Abdel-Aal, Y.A.I.; Hammock, B.D. In Bioregulators for Pest Control; Hedin, P.A., Ed.; ACS Symp. Ser. No. 276; American Chemical Society: Washington, DC, 1985; pp. 135-60.
12. Székács, A.; Hammock, B.D.; Abdel-Aal, Y.A.I.; Philpott, M.; Matolcsy, Gy. In Biotechnology in Crop Protection; Hedin, P.A.; Menn, J.J.; Hollingworth, R.M., Eds.; ACS Symp. Ser. No. 379; American Chemical Society: Washington, DC, 1988; pp. 215-27.
13. Draper, N.R.; Smith,H. Applied Regression Analysis; 2nd Ed.; Wiley: New York, 1981; pp. 307-11.
14. Draper, N.R.; Smith,H. Applied Regression Analysis; 1st Ed.; Wiley: New York, 1966; pp. 178-94.
15. Sokal, R.F.; Rohlf, F.J. Biometry; 2nd Ed.; Freeman: New York, 1981; pp. 663-71.
16. Pope, P.T. Ind. Eng. Chem. 1970, 62, 35-6.
17. Stowe R.A.; Mayer, R.P. Ind. Eng. Chem. 1969, 61, 11-6.
18. Birth, G.S. Appl. Spectrosc. 1985, 39, 729-32.
19. Bordás, B. In QSAR and Strategies in the Design of Bioactive Compounds, Proc. Fifth Eur. Symp. on QSAR; J.K. Seydell, Ed.; VCH Verlag: Weinheim, FRG, 1985; pp. 389-92.
20. Rohrbaugh, R.H.; Jurs, P.C.; Ashman, W.P.; Davis, E.G.; Lewis, J.H. Chem. Res. Toxicol. 1988, 1, 123-7.
21. Bell, A.R.; Covey, R.A.; Magee, P.S. In Proc. Brit. Crop Protec. Conf. in Weeds; PBPC Publ.: Surrey, UK, 1987; pp. 249-55.
22. Enslein, K.; Tuzzeo, T.M.; Borgstedt, H.H.; Blake, B.W.; Hart, J.B. In Proc. 2nd Internl. Workshop on QSAR in Environmental Toxicology; Kaiser, K.L.E., Ed.; D. Reidel Publ.: Dordrecht, 1987; Vol. II, pp. 91-106.
23. Takayama, C.; Kirino, O.; Hisada, Y.; Fujinami, A. Agric. Biol. Chem. 1987, 51, 1547-52.
24. Kirino, O.; Hashimoto, S.; Furuzawa, K.; Takayama, C.; Ohshio, H. J. Pestic. Sci. 1983, 8, 315-9.
25. Nishimura, K.; Ueno, A.; Nakagawa, S.; Fujita, T.; Nakajima, M. Pestic. Biochem. Physiol. 1983, 17, 271-9.
26. Takayama, C.; Fujinami, A. Pestic. Biochem. Physiol. 1979, 12, 163-71.
27. Kamoshita, K.; Ohno, I.; Fujita, T.; Nishioka, T.; Nakajima, M. Pestic. Biochem. Physiol. 1979, 11, 83-103.
28. Hammock, B.D.; Sparks, T.C. Anal. Biochem. 1977, 82, 573-9.
29. Hammock, B.D.; Roe, R.M. In Methods in Enzymology; Law, J.H.; Rilling, H.C.,Eds.; Academic Press: Orlando, 1985; Vol. 111B, pp. 487-94.

30. Hansch, C. In *Molecular Structure and Energetics*; Liebman, J.F.; Greenberg, A., Eds.; VCH Publ.: Deerfield Beach, 1987, Vol. 4, pp. 341-79.
31. Hansch, C.; Leo, A. *Substituent Constants for Correlation Analysis in Chemistry and Biology*, Wiley: New York, 1979.
32. Fujita, T.; Iwasa, J.; Hansch, C. *J. Am. Chem. Soc.* 1964, 86, 5175-80.
33. Hammett, L.P. In *Physical Organic Chemistry*; 2nd Ed.; McGraw-Hill: New York, 1956; pp. 347-90.
34. Swain, C.G.; Lupton, E.C.,Jr. *J. Am. Chem. Soc.* 1968, 90, 4328-37.
35. Taft, R.W. In *Steric Effects in Organic Chemistry*, Newman, M.S., Ed.; Wiley: New York, 1956; pp. 556-675.
36. Verloop, A.; Hoogenstraaten, W.; Tipker, J. In *Drug Design*; Ariëns, E.J., Ed.; Academic Press: New York, 1976; Vol VII, pp. 165-207.
37. Topliss, J.G.; Edwards, R.P. *J. Med. Chem.* 1979, 22, 1238-44.
38. Watanabe, K.; Fujita, T.; Takimoto, A. *Plant Cell Physiol.* 1981, 22, 1469-79.
39. Fujita, T.; Nishidoka, T. *Prog. Phys. Org. Chem.* 1976, 12, 49-89.
40. *SAS User's Guide: Statistics*, SAS Institute Inc.: Cary, NC, 1982; pp. 269-336.
41. Charton, M. In *Topics in Current Chemistry*; Charton, M.; Motoc, I., Eds.; Springer Verlag: Berlin, 1983; pp. 107-18.
42. Li, R.; Hansch, C.; Matthews, D.; Blaney, J.M.; Langridge, R.; Delcamp, T.J.; Susten, S.S.; Freisheim, J.H. *Quant. Struct.-Act. Relat.* 1982, 1, 1-7.
43. Hansch, C.; Blaney, J.M. In *Drug Design: Fact or Fantasy?*; Jolles, G.; Woolridge, K.R.H., Eds.; Academic Press: London, 1984; pp. 185-205.
44. Gupta, S.P. *Chem. Rev.* 1987, 87, 1183-253.
45. Magee, P.S. *Quant. Struct.-Act. Relat.* 1986, 5, 158-65.
46. Magee, P.S. *Chemtech* 1981, 11, 178-84.
47. Magee, P.S. In *Computer-Assisted Drug Design*; Olson, E.C.; Christoffersen, R.E., Eds.; ACS Symp. Ser. No. 112; American Chemical Society: Washington, DC, 1979; pp. 319-40.

RECEIVED March 30, 1989

Chapter 13

Conformational Analysis of Fenvalerate and an Ether-Type Pyrethroid

Yasuyuki Kurita, Kazunori Tsushima, and Chiyozo Takayama

Pesticides Research Laboratory, Takarazuka Research Center, Sumitomo Chemical Company, Ltd., 4-2-1 Takatsukasa, Takarazuka, Hyogo 665, Japan

> A candidate for the active conformer of esfenvalerate (the most insecticidally active stereo isomer of fenvalerate) has been presented based on conformational analyses using the AM1 molecular orbital method and shape comparisons with low activity pyrethroids. Esfenvalerate and a new type pyrethroid, 3-phenoxybenzyl (R)-2-(4-ethoxyphenyl)-3,3,3-trifluoropropyl ether, which has a configuration opposite to that of esfenvalerate, were reasonably superimposed.

Design of highly bioactive drugs should be assisted with the knowledge of their active conformers. If conformation of a drug can be fixed in its active form by ring closure or other methods without losing factors needed for its activity, the drug becomes more active. However, it is very difficult to know the active conformers of flexible molecules such as pyrethroids. Pyrethroids are insecticides having high activity against various insects and low toxicity to mammalia. Hopfinger et al. investigated the conformational energy maps of some ester type pyrethroids with their activities and proposed the active conformers (1). Tosi and his coworkers (2,3) presented the candidates for active conformers of chrysanthemic acid (1R-trans) and the acid component of esfenvalerate (the most insecticidally active stereo isomer of fenvalerate) with the FMFIT (Flexible Molecular Fit) method and conformational energy calculations. According to their assumption, which was originally proposed by Ohno et al. (4), the configurations at the carbon atom α to the carboxyl group are superimposable in the two molecules and the 4- and 3-positions of the phenyl ring of esfenvalerate acid correspond to the terminal part of the side chain of chrysanthemic acid. They also assumed that the dimethyl moieties at the carbon atom β to the carboxyl group are superimposable. Heritage (5) used the Venn diagram to search for the active conformers of permethrin etc. As to biphenyl pyrethroids, Plummer and his coworkers (6) got experimentally or calculated dihedral angles in the biphenyl moieties substituted or ring-closed at the ortho position(s) and correlated them with their activities. They speculated that the dihedral angle around the two benzene rings is about 50 degrees in the active form. Byberg et al. (7) offered the active conformers of both acid and alcohol components of ester type pyrethroids by using the method similar to that of Tosi et al.

Recently, Tsushima et al. synthesized a new synthetic pyrethroid, 3-phenoxybenzyl 2-(4-ethoxyphenyl)-3,3,3-trifluoropropyl ether, without a geminal dimethyl group and found the fact that the trifluoromethyl group in the more active optical isomer of the molecule is oriented in the opposite direction to that of the isopropyl group in esfenvalerate (8). Therefore, we thought it necessary to reconsider the assumption adopted by Tosi et al. and Byberg et al.

In this article, we report the results of conformational analyses of esfenvalerate, 3-phenoxybenzyl (R)-2-(4-ethoxyphenyl)-3,3,3-trifluoropropyl ether, the other two pyrethroids, and their stereo isomers. We compare them on the assumption that the position of benzene rings in the 3-dimensional space is determinant in their insecticidal activities. In other words, we assume that the overall shape of pyrethroids is the most important factor for high activity and that the bonds and substituents between benzene rings are necessary to fix the benzene rings at the position appropriate for recognition by the receptor. This assumption seems to be reasonable considering not only the fact found by Tsushima et al. but also the existance of highly active non-ester type pyrethroids such as ethofenprox, MTI-800 (9), oxime ether types (10), and alkene types (11). We propose candidates for the active conformers of the pyrethroids such as esfenvalerate and interpret the fact found by Tsushima et al.

Methods

Conformational analyses were carried out for the substructures of fenvalerate (I; SS (esfenvalerate) and SR isomers), 3-phenoxybenzyl 2-(4-ethoxyphenyl)-3,3,3-trifluoropropyl ether (II; R isomer), α-cyano-3-phenoxybenzyl 2-(4-chlorophenyl)-2-methylpropionate (III; S isomer), and deltamethrin (IV) shown in Figure 1. In these pyrethroids, compounds I(SS), II(R), and IV are highly active and compounds I(SR), I(RS), I(RR), II(S), III(S), and III(R) have low activity (8, 12-14). Conformers of the optical isomer of each substructure were obtained simultaneously without further calculations because the Schrödinger equation yields the same energies for both optical isomers. For conformational energy calculations and optimizations of molecular structure, the AM1 molecular orbital method (15) and Broyden-Fletcher-Goldfarb-Shanno optimization method (16) integrated into MOPAC (17) were used. These methods are the most reliable ones applicable to the analyses. Although ab initio molecular orbital methods are more reliable when appropriate basis sets are used, they require much more computation time and memories for integrals. Molecular mechanics (MM) methods require less computation time. However, generally, MM methods are less reliable when some hetero atoms are contained in the molecule.

First, geometry of each substructure was fully optimized except some internal coordinates described below. The bond lengths in benzene rings were fixed to 1.399 and 1.101 Å for C-C and C-H bonds, respectively. All the bond angles in benzene rings were fixed to 120

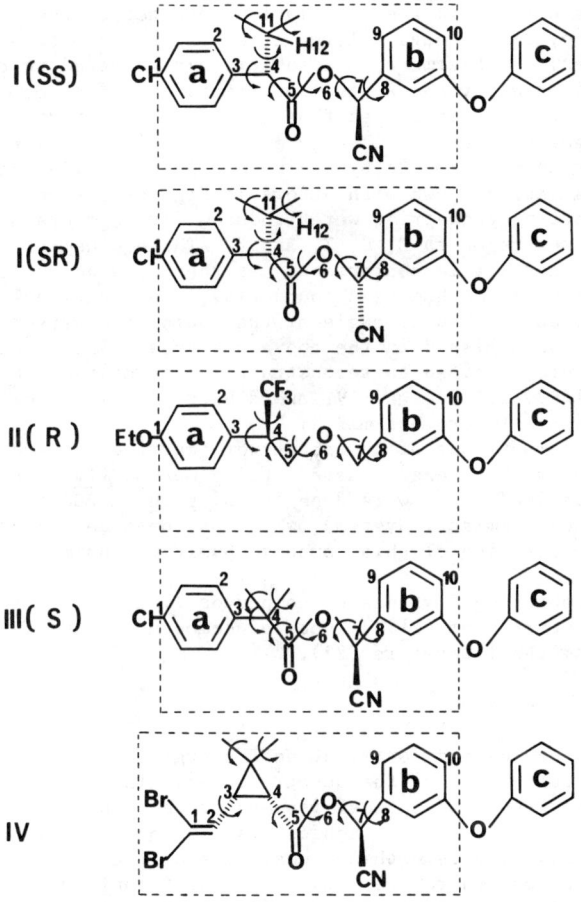

Figure 1. Structures of compounds I, II, III, and IV. Conformational analyses were carried out for the substructures enclosed with dashed lines.

degrees. Benzene rings, carbonyl groups, and 2,2-dibromoethenyl group were assumed to be flat, and the cyano groups were assumed to be linear. Ester moieties were set to <u>trans</u>-coplanar in starting geometries. The extraction of the substructures and such fixation of geometries were carried out in order to shorten the computation time. Such a treatment is reasonable since benzene rings, carbonyl groups, 2,2-dibromoethenyl group and cyano groups are relatively rigid, and ester moieties are not <u>cis</u> in low energy conformers. As the substituents omitted in the calculations are not at the <u>ortho</u> position, they do not have a large effect on conformation of the substructures. Furthermore, the substituents omitted in compounds I and III to search for the active conformers are identical.

Second, the substructures thus optimized were further divided into the segments depicted in Figure 2. Starting geometries were produced for these segments by adding 0, 120, and -120 degrees to each dihedral angle shown with arrows in Figure 2. Twenty seven (=3X3X3) starting geometries were produced for segments A and B, and nine (=3X3) for segments C, D, E, and F. These dihedral angles were optimized to get stable conformers. As there was no guarantee that all low-energy minima had been found (<u>18</u>), the rotational energy diagrams for each dihedral angle of the conformers were drawn to obtain conformers missed in the above analyses. These stable conformers were combined to reconstruct substructures I(<u>SS</u> and <u>SR</u> isomers), II(<u>R</u>), III(<u>S</u>), and IV, and dihedral angles shown with arrows in Figure 1 were optimized.

Third, comparisons of the shape were carried out among the substructures with a least square fitting method (<u>19</u>). All the analyses described above were done by using the ACACS (Advanced Computer Aided Chemistry System), which has been developed through the joint cooperation of this company, Sumitomo Pharmaceuticals, and NEC (<u>20</u>).

Hydrophobic log P values of compounds II and III were calculated with the CLOGP program (<u>21</u>). Those of compounds I and IV were obtained from the literature (<u>22</u>).

Results and Discussion

The dihedral angles and conformational energies are listed in Tables I-V for the conformers of the substructures. The sign of dihedral angles are reversed for optical isomers of the substructures. For the substructure II, all the conformers (15 conformers) have conformational energies within 6 kcal/mole of the most stable conformer, whereas there are 11, 11, 4, and 5 conformers in the same energy range for the substructures I(<u>SS</u> and <u>RR</u> isomers), I(<u>SR</u> and <u>RS</u> isomers), III, and IV, respectively. In other words, compound II is more flexible than compounds I, III, and IV. This may be one reason why the difference in activity between optical isomers of compound II is smaller than that between <u>SS</u> and <u>RS</u> isomers of compound I (<u>8,12</u>).

When comparisons are carried out among highly active compounds, it generally is a problem whether the receptor sites for them are same or not. But this problem does not tend to appear when a highly active compound is compared with a very low active one. The low activity should be attributed to either the low affinity for the receptor (e.g. the low steric complementarity to the receptor) or the low concentration around the receptor. Therefore, first, the

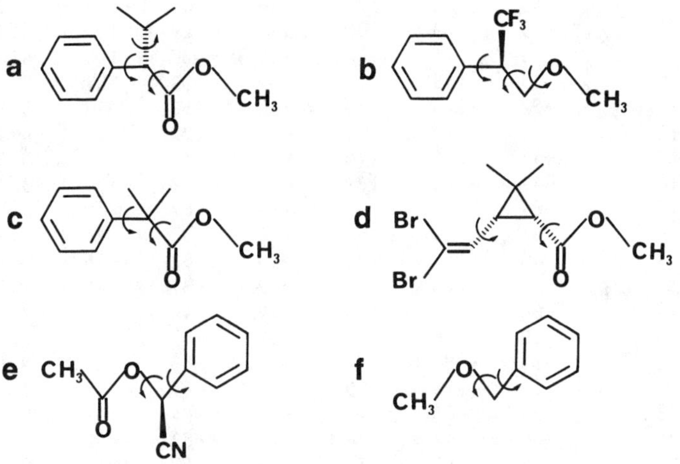

Figure 2. Segments of the substructures I, II, III, and IV.

Table I. Conformers of the substructure I(SS)

conformer	θ1	θ2	θ3	θ4	θ5	θ6	E
1	-52	-110	179	136	41	72	0.0
2	-43	-113	179	137	42	-179	0.5
3	-60	77	180	140	37	69	0.9
4	-40	-112	179	137	41	-46	1.3
5	11	-119	-180	137	41	113	3.5
6	-34	87	178	135	43	-28	3.5
7	-45	76	179	134	42	-163	3.7
8	-75	40	180	129	29	137	3.7
9	5	71	178	140	37	-165	4.3
10	25	58	179	137	39	102	5.2
11	-52	-110	173	-68	-83	72	5.8
12	-51	-110	179	-69	-37	72	6.1
13	-43	-114	174	-68	-82	-180	6.3
14	-42	-113	179	-69	-37	-179	6.5
15	-62	77	175	-68	-82	70	6.9
16	-66	34	-180	124	32	-91	6.9
17	-40	-113	175	-69	-82	-46	7.1
18	-40	-113	179	-69	-37	-46	7.3
19	-66	70	-175	-72	-43	64	7.5
20	10	-118	173	-69	-83	113	9.5
21	-33	88	171	-67	-83	-29	9.6
22	13	-118	180	-69	-37	112	9.7
23	-75	43	-174	-75	-84	139	9.8
24	-45	77	176	-69	-82	-163	9.9
25	-34	89	177	-68	-40	-29	9.9
26	-76	44	-165	-76	-43	141	10.3
27	4	72	175	-68	-81	-165	10.3
28	12	65	-175	-72	-45	-169	10.6
29	27	57	180	-71	-82	101	11.2
30	-65	40	-174	-75	-84	-90	13.0
31	-65	39	-166	-76	-43	-89	13.4

Tortional angle θi in degrees is defined as follows:
θ1, 2-3-4-5; θ2, 3-4-5-6; θ3, 4-5-6-7; θ4, 5-6-7-8; θ5, 6-7-8-9;
θ6, 12-11-4-5. The numbering of atoms is shown in Figure 1.
E is the difference in heat of formation from that of the most
stable conformer (kcal/mole).

Table II. Conformers of the substructure I(SR)

conformer	θ_1	θ_2	θ_3	θ_4	θ_5	θ_6	E
1	-52	-110	179	-137	-40	72	0.0
2	-43	-113	179	-137	-40	-179	0.4
3	-63	74	-179	-139	-37	68	1.1
4	-40	-113	-180	-138	-40	-46	1.3
5	11	-118	-179	-135	-39	113	3.6
6	-75	45	-177	-135	-35	137	3.6
7	-33	89	-180	-134	-32	-29	3.6
8	-45	76	-179	-137	-52	-163	4.1
9	7	68	-180	-142	-45	-166	4.4
10	21	59	179	-133	-33	105	5.2
11	-51	-106	-179	71	83	71	6.0
12	-51	-112	178	70	35	69	6.2
13	-42	-110	-178	71	83	-179	6.4
14	-42	-111	179	70	37	-180	6.5
15	-66	39	-176	-139	-44	-91	6.9
16	-64	75	-176	70	83	68	7.1
17	-39	-110	-175	69	83	-46	7.1
18	-65	72	179	69	34	69	7.2
19	-40	-112	-179	69	37	-46	7.5
20	-75	44	-171	67	83	138	9.4
21	7	-116	-179	71	82	114	9.6
22	-32	89	-177	70	82	-29	9.6
23	5	-115	180	70	38	116	9.7
24	-74	44	-177	68	36	139	9.8
25	-29	82	-177	70	83	-172	10.1
26	-43	78	-178	71	83	-164	10.1
27	-31	108	175	73	45	-34	10.4
28	-51	69	174	70	35	-164	10.6
29	22	58	-178	70	83	104	11.3
30	32	60	174	72	41	97	11.9
31	-66	39	-171	67	83	-91	12.6
32	-65	41	-176	67	36	-90	12.9

θ_i and E are defined in Table I.

Table III. Conformers of the substructure II($\underline{R}$)

conformer	θ1	θ2	θ3	θ4	θ5	E
1	61	112	91	70	71	0.0
2	61	113	95	74	39	0.1
3	56	140	172	70	30	0.2
4	54	145	173	70	30	0.5
5	58	-77	177	70	26	1.0
6	63	167	-106	81	40	1.2
7	60	167	-105	81	40	1.6
8	57	143	179	-65	-31	2.0
9	62	99	-89	-67	-32	2.6
10	60	115	93	177	-5	3.2
11	62	104	-82	161	10	3.3
12	59	-73	179	-65	-31	3.3
13	54	149	-94	-66	-35	3.3
14	-23	83	160	69	41	3.4
15	-22	81	177	-64	-37	5.5

θi and E are defined in Table I.

Table IV. Conformers of the substructure III($\underline{S}$)

conformer	θ1	θ2	θ3	θ4	θ5	E
1	-54	87	-180	136	33	0.0
2	-47	-86	178	138	38	0.3
3	-145	91	180	136	35	1.1
4	-152	-91	179	137	36	1.9
5	-55	86	173	-68	-83	6.1
6	-48	-87	176	-69	-82	6.2
7	-55	86	176	-68	-36	6.4
8	-49	-92	-179	-70	-38	6.5
9	-146	90	175	-69	-83	6.9
10	-146	90	178	-69	-36	7.1
11	-156	-92	178	-70	-83	7.8

θi and E are defined in Table I.

Table V. Conformers of the substructure IV

conformer	θ1	θ2	θ3	θ4	θ5	E
1	175	-153	180	137	41	0.0
2	-180	45	178	139	55	4.0
3	120	15	174	143	85	4.8
4	173	-153	174	-68	-82	5.7
5	173	-151	179	-68	-37	6.0
6	-172	-68	178	-70	-81	7.8
7	116	8	178	-70	-82	9.5
8	-170	42	171	-67	-81	9.9
9	-168	40	-175	-71	-49	10.5
10	-33	133	160	142	39	84.4
11	-33	-51	-162	105	48	85.2
12	-33	131	140	-58	-90	90.4
13	-33	131	143	-59	-40	91.5
14	-33	-48	-144	-91	-89	92.1

Tortional angle θi in degrees is defined as follows:
θ1, 1-2-3-4; θ2, 3-4-5-6; θ3, 4-5-6-7; θ4, 5-6-7-8;
θ5, 6-7-8-9. The numbering of atoms is shown in Figure 1.
E is defined in Table I.

conformers of the substructure I(SS) were compared with those of the substructure III(S) in order to search for the active conformer of the pyrethroid. Compound I is a highly active pyrethroid and the most active isomer of it has S configurations at both acid and alcohol moieties (12). The insecticidal activity of compound III is much lower than that of compound I (13). On the other hand, log P values of compounds I, II, III, and IV are 6.2, 5.8, 6.2, and 6.2, respectively. Differences in the values are not large. Hence, it was assumed that the low activity of compound III should be attributed to the 3-dimensional structure and not to the transport process from the skin of insects to the receptor site. As we assumed that the overall shape is important for the insecticidal activity of pyrethroids, we made it a criterion for the resemblance of shape that the root mean square value (RMS) of distances between atoms used to fit the substructures and the angle between benzene rings b shown in Figure 1 ($\varphi 2$) should be within 0.5 Å and 45 degrees, respectively, when the substructures were fitted with each other at the points 1, 3, 8, and 10 (Criterion 1). The restriction on $\varphi 2$ was needed because the distance between the benzene rings c shown in Figure 1 would be large if the fit at the benzene rings b is bad. On the other hand, some degree of difference in $\varphi 2$ was assumed to be acceptable, since the rotational potential for the benzene rings b around the C7-C8 axis is not so steep. For the benzene rings a, any angles between them ($\varphi 1$) were assumed to be acceptable, since distance between substituents at the para positon of benzene rings a does not change by the rotation of the rings. Considering the conformational distribution in the substructure of compound II, which is a highly active pyrethroid, we also made it a criterion that the active conformer should have a conformational energy within 6 kcal/mole of the most stable conformer (Criterion 2). In the set of the substructures I(SS) and III(S), the pairs of conformers which passed criteria 1 and 2 are listed in Table VI. None of them are considered to be candidates for the active conformer, since compound III should be expected to have high activity if the active conformer were available to it. In the substructure I(SS), only conformer 8 (Figure 3-a) remained as the candidate for the active conformer. Then, this conformer was compared with the conformers of the substructures III(R), I(SR), I(RS), and I(RR). There were not any conformers which passed criteria 1 and 2 in the latter substructures. This result is consistent with the fact that the activity of compounds I(SR), I(RS), I(RR), III(S) and III(R) is very low (12,13). Conformer 8 of the substructure I(SS) was also compared with conformers of the substructure IV. Conformer 2 of the substructure IV (Figure 3-c) has high similarity to it. RMS and $\varphi 2$ are 0.1 Å and 25 degrees, respectively, when these two conformers are fitted at the points 4, 5, 6, 7, 8, and 10 shown in Figure 1. This result is consistent with the fact that compound IV has high insecticidal activity and belongs to the same type of pyrethroids (so-called type II) as compound I (14). In Figure 4, superimposition of the two conformers is displayed. Although the benzene ring a of the substructure I(SS) and the 2,2-dibromoethenyl moiety of the substructure IV occupy almost the same position, geminal dimethyl groups separate to some extent from each other. Conformer 8 of the substructure I(SS) and conformer 2 of the substructure IV are different from the active conformers proposed by Hopfinger et al. (1), Tosi et al. (2,3), and Byberg et al. (7). This discrepancy is attributable to the different assumptions we adopted. At the present stage we cannot say whether our results or theirs are correct. Further syntheses and biological tests of compounds designed by referring to the results are needed. These conformers are also different from the ones found in crystals (23,24). The conformer in crystals is not always the same as that at receptor site (25).

Next, all the lower energy conformers of the substructure I(SS), which passed criterion 2, were compared with conformers of the substructures II(R) and II(S). The pairs which passed criterion 1 are listed in Tables VII and VIII. It can be seen from these results that the substructure II(R) has higher similarity to the substructure I(SS) in lower energy conformational space than the substructure II(S) in spite of the fact that the trifluoromethyl group in the substructure II(R) is oriented into the opposite direction to that of the isopropyl group in the substructure I(SS). Moreover, conformer 9 of the substructure II(R) (Figure 3-b), which is similar in shape to the candidate for the active conformer of the substructure I(SS) (Figure 5), has lower conformational energy than conformer 12 of the substructure II(S), which also resembles the candidate. These results are consistent with the fact that the more active isomer of compound II has an R configuration.

The conformational energy of conformer 8 of the substructure I(SS), which is the candidate for the active conformer of the substructure I(SS), is higher than that of conformer 12 of the substructure II(S). On the other hand, the substructure I(SS) has less conformers in the low energy conformational space than the substructure II(S). The substructure I(SS) has almost the same population (0.1% assuming the Boltzmann distribution at 298.15 K) in an active form as the substructure II(S). This seems to be in

Table VI. Conformer pairs between the substructures I(SS) and III(S) which passed criteria 1 and 2

conformer of I(SS)	conformer of III(S)	RMS	φ1	φ2
1	2	0.4	20	7
1	4	0.3	89	7
2	2	0.4	16	8
2	4	0.3	80	8
3	1	0.1	8	2
3	3	0.2	81	6
4	2	0.4	15	8
4	4	0.3	78	8
5	2	0.5	49	9
5	4	0.4	31	9
6	1	0.1	20	6
6	3	0.1	71	3
7	1	0.2	4	4
7	3	0.2	88	7
9	1	0.2	52	7
9	3	0.2	39	11
10	1	0.4	63	14
10	3	0.4	31	19
11	1	0.5	56	30

RMS (Å), φ1, and φ2 (degrees) are defined in text.

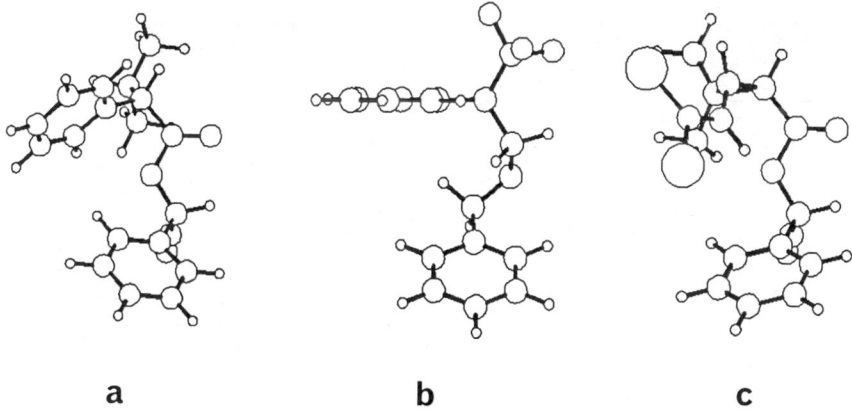

a b c

Figure 3. Candidates for the active conformers of the substructures: (a), I(SS); (b), II(R); (c), IV.

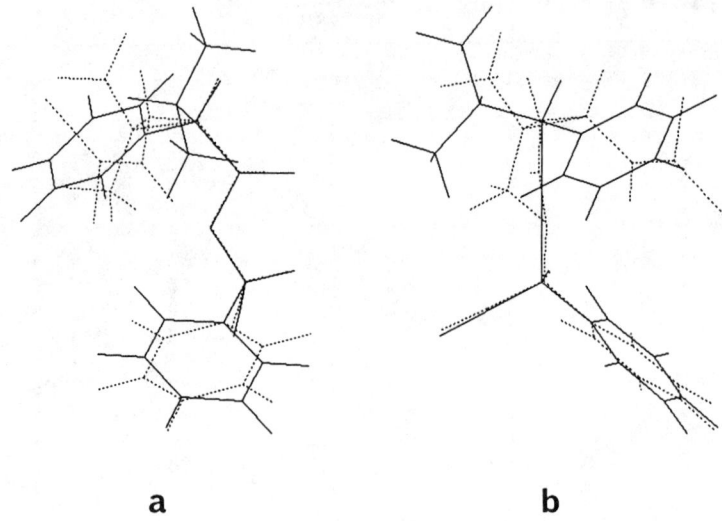

                    a   b

Figure 4. Superimposition of conformer 2 of the substructure IV (dotted line) on conformer 8 of the substructure I(SS) (solid line). (a) and (b) show front and side views, respectively.

Table VII. Conformer pairs between the substructures I(SS) and II(R) which passed criteria 1 and 2

conformer of I(SS)	conformer of II(R)	RMS	$\varphi 1$	$\varphi 2$
1	5	0.5	74	23
2	5	0.5	84	23
3	3	0.4	81	13
3	4	0.5	81	10
4	5	0.5	86	23
5	5	0.5	44	19
6	3	0.3	72	19
6	4	0.4	73	16
6	13	0.5	69	4
7	3	0.5	89	13
8	9	0.5	42	13
9	3	0.5	42	7

RMS (Å), $\varphi 1$, and $\varphi 2$ (degrees) are defined in text.

Table VIII. Conformer pairs between the substructures I(SS) and II(S) which passed criteria 1 and 2

conformer of I(SS)	conformer of II(S)	RMS	φ1	φ2
3	13	0.4	57	44
6	13	0.3	84	37
7	13	0.4	67	44
8	12	0.5	19	17
10	12	0.5	63	44

RMS (Å), φ1, and φ2 (degrees) are defined in text.

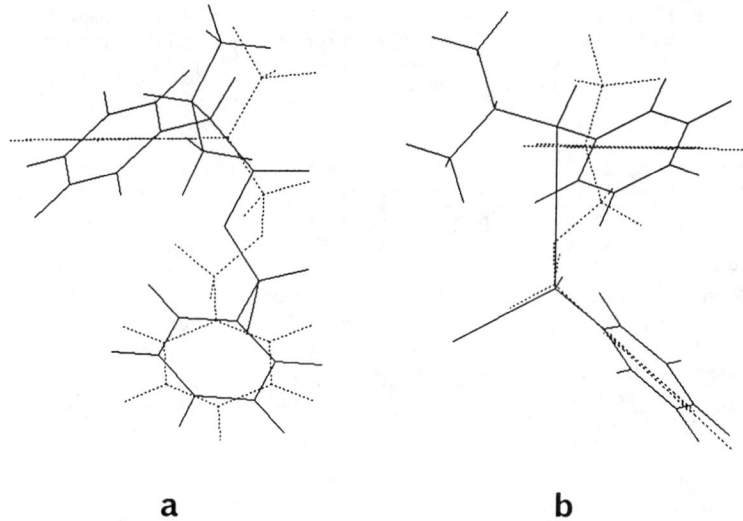

a   b

Figure 5. Superimposition of conformer 9 of the substructure II(R) (dotted line) on conformer 8 of the substructure I(SS) (solid line). (a) and (b) show front and side views, respectively.

conflict with the fact that compound I(SS) has higher insecticidal activity than compound II(S) (Tsushima, K.; Yano, T.; Takagaki, T.; Matsuo, N.; Hirano, M.; Ohno, N., Sumitomo Chemical Co., Ltd., unpublished data.). Other factors, e.g. dipole moment or hydrogen bond, may have an effect on binding to the receptor of pyrethroids.

In the discussion described above, it is assumed that active conformers are in energy minima in vacuum. This assumption, however, may not be correct. Although the more general way to search for an active conformer may be by using Venn diagram (26), a multi-dimensional diagram is needed to treat flexible molecules as a whole, and it is very difficult to carry out such an analysis. Also, the Venn diagram is useless in the case where only the positions of benzene rings in 3-dimensional space are important and bonds between them are not directly important. For example, bonds between benzene rings are not fitted well with each other as shown in Figure 5. Obviously, the conclusions we have obtained depend on the assumptions adopted. Whether the active conformers proposed by us are real ones can be confirmed only by synthesizing a rigid molecule which is fitted well to our model without losing other factors needed for activity.

Acknowledgments

We wish to thank Dr. N. Ohno and Dr. N. Matsuo of this company for their valuable discussions. We also express appreciation to Mr. H. Katsumi of this company for installing the CLOGP program.

Literature Cited

1. Hopfinger, A. J.; Malhotra, D.; Battershell, R. D.; Ho, A. W. J. Pesticide Sci. 1984, 9, 631.
2. Barino, L. Comput. Chem. 1981, 5, 85.
3. Tosi, C.; Barino, L.; Castellani, G.; Scordamaglia, R. THEOCHEM 1982, 87, 315.
4. Ohno, N.; Fujimoto, K.; Okuno, Y.; Mizutani, T.; Hirano, M.; Itaya, N.; Honda, T.; Yoshioka, H. Pestic. Sci. 1976, 7, 241.
5. Heritage, K. J. Biochem. Soc. Trans. 1982, 10, 310.
6. Plummer, E. L. In Pesticide Synthesis Through Rational Approaches; Magee, P. S.; Kohn, G. K.; Menn, J. J., Eds.; ACS Symposium Series No. 255; American Chemical Society: Washington, DC, 1984; p 297.
7. Byberg, J. R.; Jorgensen, F. S.; Klemmensen, P. D. J. Comput.-Aided Mol. Design 1987, 1, 181.
8. Tsushima, K.; Yano, T.; Takagaki, T.; Matsuo, N.; Hirano, M.; Ohno, N. Agric. Biol. Chem. 1988, 52, 1323.
9. Udagawa, T.; Numata, S.; Oda, K.; Shiraishi, S.; Kodaka, K.; Nakatani, K. In Recent Advances in the Chemistry of Insect Control; Janes, N. F., Ed.; Royal Society of Chemistry: London, 1985; p 192.
10. Bull, M. J.; Davies, J. H.; Searle, R. J. G.; Henry, A. C. Pestic. Sci. 1980, 11, 249.
11. Tsushima, K.; Matsuo, N.; Nishida, S.; Yano, T.; Hirano, M. Japanese Patent 60-115545A, 1985.
12. Nakayama, I.; Ohno, N.; Aketa, K.; Suzuki, Y.; Kato, T.; Yoshioka, H. In Advances in Pesticide Science; Geissbuhler, H., Ed.; Pergamon Press: Oxford, 1979; Part 2, p 174.

13. Numata, S. In Design of Bioactive Molecules; Yoshioka, H.; Shudo, K., Eds.; Soft Science Publications: Tokyo, 1986; p 338.
14. Gammon, D. W.; Brown, M. A.; Casida, J. E. Pestic.Biochem. Physiol. 1981, 15, 181.
15. Dewar, M. J. S.; Zoebisch, E. G.; Healy, E. F.; Stewart, J. J. P. J. Am. Chem. Soc. 1985, 107, 3902.
16. Shanno, D. F. J. Optim. Theo. Appl. 1985, 46, 87.
17. Stewart, J. J. P.; Seiler, F. J. QCPE 1987, 455 (Version 4.00).
18. Lipton, M.; Still W. C. J. Comput. Chem. 1988, 9, 343.
19. Kabsch, W. Acta Cryst. 1976, A32, 922.
20. Yoshida, M.; Ueda, A.; Kikuzono, Y.; Takayama, C.; Mizutani, T.; Motoki, T.; Morooka, S.; Shimojyu, T.; Kanaoka, S.; Yokota, A. Proc. Int. Chem. Cong. Pacific Basin Soc., 1984, 08A23.
21. Medicinal Chemistry Project, Release 3.42; Pomona College: Claremont, 1986.
22. Hansch, C.; Leo, A. The Log P Database (Medicinal Chemistry Project), Issue 26; Pomona College: Claremont, 1985.
23. Yanagi, K.; Moriguchi, K.; Uemura, Y.; Aoyagi, M. J. Sumitomo Chemical 1987, 1987-II, 4.
24. Owen, J. D. J. Chem. Soc. Perkin I 1975, 1865.
25. Marshall, G. R.; Motoc, I. In Topics in Molecular Pharmacology; Burgen, A. S. V.; Roberts, G. C. K.; Tute, M. S., Eds.; Elsevier: Amsterdam, 1986; Vol. 3, p 115.
26. Richards, W. G. Quantum Pharmacology (Second Edition); Butterworths: London, 1983; p 178.

RECEIVED March 21, 1989

Chapter 14

Molecular Design and Target Site Analysis in Fungicide Development

Hugh D. Sisler[1] and Nancy N. Ragsdale[2]

[1]Department of Botany, University of Maryland, College Park, MD 20705
[2]Cooperative State Research Service, U.S. Department of Agriculture, Washington, DC 20250-2200

> Antifungal compounds that specifically block sterol biosynthesis, microtubule assembly, succinate oxidation or polyketide melanin biosynthesis are discussed. These compounds are considered in respect to structure-activity relationships as they are influenced by target site characteristics and cellular biochemistry.

Most fungicides developed for plant disease control before 1965 are multisite biochemical inhibitors that lack the chemical properties and biological specificity required for internal therapeutic action in higher plants. This group of plant protectants include such compounds as the inorganic copper fungicides, captan, chlorothalonil, dithiocarbamates and chlorinated quinones. Although these compounds are usually less effective than some of the newer systemic fungicides, they are inexpensive and have encountered very few problems with fungal resistance; therefore, they continue to play an important role in crop protection. They are particularly valuable for use in programs designed to prevent or manage problems of fungal resistance to biochemically specific fungicides.

Since 1965 most of the new fungicides adopted for practical use are biochemically specific compounds that have local or systemic internal therapeutic activity. This development has extended the disease control spectrum of available fungicides and has made possible the use of a curative as well as a preventative strategy in disease control programs. However, not all has gone well because serious problems have been encountered with the development of fungal resistance to several of these fungicide mechanism groups (1).

Various studies have indicated that most systemic fungicides act at a single target site and that most cases of fungicide resistance result from mutations that lead to loss of target affinity (2). There has been intense interest in devising ways to

counter these resistance problems. While using mixtures or
alternating different fungicide mechanism groups are important
measures for reducing the risk of resistance development, the
design of chemical structures to fit the mutated target sites in
the resistant organisms and accelerated development of fungicides
with new modes of action are ways in which the organic chemists,
fungal physiologists, biochemists and enzymologists can help to
combat the fungicide resistance problems and improve chemical
control of fungal diseases in other respects. For existing
fungicides, it is important to know and characterize their target
sites not only because this knowledge can help in the design of new
analogues to counter resistance problems, but because it may
provide fundamental information about the molecular design
necessary to broaden the antifungal spectrum of a fungitoxic
mechanism group.

The identification and characterization of target sites are
critically important for programs aimed at the deliberate design of
new fungicide mechanism groups because these programs are likely to
depend on mimicking the structure and conformation of a substrate,
a high energy or transition state intermediate or an allosteric
regulator. Selection of candidate targets on which selective
toxicity is to be based can be a risky endeavour because factors
such as uptake, metabolism, and compensation can render an
excellent inhibitor at the subcellular target site ineffective in
the intact target organism. A reaction or pathway characteristic
of the fungal pathogen but not of the plant host or mammals would
ordinarily be of great interest in the development of selective
fungicides. Chitin biosynthesis is often suggested as an example
of such a pathway. On the other hand, most targets of existing
fungicides also occur in non-target organisms and still, good
selectivity exists.

The following discussion concerning the mode of action of
selected fungicides now in use will serve to illustrate the
importance of some of the points mentioned above in guiding the
development of new fungicides.

Ergosterol Biosynthesis Inhibitors

The triterpenoid pathway leading to the biosynthesis of ergosterol
(Figure 1) is the target of 3 fungicide mechanism groups used to
control fungal pathogens of plants and animals. These groups
consist of: (A) squalene epoxidase inhibitors (allylamines and
thiocarbanilates) which block conversion of squalene to 2,3-
oxidosqualene; (B) sterol C-14 demethylase inhibitors
(piperazines, pyridines, pyrimidines, imidazoles and triazoles)
which block conversion of intermediate b to c (Figure 1); and (C)
sterol Δ^8 ---> Δ^7 isomerase inhibitors and/or sterol Δ^{14} reductase
inhibitors (morpholines) which block conversion of intermediate c
to d and/or f to g (Figure 1). It is an interesting phenomenon
that these 3 fungicide mechanism groups discovered by routine
screening procedures should all prove to be inhibitors of the
triterpenoid pathway to ergosterol. This would suggest that other
enzymes in the pathway like 2,3-oxidosqualene cyclase or sterol C-4
demethylase might also be good prospective targets for fungicide
development.

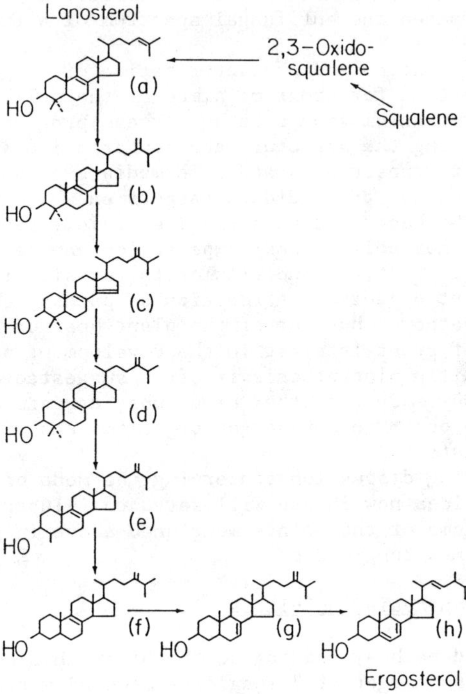

Figure 1. Pathway of ergosterol biosynthesis from squalene.

Thus far, compounds in mechanism group A have not been adopted for practical use in plant protection, but are used for the control of fungal pathogens of humans (3,4). On the other hand, compounds of group B are used to control a variety of fungal pathogens of both plants and animals. Compounds in group C are used only for the control of plant pathogens; however, their practical disease control spectrum is narrower than that of the compounds in group B.

Inhibitors of Sterol C-14 Demethylation. The greatest variety of fungicides in a single mechanism group used to control pathogens of plants and animals are inhibitors of sterol C-14 demethylase, a cytochrome P-450 ($P-450_{14DM}$) enzyme. Typically, these sterol demethylation inhibitors (DMI) consist of a pyrimidine, imidazole, triazole or pyridine heterocycle attached to a single lipophilic substituent (5-7). An exception to this general structure is that of the piperazine, triforine. Structures of representative compounds of this mechanism group are shown in Figure 2.

These inhibitors produce type II difference spectra with the cytochrome P-450 monooxygenase enzymes which indicate binding of an N atom of the heterocycle at the 6th coordination position of the iron atom of the P-450 heme prosthetic group, the position at which O_2 or CO would normally bind (Figure 3). The lipophilic substituent of the fungicide binds concurrently to a nearby lipophilic region of the apoenzyme normally occupied at least in part by a 14α-methyl sterol (8) (Figure 3). Consequently, affinity of these inhibitors for the P-450 enzyme is determined by the binding of the N atom to the heme iron and the lipophilic substituent to the apoenzyme (9). Extensive structural-activity data indicate that the lipophilic substituent is primarily responsible for the selectivity of DMI fungicides.

The low specificity and open nature of the $P-450_{14DM}$ apoenzyme in the vicinity of the heme prosthetic group very likely explains why it is possible to have so much structural variability among the many pyrimidines, triazoles, imidazoles and pyridines that have been successfully targeted on this enzyme in fungi. Structural studies indicate that the environment and topography in the vicinity of the O_2 binding site is very similar in all cytochrome P-450 enzymes (10). It may seem surprising, therefore, that $P-450_{14DM}$ from rat liver is much less sensitive to ketoconazole and miconazole than the same enzyme from yeast (11). Some detailed differences obviously account for fungicide selectivity between these enzymes of the same class and function. The selectivity of DMI fungicides for the target enzyme in a nontarget species has been demonstrated with maize sterol C-14 demethylase. The DMI fungicides triadimefon and propiconazole are moderate (IC_{50} 8 μM) or good (IC_{50} 2 μM) inhibitors of maize $P-450_{obt\ 14DM}$, whereas the experimental DMI fungicides lab 158241F and lab 170250F are highly potent inhibitors of this enzyme with respective IC_{50} values of 0.3 and 0.05 μM (12). These observations indicate that while there may be much similarity in various cytochrome P-450 enzymes, small differences in structural designs of DMI can be used to exploit those that do exist for purposes of fungal disease control. However, side effects are known to occur with the use of these fungicides. One commonly observed in controlling plant diseases

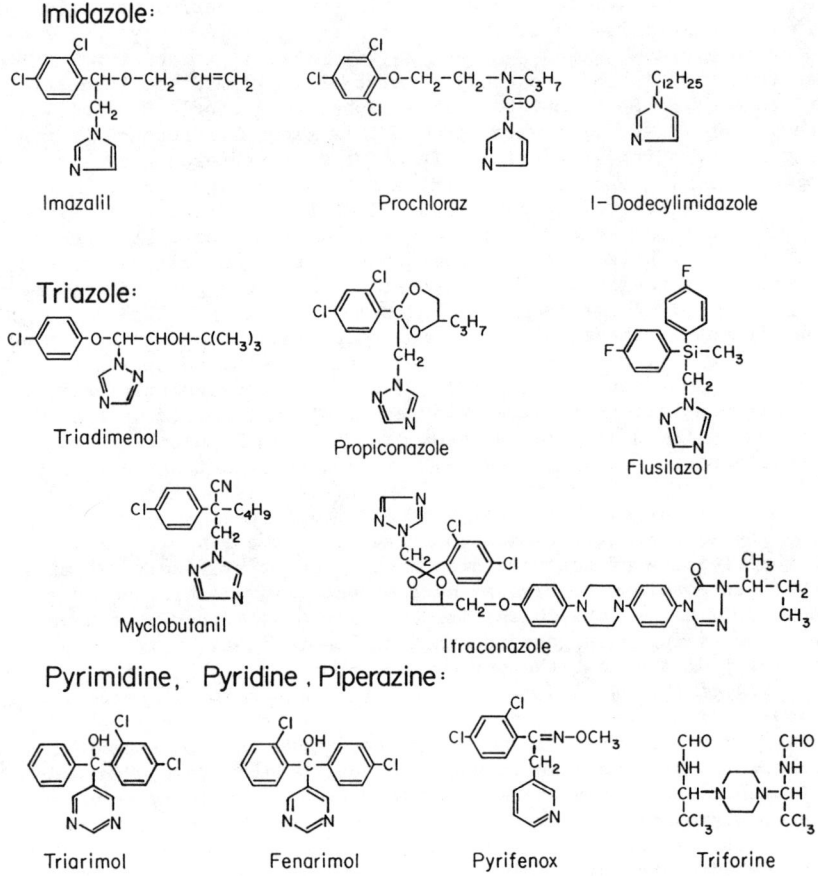

Figure 2. Structures of sterol C-14 demethylation inhibitors.

with DMI is the growth regulator effect (13) that results from inhibition of the P-450 enzyme involved in oxygenation of the C-19 methyl group of kaur-16-ene (14), a precursor of gibberellic acid. When high levels of ketoconazole are used medicinally, there may be interference with the cytochrome P-450 17,20-lyase involved in the pathway of conversion of 17α-hydroxy progesterone to testosterone (15). With careful selection of fungicide structure and proper regulation of dosage, such side effects can largely be avoided.

Using structure-activity correlations and analyses of sterochemistry, lipophilicity, conformation and other physiochemical parameters, various investigators (16-19) have constructed cytochrome P-450 binding-site models and have attempted to predict inhibitor structures with optimal fungicidal activity. It is apparent from examining structures of sterol C-14 demethylation inhibitors (Figure 2) that many would not fully occupy the sterol binding site on the apoenzyme and that some such as itraconazole, probably extend outside the sterol binding site. Moreover, the regions of contact with the apoenzyme must differ appreciably among these derivatives. Steric fit comparisons by computer graphics of diniconazole with lanosterol when the N-4 atom of the fungicide is in close proximity of the lanosterol C-14 methyl group, show overlap of these molecules mainly in the region of the "D" ring and side chain of the sterol (18). Using computer graphic procedures it has been possible to align the fungicide RR-paclobutrazol so that it has good overlap with the 24-methylene-24,25-dihydrolanosterol substrate (17). Fujimoto et al. (19) have recently constructed a folded conformation model of myclobutanil that resembles a portion of the active conformation of the lanosterol molecule. Advancement of these techniques may eventually guide synthesis of entirely new fungicides rather than the modification or improvement of fungicides already discovered as is now the case.

As mentioned earlier, the apoenzyme region surrounding the heme prosthetic group of cytochrome P-450 enzymes is one of relatively low specificity. However, there are apparently significant differences in $P-450_{14DM}$ even among various fungi as suggested by the optimal inhibitory structure of DMI for different species. While most single gene mutations affecting this region of the enzyme would probably not result in drastic loss of fungicide effectiveness and still allow the organism to retain good fitness, the change in sensitivity might be as great as that presently existing among certain fungal species. It seems likely, therefore, that if a target change leads to decreased sensitivity, this could be countered by a DMI fungicide presently available or one that could be readily synthesized. Quantitative structure activity relationships should be helpful in designing the proper derivatives to combat such target site resistance. Development of resistance to DMI has been less rapid than expected, possibly because of the characteristics of the P-450 target site described above. With these fungicides, it may be the efflux system described by DeWaard and Van Nistelrooy (20) or some other mechanism rather than target site change that will present the greatest hazard for practical resistance.

Inhibitors of Sterol Δ^8 ---> Δ^7 Isomerization. Initial mode of action studies indicated that the morpholine fungicide tridemorph blocked sterol Δ^8 ---> Δ^7 isomerization (21-23) or sterol Δ^{14} reduction (24). Subsequently, it was shown that morpholines may block both reactions or only the former, depending on the organism and morpholine derivative involved (25). Detailed mode of action studies have focused mainly on inhibition of the Δ^8 ---> Δ^7 isomerization by tridemorph. An analysis of the mechanism of sterol Δ^8 ---> Δ^7 isomerization (26) indicates that the enzymatic reaction involves first an α-protonation of the Δ^8 double bond leading to a high energy intermediate (HEI) with a carbocation at C-8, followed by an elimination of the C-7 proton to give a Δ^7-double bond (Figure 4). Studies in the laboratory of Benveniste (27-29) have led to the proposal that since the pKa of N-substituted morpholines like tridemorph is between 7 and 8, some morpholinium cations exist at physiological pH values and these can act as an inhibitory carbocationic mimics of the sterol HEI involved in the isomerase reaction. This reasoning can be understood by comparing the structure of the sterol carbocationic HEI with tridemorph and an 8-azadecalin like the derivative shown in Figure 4 which is a potent inhibitor of Δ^8 ---> Δ^7 isomerase (28).

Taton et al. (28) have pointed out that several reactions in the sterol biosynthetic pathway involve cationic HE (high energy) or transition state (TS) intermediates with a high dipole moment. Some of these are indicated in Table I. These intermediates can serve as models for design of stable analogues that might be effective fungicides.

Table I. Target Enzymes, TS or HE Intermediates and Corresponding Inhibitor Analogues

Target Enzyme	Intermediate	Inhibitor Analogue
2,3-Oxidosqualene cyclase	TS electron deficient C-2 squalene	2-amino, 2N-oxide(27) and 2,3-imino(30) squalene
Sterol Δ^8 ---> Δ^7 isomerase	HE C-8 carbocationic sterol	Cationic morpholines(21,29) and 8 azadecalins (28)
Sterol C-24 methyl transferase	HE C-25 carbocationic sterol	C-25 cationic sterols (amine, sulfonium, arsonium groups) (27)
Sterol Δ^{14} reductase	HE C-15 carbocationic sterol	Cationic morpholines(31)

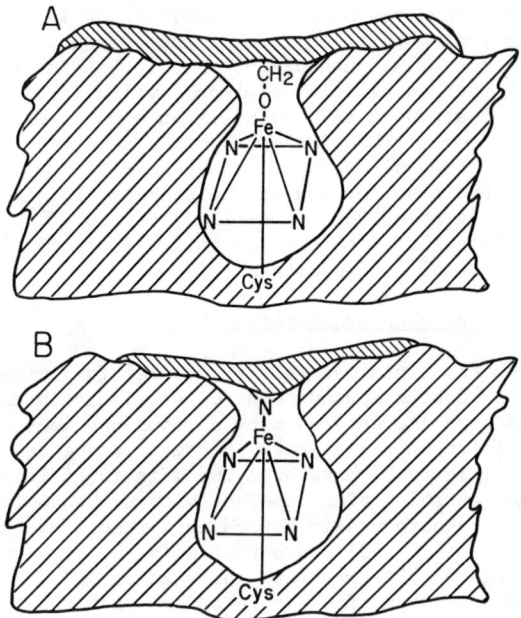

Figure 3. Diagrams portraying the interaction of 14α-methyl sterol (A) and a sterol C-14 demethylation inhibitor (B) with cytochrome P-450$_{14DM}$ in the region of the heme prosthetic group. An atom of oxygen has already been inserted into the sterol C-14 methyl group (A). A nitrogen atom of the fungicide heterocycle is shown interacting with the heme prosthetic group of the enzyme (B).

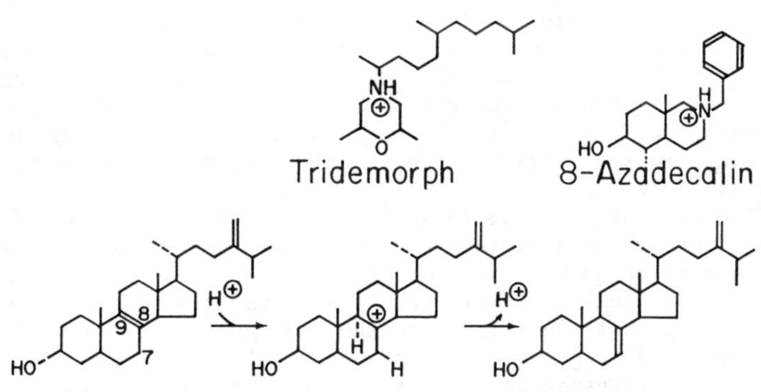

Figure 4. Pathway showing mechanism of sterol Δ^8 ---> Δ^7 isomerization and structural similarity of protonated tridemorph and an 8-azadecalin to the C-8 carbocationic HEI sterol.

The 2,3-oxidosqualene cyclization reaction is one of particular interest as a fungicide target since it is quite complicated and involves more than one TS or HE intermediate. It is surprising that more inhibitors of the enzyme catalyzing this reaction have not been synthesized. A point of interest is the unusual overlap of specificity of 1-dodecyclimidazole (Figure 2) as an inhibitor of sterol C-14 demethylation and 2,3-oxidosqualene cyclization (12,32). The simple aliphatic substituent of this imidazole may allow the compound to act as a TS analogue of 2,3-oxidosqualene and also to bind to the sterol C-14 demethylase and interact with the heme iron. A similar overlap of specificity apparently exists also with the morpholines for sterol $\Delta^8 \dashrightarrow \Delta^7$ isomerase and sterol Δ^{14} reductase (25).

Benzimidazoles and Phenylcarbamates

Benzimidazoles came into use as agricultural fungicides in the early 1970's. These compounds control a relatively broad spectrum of fungi, but have encountered many serious problems with the development of fungal resistance. Benomyl, carbendazim and thiabendazole (Figure 5) have been the main fungicides used for plant disease control. Since the activity of benomyl can be attributed to the carbendazim formed as a result of the loss of the butylcarbamoyl moiety (33), carbendazim is the structure of basic interest in regard to target site interaction.

Benzimidazoles specifically interfere with the formation of microtubules which function in a variety of cellular processes including mitosis (34). Microtubules are formed under appropriate conditions by assembly of tubulin, a heterodimeric protein of which the subunits, are usually designated as α- and β-tubulin. Biochemical and genetic studies have clearly shown that the target site of carbendazim and related benzimidazole fungicides is the colchicine binding site on tubulin, and that the β-tubulin subunit is usually the primary determinant of binding affinity (34). In regard to target site selectivity of benzimidazoles, carbendazim is at least 300 times more effective in preventing assembly of yeast tubulin than pig brain tubulin (35). On the other hand, substitution of the benzene ring of carbendazim as in nocodazole leads to about a 200 fold increase in effectiveness for blocking assembly of pig brain tubulin but only a 4-fold increase for blocking assembly of yeast tubulin. Thus, this substituent largely eliminates the selectivity shown by carbendazim for the yeast and the mammalian target sites.

With the emergence of fungal resistance to benzimidazoles, there has been intense interest in ways to counter this problem. Since most cases of benzimidazole resistance are believed to be due to changes in tubulin affinity for these fungicides, one approach to combatting resistance is to design analogues with high affinity for the mutated target sites. The potential for success of this approach was suggested by the observations of Leroux and Gredt (36) that benzimidazole resistant mutants of *Botrytis cinerea* and *Penicillium expansum* are much more sensitive to phenyl carbamate herbicides than the benzimidazole sensitive strains of these fungi. Following this observation, it was shown that methyl N-(3,5-

dichlorophenyl) carbamate (MDPC) and isopropyl N-(3,4-diethoxyphenyl) carbamate (diethofencarb) are highly toxic to benzimidazole carbamate resistant strains of *B. cinerea*, but not to benzimidazole carbamate sensitive strains (37,38). These compounds (Figure 5) also show little phytotoxicity. Mixtures of benzimidazole carbamates and diethofencarb are presently being used in France to control *Botrytis* disease of grape in regions where benzimidazole carbamates alone are no longer effective. The extent to which this approach can be used to control benzimidazole resistance in other fungal species is unclear. It is known, for example, the MDPC does not control some benzimidazole resistant strains of *Cercospora beticola* (39). Whether a practical problem from double resistance to the paired inhibitors or difficulties with toxicity to nontarget organisms prove to be limiting factors must await further experience.

Recent studies have shown that a strain of *Neurospora crassa* resistant to benzimidazole carbamates but sensitive to the phenylcarbamates, MDPC and diethofencarb, resulted from a mutational change in the β-tubulin gene (40). The change of a single amino acid from glutamic acid to glycine in β-tubulin apparently accounted for benzimidazole resistance in this mutant strain. Revertants of this mutant were resistant to diethofencarb but exhibited the same benzimidazole sensitivity as the wild type.

Extensive analyses of structure activity relationships (SAR) of fungicidal methyl N-phenyl carbamates indicate that receptor regions corresponding to the O and M-substituents on the benzene ring are highly hydrophobic whereas that for the P substituent is moderately hydrophobic (38). Steric interaction of the M-substituent with the receptor as well as hydrogen bonding by the P substituent appeared to be necessary for high activity. Appreciable modification of potency of phenylcarbamates could be produced by changing the nature of the alcohol of the ester moiety. The isopropyl ester for example was highly fungitoxic whereas the isobutyl ester showed very low toxicity (41).

Studies of the SAR of benzimidazoles and phenylcarbamates have shown that target site selectivity between fungal strains and among fungal species, higher plants and mammals can be affected by rather simple structural changes. This not only points out opportunities for combatting fungal resistance to these derivatives and for selective control of various pests, but also indicates the potential hazard for nontarget species that might result because of the close similarity of tubulin among diverse eukaryotic organisms.

Carboxamides

Carboxamides are a group of fungicides that control diseases caused by Basidiomycete type fungi (42). The best known member of this group is carboxin (Figure 6). Carboxamides specifically block membrane bound succinate-ubiquinone oxidoreductase activity in the mitochondrial electron transport chain (43,44). The carboxin receptor in the succinic dehydrogenase complex (SDC) is believed to be the iron-sulfur cluster S_3 complexed with small coenzyme Q binding polypeptide(s) in a phospholipid environment (45,46).

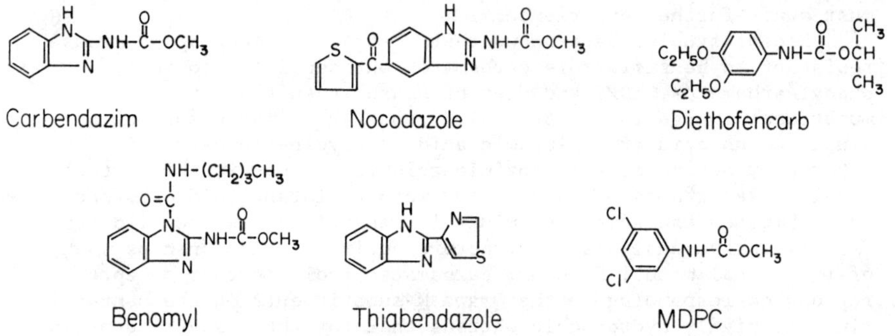

Figure 5. Structures of benzimidazole and phenylcarbamate fungicides.

Figure 6. Structure of carboxin.

Extensive studies of structural-activity relationships of carboxamides have been made, particularly in regard to the molecular designs that inhibit carboxin resistant mutants of *Ustilago maydis* (45,47,48). These investigations have shown that it is possible to produce analogs of carboxin which are good inhibitors of the mutated succinic dehydrogenase complex (SDC) in carboxin resistant strains that are also good inhibitors of growth of these strains. There are indications that with proper structural design, succinic dehydrogenase inhibitors (SDI) can be produced which control fungal pathogens in mycological groups other than Basidiomycetes (49,50). It is puzzling, therefore, why the practical usefulness of carboxanilides remains essentially confined to the control of Basidiomycete type fungi. Whether this is due to inadequate target site affinity, cellular permeability, metabolism or other factors is not known. There appears to be an interesting and perhaps an economically rewarding challenge to design effective SDI for nonBasidiomycete fungi, but it is possible that blocking of the SDC complex may not have the same consequences in these fungi as in the sensitive Basidiomycetes. While it is believed that primary cellular toxicity of SDC inhibitors results from failure of the citric acid cycle to operate to produce ATP and biosynthetic intermediates (51), the destruction of mitochondrial membranes and enzymes by reactive oxygen radicals resulting from the blocking of the SDC could actually be the primary toxic mechanism in sensitive Basidiomycetes. If this is the case, then fungi of other mycological groups may be less sensitive because they are more capable than Basidiomycetes of eliminating these reactive oxygen products. Evidence for a radical type of toxic mechanism is suggested by the observation that carboxin treatment leads to structural damage to mitochondria (52) and destruction of the S_3 center of the SDC (44).

Melanin Biosynthesis Inhibitors. Among the most desirable targets for selective fungitoxic action are those uniquely associated with fungal pathogenicity such as the polyketide pathway to melanin in the fungus *Pyricularia oryzae* which causes rice blast disease. This pathway of secondary metabolism (Figure 7), which is not required for growth of the fungus as a saprophyte, is induced in the appressoria (penetration structures) formed by germ tubes of spores on a plant epidermal surface or other barriers. This pathway leads to appressorial wall melanization and is necessary for appressorial penetration of plant epidermal walls by *P. oryzae* (53,54) and *Colletotrichum* species (55). Genetic or chemical blocks in the pathway that result in the inhibition of melanin biosynthesis, prevent appressorial penetration of plant epidermal barriers. A genetic or a chemical block (by cerulenin) prior to pentaketide cyclization leads to albino appressoria that fail to penetrate epidermal or cellulosic barriers (56,57). Melanization and penetration capacity in *P. oryzae* can be restored in these appressoria by adding the melanin precursors scytalone, vermelone or 1,8-dihydroxynaphthalene (1,8-DHN) (57). The compounds shown in Figure 8, that have been developed for the control of rice blast disease, block NADPH dependent reductase reactions in the pathway at sites indicated by asterisks in Figure 7. Yellow or reddish

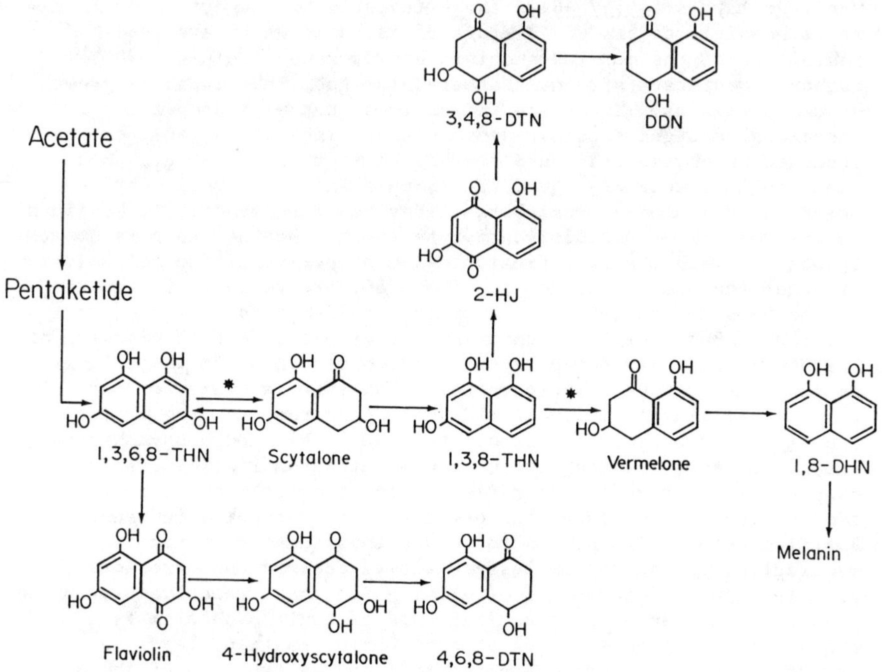

Figure 7. Polyketide pathway of fungal melanin biosynthesis showing branch pathways leading to synthesis of 2-hydroxyjuglone (2HJ), flaviolin and other shunt metabolites.

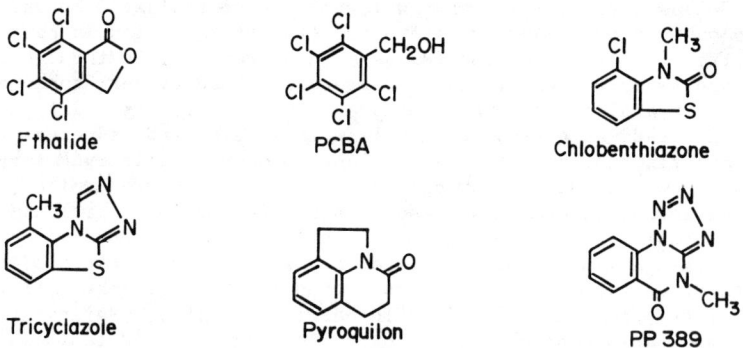

Figure 8. Structures of compounds that block reductase reactions in the polyketide pathway to melanin at points indicated by asterisks in Figure 7.

brown pigments are formed instead of black melanin. Vermelone and 1,8-DHN, but not scytalone, can restore melanization and partially restore penetration capacity in appressoria treated with these compounds (57).

Utilizing knowledge concerning the structural relationship of known melanin biosynthesis inhibitors (MBI) to 1,3,8-trihydroxynaphthalene (1,3,8-THN), the substrate of a target enzyme in this pathway, Omata et al. (58) have recently designed phthalazine derivatives that are good MBI and rice blast control agents.

While it is known that blocking of the melanin biosynthetic pathway in P. oryzae by antipenetrants like tricyclazole, fthalide, pyroquilon and cerulenin, results in failure of appressorial penetration, it remains unclear why this is so. Among the various causes that have been suggested are lack of appressorial wall rigidity (59), poor adhesion (60) or the accumulation of toxic polyketide shunt products such as 2-hydroxyjuglone (2HJ) (61).

Control of a major fungal disease of plants by MBI illustrates how specific knowledge concerning fungal pathogenicity can be exploited for plant protection. A better understanding of fungal pathogenicity mechanisms and of host-pathogen interactions will no doubt reveal other specific targets that can be used to control fungal diseases. However, there may be limited economic value of inhibitors of a particular target if it is restricted to a single species or a very narrow mycological spectrum. This is the case with the MBI. The polyketide pathway to melanin appears to be critical only for pathogenicity of *P. oryzae* and *Colletotrichum* species and at present, the MBI are used only for control of rice blast disease caused by *P. oryzae*.

Conclusions

We have examined information on molecular design in relation to target site fit and antifungal activity. This knowledge has aided in the development of derivatives that broaden the antifungal spectrum of existing fungicide groups to include species or resistant strains not adequately controlled by the original compounds discovered by conventional screening procedures.

Accelerated development of fungicides based on rational designs will be highly dependent on the ability to identify and precisely analyze the structural features of promising target sites. High potency in the laboratory can only be suggestive of practical success. These fungicides will need to exhibit properties of stability, selectivity and mobility in the plant similar to those of successful fungicides discovered by conventional proecedures.

Literature Cited

1. Dekker, J. In Modern Selective Fungicides; Lyr, H., Ed.; Longman: London, 1987; pp 39-52.
2. Sisler, H. D. In Fungicide Resistance in North America; Delp, C. J., Ed.; Amer. Phytopathological Soc. Press: St. Paul, 1988; pp 6-8.

3. Ryder, N. S. Pestic. Sci. 1987, 21, 281-288.
4. Ryder, N. S.; Frank, I.; Dupont, M. C. Antimicrob. Agents Chemother. 1986, 29, 858-860.
5. Sisler, H. D.; Ragsdale, N. N. In Mode of Action of Antifungal Agents; Trinci, A. J. P.; Ryley, J. F., Eds.; Cambridge Univ. Press: London, 1984; pp 257-282.
6. Scheinpflug, H.; Kuck, K. H. In Modern Selective Fungicides; Lyr, H. Ed.; Longman: London, 1987; pp 173-204.
7. Buchenauer, H.; In Modern Selective Fungicides; Lyr, H., Ed.; Longman: London, 1987; pp 205-231.
8. Gadher, P.; Mercer, E. I.; Baldwin, B. C.; Wiggins, T. E. Pestic. Biochem. Physiol. 1983, 19, 1-20.
9. Vanden Bossche, H. In Recent Trends in the Discovery, Development and Evaluation of Antifungal Agents; Fromtling, R. A., Ed.; J. R. Prous: Barcelona, 1987; pp 207-221.
10. Poulos, T. L. In Cytochrome P-450: Structure, Mechanism and Biochemistry; Oritz deMontellano, P. R., Ed.; Plenum: New York, 1986; pp 505-523.
11. Isaacon, D. M.; Tolman, E. L.; Tobia, A. J.; Rosenthale, M. E.; McGuire, J. L.; Vanden Bossche, H.; Janssen, P. A. J. J. Antimicrob. Chemother. 1988, 21, 333-343.
12. Taton, M.; Ullmann, P.; Benveniste, P.; Rahier, A. Pestic. Biochem. Physiol. 1988, 30, 178-189.
13. Shive, J. B.; Sisler, H. D. Plant Physiol. 1976, 57, 640-644.
14. Coolbaugh, R. C.; Swanson, D. I.; West, C. A. Plant Physiol. 1982, 69, 707-711.
15. Santen, R. J.; Vanden Bossche, H.; Symoens, J.; Brugmans, J.; Decoster, R. J. Clin. Endocrinol. Metab. 1983, 57, 732-736.
16. Marchington, A. F.; Lambros, S. A. In Modern Selective Fungicides; Lyr, H., Ed.; Longman: London, 1987; pp 325-336.
17. Worthington, P. A. In Synthesis and Chemistry of Agrochemicals; Baker, D. R.; Fenyes, J. G.; Moberg, W. K.; Cross, B., Eds. ACS Symposium Series No. 355; American Chemical Society, Washington, DC, 1987; pp 302-317.
18. Katagi, T.; Mikami, N.; Matsuda, T.; Miyamoto, J. In Synthesis and Chemistry of Agrochemicals; Baker, D. R.; Fenyes, J. G.; Moberg, W. K.; Cross, B., Eds.; ACS Symposium Series No. 355; American Chemical Society, Washington, DC, 1987; pp 340-352.
19. Fujimoto, T. T.; Quinn, J. A.; Egan, A. R.; Shaber, S. H.; Ross, R. R. Pestic. Biochem. Physiol. 1988. 30, 199-213.
20. DeWaard, M. A.; Van Nistelrooy, J. G. M. Experimental Mycol. 1987, 11, 1-10.
21. Kato, T.; Shoami, M.; Kawase, Y. J. Pestic. Sci. 1980, 5, 69-79.
22. Leroux, P.; Gredt, M. Agronomie 1983, 3, 123-130.
23. Berg, L. R.; Patterson, G. W.; Lusby, W. R. Lipids 1983, 18, 448-452.
24. Kerkenaar, A.; Uchiyama, M.; Versluis, G. G. Pestic. Biochem. Physiol. 1981, 16, 97-104.
25. Baloch, R. I.; Mercer, E. I.; Wiggins, T. E. and Baldwin, B. C. 1984. Proc. Brit. Crop Prot. Conf. Pests Dis. 1984, 3, 893-898.

26. Wilton, D. C.; Rahimtula, A. D.; Akhtar, M. Biochem. J. 1969, 114, 71-73.
27. Rahier, A.; Taton, M.; Bouvier-Navé, P.; Benveniste, P.; Schuber, F.; Narula, A. S.; Cattel, L.; Anding, C.; Place, P. Lipids 1986, 21, 52-62.
28. Taton, M.; Benveniste, P.; Rahier, A. Pure and Appl. Chem. 1987, 59, 287-294.
29. Rahier, A.; Schmitt, P.; Huss, B.; Benveniste, P.; Pommer, E. H. Pestic. Biochem. Physiol. 1986, 25, 112-124.
30. Pinto, W. J.; Nes, W. R. J. Biol. Chem. 1983, 258, 4472-4476.
31. Kerkenaar, A. Recent Trends in The Discovery and Evaluation of Antifungal Agents; Fromtling, R. A., Ed.; J. R. Prous: Barcelona, 1987 p 523-542.
32. Henry, M. J.; Sisler, H. D. Antimicrob. Agents Chemother. 1979, 15, 603-607.
33. Clemons, G. P.; Sisler, H. D. Phytopathology 1969, 59, 705-706.
34. Davidse, L. C. Annu. Rev. Phytopathology 1986, 24, 43-65.
35. Kilmartin, J. V. Biochemistry 1981, 20, 3629-3633.
36. Leroux, P.; Gredt, M. C. R. Acad. Sci. 1979, 289, 691-693.
37. Suzuki, K.; Kato, T.; Takahishi, J.; Kamoshita, K. J. Pestic. Sci. 1984, 9, 497-501.
38. Takahishi, J.; Kirino, O.; Takayama, C.; Nakamura, S.; Noguchi, H.; Kato, T., Kamoshita, K. Pestic. Biochem. Physiol. 1988, 30, 262-271.
39. Demakopoulou, M. G.; Georgopoulos, S. G. Abstracts, 6th Int. Congr. Pesticide Chem. 1986, 3E-05.
40. Fujimura, M.; Oeda, K.; Inoue, H.; Kato, T. Abstracts, 5th Int. Congr. Plant Pathol. 1988, IX-3-4, p 308.
41. Takahishi, J.; Nakamura, S.; Noguchi, H.; Kato, T.; Kamoshita, K. J. Pestic. Sci. 1988, 13, 63-69.
42. Kulka, M.; Von Schmeling, B. In Modern Selective Fungicides; Lyr, H., Ed.; Longman: London, 1987; pp 119-131.
43. Ulrich, J. T.; Mathre, D. E. J. Bacteriol. 1972, 110, 628-632.
44. Mowery, P. C.; Ackrell, B. A. C.; Singer, T. P.; White, G. A.; Thorne, G. D. Biochem. Biophys. Res. Commun. 1976, 71, 354-361.
45. White, G. A.; Thorne, G. D.; Georgopoulos, S. G. Pestic. Biochem. Physiol. 1978, 9, 165-182.
46. Schewe, T.; Lyr, H. In Modern Selective Fungicides; Lyr, H., Ed.; Longman: London, 1987; pp 133-142.
47. White, G. A.; Thorne, G. D. Pestic. Biochem. Physiol. 1980, 14, 26-40.
48. White, S. G. Pestic. Biochem. Physiol. 1988, 31, 129-145.
49. Edgington, L. V.; Barron, G. L. Phytopathology 1967, 57, 1256-1257.
50. White, G. A.; Georgopoulos, S. G. Pestic. Biochem. Physiol. 1986, 25, 188-204.
51. Ragsdale, N. N.; Sisler, H. D. Phytopathology 1970, 60, 1422-1427.
52. Lyr, H.; Ritter, G.; Casperson, G. Z. Allg. Mikrobiol. 1972, 12, 271-280.

53. Woloshuk, C. P.; Sisler, H. D.; Tokousbalides, M. C.; Dutky, S. R. Pestic. Biochem. Physiol. 1980, 14, 256-264.
54. Woloshuk, C. P.; Sisler, H. D. J. Pestic. Sci. 1982, 7, 161-166.
55. Kubo, Y.; Suzuki, K.; Furusawa, I.; Ishida, N.; Yamamoto, M. Phytopathology 1982, 72, 498-501.
56. Kubo, Y.; Katoh, M.; Furusawa, I.; Shishiyama, J. Exp. Mycol. 1986, 10, 301-306.
57. Chida, T.; Sisler, H. D. J. Pestic. Sci. 1987, 12, 49-55.
58. Omata, K.; Tomita, H.; Nakajima, T.; Natsume, B. Abstract 103, Div. Agrochemicals, 196th National Meeting, Amer. Chemical Soc., 1988, Los Angeles.
59. Woloshuk, C. P.; Sisler, H. D.; Vigil, E. L. Physiol. Plant Pathol. 1983, 22, 245-259.
60. Inoue, S.; Kato, T.; Jordan, V. W. L.; Brent, K. J. Pestic. Sci. 1987, 19, 145-152.
61. Yamaguchi, I.; Sekido, S.; Seto, H.; and Misato, T. J. Pestic. Sci. 1983, 8, 545-550.

RECEIVED June 28, 1989

Chapter 15

Modeling of Photosystem II Inhibitors of the Herbicide-Binding Protein

Inhibitory Pattern, Quantitative Structure–Activity Relationships, and Quantum Mechanical Calculations of New Hydroxyquinoline Derivatives

W. Draber[1], B. Pittel[2], and A. Trebst[3]

[1]Agrochemical Research Monheim, Bayer AG, 5090 Leverkusen, Federal Republic of Germany
[2]Central Research, Bayer AG, 5090 Leverkusen, Federal Republic of Germany
[3]Department of Biology, Ruhr-University Bochum, 4630 Bochum, Federal Republic of Germany

Substituted hydroxyquinolines are inhibitors of photosynthetic electron flow on the Q_B acceptor site of photosystem II. They displace radioactive metribuzin from its binding site on the D-1 reaction center polypeptide of photosystem II. They neither loose inhibitory potency in tris-treated membrane preparations, nor in a metribuzin resistant mutant of Chlamydomonas where serine 264 has been exchanged in the D-1 polypeptide. A QSAR study of 15 substituted hydroxyquinolines suggests a dependency on two steric parameters and an electronic parameter for position 6.
In their functional inhibitory pattern and their physicochemical parameters for high inhibitory potency the hydroxyquinolines resemble the phenol, but not the urea/triazine family of PS II inhibitors.
This is substantiated by charge distributions of hydroxyquinolines and of the tautomeric quinolones obtained from quantum mechanical calculations. It shows a negative π-charge at a particular essential atom. The comparison of these with the charge distribution on chromones and napththoquinones revealed the gradual shift from the phenol- to the urea-type inhibitors.

Two families of inhibitors interfere with the plastoquinone Q_B or herbicide binding site on the D-1 polypeptide, i.e. on one of the reaction center subunits of PS II. The phenol and urea/triazinone family of PS II inhibitors are different in their functional inhibitory pattern (reviewed in [1]), although they both bind to the D-1 polypeptide and displace each other from the binding site (1). Both QSAR studies (2) and - more refined - quantum mechanical calculations (3,4) indicate differences in the physicochemistry of the interaction of these two families with the amino acids in the binding niche on the D-1 polypeptide.

Many of the PS II inhibitors are herbicides like ureas, triazines, and aminotriazinones (reviewed in (5) and (6)). This class shares a common structural element essential for inhibitory potency, and their inhibitory potency depends on electronic, lipophilic and steric properties of the substituents.

Another class of PS II inhibitors are phenol-derivatives, like the herbicides ioxynil and dinoseb where the QSAR is governed by steric parameters (2), as well as hydroxypyridines (7,8) and ketonitriles (9). The two inhibitory families may also be designated as a serine and a histidine families, named after the amino acid in the D-1 polypeptide to which the inhibitor is predominantly oriented (though not necessarly bound) (10).

We wish to report on hydroxyquinoline derivatives which are very potent inhibitors and appear to belong to the histidine or phenol family, as the inhibitory pattern suggests. QSAR studies with a selected group of hydroxyquinolines also indicate a similarity to the phenol-type family of inhibitors. Furthermore, quantum chemical calculations of the hydroxyquinolines (or of the tautomeric quinolones) are compared with those of chromones and naphthoquinones, also inhibitors of PS II. It will become evident that, depending on substitution, within these three classes of compounds the charge distributions gradually change from the one to the other family of PS II inhibitors.

Materials and Methods

Thylakoid membranes were prepared from spinach leaves according to well known standard procedures. Leaves were homogenized in 0.4 M NaCl, 20 mM tricine-NaOH buffer pH 8.0, and 20 mM $MgCl_2$. After centrifugation the chloroplasts were osmotically shocked in 20 mM tricine pH 8.0, 15 mM NaCl, and 5 mM $MgCl_2$ and the thylakoid membranes were centrifuged down.

PS II particles were prepared according to Berthold et al. (11) by triton-treatment of the thylakoids (2 mg chlorophyll/ml) in a medium containing 5 mM NaCl, 20 mM hepes buffer pH 7.5, and 50 mg/ml triton X-100 for 30 min. at 4°C and centrifuged at 40.000 g. Activity was measured in a MDBQ (methylenedioxy-dimethyl-benzoquinone) system.

Tris-treated thylakoid membranes were prepared according to Yamashita and Butler (12) by suspending thylakoid membranes (5 mg chlorophyll) in 10 ml of 1 M tris buffer pH 9.0 and centrifuged down after 10 min. at 0°C. They are resuspended in MES buffer pH 6.5, 5 mM $MgCl_2$, and 15 mM NaCl. Photosynthetic activity was measured in the same medium in a DCP -> DCPIP (10 mM) system.

Photosynthetic activities were measured with an oxygen electrode in a methylviologen (0.1 mM) or MDBQ (0.1 mM) system or spectrophotometrically in a K-ferricyanide system (420 nm) or DCPIP (600 nm) system in a medium containing 80 mM tris buffer pH 8.0, 10 mM $MgCl_2$, 0.3 mM Na-azide, 10 µg gramicidin, and thylakoid membranes equivalent to 50 µg chlorophyll.

Herbicide sensitivity in thylakoids from metribuzin resistant *Chlamydomonas reinhardii* was investigated in the system described by

Wildner et al. (13). Displacement of radioactive metribuzin from its binding site on the membrane was performed according to Oettmeier et al. (14).

The compounds were synthesized according to known methods. The Sterimol parameters by Verloop et al. (16,17) proved very useful in the QSAR work. CNDO/2 calculations were carried out by using a QCPE program (15) which was modified inhouse. ^{13}C studies were made in $CDCl_3$ with $SiMe_4$ as internal standard.

Results

A number of derivatives of 2-trifluoromethyl-hydroxyquinolines were synthesized and photosynthesis inhibition determined with isolated thylakoids in the water -> methylviologen electron flow system. The inhibition is expressed as the pI_{50}-value, i.e. the negative logarithm of the molar concentration inhibiting 50%. Table I gives the values for 17 compounds, among them 11 3-bromo derivatives.

Table I. Inhibitory potency of substituted 2-trifluoromethyl-4-hydroxy-quinolines on photosynthetic oxygen evolution in spinach thylakoid membranes
pI_{50} = negative logarithm of the concentration for 50% inhibition

number	substituent	pI_{50}
1	H	4.1
2	6-Cl	4.5
3	6-N(CH$_3$)$_2$	4.7
4	3-Br,8-CH$_3$	4.8
5	8-Cl	4.9
6	6-CF$_3$	4.9
7	6-OCF$_3$	4.9
8	3-Br	5.1
9	3-Br,6-N(CH$_3$)$_2$	5.3
10	3-Br,8-CF$_3$	5.4
11	3-Br,6-CH$_3$	5.6
12	3-Br,6,8-Cl$_2$	5.9
13	3-Br,7,8-Cl$_2$	6.4
14	3-Br,6-CF$_3$	6.6
15	3-Br,6-OCF$_3$	6.7
16	3-Br,5,8-Cl$_2$	6.7
17	3-Br,8-SCF$_3$	7.0

The most active compounds (No.s 14-17) are inhibiting 50% of electron flow already at 0.1 μmolar concentrations. It is clear that a bromine in 3-position enhances the inhibitory activity.

In Table II the localization of the inhibition site of compound 14 (Table I) is shown. Not only the water -> methylviologen electron transport involving both photosystems, but also a Hill reaction that is driven by photosystem II only (water -> MDBQ) is inhibited by the hydroxyquinoline. On the other hand a PS I donor system that includes

Table II. Inhibitory Potency of a Hydroxyquinoline on Part Sequences of the Electron Flow System, Involved either in PS II and/or PS I

| | pI_{50}-Values of Non-cyclic Electron Flow | | |
| | $H_2O \rightarrow MV$ | $H_2O \rightarrow MDBQ$ | $DQH_2 \rightarrow MV$ |
Photosystem	I+II	II	I
DCMU	7.1	7.5	<3
quinoline	6.8	7.0	<3
DNP-INT	7.2	7.3	5.4

quinoline = 3-bromo-2,6-di-trifluoromethyl-hydroxyquinoline (= 14)
DNP-INT = dinitrophenylether of iodonitrothymole
MV = methylviologen
MDBQ = methylenedioxy-dimethylbenzoquinone
DQH_2 = duroquinol

PS I and the cytochrome b_6/f-complex ($DQH_2 \rightarrow MV$) is not impaired by compound 14. The displacement of ^{14}C metribuzin (14), a well known inhibitor of the acceptor side of PS II, by hydroxyquinolines (Figure 1) shows directly that the new compounds share the same binding site on the D-1 polypeptide.

Nevertheless, there is a functional difference between metribuzin or DCMU when compared with a phenol inhibitor or the hydroxyquinolines. This is seen in tris-treated chloroplasts where a modification of the donor side of the herbicide binding protein affects the

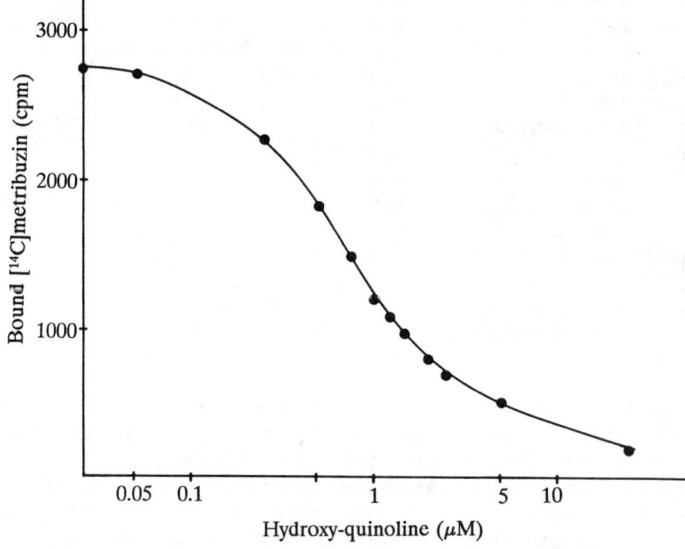

Figure 1. Displacement of ^{14}C metribuzin by 3-bromo-2.6-di-trifluoromethyl-4-hydroxyquinoline (No. 14 in Table I)

inhibitory efficiency of metribuzin and DCMU (see 7,18), but does not change that of the phenol and hydroxyquinoline (Table III). (In this table DCPIP was used as electron acceptor for the photosynthetic system. Inhibitory potency is experimentally higher in this system than in the methylviologen system of Table I).

Table III. Effect on the Inhibitory Potency of Different PS II Inhibitors by Tris-treatment of the Spinach Thylakoid Membrane and of the Amino Acid Change (Ser 264 to Ala) in Metribuzin Tolerant *Chlamydomonas reinhardii*

	pI_{50}-Values			
	Thylakoids		*Chlamydomonas*	
	Control	Tris-treated	Wild-type	Mutation Ser 264 to Ala
Metribuzin	7.3	6.3	7.5	3.5
DCMU	7.3	6.7	7.7	5.4
Bromonitrothymol	7.9	7.9	7.3	7.8
Compound 14 in Table I	7.1	7.1	7.0	7.0

Compound 14 = 3-bromo-2.6-di-trifluoromethyl-hydroxyquinoline

The notion that DCMU/metribuzin and phenol/hydroxyquinoline react with the same binding area on the acceptor side of photosystem II, but do not use identical binding interactions with the amino acids in the binding niche, is supported by the behaviour of the hydroxyquinolines in a metribuzin tolerant mutant of *Chlamydomonas* where Ser 264 is replaced by Ala (13,19) as shown in Table III. In this mutant there is no decrease of the inhibitory potency of hydroxyquinolines contrary to that of metribuzin and DCMU, but in line with other phenol-type inhibitors (13,19). This is further explored in ref. (13) with other mutations as well.

The inhibitory potency of naphthoquinones on PS II has been described earlier (20,21). Here we add (Table IV) a few more hydroxy derivatives, which are relevant for the quantum mechanical calculations, discussed below. Natural chromone derivatives have been described as inhibitors of PS II (22), as well as pyrones (23).

In Table V we have listened the observed pI_{50}-values of PS II inhibition of a set of 15 2-CF_3-substituted 4-hydroxyquinolines together with the calculated values obtained by regression analysis (using Equation 1 of Table VI, some less satisfactory examples are Equations 2-4). The occurrence of linear and quadratic L_8 terms seems to indicate a need for an optimum length of the substituent in position 8. This is confirmed by Equations 3 and 4 that are less significant.

The three-parameter Equation 2 demonstrates that lack of a parameter referring to position 8 in results in only minor decrease of statistical quality.

Table IV. Inhibition of Photosystem II Electron Flow by Substituted Naphthoquinones
pI_{50} = negative logarithm of the concentration for 50% inhibition

Number	R	pI_{50}
1	$-CH=CH(CH_2)_9CH_3$	5.8
2	$-(CH_2)_3-3-CF_3C_6H_4$	5.8
3	$-CH=CH(CH_2)_8CH_3$	5.7
4	$-4-(CH_3)_3C-C_6H_4$	5.6
5	$-CH=CH(CH_2)_4CH_3$	5.5
6	$-3-CF_3C_6H_4$	5.3
7	$-CH=CHCH_2-C_6H_5$	5.2
8	$-3-NO_2C_6H_4$	5.1
9	$-CH=CHC_2H_5$	4.5
10	$-CH_3$	3.3

Table V. QSAR of Hydroxyquinolines.
Comparison of Experimentally Measured to Calculated Inhibitory Potency of Substituted 2-trifluoromethyl-hydroxyquinolines Using Equation 1 of Table VI

No.	Substitution	pI_{50}exp.	pI_{50}calc.	pI_{50}exp.-pI_{50}calc.
1	H	4.10	4.07	-0.03
2	6-Cl	4.50	4.54	-0.04
3	6-NMe$_2$	4.70	4.39	0.31
4	3-Br,8-Me	4.80	4.84	-0.04
5	8-Cl	4.90	4.59	0.31
6	6-CF$_3$	4.90	5.16	-0.26
7	6-OCF$_3$	4.90	5.35	-0.45
8	3-Br	5.10	5.26	-0.16
9	3-Br,6-NMe$_2$	5.30	5.58	-0.28
10	3-Br,8-CF$_3$	5.40	5.33	0.07
11	3-Br,6-Me	5.60	5.57	0.03
12	3-Br,6,8-Cl$_2$	5.90	6.25	-0.35
14	3-Br-6-CF$_3$	6.60	6.27	0.33
15	3-Br,6-OCF$_3$	6.70	6.54	0.16
17	3-Br,8-SCF$_3$	7.00	6.77	0.23

Table VI. Regression Equations for the Inhibitory Potency of Hydroxyquinolines

---

1. $pI_{50} = 1.19D_3 + 0.40B5_6 + 0.63\sigma_6 - 7.20L_8 +$
   $\phantom{pI_{50} =}(6.88) (4.21) (3.26) (-1.96)$
   $\phantom{pI_{50} =}1.35L_8^2 + 12.7$
   $\phantom{pI_{50} =}(2.04)$

   $n = 15 \quad F = 19.9 \quad r^2 = 0.92 \quad s = 0.31$

2. $pI_{50} = 1.12D_3 + 0.38B5_6 + 0.67\sigma_6 + 3.81$
   $\phantom{pI_{50} =}(5.75) (4.12) (2.87)$

   $n = 15 \quad F = 21.1 \quad r^2 = 0.85 \quad s = 0.37$

3. $pI_{50} = 1.08D_3 + 0.47B5_6 + 0.64\sigma_6 - 0.29L_8^2 + 2.96$
   $\phantom{pI_{50} =}(5.73) (4.43) (2.91) (1.48)$

   $n = 15 \quad F = 18.6 \quad r^2 = 0.88 \quad s = 0.35$

4. $pI_{50} = 0.09D_3 + 0.51B5_6 + 0.06L_8^2 + 3.21$
   $\phantom{pI_{50} =}(4.51) (3.82) (1.39)$

   $n = 15 \quad F = 12.9 \quad r^2 = 0.78 \quad s = 0.45$

The t-values are given in parentheses below the coefficients. Equation 1 has been used in Table V.

The correlation matrix for the parameters employed in Table VI is shown in Table VII. From this it is evident that the indicator variable D_3 which stands for the presence of Br in 3-position is indispensible.

The QSAR indicate a similarity of the hydroxyquinolines to the phenols, but not to the classical inhibitors of the triazine-type.

This difference is even more clearly observed in the charge distribution of essential elements (3,4).

Table VIII shows a selection of results obtained by CNDO/2 calculations on three slightly different classes of compounds that are

Table VII. Correlation matrix for the equations in Table VI

	$B5_6$	$B5_6^2$	L_8	L_8^2	$s(p)_6$	$s(p)_6^2$	D_3	pI_{50}
$B5_6$	1.00							
$B5_6^2$	0.98	1.00						
L_8	-0.55	0.52	1.00					
L_8^2	-0.52	-0.49	0.99	1.00				
$s(p)_6$	-0.02	-0.02	0.03	0.02	1.00			
$s(p)_6^2$	0.65	0.60	-0.40	-0.37	-0.58	1.00		
D_3	-0.14	-0.14	0.27	0.27	-0.04	-0.13	1.00	
pI_{50}	0.17	0.17	0.42	0.46	0.25	-0.06	0.69	1.00

Table VIII. π-Densities and Net Charges of Quinolones, Chromones and Naphthoquinones

4-Quinolones

X	Cl		Cl		OH		OH	
Y	CH$_3$		CF$_3$		CH$_3$		CF$_3$	
	π	net	π	net	π	net	π	net
1	-.19	-.18	-.11	-.03	-.11	+.04	-.03	-.10
2	+.16	+.28	+.15	+.28	+.15	+.24	+.15	+.24
3	-.07	-.07	-.07	-.06	-.07	-.07	-.08	-.07
4	-.36	-.34	-.34	-.33	-.37	-.36	-.34	-.34

4-Chromones

X	Cl		Cl		OH		OH	
Y	CH$_3$		CF$_3$		CH$_3$		CF$_3$	
	π	net	π	net	π	net	π	net
1	-.15	-.08	-.05	-.01	-.08	+.05	+.02	+.13
2	+.15	+.28	+.14	+.27	+.15	+.24	+.14	+.24
3	-.05	-.07	-.06	-.07	-.06	-.07	-.06	-.07
4	-.31	-.31	-.27	-.29	-.32	-.33	-.28	-.29

1,4-Naphthoquinones

X	Cl		Cl		OH		OH	
Y	CH$_3$		CF$_3$		CH$_3$		CF$_3$	
	π	net	π	net	π	net	π	net
1	-.03	-.02	-.07	-.08	-.04	+.15	+.13	+.21
2	+.13	+.22	+.12	+.22	+.13	+.19	+.12	+.18
3	-.01	-.02	-.01	-.01	-.01	-.02	-.02	-.01
4	-.18	-.23	-.16	-.20	-.19	-.23	-.16	-.21

all inhibitors of either class: 4-quinolones (which are tautomeric with 4-hydroxyquinolines), 4-chromones and 1,4-naphthoquinones. They have several features in common: 1. All are benzo derivatives with a carbonyl group in the condensed ring; 2. In p-position to the carbonyl groups is either a nitrogen, an oxygen, or a second carbonyl group. The π-densities (minus the formal charge) and the total net charges are shown on three carbon and one oxygen atom that we consider as being significant for the type of protein to which these potential PS II inhibitors bind.

In the literature there are controversial arguments concerning the structure of the quinolones, whether or not they exist in the tautomeric form as 4-hydroxyquinolines. Careful ^{13}C-NMR spectroscopy of some 2-CF_3-substituted compounds has revealed that they actually exist (in diluted $CHCl_3$-solution) as 4-hydroxyquinolines. Therefore we carried out CNDO/2 calculations on 2-CF_3-3-Cl-4-quinolone and the corresponding hydroxyquinoline (Table IX).

Table IX. Comparison of the Charges on a Selected 4-Quinolone vs the Tautomeric 4-Hydroxyquinoline

position	π	net	π	net
1	-.11	-.03	-.11	-.03
2	+.12	+.24	+.15	+.28
3	-.05	-.04	-.07	-.06
4	+.10	-.25	-.34	-.33

Accordingly the tautomeric forms do not differ significantly from each other except for the charge on the oxygen (position 4) which is not at all surprising since this atom is in different binding states in the two forms. Although we assume the hydroxy form to be present in the protein-bound state, all compounds were calculated as quinolones for better comparison with chromones and naphthoquinones in Table VIII.

Discussion

Numerous inhibitors of quite different chemical structure inhibit at the acceptor side of PS II (5,6). On general terms they bind on the herbicide binding niche on the D-1 polypeptide reaction center subunit of PS II in the same binding domain, but with specific, though overlapping binding sites. The concept was developed several years ago (2,24,25). Amino acid changes in the D-1 polypeptide identified in herbicide tolerant mutants, provide support for a folding model of the amino acid sequence of the D-1 polypeptide (10,18), based on the X-ray structure of the equivalent and homologous subunits of the bacterial reaction center (27). This allowed to phrase the concept of overlapping binding sites on more molecular terms and on interactions with specific amino acids. It was concluded that the phenol-type herbicides bind to the same stretch of amino acids (AS 210-AS 275) in the D-1 polypeptide as the urea/triazine family does, but with a different orientation in the space that is provided by the folding of that sequence. We proposed a serine (264) and histidine (215) family depending towards which amino acid the inhibitors were predominantly oriented (10) (but not necessarily bound to).

Many herbicides, like ureas and triazines of the serine family share a common substructure: a sp^2 carbon attached to a nitrogen with a free electron pair and a positive π-charge (2,18,28). Their QSARs show usually a dependence on electronic and lipophilicity parameters. Individual compounds, chemically different, displace each other from the membrane (14,29). This family looses inhibitory potency in tris-treated chloroplast membranes (7,18). Cross resistance studies of chloroplasts in triazine/triazinone or DCMU tolerant plants and algae have indicated subfamilies (reviewed in 13,18). None of these mutants are tolerant to phenol-type inhibitors.

The second group of inhibitors, the phenol-type (2) or histidine family (10) have been shown also to displace each other as well as compounds of the diuron/atrazine family from the membrane (14,29). But these phenol-type inhibitors lack the essential substructure of the diuron family, as well as the positive π-charge (3,4). Their QSARs point to a large dependence on steric parameters (2). The difference in inhibitory pattern between the two families is reviewed in references (1,10) and an example is given in Table III.

This paper on the properties of inhibitory potency of hydroxyquinolines reports on another large group of phenol-type inhibitors, as judged from their inhibitory pattern in photosynthetic electron flow. In accordance with this grouping is also the QSAR of the inhibitory potency of these hydroxyquinolines. Equation 1 (of Table VI) on the correlation of physicochemical parameters to the inhibitory value comprises two Sterimol parameters. One electronic parameter accounts for effects of substituents in position 6. The Sterimol parameters refer to the 6- and 8-position of the benzene ring (16,17). This QSAR dependence on steric parameters is like that of phenols (2).

In our earlier equation in ref. (2) dealing with p-nitrophenols (pI_{50} = -0.39 + 2.01B1-1 + 0.97B3-2) two Sterimol parameters provided excellent statistics (n = 33, r = 0.97, s = 0.25, f = 217, p <0.0001), whereas in this case we need an additional electronic parameter (σ-p). This, however, depends largely on the composition of our present data set. The p-nitrophenol series consisted of compounds which had some electronic variability ortho to the OH-group (H, Cl, Br, J, NO_2), but only there. The other ortho position carried electronically neutral groups. The hydroxyquinoline case looks different since there are two positions more available (although n is much lower, hence the worse statistical quality). Compounds number 13 and 16 were not included in the regressions because they are singularities.

In some earlier papers (3,4,) we have already emphasized the importance of quantum chemical calculations for a better understanding of the binding of PS II inhibitors to the D-1 polypeptide subunit and for the differentiation of the two inhibitor families. Calculations of the electron densities in combination with quantitative structure activity correlations enabled us to classify several groups of compounds with respect to their binding behaviour (2,3,4,18). Some other authors (9,30,31) have also employed quantum chemical calculations for the interpretation of PS II inhibitor binding though they did not use them in combination with structure activity analyses.

Here we present CNDO/2 calculations of three classes of compounds: 4-quinolones, 4-chromones, and 1,4-naphthoquinones that are all inhibitors of PS II. We compared quinolones with the tautomeric hydroxyquinolines, the latter form likely to be the form bound to the membrane as already discussed. The tautomers do not differ significantly, hence we used for symmetry reasons the quinolone form in the comparison with the chromones and naphthoquinones

As we have previously shown (3,4) the charge distributions for p-nitro-phenols and DCMU-like PS II inhibitors differ markedly: whereas the carbon atoms in ortho-position of the phenols bear a negligible charge, for good inhibitory potency a positive charge on one of the carbons or nitrogens (in 1,2,4-triazinones) in α-position to the carbonyl group is required (4,18,28). From Table VIII one can take that this charge distribution holds true also for the quinolones (hydroxyquinolines) with the probable exception of the $2-CF_3-3-OH$ compound. The chromones look very similar, but the naphthoquinones are different. Here the charges at the carbon adjacent to the carbonyl are slightly to markedly positive, particularly in the OH-derivatives. Indeed naphthoquinones may be more likely grouped among the classical inhibitors (see 21).

These quantum mechanical considerations for the two groups of PS II inhibitors fit the biochemical inhibitory pattern in the thylakoid systems. The hydroxyquinolines follow the pattern of the phenols (2), hydroxypyridines (7) and the ketonitriles (9. Furthermore, the MO calculations of the hydroxyquinolines - in comparison with the chromones and naphthoquinones - indicate the two basic elements in the two groups. They support the grouping of the inhibitors in a serine- and histidine-type (10) to indicate the orientation in the three-dimensional folding of the D-1 polypeptide either towards serine 264 or towards histidine 215. As is summarized in Figure 2, the first group has hydrogen bridges to the nitrogen backbone of the amino acid sequence of the D-1 protein and to Ser 264. It carries a positive π-charge in ortho to the carbonyl group. This group can dispense of the part below the dashed line altogether, if additional binding affinities towards not yet clearly defined amino acid residues via R_1 can compensate, like in the ureas. The basic element for the urea class proposed in (4,28) is shown in Figure 2 left. This element has recently been established very clearly also by a quite different approach via free energy differences for electron transfer between the Q_A and Q_B sites (32).

The second group of inhibitors, the phenols and hydroxyquinolines, with the hydroxyl group is pushed away from serine 264 towards histidine 215 of the D-1 protein and his 215 may undergo interactions with X for high affinity binding. Their basic element is shown in Figure 2 right.

Clearly the two basic elements in Fig. 2 reflect the quinone and hydroxyquinone form of the plasto(hydro)quinone, i.e. the natural Q_B which the inhibitors displace in the binding niche in the D-1 poly-

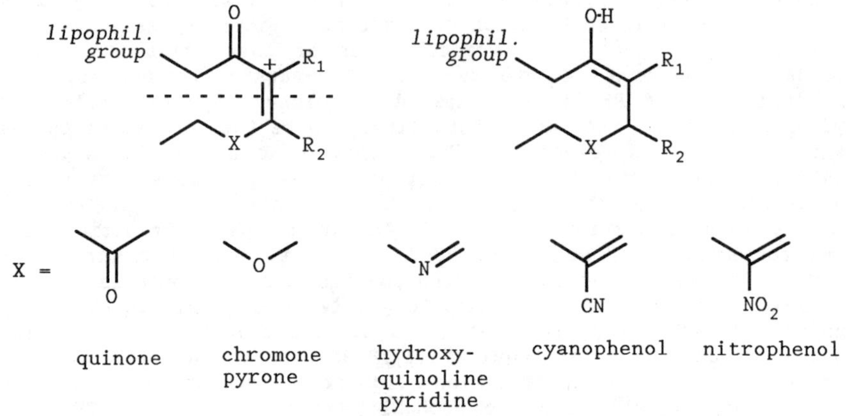

Figure 2. The essential chemical element for the two classes of photosystem II inhibitors
The upper half of the structure is oriented towards serine 264, the lower towards histidine 215 of the D-1 polypeptide (see 10).
The essential element in the serine group can be further shortened, for example in urea derivatives not even the part below the dashed line is present.

peptide. The plastoquinone is likely oriented in the binding niche of the D-1 protein with the carbonyl group near the isoprene side chain towards serine 264. The difference to the inhibitors is that plastohydroquinone looses binding efficiency to the Q_B site entirely and is expelled (which is functionally necessary in photosynthesis in order to carry the electrons from PS II to PS I). Whereas the inhibitors of the histidine-type remain tightly bound, particularly when they contain heteroatoms at the appropriate position (at X) to interact with his 215 of the D-1 polypeptide. The quinone-type inhibitors (naphthoquinones) may fall into a third class, as suggested by Oettmeier et al. (21) because the redox systems of PS II may produce the hydroquinone form with consequences on binding affinities and specificities.

Certain hydroxyquinolines have already been reported to be herbicides in the patent literature (33,34), although their mode of action had not been established. The well known inhibitor (35,36) of electron flow systems, hydroxyquinoline-N-oxide, is, of course, also a hydroxyquinoline derivative, although with a different substitution pattern to those reported here. Therefore the compound may be oriented in a turned around way in the binding niche (see 37), placing it in the serine family.

Acknowledgments

The authors are grateful to Dr. K. Ditgens, Agrochemical Research, Wuppertal, Germany, for synthesis of the hydroxyquinolines and efficient cooperation. We are also indebted to Dr. W. Lindner, Agrochemical Research Monheim, Bayer AG, for some naphthoquinones, Dr. B. Weber, Analytical Laboratories, Bayer AG, for the NMR-studies and H. Wietoska, Bochum, for biochemical measurements. Work at Bochum was supported by Fonds der Chemischen Industrie.

Literature Cited

1. Oettmeier, W.; Trebst, A. In The Oxygen Evolving System of Photosynthesis; Inoue, Y., Crofts, A.R., Govindjee, Murata, N., Renger, G. and Satoh, K., eds.; Academic Press: Tokyo, 1983, pp 411-420.
2. Trebst A.; Draber, W. In Advances in Pesticide Science; Geissbühler, Ed.; Pergamon Press: Oxford, New York, 1979; Part 2, pp 223-234.
3. Trebst, A.; Draber; Donner, W.T. In IUPAC, Pesticide Chemistry. Human Welfare and the Environment; Miyamoto, J. et al., Eds.; Pergamon Press: Oxford, 1983; pp. 85-90.
4. Trebst, A.; Draber, W.; Donner, W. Z. Naturforsch. 1984, 39c, 405-411.
5. Fedtke, C. Biochemistry and Physiology of Herbicide Action; Springer-Verlag: Berlin, Heidelberg, New York, 1982.
6. Pfister, K.; Urbach, W. Encyclopedia of Plant Physiology. Physiological Plant Ecology IV; Lange, O.L.; Nobel, P.S.; Osmond, C.B.; Ziegler, H. Eds.; Springer-Verlag: Berlin, Heidelberg, 1983; New Series, Vol. 12D, p 329.
7. Trebst, A.; Depka, B.; Ridley, S.M.; Hawkins, A.F. Z. Naturforsch. 1985, 40c, 391-399.
8. Asami, T.; Yoshida, S.; Takahashi, N. Agric. Biol. Chem. 1986, 50, 469-474.
9. Bühmann, U.; Herrmann, E.C.; Kötter, C.; Trebst, A.; Depka, B.; Wietoska, H. Z. Naturforsch. 1987, 42c, 704-712.
10. Trebst, A. Z. Naturforsch. 1987, 42c, 742-750.
11. Berthold, D.A.; Babcock, G.T.; Yocum, C.F. FEBS Lett. 1981, 134, 231.
12. Yamashita, T.; Butler, W.L. Plant Physiol., 1968, 43, 1978.
13. Wildner, G.F.; Heisterkamp, U.; Trebst, A. Photosynthesis Res. 1989, in press.
14. Oettmeier, W.; Masson, K. Pestic. Biochem. Physiol. 1980, 14, 96.
15. Quantum Chemical Program Exchange, No. 141.
16. Verloop, A.; Hoogenstraten, W.; Tipker, J. In Drug Design,; Ariens, E.J. Ed; New York, San Francisco, London 1978; Vol. VII, p 165.
17. Tipker, J; Verloop, A. In Pesticide Synthesis Through Rational Approaches; ACS Symposium Series 255; Magee, P.S.; Kohn, G.K.; Menn, J.J., Eds.; American Chemical Society: Washington D.C. 1984; pp 279-296.
18. Trebst, A.; Draber, W. Photosynthesis Res. 1986, 10, 381-392.

19. Janatkova, H.; Wildner, G.F. Biochim. Biophys. Acta 1982, 682, 227.
20. Pfister, K.; Lichtenthaler, K.H.; Burger, G.; Musso, H.; Zahn, M. Z. Naturforsch. 1981, 36c, 645-655.
21. Oettmeier, W.; Dierig, C.; Masson, K. Quant. Struct.-Act. Relat. 1986, 5, 50-54.
22. Oettmeier, W.; Godde, D.; Kunze, B.; Höfle, G. Biochim. Biophys. Acta 1981, 807, 216-219.
23. Kuwabara, M.; Yoshida, S.; Takahashi, N.; Fujita, Y. Plant & Cell Physiol. 1980, 21, 745-753.
24. Pfister, K.; Arntzen, C.J. Z. Naturforsch. 1979, 34c, 996-1009.
25. Renger, G. Z. Naturforsch. 1979, 34c, 1010-1014.
26. Pfister, K.; Steinback, K.E.; Gardner, G.; Arntzen, C.J. Proc. Natl. Acad. Sci. USA 1981, 78, 981.
27. Deisenhofer, J.; Epp, O.; Miki, K.; Huber, R.; Michel, H. Nature 1985, 318, 618-624.
28. Draber, W.; Fedtke, C. In Advances in Pesticide Science 3; Geissbühler, H., Ed; Pergamon Press: Oxford, New York 1979; p. 475.
29. Tischer, W.; Strotmann, H. Biochim. Biophys. Acta 1977, 460, 143.
30. van Assche, C.J.; Carles, P.M. In Biochemical Responses Induced by Herbicides. ASC Symposium Series 181; Moreland, D.E.; StJohn, J.B.; Hess, F.D., Eds.; American Chemical Society: Washington D.C. 1982; pp. 1-21.
31. Shipman, L.L. J. Theor. Biol. 1981, 90, 123-148.
32. Giargiacomo, K.M.; Dutton, P.L. Proc. Natl. Acad. Sci. USA, 1989, in press.
33. Pfizer Inc. New York/USA; British Patent Specification; 1 419 788; The Patent Office: London 1975.
34. Pfizer Inc. New York/USA; British Patent Specification; 1 419 789, The Patent Office: London 1975.
35. Avron, M.; Shavit, N. Biochim. Biophys. Acta 1965, 109, 317.
36. Clark, R.D.; Hind, G. Proc. Natl. Acad. Sci. USA 1983, 80, 6249.
37. Trebst, A. In Progress in Photosynthesis Research; Biggins, J., Ed; Martinus Nijhoff Publishers: Dordrecht 1987; Vol. II, pp. 109-112.

RECEIVED July 10, 1989

DRUG MECHANISMS

Drug Mechanisms: Introduction

The rational design of biologically active molecules will include investigations into the mechanism of action of the molecule. The following underlying assumption holds for all bioactive substances: *These are chemicals which bind with a receptor, itself a chemical or chemical complex. This binding initiates a series of biochemical events resulting in a biological response.* Examples of the end result include the death of an insect, parasite, bacterium, fungus, or virus; correction of a pathological state such as hypertension, cholesterolemia, gastric ulcer, schizophrenia, or epilepsy; and alleviation of a condition including pain, inflammation, allergy, cardiac arrythmia or hypertension. An overview of methodologies used in discovering or designing bioactive molecules, including examples, will be found in the first chapter of this book. This section contains six chapters describing current projects using a variety of approaches in order to elucidate some aspect of the mechanism of action including the classification of receptor subtypes, fitting the drug to the receptor, or enhancing a specific mechanism of drug action.

One of the more recently discovered receptors is one that binds the ubiquitous purine as its nucleoside, adenosine. It is now realized that adenosine acts both peripherally and centrally. Further work has shown that there are at least two receptor subtypes, A_1 and A_2. Before chemical agents can be designed to initiate a desired pharmacological response, the receptor subtype found on a particular tissue must be determined. Using a series of substituted adenosines, Tom Murray and his coworkers describe how a subtype od the adenosine receptor found in atrial tissue was classified.

The role of the parathyroid hormone (PTH) as one of the regulators of blood calcium levels has been appreciated for several years. The absence of both PTH and 1,25-dihydroxy cholecalciferol leads to hypocalcemia. With this information, Mark Goldman and Michael Rosenblatt report on the development of a series of parathyroid hormone antagonists that have the potential to reverse hypercalcemia. In so doing they were able to obtain information regarding those structural attributes of PTH responsible for both agonist and antagonist activities.

The fact that serotonin (5-hydroxytryptamine; 5-HT) is a neurotransmitter has been known for many years. Of course that means there is a serotonin receptor. Careful studies of this receptor has led to the discovery of receptor subtypes known as 5-HT_1, 5-HT_2, and 5-HT_3. Each of these receptors has its own structural requirements and, when the ligand-receptor interaction occurs, will cause a specific biological response. Richard Glennon and Mark Seggel describe the use of quantitative structure activity relationships (QSAR) as a tool to elucidate those structural attributes of phenylisopropylamines which will facilitate binding to the 5-HT_2 receptor.

The inhibitory neurotransmitter, 4-aminobutyric acid (γ-aminobutyric acid; GABA) has been studied for several years. Many studies have shown that antagonists cause convulsions whereas agonists and agents which may promote the binding of GABA to its receptor can cause a placid response. Literally hundreds of compounds have been synthesized in an attempt to gain an understanding of structural requirements of the GABA receptor. Phil Magee and Jim King have carried out a structure activity relationship analysis of binding at GABA receptor sites. Interestingly, they speculate that the search for synthetic GABA analogues likely will not lead to more active agonists.

In the continuing search for cytotoxic agents for the treatment of cancers, the overriding goal is to design molecules that show good selective toxicity for the malignant cells. One approach is to target some metabolic process associated more with the tumor cells relative to normal tissue. William Denny and his coworkers have authored a chapter illustrating how QSAR can be used in the design of hypoxia-selective drugs.

The final chapter is this section is somewhat different from the others. In the search for improved antibiotics, several hundred compounds based on the nalidixic acid structure have been synthesized in a number of laboratories. This collection of agents lends itself to a retrospective QSAR study. The authors show that sometimes even large data sets do not guarantee obtaining a clean set of equations.

It should become obvious after reading the chapters in this section that the approaches used to study drug mechanisms are no different from those used in investigating agricultural mechanisms (the preceding section) or toxicity mechanisms (the following section). After all, the chemical principles governing ligand-receptor interactions are the same independent of the biological response. What is important in the design of bioactive molecules is developing chemicals which will interact with specific receptors in order to produce a defined effect. Taken as a whole the chapters in this book should help researchers in the field of molecular design realize these goals.

JOHN H. BLOCK
College of Pharmacy
Oregon State University
Corvallis, OR 97331–3507

Chapter 16

A_1 Adenosine Receptors in the Heart

Functional and Biochemical Consequences of Activation

T. F. Murray[1], T. A. Blair[1], M. Leid[1], P. H. Franklin[1], and J. F. Siebenaller[2]

[1]College of Pharmacy, Oregon State University, Corvallis, OR 97331
[2]Department of Zoology and Physiology, Louisiana State University, Baton Rouge, LA 70803

Adenosine and its analogs exert negative inotropic, chronotropic and dromotropic effects in a variety of cardiac preparations. The cardioinhibitory effects of adenosine are mediated via an interaction with adenosine receptors of the A_1 subtype. Activation of A_1 adenosine receptors expresses negative chronotropy in embryonic chick atria. The appearance of A_1 adenosine receptors precedes the onset of physiologic sensitivity to adenosine analogs in the embryonic chick heart. These results suggest that there are physiologically inactive A_1 receptors in embryonic day 5 through day 9 chick hearts. In contrast the functional coupling of A_1 adenosine receptors to the inhibition of adenylyl cyclase is complete by embryonic day 5. The lack of temporal correlation between A_1 adenosine receptor coupling to adenylyl cyclase and the responsiveness of isolated atria to adenosine analog-induced negative chronotropy argues against the involvement of cyclic AMP in the negative chronotropic effects of adenosine. The agonist radioligand $(-)N^6-[^{125}I]$-p-hydroxyphenylisopropyladenosine ($[^{125}I]$HPIA) has been used to characterize adenosine recognition sites in porcine atrial membranes. $[^{125}I]$HPIA showed saturable binding to a homogeneous population of sites with a maximum binding capacity of 35 ± 3 fmol/mg of protein and an equilibrium dissociation constant of 2.5 ± 0.4 nM. The kinetics of $[^{125}I]$HPIA binding were consistent with those expected for a simple bimolecular reaction. Guanyl nucleotides negatively modulated $[^{125}I]$HPIA binding by increasing its rate of dissociation. Thus, porcine atria appear to represent an acceptable source for further characterization of cardiac A_1 adenosine receptors.

Adenosine regulates numerous physiological processes including platelet aggregation, lipolysis, coronary vasodilation, cardiac automaticity and contractility, and neuronal function in the central nervous system (1-3). In the brain the neurophysiologic actions of adenosine are largely inhibitory and appear to involve both presynaptic and postsynaptic sites of action (4,5). Adenosine influences adenylyl cyclase activity via an interaction with at least two distinct membrane-associated receptors. The A_1 adenosine receptor mediates an inhibition of adenylate cyclase, whereas A_2 receptor activation results in the stimulation of this enzyme (6). These receptor subtypes have been defined further by their structure-activity profiles of agonists (7-9). At A_1 adenosine receptors the rank order potency of adenosine analogs is N^6-(R-phenylisopropyladenosine (R-PIA) > N'-ethylcarboxamidoadenosine (NECA) $\geq$ 2-chloroadenosine (2-ClA) > S-PIA. In contrast at A_2 adenosine receptors the potency series for these agonists is NECA > 2ClA > R-PIA > SPIA (7,9). Adenosine analogs acting on A_1 receptors have been shown to inhibit adenylyl cyclase in brain tissue. Moreover, the adenosine-induced depression of neuronal activity in the rat hippocampus has been shown to involve an interaction with A_1 receptors (10).

The use of radioligands has recently permitted the direct labeling of A_1 adenosine receptors in brain (11), testes (12), fat cells (13) and heart (14). In these studies tritiated agonist radioligands were demonstrated to label a recognition site with A_1 adenosine receptor selectively. Analogous to numerous other neurotransmitter receptors which are G protein coupled, Al adenosine receptors can exist in two affinity states for agonists (15). Using the adenosine agonist [^{3}H]cyclohexyladenosine ([^{3}H]CHA) as a radioligand, A_1 adenosine receptors have been shown to have a broad phylogenetic distribution in brains of both primitive and advanced vertebrates (16). The presence of substantial amounts of specific [^{3}H]CHA binding in the brain of primitive vertebrate species such as the hagfish suggested a functional role for adenosine throughout the course of vertebrate evolution.

The transduction pathways and effector systems that underlie the various physiological actions of adenosine remain to be defined. The intent of this article is to provide a discussion of our attempts to understand more fully the biochemical and functional consequences of A_1 adenosine receptor mechanisms in the heart.

Ontogeny of Pharmacologic Sensitivity to Adenosine Analogs in Embryonic Chick Heart
The negative chronotropic properties of adenosine were first reported by Drury and Szent-Gyorgy in 1929 (17). Additional cardiovascular effects of adenosine include negative inotropic and dromotropic activities as well as vasodilation of coronary arteries. These effects are presumably mediated by a complex series of interactions subsequent to adenosine binding to membrane receptors. Physiological and pharmacological evidence suggests that the negative chronotropic response to adenosine and related congeners in various species is mediated by an activation of the A_1 subtype of adenosine receptor (18).

While the pharmacologic profile for the negative chronotropic response indicates the involvement of A_1 adenosine receptors, the transmembrane signaling mechanisms which underlie these physiological effects remain unclear. A number of events have been reported to occur in cardiac tissue in response to stimulation by adenosine agonists such as inhibition of basal and catecholamine-stimulated adenylyl cyclase activity (14,19), activation of an inwardly rectifying K^+ channel (20,21) and inhibition of a slow inward CA^{++} channel (22,23). In attempts to investigate the role of regulation of adenylyl cyclase activity in adenosine receptor mediated cardiac responses, some investigators have shown stimulation of adenylyl cyclase (24) whereas others have reported an inhibition of the enzyme (25-27) or no influence (28). To better define the molecular mechanisms which mediate the negative chronotropic response of adenosine analogs, we have utilized the developing embryonic chick heart which has been used extensively to study the biochemical events involved in the development and functioning of various neuroreceptors (29).

In isolated atria from embryos 12 days in ovo the adenosine analogs R-phenylisopropyladenosine, 5'-N-ethylcarboxamidoadenosine, 2-chloroadenosine and S-phenylisopropyladenosine all produced a concentration dependent negative chronotropy. The maximal effect achieved with all four adenosine analogs was a complete cessation of spontaneous beating in embryonic day 12 atria. The EC_{50} values derived from concentration-response data presented in Table I indicate that the rank order of potency was R-PIA > NECA > 2-ClA > S-PIA. R-PIA was approximately 17 fold more potent than S-PIA as an inhibitor of spontaneous atrial beating rate. This rank order of potency and diasteriomeric selectivity for R- and S-PIA is characteristic of a response mediated by an A_1 rather than an A_2 adenosine receptor.

Table I. Relative Potencies and Maximum Effects of Adenosine Analogs as Inhibitors of Spontaneous Beating Rate in Embryonic Day 12 Chick Atria

Compound	EC_{50}	Maximum Effect
	μM	%
R-PIA	0.176 ± 0.065	99.7 ± 14.2
NECA	0.516 ± 0.009	95.8 ± 0.49
2-ClA	0.789 ± 0.079	100.7 ± 2.88
S-PIA	2.94 ± 1.74	116.8 ± 20.1

In order to further verify that this negative chronotropic effect of adenosine analogs was mediated via a cell surface adenosine receptor the effects of the adenosine receptor antagonist (8-parasulfophenyl-theophylline (8-pSPT) on the response to 2-ClA was investigated. Because most methylxanthines are permeant to cell membranes and exert secondary effects that might modify actions at surface receptors, 8-pSPT, a polar methylxanthine, was chosen for these beating rate experiments. In the presence of 8-pSPT, the 2-ClA concentration-response curve was shifted to the right without

a change in slope. This shift in the concentration-response curve
is consistent with a competitive antagonism of the response to 2-
ClA by 8-pSPT (Figure 1).
 The development of an adenosine-receptor mediated negative
chronotropic response was investigated in 4- through 16-day old
embryos by measuring 2-ClA-induced inhibition of spontaneous
beating in isolated atria. Atria isolated from 4-, 5- and 6-day
old embryos were essentially unresponsive to the negative
chronotropic effects of 2-ClA using concentrations as high as 30
μM. A gradual increase in the maximum negative chronotropic
response occurred from day 5 to day 14 in ovo with day 14 atria
being fully responsive (Figure 2).
 Companion studies have employed the A_1 selective antagonist
radioligand 8-cyclopentyl-1,2[^{3}H]dipropylxanthine ([^{3}H]DPCPX) to
monitor the ontogenesis of adenosine receptors in embryonic chick
hearts. These studies have revealed that the density of A_1
adenosine receptors increases approximately 2.5 fold between
embryonic day 5 and 9 and then remains relatively stable through
day 14 in ovo (Figure 2). Considered together these data indicate
that the appearance of A_1 adenosine receptors precedes the onset
of physiologic sensitivity to adenosine analogs in the embryonic
chick heart. These results suggest that there are physiologically
inactive A_1 adenosine receptors in hearts from day 5 through day 9
embryos. The developmental change in the A_1 receptor-mediated
negative chronotropic response lagged behind the increase in the
density of [^{3}H]DPCPX recognition sites until embryonic day 12.
Thus, it is reasonable to infer that a large fractional occupancy
of chick atrial A_1 adenosine receptors is required to express
negative chronotropy during this period of embryonic development.
 The appearance of physiologically inactive A_1 adenosine
receptors in day-5 embryonic chick hearts suggested that a defect
in the functional coupling of A_1 receptors to a relevant effector
system may underlie the lack of responsiveness of atria to 2-ClA
at early embryonic ages. Given the ability of A_1 adenosine
receptor activation to affect a G_i transduced inhibition of
adenylyl cyclase, we characterized the sensitivity of adenylyl
cyclase to inhibition to CPA as a function of embryogenesis. CPA
inhibited basal adenylyl cyclase activity to a similar maximal
extent from embryonic day 5 through day 16. Thus, the functional
coupling of A_1 adenosine receptors to a GTP-dependent inhibition
of adenylyl cyclase was similar in unresponsive and responsive
embryonic hearts. The efficacy of CPA as an inhibitor of adenylyl
cyclase activity was, therefore, stable during a developmental
period when A_1 receptor density increased approximately 2.5 fold.
Hence, only a fraction of the A_1 receptors present during
embryogenesis need to be coupled to produce a maximum response
with respect to adenylyl cyclase inhibition, which is indicative
of the presence of spare receptors. These results demonstrate that
the development of sensitivity to A_1 receptor mediated negative
chronotropy is not paralleled by developmental changes in
adenosine agonist inhibition of adenylyl cyclase. Although the
negative chronotropic effect of adenosine has been suggested to be
mediated by an inhibition of adenylyl cyclase activity (26), the
lack of temporal correlation between A_1 adenosine receptor
coupling to adenylyl cyclase and the responsiveness of isolated

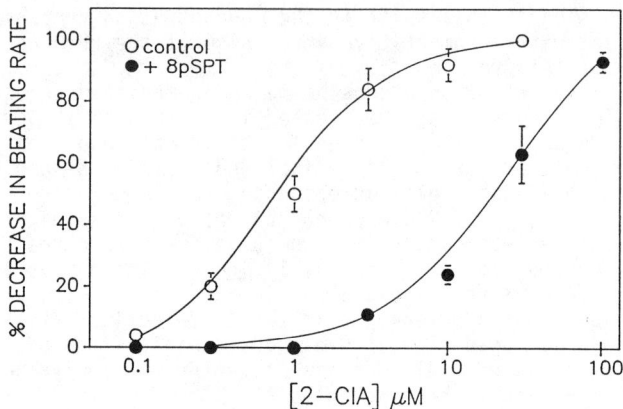

Figure 1: Antagonism of the 2-ClA-induced inhibiton of atrial beating rate by 8-p-sulfophenyltheophylline (8pSPT). Spontaneous beating rates were determined in the presence of increasing concentrations of 2-ClA alone (O) or 2-ClA and 10 μM 8pSPT (●). Each value represents the mean ± S.E. percentage inhibition of beating rate of 5-6 atria from embryos 16-days in ovo.

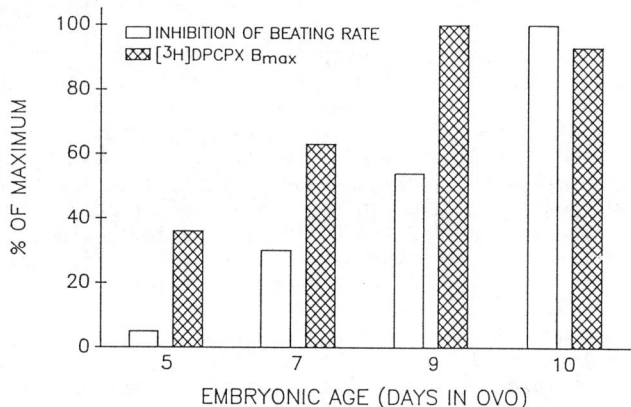

Figure 2: Relationship between the developmental profiles for the maximal sensitivity to 2-ClA-induced inhibition of atrial beating rate and the maximum number (Bmax) of A_1 adenosine receptors labeled by [^{3}H]DPCPX in embryonic chick heart membranes. Values for 2-ClA-induced suppression of atrial beating rate and [^{3}H]DPCPX Bmax were normalized to the percentage of the maximal value obtained for each parameter during embryogenesis. The normalized values for [^{3}H]DPCPX Bmax for each embryonic age depicted were calculated as the percentage of the value obtained on embryonic day 9 (74.8 ± 6.5 fmol/mg protein), while values for sensitivity to 2-ClA-induced negative chronotropy are the percentages of the maximum response which was a complete suppression of beating rate.

atria to adenosine analog-induced negative chronotropy argues against this proposal.

It has recently been demonstrated that adenosine receptors are coupled to K^+ channels via a guanine nucleotide regulatory protein in guinea pig atrial tissue (30). Moreover, adenosine and muscarinic receptors have been shown to share the same pool of cardiac K^+ channels in a single cell. The changes in physiological sensitivity of the chick atrium to adenosine agonists during embryogenesis may therefore be related to the development of functional coupling between A_1 receptors and K^+ channels via guanine nucleotide regulatory proteins. The appearance of physiologically inactive A_1 adenosine receptors on embryonic day 5 through day 9 may reflect the absence of functional coupling of these recognition sites to K^+ channels via a guanine nucleotide binding protein during this developmental period. Ongoing investigations of the developing chick heart are exploring these questions in an effort to understand more fully the molecular mechanisms which underlie the cardioinhibitory effects of adenosine.

Characterization of the Porcine Atrial A_1 Adenosine Receptor

We have undertaken a biochemical characterization of the A_1 adenosine receptor using pig heart as a tissue source. The porcine heart and circulatory system resembles that of man in many respects and this system may provide sufficient material for studies on the solubilization and purification of the receptor protein. The agonist radioligand (-)-N^6-[^{125}I]-p-hydroxyphenylisopropyladenosine ([^{125}I]HPIA) has been used to characterize adenosine recognition sites in porcine atrial membranes (31). [^{125}I]HPIA bound saturably, reversibly and with high affinity to an apparently homogenous population of recognition sites in porcine atrial membranes. The number of porcine atrial membrane recognition sites labeled by [^{125}I]HPIA is very similar to that previously reported for the adenosine receptor antagonist [^{3}H]DPCPX (35 ± 3 and 32 ± 1 fmol/mg protein, respectively) (32). Thus, the labeling of equivalent numbers of recognition sites in porcine atrial membranes by [^{125}I]HPIA and [^{3}H]DPCPX suggests that the presence of 5 mM $MgCl_2$ in the incubation medium resulted in a quantitative conversion of adenosine receptors from a low to a high affinity state. Adenosine agonists inhibited the specific binding of [^{125}I]HPIA in a manner consistent with the labeling of an A_1 adenosine receptor (Figure 4). This hypothesis is supported by the rank order potency of adenosine analogs as inhibitors of [^{125}I]HPIA binding (R-PIA > NECA ≥ S-PIA), and the finding that CV-1808 (an A_2-selective ligand) is approximately three orders of magnitude less potent than A_1-active ligands (Table II).

Toward the goal of understanding potential transduction mechanisms associated with adenosine receptor activation in porcine atria, the effect of the A_1-selective agonist cyclopentyladenosine (CPA) on adenylyl cyclase activity was evaluated. CPA (1 μM) inhibited basal adenylyl cyclase activity in a GTP-dependent manner with maximal inhibition by CPA occurring at a GTP concentration of 100 μM (Figure 3). The inhibition of

Table II. Adenosine Receptor Agonists Inhibiting Specific Binding of [^{125}I]HPIA to Membrane-Bound and Solubilized Porcine Atrial Adenosine Receptors [a]

Compound	Membrane-Bound Receptor		Solubilized Receptor	
	K_I (nM)	slope factor	K_I	slope factor
(R)-PIA	1.38 ± 0.60	0.82 ± 0.17	2.2 ± 0.5	1.01 ± 0.005
NECA	27 ± 12	0.83 ± 0.14	49 ± 11	1.05 ± 0.06
(S)-PIA	30 ± 12	0.83 ± 0.14	94 ± 20	0.95 ± 0.04
CV-1808	2310 ± 355	0.74 ± 0.06	5640 ± 1940	1.00 ± 0.1

[a] Receptor and [^{125}I]HPIA concentration were approximately 40 and 750 pM, respectively. Parameter estimates were obtained using EBDA (Elseview-Biosoft, Cambridge, UK) and represent the mean ± S.E.M. of 3-5 Experiments.

porcine atrial adenylyl cyclase activity by CPA was reversed by the A_1-selective antagonist DPCPX. Although the magnitude of adenylyl cyclase inhibition induced by CPA was modest and varied between 5 to 20% for individual membrane preparations, both the GTP-dependence and DPCPX reversal of CPA-mediated inhibition of adenylyl cyclase were highly reproducible. These findings, therefore, provide functional evidence for the existence of A_1 adenosine receptors in porcine atrial membranes. The porcine atrial A_1 adenosine receptor has been solubilized using a mixed detergent system (0.4% w/v digitonin and 0.08% w/v cholate) and biochemically characterized (33). Solubilization in this mixed detergent system resulted in a 2.5 fold enrichment of A_1 adenosine receptor specific activity over that observed in experiments with the membrane-bound receptor of porcine atria. The association of [^{125}I]HPIA with solubilized cardiac A_1 adenosine receptors apparently involves a simple bimolecular reaction which is consistent with kinetic behavior observed in experiments using membrane-bound receptors (31). The pharmacological signature of the solubilized cardiac adenosine receptor was assessed in equilibrium competition assays of [^{125}I]HPIA binding (Figure 4). In this series of experiments the A_1 adenosine receptor-selective agonist R-PIA was more potent as an inhibitor of [^{125}I]HPIA binding to the solubilized receptor than its less active diastereomer, S-PIA. This agonist rank order potency diastereomeric affinity ratio is consistent with the labeling of an A_1 adenosine receptor subtype. Thus, these findings suggest that the pharmacological profile of the porcine atrial A_1 receptor is well preserved in this detergent system.

The affinity differences for the diastereomers of PIA are often used as criteria for adenosine receptor subtype classification. The results summarized in Table III demonstrate that the diastereomeric selectivity for R- and S-PIA at A_1 adenosine receptors is remarkably stable in a variety of radioligand binding and physiological assays. The largest difference in affinity was observed in the solubilized porcine atrial A_1 adenosine receptor where R-PIA was approximately 43 fold more potent than S-PIA as an inhibitor of [^{125}I]HPIA binding. These findings suggest that the lipid milieu in which the receptor

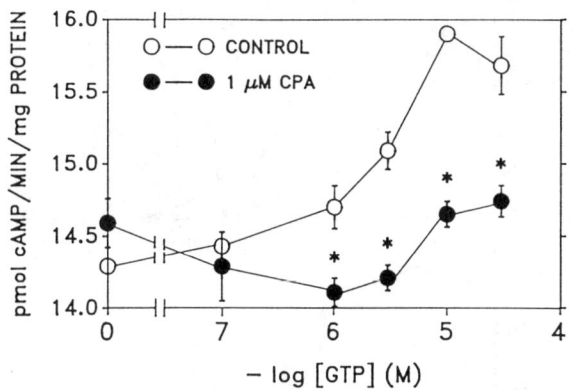

Figure 3: Inhibition of basal porcine atrial adenylyl cyclase activity by 1 μM CPA as a function of GTP concentration. Using α-[^{32}P]ATP as a substrate, reactions were carried out for 20 min at 30°C and terminated by addition of a stopping solution (2% SDS, 45 mM ATP and 1.3 mM cAMP) followed by boiling samples for 3 min. [^{32}P]cAMP was separated from α-[^{32}P]ATP by sequential chromatography on Dowex A6 50W-X4 (400 mesh) and alumina columns. Recovery of [^{32}P]cAMP was monitored by addition of ≈ 10,000 cpm [^{3}H]cAMP to samples prior to boiling. Radioactivity was quantified by use of a Beckman 6800 LSC.

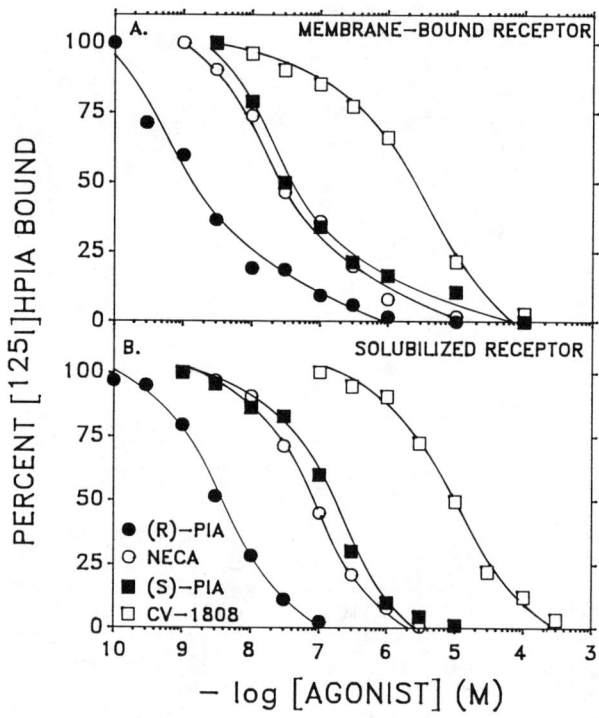

Figure 4: Adenosine receptor agonist inhibition of the specific binding of [^{125}I]HPIA to membrane bound (A) and solubilized (B) porcine atrial adenosine receptor preparations. Each point represents the mean of 3-9 experiments and the lines drawn are the best fits derived from non-linear least squares regression analysis using EBDA software.

protein is embedded is an important determinant of the R- and S-PIA diastereomeric affinity ratio.

Table III. Comparison of the Affinity Ratios for R- and S-Phenylisopropyladenosine (PIA) Derived from Both Radioligand Binding Assays and Functional Measures of A_1 Adenosine Receptor Activation

Assay	S-PIA/R-PIA Affinity Ratio
[^{125}I]HPIA binding assay	
Bovine brain membranes	15.6
Piscine brain membranes	15.5
Porcine atrial membranes	21.7
Porcine atrial extract	42.7
Physiological assay	
Inhibition of adenylyl cyclase in chick heart	21.5
Negative chronotropy in chick atria	16.7
Intracerebral anticonvulsant effect	10.2

Conclusions

The results of both radioligand binding and functional assays indicate that the adenosine receptor of atrial tissue displays a pharmacologic profile characteristic of the A_1 subtype. Activation of these A_1 adenosine receptors expresses negative chronotropy in embryonic chick atria. The A_1 receptors in both avian and porcine atria are negatively coupled to adenylyl cyclase; however, inhibition of adenylyl cyclase does not appear to mediate the effects of adenosine analogs on atrial beating rate. The elucidation of the particular transduction systems and ion channel coupling which underlie the negative chronotropic influence of A_1 adenosine receptor activation awaits further research in this area.

Acknowledgments

The excellent word processing assistance of Ms. Elaine Luttrull is gratefully acknowledged. This work was supported by a grant from the Oregon Affiliate of the American Heart Association.

Literature Cited

1. Burnstock, G.; Brown, G.M. In: Purinergic receptors. Burnstock, G., Ed.; Chapman and Hall. London New York, 1977; p 1045.
2. Phillis, J.W.; Wu, P.H. Prog. Neurobiol. 1981, 16, 187-239.
3. Snyder, S.H. Ann. Rev. Neurosci. 1985, 8, 103-124.
4. Fredholm, B.B.; Hedqvist, R. Biochem. Pharmacol. 1980, 29, 1635-1643.
5. Lee, K.S.; Schubert, P.; Heinemann, U. Brain Res. 1984, 321, 160-164.
6. Daly, J.W.; Bruns, R.F.; Snyder, S.H. Life Sci. 1981, 28, 2083-2097.
7. Daly, J.W. In: Physiology and Pharmacology of Adenosine Derivatives Daly, J.W.; Kuroda, Y.; Phillis, J.W.; Shimizu, H.; Ui, M., Eds. Raven, New York; 1983, p 275.

8. Wolff, J.; Londos, C.; Cooper, D.M.F. Adv. Cyclic Nucleotide. Res. 1981, 14, 199-214.
9. Stone, T.W. In: Purines: Pharmacology and Physiological Roles, Stone, T.W. Ed.; VCH Publishers, Deerfield Beach, Florida, 1985; p 1.
10. Reddington, M.; Lee, K.; Schubert, P. Neurosci. Lett. 1982, 28, 275-279.
11. Bruns, R.F.; Daly, J.W.; Snyder, S.H. Proc. Natl. Acad. Sci. USA. 1980, 77, 5547-5551.
12. Murphy, K.M.M.; Snyder, S.H. Mol. Pharmacol. 1982, 22, 250-257.
13. Trost, T.; Schwabe, U. Mol. Pharmacol. 1981, 19, 228-235.
14. Hosey, M.M.; McMahon, K.G.; Green, R.D. J. Mol. Cell. Cardiol. 1984, 16, 931-942.
15. Lohse, M.J.; Lenschow, U.; Schwabe, U. Mol. Pharmacol. 1984, 26, 1-9.
16. Siebenaller, J.F.; Murray, T.F. Biochem. Biophys. Res. Commun. 1986, 137, 182-189.
17. Drury, A.N.; Szent-Gyorgyi, A. J. Physiol. (Lond.) 1929, 68, 213-226.
18. Haleen, S.; Evans, D. Life Sci. 1985, 36, 127-137.
19. Schutz, W.; Freissmuth, M.; Haussleithner, V.; Tuisl, E. Naunyn-Schmiedeberg's Arch. Pharmacol. 1986, 333, 156-162.
20. Jochem, G.; Nawrath, H. Experientia. 1983, 39, 1347-49.
21. Belardinelli, L.; Isenberg, G. Am. J. Physiol. 1983, 244, H734-H737.
22. Isenberg, G.; Belardinelli, L. Circ. Res. 1984, 55, 309-325.
23. Caparrotta, L.; Fassina, B.; Froldi, G.; Poja, R. Br. J. Pharmacol. 1987, 90, 23-30.
24. Anand-Srivastava, M. Arch. Biochem. Biophys. 1985, 243, 439-446.
25. Schrader, J.; Baumann, G.; Gerlach, E. Pflugers Arch. 1977, 372, 29-35.
26. Leung, E.; Johnston, C.; Woodcock, E. J. Cardiovas. Pharmacol. 1986, 8, 1003-1008.
27. Martens, D.; Lohse, M.; Rauch, B.; Schwabe, U. Naunyn-Schmiedeberg's Arch. Pharmacol. 1987, 336, 342-348.
28. Bruckner, R.; Fenner, A.; Meyer, U.; Nobis, T.; Schmitz, W.; Scholz, H. J. Pharmacol. Exp. Ther. 1985, 234, 766-774.
29. Pappano, J.W.; A. Pharmacol. Rev. 1977, 29, 3-33.
30. Kurachi, Y.; Nakajima, T.; Sugimoto, T. Pflugers Arch. 1986, 407, 264-274.
31. Leid, M.; Schimerlick, M.; Murray, T. Mol. Pharmacol. 1988, 34, 334-339.
32. Leid, M.; Franklin, P.; Murray, T. Eur. J. Pharmacol. 1988, 147, 141-144.
33. Leid, M.; Schimerlick, M.; Murray, T. Mol. Pharmacol. in press.

RECEIVED March 21, 1989

Chapter 17

Therapeutic Potential for Parathyroid Hormone Antagonists

Mark E. Goldman and Michael Rosenblatt

Department of Biological Research and Molecular Biology, Merck Sharp and Dohme Research Laboratories, West Point, PA 19486

> Due to a lack of selective pharmacological agents, therapy of hyperparathyroid hypercalcemia and other hypercalcemic disorders has largely been considered a medical problem managed symptomatically or surgically by clinicians. Based upon recent gains in the understanding of parathyroid hormone (PTH) structure-activity relationships and the pathogenesis of hyperparathyroidism and hypercalcemia, the rational development of drugs for treating these disorders should now be considered. Potent and selective PTH antagonists, when available, may prove useful in treating not only primary hyperparathyroidism and post-renal transplant secondary hyperparathyroidism, but also hypercalcemia of malignancy and osteoporosis.

Hypercalcemic disorders caused by increased circulating levels of PTH or PTH-like peptides are common, adversely affect most organs and are potentially life-threatening. Selective therapeutic agents, however, are not currently available. To expedite the development of such agents, an understanding of physiological processes regulating blood calcium levels as well as endocrine changes during hypercalcemic states is required. Structure-activity relationship studies have been initiated based upon knowledge of the amino acid sequences of PTH and PTH-like peptides. As a result of peptide truncation, substitution of natural and synthetic amino acids and combination of various motifs from PTH and PTH-like peptides, novel PTH antagonists possessing PTH inhibitory activities _in vitro_ and _in vivo_ have been developed.

NORMAL CALCIUM HOMEOSTASIS
The normal maintenance of blood calcium levels within a narrow range is regulated by two hormones in humans: PTH and 1,25 dihydroxy vitamin D_3 (1,25(OH)$_2$VitD$_3$). In the absence of these hormones, blood calcium levels fall. The integrated actions of PTH and 1,25(OH)$_2$VitD$_3$ are responsible for maintaining calcium levels in the normal range. Calcitonin's actions in many aspects oppose those

of PTH. Although it is responsible for decreasing calcium levels in lower species, the physiological role of calcitonin in humans is uncertain.

As blood calcium levels fall, PTH is released from the parathyroid gland and acts directly upon the kidney and bone as well as indirectly on the intestines. The renal actions of PTH, mostly mediated by cAMP, are summarized in Table I. These direct renal PTH actions, therefore, elevate blood calcium levels by enhancing calcium reabsorption and by lowering blood phosphate concentrations and pH. The latter two actions of PTH allow calcium levels to rise to a greater extent without precipitating calcium phosphate.

Table I. Renal Actions of Parathyroid Hormone

1. $1,25(OH)_2VitD_3$-dependent stimulation of renal tubular calcium reabsorption
2. Inhibition of phosphate reabsorption and stimulation of proximal tubular phosphate excretion
3. Inhibition of sodium-hydrogen exchange in the proximal tubules resulting in the inhibition of hydrogen ion secretion and enhanced excretion of bicarbonate.
4. Activation of proximal convoluted tubule $25(OH)VitD_3$ 1-α-hydroxylase

Bone is the body's calcium reservoir. PTH stimulates bone resorption leading to the dissolution of hydroxyapatite and release of calcium and phosphate into the blood. This action of PTH appears to be the major mechanism for the rapid elevation of blood calcium levels. PTH also maintains blood calcium levels by promoting calcium reabsorption from the renal tubules.

Indirect calcium level-increasing actions of PTH on the gut are mediated through the vitamin D system. Following synthesis in the skin and hydroxylation by the liver, $25(OH)VitD_3$ is converted to its biologically-active form, $1,25(OH)_2VitD_3$, by renal 1-α-hydroxylase (1). This is the rate limiting step in $1,25(OH)_2VitD_3$ synthesis and is mediated by the stimulation of adenylate cyclase-linked PTH receptors located in the proximal convoluted tubule. $1,25(OH)_2VitD_3$ acts upon the intestines to increase calcium absorption through stimulation of a vitamin D-dependent calcium binding protein. Although vitamin D production is not directly regulated by acute changes in calcium homeostasis, the $1,25(OH)_2VitD_3$ and PTH systems are intertwined. $1,25(OH)_2$-$VitD_3$ modulates the calcium set-point for PTH secretion by stimulating parathyroid gland vitamin D receptors which decrease PTH gene transcription and modulate PTH release.

PTH is synthesized in the parathyroid gland as an 115-amino acid gene product (preproPTH) and processed by the endoplasmic reticulum and golgi to an 84-amino acid linear biologically-active peptide (for a review, see (2)). PreproPTH synthesis is regulated at both transcriptional and post-transcriptional levels by extracellular calcium and $1,25(OH)_2VitD_3$ (1,3,4). In the presence of elevated concentrations of calcium, the intracellular degradation of PTH in the parathyroid gland is greater than at low calcium concentrations. Release of various endogenous fragments of PTH-(1--84) from the parathyroid gland is partially responsible for the heterogeneity of circulating PTH molecules. PTH is further metabolized upon release from the parathyroid gland by the kidney and liver. The major circulating forms of PTH in the bloodstream are biologically inactive.

The classical biological activities of PTH reside in the N-terminal 34 amino acids (5). The remainder of the PTH molecule may be responsible for prolonging the half-life of PTH in vivo. In certain disease states, mid-portion and C-terminal fragments may accumulate in the circulation and have been hypothesized to inhibit erythropoiesis, cardiac function, nerve conduction, red blood cell survival time and glucose tolerance (6).

The main signal transduction mechanism for PTH receptor stimulation is the cAMP system. Stimulation of renal or bone PTH receptors causes an elevation of membrane-bound and guanyl nucleotide-dependent adenylate cyclase activity resulting in the enhanced formation of cAMP from ATP. Recent studies suggest that PTH may also stimulate inositol triphosphate and diacylglycerol production in renal tissue (7).

PATHOPHYSIOLOGY OF HYPERPARATHYROID HYPERCALCEMIA

Symptoms and Side Effects of Hypercalcemia

Hypercalcemia affects most organ systems. Depending upon the age of the patient, symptoms may easily be recognized as an indication of hypercalcemia or may be confused with other disease states.

Central nervous system symptoms of hypercalcemia include lethargy, decreased cognitive functions, depression, confusion and irritability. The changes in cognitive abilities can vary from simple memory loss to dementia or psychosis. These hypercalcemia-induced personality changes may be interpreted wrongly as normal age-related symptoms, especially if they are slow in appearing. Psychiatric symptoms were more frequent in elderly patients (8). During severe hypercalcemic crises, such as in advanced malignancies, stupor or coma is not uncommon.

There is a strong correlation between hypercalcemia and hypertension (9,10). Black males may have a higher risk for this side effect (11). Caution must be used when treating hypercalcemia-induced hypertension since thiazide diuretics can transiently produce hypercalcemia. With long-standing hypercalcemia, calcium may be deposited on the cardiac valves and coronary arteries (12). Another cardiovascular side effect of hypercalcemia is lengthening of the ECG QT interval.

Demineralization of bone accompanies hypercalcemia, producing osteopenic skeletal changes. Manifestations include osteoporosis, osteomalacia and osteitis fibrosa et cystica (13). As a result of decreased bone mass, fractures of the femoral neck and spontaneous fractures of the vertebrae are most frequent. Morphometric changes in metacarpal bones provide quantifiable data for the progression and treatment of bone disease (14,15). Other skeletal effects include arthralgia, periarticular calcification, gout or pseudogout and loose teeth.

Renal effects of hypercalcemia include reduced glomerular filtration rate (GFR), polyuria, nephrocalcinosis, and renal stone disease. Hypercalcemia causes renal vasoconstriction which may contribute to decreased GFR. The hypercalcemia-induced polyuria results from 1) an impairment of active transport of NaCl in the loop of Henle, distal tubule and collecting duct and 2) an inhibition of vasopressin-facilitated absorption of water in the distal nephron. As a direct result of the polyuria, many side effects including polydipsia, thirst, nocturia and dehydration are common. Precipitation of calcium salts within the kidney leads to chronic inflammatory reactions (nephrocalcinosis), fibrosis, renal impairment, nephrolithiasis and urolithiasis. Further renal damage may occur indirectly from hypertension.

The gastrointestinal manifestations of hypercalcemia include abdominal discomfort as a result of peptic ulceration or pancreatitis, pancreatic calcification, vomiting, anorexia, constipation and weight loss.

Neuromuscular symptoms vary from fatigue to hypotonia and ataxia. These symptoms may increase the risk of bone fracture.

As with other organ systems such as the kidney and pancreas, elevated calcium levels may cause calcium deposition in the eye as demonstrated by conjunctival calcium deposition, conjunctivitis and band keratinopathy.

Causes of Hyperparathyroid Hypercalcemia

Primary Hyperparathyroidism (1HPT). Prior to the routine analysis of serum calcium levels, 1HPT was usually diagnosed by the radiologic presence of osteitis fibrosa et cystica or the presence of renal stones. During the last two decades, 1HPT has become recognized as the most common form of hypercalcemia in the general population and is usually diagnosed at the asymptomatic stage. The most common causes of 1HPT are parathyroid adenomas (80%), hyperplasia (15%) and carcinoma (1-5%) (16,17).

In the hyperparathyroid state, the parathyroid gland does not become autonomous, although the regulation of PTH secretion by calcium is altered. Murray and co-workers (18) demonstrated that modulation of serum calcium levels in 1HPT patients by EDTA or calcium infusion still resulted in the appropriate changes in PTH secretion. In addition, studies with normal and adenomatous parathyroid cells in culture have shown that 1HPT may cause changes in the calcium inhibitory set point as well as an inability for calcium to completely suppress PTH secretion in spite of reduced PTH mRNA levels (3,19).

Multiple Endocrine Neoplasia (MEN). Several polyglandular disorders are known which result in an autonomous hyperfunction of two or more endocrine glands. These disorders are inherited as autosomal dominant traits. Depending upon which glands are dysfunctioning at the time of diagnosis, the clinical presentation is variable.

In MEN type 1 (MEN1), multiple tumors of the parathyroid gland, anterior pituitary gland and pancreatic islets are common. Parathyroid gland involvement in MEN1 is the most common manifestation and is seen in 90-95% of patients.

MEN types 2 and 3 (MEN2, MEN3) are entirely distinct syndromes from MEN1. MEN2 is most frequently characterized by medullary carcinoma of the thyroid gland. MEN3 resembles MEN2, although there are several differences.

Secondary Hyperparathyroidism (2HPT). As with 1HPT, 2HPT is a metabolic disorder involving enhanced secretion of PTH. In the case of 2HPT, however, the HPT is a compensatory parathyroid gland adaptation resulting from prolonged tendency toward hypocalcemia or relative resistance to the metabolic actions of PTH; it occurs in an attempt to protect calcium homeostasis. Another distinction between the 2 forms of HPT is that 1HPT is apparently irreversible whereas 2HPT is usually reversible (given sufficient time) after removal of the initial stimulus.

Although 2HPT may occur with osteomalacia, vitamin D deficiency or pseudohypoparathyroidism, by far, the most common cause is chronic progressive renal disease. Loss of functional nephrons is a common occurrence in the aging process and following hypertensive, inflammatory or infectious renal injury. When the GFR falls below 50-75% of normal, a sequence of events resulting in early 2HPT has been hypothesized to occur (for reviews, see [20-22]). This sequence can be briefly summarized as follows: 1) decreased GFR causes a reduced elimination of phosphate, 2) as a result of elevated intracellular and extracellular phosphate levels, there is a stoichiometric decrease in serum ionized calcium levels and a decrease in the 1-α-hydroxylation of $25(OH)VitD_3$, 3) this hypocalcemia and hypovitaminemia D then induce the enhanced secretion of PTH which 4) reduces the tubular reabsorption of phosphate, thus increasing phosphate excretion leading to 5) changes towards normal in serum phosphate, $1,25(OH)_2VitD_3$ and calcium levels, although mild hypocalcemia usually persists.

As renal disease progresses, there is a proportional increase in serum PTH levels by several mechanisms. First, as a direct response to hypocalcemia, PTH secretion is enhanced. Second, since the kidney is the sole source of 1-α-hydroxylase, progressive renal disease reduces the formation of $1,25(OH)_2VitD_3$ leading to decreased intestinal absorption of calcium. Third, decreased $1,25(OH)_2VitD_3$ levels may result in elevated PTH secretion by preventing $1,25(OH)_2VitD_3$ feedback suppression of PTH synthesis. Fourth, since the kidney is a principal organ for PTH metabolism, renal failure may also contribute to elevated serum PTH levels by decreasing PTH breakdown ([20,21]).

Recent studies suggest that decreased levels of $1,25(OH)_2$-$VitD_3$ may be as significant as hypocalcemia for inducing and maintaining 2HPT ([22,23]). Intravenous $1,25(OH)_2VitD_3$ therapy has been shown to be beneficial for reducing the bone manifestations of 2HPT including metastatic calcification, osteitis fibrosa, renal osteodystrophy and spontaneous fractures. In advanced 2HPT, however, this treatment may cause vitamin D intoxication leading to hypercalcemia, myocardial and pulmonary calcinosis, arterial calcification, further metastatic calcification and death.

Many endocrine tissues possess the ability to store large quantities of hormone for release upon demand. In contrast, the parathyroid gland stores little PTH. As a result, the turnover rate of PTH is high even under basal conditions. In 2HPT, therefore, parathyroid hyperplasia results as the parathyroid gland attempts to keep up with demands for enhanced hormone production. As the parathyroid tissue enlarges, changes in cellular control mechanisms have been demonstrated, including an increase in the value of the calcium set-point and a decreased maximal suppression by calcium of PTH secretion (21,24).

The obvious mechanism to reverse 2HPT is to remove the stimuli that caused the syndrome. Although hemodialysis coupled with carefully controlled management of the patient will provide short-term benefit, renal transplantation is eventually required in most cases. Following successful transplantation, serum phosphate and $1,25(OH)_2VitD_3$ levels will often return to normal very rapidly. At this stage, however, the hyperplastic parathyroid gland continues to secrete large quantities of PTH. Now the increased PTH secretion produces frank hypercalcemia which threatens the newly transplanted kidney.

Since parathyroid gland hyperplasia in 2HPT is the direct result of increased and continuous demand for PTH secretion, gradual involution of the gland will usually occur upon restoration of normal renal function following the transplant. The hypoplastic transformation, however, is a slow process since parathyroid cells do not turnover rapidly (25). Post-renal transplant hypercalcemia is common and is usually resolved within 1-2 years. If the hypercalcemia is mild to moderate and does not cause further skeletal or renal deterioration, no steps are taken to manage this problem. In cases where the pretransplant control of 2HPT was poor, there is a great chance for severe hypercalcemia with possible damage to the renal allograft (25). In this situation or when spontaneous resolution of the PTH-induced hypercalcemia does not occur within 2-3 years, parathyroidectomy must be considered.

Humoral Hypercalcemia of Malignancy

Hypercalcemia of malignancy is a common occurrence in solid tumors of the lung and breast as well as multiple myeloma and adult T-cell lymphoma/leukemia (26). The hypercalcemia associated with breast cancer is usually seen in late stages of the disease in patients with extensive bone metastases. In squamous cell carcinoma of the lung or kidney, however, hypercalcemia is not correlated with disease stage and is not necessarily associated with bone metastases. The hypercalcemia results from increased bone resorption, decreased bone formation and increased renal tubular calcium reabsorption. These findings suggest that some tumors may secrete humoral factors with PTH-like actions.

The humoral hypercalcemia of malignancy hypothesis states that an osteolytic non-PTH substance is secreted by certain tumors and, in an endocrine manner, is transported from tumor to bone through the bloodstream. The evidence for this hypothesis is that 1) bone destruction occurs in patients without bone metastases, 2) serum PTH levels in these patients were usually normal, 3) PTH mRNA was absent in tumors (27), and 4) tumor extracts from hypercalcemic patients enhanced bone cell adenylate cyclase activity and phosphate transport in kidney epithelial cells (28-30).

A 141-amino acid tumor-secreted, PTH-like peptide has recently been identified from human carcinomas based upon partial sequence analysis and cDNA cloning (31-33). Further studies demonstrated the presence of two forms of mRNA encoding for this protein which originate from a single gene by alternate splicing mechanisms (34). The 3' untranslated region is homologous to the corresponding domain of the c-myc proto-oncogene (34).

Besides tumors, the only other tissues known to produce this hypercalcemia factor (HCF) are human keratinocytes, fetal parathyroid gland, placenta and mammary gland (35-38). The mRNA coding for this protein were expressed in rat mammary tissue only during lactation and the response changed rapidly as a function of the suckling stimulus (35). In the mammary gland, HCF may play a physiologic role in the mobilization and/or transfer of calcium into milk.

The N-terminal portion of the 141 amino acid human HCF (hHCF) bears sequence homology with the corresponding regions of human and bovine PTH (hPTH, bPTH):

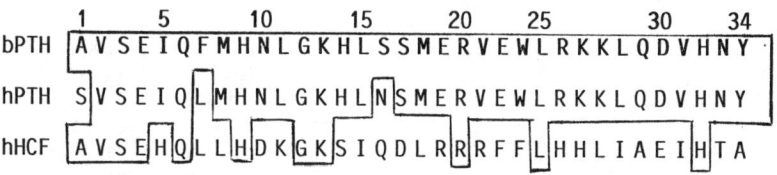

Within the first 13 amino acids, there is 60% sequence homology between PTH and hHCF. After this region, the homology is minimal.

To determine if HCF is responsible for mediating some or all of the components of the humoral hypercalcemia of malignancy syndrome, an N-terminal fragment of this peptide (hHCF-(1--34)NH$_2$) was synthesized and its biological properties investigated. The tumor factor caused a concentration-dependent inhibition of radiolabeled PTH binding to bovine renal cortical membranes, in vitro, with a potency similar to 34-amino acid PTH fragments (39). This peptide also enhanced adenylate cyclase activity in renal membranes, bone cells and isolated, perfused kidneys (39,40). Using bone cells, dexamethasone pretreatment modulated the actions of both hHCF and PTH similarly (41). Both peptides also inhibited bone cell alkaline phosphatase activity in vitro (41).

In vivo, using the thyroparathyroidectomized rat model system, hHCF-(1--34)NH$_2$ caused hypercalcemia with an apparent potency 6-10-fold that of bPTH-(1--84) (39). hHCF-(1--34)NH$_2$ also reduced serum phosphate and elevated 1,25(OH)$_2$VitD$_3$ levels. Following a 48 hr infusion of hHCF-(1--34)NH$_2$, histological changes in bone consistent with the stimulation of bone PTH receptors were evident (42). Similarly, using the thyroparathyroidectomized/nephrectomized/low dietary calcium rat bone model, hHCF-(1--34)NH$_2$ was equipotent with bPTH-(1--34) and more potent than bPTH-(1--84) for causing hypercalcemia (39).

Further support for the PTH-like actions of hHCF were gained using a PTH antagonist. The prototypical PTH antagonist, [Tyr^{34}]bPTH-(7--34)NH$_2$ blocked completely the adenylate cyclase-enhancing actions of hHCF-(1--34)NH$_2$ on both renal membranes and bone cells in vitro (43). This PTH antagonist also prevented the actions of hHCF-(1--34)NH$_2$ in vivo (Horiuchi et al., submitted).

In summary, these results suggest that at least some and perhaps all of the actions of hHCF are mediated through stimulation of the PTH receptor. Indeed, recent studies using bone cells indicate that both hHCF and PTH recognize the same plasma membrane receptor (44,45). As the actions of these structurally-distinct peptides are compared on other tissues and biological parameters, it is possible that PTH/HCF receptor subtypes may be identified.

Other possible humoral mediators of bone resorption fall into several categories including transforming growth factors, prostaglandin E's, cytokines, 1,25(OH)$_2$VitD$_3$ and colony stimulating factors (for a review, see (26)). With the exception of PTH-like peptides, evidence for a direct role of any individual factor is limited. It is possible that several factors may act together to cause humoral hypercalcemia of malignancy.

Osteoporosis and Hyperparathyroidism

Osteoporosis is a specific form of osteopenia in which the bone histology is normal but there is a quantitative decrease in the amount of bone (13). Osteoporotic reduction of bone mass, especially in the elderly, is a common problem resulting in an increased frequency of pathological fractures and sometimes death from ensuing complications. Two osteoporotic syndromes have been hypothesized. Type I or postmenopausal osteoporosis is 6-fold more common in women than men and results mainly in trabecular bone loss. Type I osteoporosis is due primarily to estrogen deficiency beginning at menopause. Although type I osteoporosis has not been linked directly to elevated parathyroid function, it is possible that certain osteoporotic patients may secrete factors(s) which potentiate the actions of PTH (46). Based upon the recent identification of hHCF (see above), it is conceivable that PTH-like factors may be released which stimulate PTH receptors in certain tissues.

Type II, involutional or senile osteoporosis occurs later in life than type I, has a closer ratio between sexes, results in both trabecular and cortical bone loss and is related to increased PTH secretion (26). Under these circumstances, osteoporosis may essentially be a side-effect of 1HPT or 2HPT (15,26,46-48). Type II osteoporosis is managed, therefore, by correcting the underlying cause of the HPT.

Incidence of Hypercalcemia and Hyperparathyroid Disorders

Since the introduction of modern procedures for the routine automated analysis of serum calcium levels, hypercalcemia has become recognized as a relatively common clinical problem. 1HPT is the most common cause of hypercalcemia in the general population and the incidence increases with age (26). In men or women under the age of 40, the annual incidence of 1HPT was found to be 10 cases/10,000 population (49). This rate rose steadily with age to 92 and 188 cases/10,000 in men and women, respectively, over 60 years of age.

In another study of 207 hypercalcemic individuals, 111 (54%) were diagnosed as having 1HPT and again, the greatest risk group was elderly women (50). Hypercalcemia identified from routine serum calcium monitoring was the initial indication of 1HPT in over 50% of the patients (49-51). The annual age adjusted incidence of 1HPT was 27.7/100,000 population of the U.S. (49). In Sweden, the prevalence of 1HPT has been estimated, in one study, to be 520 cases/100,000 population (52). Extrapolation to the U.S. predicts that over 1 million cases of 1HPT may exist (49).

In hospitalized patients, hypercalcemia of malignancy is the major cause of hypercalcemia (26,51,53,54). In two separate studies, 9% of all cancer patients had hypercalcemia of malignancy (55,56). Hypercalcemia of malignancy is most common in solid tumors such as carcinomas of the lung, breast, kidney, pancreas and ovary, but also occurs with multiple myeloma and adult T-cell lymphoma/leukemia. It has been estimated that there is approximately an equal distribution between humoral factors and metastatic bone tumors for causing hypercalcemia of malignancy (26,57,58).

With advanced age, there is an increased impairment of renal function which has been shown to correlate with elevated PTH levels and radiologic bone disease (15). When the impairment of function reaches the renal failure stage, 2HPT has been found in 80-94% of patients (59,60). While some patients are asymptomatic, osteodystrophy is often manifested as spontaneous fractures (femoral neck, vertebrae) or bone and joint pains (59). Chronic hemodialysis may cause an exacerbation of the 2HPT and increase the risk of bone disease (61,62).

Twenty to 66% of post-renal transplant patients displayed persistent hypercalcemia and/or 2HPT (14,63-66). In most cases, the patients became normocalcemic and euparathyroid within 1-3 years. Only a small percentage (3.2-7%) required parathyroidectomy as a result of impaired renal function or progression of bone disease (14,60,63,65,66).

Following initiation of antihypertensive therapy with thiazide diuretics, transient hypercalcemia has been seen in over one-third of patients (67). Two percent of patients receiving long-term thiazide diuretics administration had persistent hypercalcemia (68). In the elderly (especially women), combined administration of thiazides with vitamin D supplements (for osteoporosis) can have synergistic effects on the elevation of serum calcium levels resulting in severe hypercalcemia (69). Similarly, if the patient is predisposed to hypercalcemia (1HPT, 2HPT or immobilization), thiazides can precipitate significant and sustained hypercalcemia (68,70).

CURRENT MANAGEMENT OF HYPERCALCEMIA AND HYPERPARATHYROIDISM

Surgical Management

The ultimate therapeutic goal of managing hyperparathyroid hypercalcemia is returning blood PTH, calcium and phosphate levels to their normal range. In 1PTH, when active bone, renal or GI symptoms are present, surgery is the most effective treatment modality when the offending parathyroid adenoma can be identified. In MEN1 or hyperplasia, controversy exists as to whether total parathyroidectomy with autotransplantation to the forearm or subtotal parathyroidectomy should be performed. In the case of parathyroid carcinoma, however, attempts should be made to remove the entire carcinoma (with capsule intact), as well as metastases, if possible. The use of non-invasive, preoperative localization techniques (high resolution cervical ultrasonography, mediastinal computed tomography, magnetic resonance imaging) may aid in the identification of parathyroid tissue, especially in cases of recurrent or persistent 1HPT (71-73).

At present, the benefit of parathyroid gland surgery for mild, asymptomatic 1HPT is not clear (11,74). Since there is a 50% chance that patients with mild 1HPT will eventually develop complications (75,76), and since some of the renal or bone impairment may not be reversible, these patients should be monitored for disease progression (11). However, parathyroidectomy may be contraindicated in older patients, patients with increased risk of anesthesia-induced complications, and patients with severe concomitant disease that is not exacerbated by 1HPT (75).

In chronic renal failure treated by renal transplantation, hyperparathyroidism may persist, as discussed above. In most cases, the parathyroid glands involute to normal within 3 years (14,59,63, 65,66). However, since severe or prolonged hypercalcemia impairs renal function and may cause permanent renal graft injury, subtotal parathyroidectomy must be considered for patients with persistent hypercalcemia or acute hypercalcemic crisis.

Medical Management

The goal of the medical management of hyperparathyroidism is directed at lowering blood calcium levels (for reviews, see 77,78). In 1HPT, oral phosphate administration may be beneficial for lowering plasma calcium levels but the long-term effects of this treatment are uncertain. In 2HPT, however, attempts are made to reduce phosphate absorption using oral phosphate binding gels (aluminum hydroxide, aluminum carbonate) and dietary protein restriction (62).

Acutely, the use of saline infusion accompanied by administration of loop diuretics enhances urinary calcium excretion. Calcitonin, mithramycin and corticosteroids decrease calcium movement from bone. Reduced intake of calcium and corticosteroids decrease intestinal absorption of calcium. Short-term hemodialysis or peritoneal dialysis is effective for the rapid removal of calcium from the blood in crisis situations, especially in patients with renal failure or congestive heart failure. Prolonged hemodialysis, however, is not a therapeutic solution because of its impracticality and high complication rate.

Further supportive measures are essential for preventing additional complications of hypercalcemia. These treatments include normal saline infusion for restoration of euvolemia and electrolyte infusion for correction of hypokalemia or hypomagnesemia. In addition, medications which may exacerbate hypercalcemia (ex: thiazide diuretics) should be avoided.

POTENTIAL CLINICAL USES OF PTH ANTAGONISTS

In 1HPT and MEN, PTH antagonists may be effective for managing mild hypercalcemia, for treating acute hypercalcemic crises or for treating patients unfit for surgery (79). Such agents also may be effective for normalizing blood calcium levels in preparation for parathyroidectomy, subsequent to unsuccessful surgery or for treating hypercalcemia caused by metastatic parathyroid carcinoma. Similarly, PTH antagonists may block the hypercalcemic actions of PTH-like peptides secreted by non-parathyroid malignancies.

PTH antagonists probably would not be effective in managing 2HPT prior to renal transplantation, but may be quite beneficial for the transient and subacute hyperparathyroid hypercalcemia present after transplantation. The use of PTH antagonists after renal transplantation may reduce the risk of hypercalcemia-associated graft injury.

PTH antagonists may also have diagnostic value. Since PTH plays a role in the minute-to-minute regulation of blood calcium levels, short-term administration of a PTH antagonist may cause a rapid reduction in blood calcium levels in hypercalcemic patients if the hypercalcemia is due to elevated PTH levels. Similarly, short-term amelioration of possible symptoms of hypercalcemia (such as lethargy, cognitive defects) may help pinpoint the etiology of the symptoms (i.e. hyperparathyroidism vs. other disorders).

If PTH contributes to the pathogenesis of osteoporosis, PTH antagonists may be effective for its treatment. Until such agents are tested in osteoporotic patients, however, this use is quite speculative.

Potential mechanism-based side effects of PTH antagonists may be related to hypocalcemia or hypovitaminemia D. Under severe conditions, PTH antagonist-induced hypocalcemia could lead to tetany, seizures or death. Since: 1) vitamin D-dependent calcium absorption from the gut is essential for maintaining calcium homeostasis, 2) $1,25(OH)_2VitD_3$ is required for inhibition of PTH synthesis/secretion and 3. renal and bone transport of calcium is vitamin D-dependent, vitamin D deficiency would not be desirable. Oral and injectable preparations of vitamin D, however, are commercially available.

DESIGNING PTH ANTAGONISTS

Methodological considerations

Prerequisite to the rapid evaluation of drug candidates is the development of methods that will allow the quantitative comparison of a large series of structurally-related compounds. These assays must be: 1) easy to establish, 2) performed rapidly on a regular basis, and 3) simple to interpret. Secondary assays should also be available to confirm initial results and measure both intrinsic activities and receptor affinities.

The inhibition of binding of labeled PTH analogs to renal membranes in vitro is a standard method for the initial characterization of PTH agonists and antagonists (80-84). This procedure is advantageous because complete dose-response curves can be generated with microgram quantities of compound in a rapid and routine manner.

Using sucrose density gradient-purified renal cortical membranes, the degradation of compound (and ligand) is minimized thereby providing a precise estimate of the molar affinity.

Recently, this screening procedure has been optimized by employing HPLC-purified radioligand ([mono^{125}I-Tyr^{34},Nle^{8,18}]-bPTH-(1--34)NH$_2$) and bovine renal cortical tissue (84). The use of chromatographically-pure radioligand is advantageous since non-iodinated peptide, di-iodinated peptide and unreacted ^{125}INa can be quantitatively resolved from [mono^{125}I-Tyr^{34},Nle^{8,18}]-bPTH-(1--34)NH$_2$ which has the best assay characteristics. This chromatographic procedure yields a radioligand that is less susceptible to non-specific radiolysis during storage. In addition, by storing in 50 mM TrisHCl (pH 7.4)/2% BSA at -70°C, the radioligand remains stable for approximately 2 months. Bovine kidneys were chosen as the tissue source for preparing renal cortical membranes since one bovine kidney will yield a sufficient quantity of membranes for 2-3 months of screening.

Once the affinity of a compound is determined in the binding assay, intrinsic activity is quantified using a renal membrane adenylate cyclase assay (80,84-86). Potential PTH agonist activity is evaluated by examining the ability of each compound to enhance adenylate cyclase activity in a PTH antagonist reversible manner. Antagonist activity is quantified by determining the potential dose-dependent inhibition of [Nle^{8,18},Tyr^{34}]bPTH-(1--34)NH$_2$-stimulated adenylate cyclase activity.

Based upon differences between kidney and bone structure-activity relationships among PTH analogs, compounds should also be evaluated in a bone-based assay. Bone cell lines that contain PTH receptors such as ROS 17/2.8 cells or UMR-106 cells, are advantageous because they consist of a single bone cell type, can be maintained in culture and have also been used for both PTH binding and adenylate cyclase studies (41,43). Recently, evidence has been provided that the ROS 17/2.8 cell adenylate cyclase assay may also be used to predict weak agonist activity of PTH analogs in vivo (41,43).

Following in vitro characterization, compounds that display potent activity either as agonists or antagonists should be evaluated in vivo to determine the influences of metabolism, pharmacokinetics etc. on biological activity. Several models using rat, chick and dog are currently available which provide information on bone and renal PTH receptor effects (39,87-90) or selective bone PTH receptor effects (87). In vivo models are critical since one potent and promising in vitro PTH antagonist ultimately showed agonist properties in vivo (see below).

Structure-activity relationships of peptide analogs

As a preface towards the rational design of potent and selective PTH antagonists, an understanding of structure-activity relationships responsible for PTH receptor occupation and activation is essential. Since full biological activity of the 84-amino acid peptide was shown to reside in the N-terminal portion (positions 1-34) of the bovine PTH molecule (5), most synthetic efforts have involved modifications, substitutions and deletions in this region. For example, replacement of the Phe[34] residue of bPTH-(1--34)OH with a Tyr and substitution of a carboxyamide for the C-terminal carboxylic acid resulted in enhanced biological activity (91,92). In addition, substitutions of the oxidation-sensitive methionine residues in positions 8 and 18 of this analog with sulfur-free norleucines (Nle) yielded a stable molecule, [Nle[8,18],Tyr[34]]bPTH-(1--34)NH$_2$, possessing full PTH agonist activities both in vitro and in vivo (87,92,93).

To understand the regions of bPTH-(1--34) responsible for biological activity, the effects of both C- and N-terminal truncations were studied. Stepwise deletions of C-terminal amino acids to position 25 resulted in a steady decline in biological activity (94,95). Further C-terminal truncations prevented the demonstration of biological activity or binding affinity.

Similar truncations from the N-terminus reduced activity but much more dramatically (80,95). Removal of Ala[1] caused a rapid decrease in activity. One further deletion yielding bPTH-(3--34) resulted in a loss of agonist activity. As a result of these studies, the minimum sequence for agonist activity is considered to be the 2-25 region of PTH.

Although bPTH-(3--34) was a weak antagonist, [Nle[8,18],-Tyr[34]]bPTH-(3--34)NH$_2$ was a potent, selective and competitive antagonist using renal and certain bone assays in vitro (96-98). In vivo, however, [Nle[8,18],Tyr[34]]bPTH-(3--34)NH$_2$ was not an antagonist. Instead, this agent was a weak agonist displaying all of the properties of PTH at approximately 0.3-1% the potency of PTH (89,90,99). Similarly, this peptide was a partial agonist in the ROS 17/2.8 cell adenylate cyclase assay as well as the GppNHp-amplified canine renal adenylate cyclase assay (28,43).

In an attempt to develop an in vitro and in vivo antagonist, the effects of further N-terminal truncations of [Nle[8,18],Tyr[34]]-bPTH-(3--34)NH$_2$ were studied (96,98). The 7-34 analog was 10-100 fold weaker than the 3-34 analog but still showed significant affinity for the PTH receptor. Following replacement of methionines in positions 8 and 18, [Tyr[34]]bPTH-(7--34)NH$_2$ was found to be a potent PTH antagonist both in vitro and in vivo without possessing partial agonist properties (43,87,100).

Using [Tyr[34]]bPTH-(7--34)NH$_2$ as the prototypical antagonist, the goal of present studies is to develop more potent and long-acting in vivo PTH antagonists. This objective is being addressed using several approaches including 1) substituting amino acids that may enhance resistance to proteolytic degradation, 2) understanding the conformational features of both agonists and antagonists that are required for biological activity so that rational changes can be made in the antagonist sequence which increase PTH receptor affinity, and 3) the synthesis of hybrid molecules of PTH and hHCF to identify new directions for PTH antagonist design. Once optimal substitutions from each approach are identified, hybrid molecules containing combinations of structural features can be synthesized.

In an effort to protect [Tyr[34]]bPTH-(7--34)NH$_2$ against proteolytic degradation by aminopeptidases, the effects of 3 N-terminal structural modifications on antagonist potency were investigated (84). First, the amino group of Phe[7] was removed

(α-desamino[Tyr^{34}]bPTH-(7--34)NH$_2$). Second, Phe^7 was replaced by N-MePhe^7([N-MePhe^7,Tyr^{34}]bPTH-(7--34)NH$_2$). Third, by replacing Met^8 with N-MeMet^8 an N-methylated peptide bond between residues 7-8 was formed ([N-MeMet^8,Tyr^{34}]bPTH-(7--34)NH$_2$). Using renal cortical membrane binding and adenylate cyclase assays, these three structural modifications were found to be tolerated in terms of preservation of biological activity in vitro (i.e. inhibition of PTH binding and PTH-stimulated adenylate cyclase activity), suggesting that a charged N-terminal amino group is not required for antagonist activity. In order to validate the enhanced stability of these analogs to aminopeptidases, they will be evaluated in vivo.

The second approach for developing potent peptide PTH antagonists is based on obtaining an understanding of the conformational requirements for receptor occupation and activation. Analogs which promote receptor-favored conformational features can then be synthesized. In the PTH region of residues 12-15, for example, a β-turn was predicted by the Chou-Fasman algorithm (101). Position 12-substituted analogs of PTH-(1--34)NH$_2$ and PTH-(7--34)NH$_2$ were, therefore, synthesized to test this hypothesis (102,103). Replacement of Gly^{12} with amino acids that favored the formation of an α-helix resulted in the retention of either agonist or antagonist properties (Table II). Substitution of Pro (a known α-helix breaker) in this position, however, caused a marked diminution of activity. Taken together, these studies favor the presence of an α-helix in this region and suggested that position 12 would be a relevant site for further substitution studies.

D-Trp was chosen as a position 12 replacement in [Tyr^{34}]bPTH-(7--34)NH$_2$ and [Nle^{8,18},Tyr^{34}]bPTH-(7--34)NH$_2$ since this non-natural amino acid was shown to increase analog potencies in other peptide systems (104-107). [D-Trp^{12},Tyr^{34}]bPTH-(7--34)NH$_2$ and [Nle^{8,18},D-Trp^{12},Tyr^{34}]bPTH-(7--34)NH$_2$ were competitive inhibitors of [mono^{125}I-Tyr^{34},Nle^{8,18}]bPTH-(1--34)NH$_2$ binding to renal cortical membranes and bone cells in vitro, with potencies 10-20-fold greater than their non-position 12-substituted counterparts (Table III) (108). These new analogs were also more potent inhibitors of [Nle^{8,18},Tyr^{34}]bPTH-(1--34)NH$_2$-stimulated adenylate cyclase activity in both renal- and bone-based tissue preparations. In addition, [D-Trp^{12},Tyr^{34}]bPTH-(7--34)NH$_2$ inhibited bPTH-(1--84)- and hHCF-(1--34)NH$_2$-stimulated adenylate cyclase activity. In contrast to [Nle^{8,18},Tyr^{34}]-bPTH-(3--34)NH$_2$ (43), however, these agents did not possess partial agonist properties in vitro, suggesting that they will not be weak agonists in vivo.

The third approach towards PTH antagonist development involves using structure-activity experiments of nature to identify new PTH antagonist leads. Based upon precedents from the PTH system, truncation of hHCF to hHCF-(7--34)NH$_2$ was hypothesized to generate a PTH antagonist. This fragment inhibited PTH binding and PTH-stimulated adenylate cyclase activity using renal membranes and bone cells in vitro (43). Although hHCF-(7--34)NH$_2$ was more potent than [Tyr^{34}]bPTH-(7--34)NH$_2$ on bone cells (but not renal membranes), the hHCF fragment displayed partial agonist properties, suggesting that it may be a weak agonist in vivo, like [Nle^{8,18},Tyr^{34}]bPTH-(3--34)NH$_2$ (43).

Table II. Biological Activities of PTH Analogs Substituted in Position 12

Analog	Biological Activity		
	Binding K_i	Adenylate Cyclase	
		K_m (Agonist)	K_i (Antagonist)
	nM		
[Tyr^{34}]hPTH-(1--34)NH$_2$	0.7 ± 0.3	0.7 ± 0.1	
[Tyr^{34}]bPTH-(1--34)NH$_2$	1.1 ± 0.3	1.1 ± 0.4	
[Tyr^{34}]hPTH-(7--34)NH$_2$	257 ± 36		842 ± 182
[Tyr^{34}]bPTH-(7--34)NH$_2$	75.0 ± 8.3		835 ± 65
[Nle^{8,18},Tyr^{34}]bPTH-(7--34)NH$_2$	144.0 ± 9.0		1550 ± 33
[Ala^{12},Tyr^{34}]hPTH-(1--34)NH$_2$	1.0 ± 0.04	1.5 ± 0.2	
[D-Ala^{12},Tyr^{34}]hPTH-(1--34)NH$_2$	0.8 ± 0.1	1.4 ± 0.1	
[Aib^{12},Tyr^{34}]hPTH-(1--34)NH$_2$	0.8 ± 0.1	0.6 ± 0.2	
[Pro^{12},Tyr^{34}]hPTH-(1--34)NH$_2$	587 ± 196	2448 ± 769	
[Ala^{12},Tyr^{34}]hPTH-(7--34)NH$_2$	114 ± 32		413 ± 67
[D-Ala^{12},Tyr^{34}]hPTH-(7--34)NH$_2$	113 ± 4		612 ± 116
[Pro^{12},Tyr^{34}]hPTH-(7--34)NH$_2$	471 ± 50		1400 ± 668
[Aib^{12},Tyr^{34}]bPTH-(7--34)NH$_2$	51.0 ± 8.7		536 ± 144
[β-Ala^{12},Tyr^{34}]bPTH-(7--34)NH$_2$	128 ± 21		304 ± 74
[Sar^{12},Tyr^{34}]bPTH-(7--34)NH$_2$	503 ± 91.4		2506 ± 732

Table III. Relative potencies of PTH antagonists in kidney (bovine renal cortical membrane) and bone (ROS 17/2.8 cell) systems

Analog [x]bPTH(7-34)NH$_2$ x=	Kidney Membranes		Bone Cells	
	Binding K_b	Adenylate Cyclase K_i	Binding K_b	Adenylate Cyclase K_i
		nM		
Tyr^{34}	80 ± 7	879 ± 68	767 ± 199	5620 ± 1670
Nle^{8,18},Tyr^{34}	145 ± 13	1631 ± 350	964 ± 170	1550 ± 361
D-Trp^{12},Tyr^{34}	7 ± 1	69 ± 5	60 ± 20	211 ± 116
Nle^{8,18},D-Trp^{12},Tyr^{34}	15 ± 1	125 ± 7	182 ± 32	69 ± 17

Values are the mean ± S.E.M. of at least three experiments.

By preparing hybrid molecules containing various motifs from PTH and hHCF, it may be possible to identify regions of the hHCF molecule that are responsible for the partial agonist activity of hHCF-(7--34)NH$_2$ as well as determine new directions for antagonist design. For example, since there is less sequence homology between PTH and HCF after the first 13 amino acids, the 7-13 region of each peptide could be coupled with the 14-34 region of the other (ex: bPTH-(7--13) + hHCF-(14--34)NH$_2$ or hHCF-(7--13) + bPTH-(14--34)-NH$_2$). More subtle hybrid molecules could help pinpoint the relative importance and role of various amino acids for PTH antagonist and partial agonist activities.

Future prospects for non-peptide PTH antagonists

The ultimate goal of PTH antagonist drug design is the development of orally effective non-peptide (or pseudo-peptide) agents. Approaches towards accomplishing this objective include conversion of peptide agents to non-peptide agents and direct identification of non-peptide agents from natural products.

Several avenues have been used for the conversion of peptides to non-peptides. One approach involves first identifying a minimal amino acid sequence required for biological activity, then synthesizing analogs possessing altered peptide bonds, thus decreasing susceptibility to proteolysis. Replacement of specific peptide bonds with CH$_2$-NH in tetragastrin, for example, resulted in the synthesis of a gastrin antagonist from the parent agonist molecule (109). Another approach involves the use of peptide structure-activity relationships to develop hypothetical models for the mechanisms of binding of an antagonist to its target (receptor or enzyme), then synthesizing small molecules that interact with the active site of the target. For example, based upon an understanding of the interaction of nonapeptides with angiotensin converting enzyme (ACE) and the similarities between ACE and carboxypeptidase A, small, orally-active and clinically-effective molecules were prepared that specifically inhibit ACE activity (110-112).

The isolation of pharmacological substances from natural product extracts provides a potentially unlimited source of new therapeutic entities. In recent years, this approach has led to the identification of many drugs including ion channel inhibitors (avermectins for treating parasitic infections), enzyme inhibitors (the HMGCoA reductase inhibitor, lovastatin, for treating hyperlipidemia) and peptide receptor antagonists (asperlicin as an inhibitor of cholecystokinin). In attempts to identify peptide hormone receptor antagonists, this process involves screening crude natural product extracts for receptor binding inhibitory activity, confirming activity in a secondary (functional) assay, determining receptor specificity of the extract or natural product, and using appropriate assays to monitor isolation and purification efforts. For example, the isolation of the selective peripheral cholecystokinin antagonist asperlicin from Aspergillus alliaceus (113), led to the synthesis of benzodiazepines possessing selectivity and enhanced potency for either the peripheral cholecystokinin receptor (L-364,718, MK-329) (114,115) or the gastrin/central cholecystokinin receptor (L-365,260) (116; Lotti, V.J.; Chang, R.S.L. Eur. J. Pharmacol. 1989, in press).

Literature Cited

1. DeLuca, H.F. FASEB J. 1988, 2, 224-236.
2. Rosenblatt, M. Mineral Electrolyte Metab. 1982, 8, 118-129.
3. Farrow, S.M.; Karmali,R.; Gleed, J.H.; Hendy, G.N.; O'Riordan, J.L.H. J. Endocrin. 1988, 117, 133-138.
4. Nygren, P.; Gylfe, E.; Larsson, R.; Johansson, H.; Juhlin, C.; Klareskoq, L.; Akerstrom, G.; Rastad, J. Biochim. Biophys. Acta. 1988, 968, 253-260.
5. Potts, J.T. Jr.; Tregear, G.W.; Keutmann, H.T.; Niall, H.D.; Sauer, R.; Deftos, L.J.; Dawson, B.F.; Hogan, M.L.; Aurbach, G.D. Proc. Natl. Acad. Sci. 1971, 68, 63-67.
6. Rosenblatt, M.; Kronenberg, H.M.; Potts, J.T. Jr. Endocrinology; DeGroot, L.J., Ed.; Philadelphia, 1989; Vol. 2, Chap. 54.
7. Hruska, K.A.; Moskowitz, D.; Esbrit, P.; Civitelli, R.; Westbrook, S.; Huskey, M. J. Clin. Invest. 1987, 79, 230-239.
8. Joborn, C.; Hetta, J.; Palmer, M.; Akerstrom, G; Ljunghall, S. Upsala J. Med. Sci. 1986, 91, 77-87.
9. Christensson, T.; Hellstrom, K.; Wengle, B. Eur. J. Clin. Inves. 1977, 7, 109-113.
10. Rapado, A. Am. J. Nephrol. 1986, 6, 49-50.
11. Rubinoff, H.; McCarthy, N.; Hiatt, R.A. J. Chron. Dis. 1983, 36, 859-868.
12. Roberts, W.C.; Waller, B.F. Am. J. Med. 1981, 71, 371-384.
13. Bauwens, S.F.; Drinka, P.J.; Boh, L.E. Clin. Pharmacy 1986, 5, 639-659.
14. Alfrey, A.C.; Jenkins, D.; Groth, C.G.; Schorr, W.S.; Gecelter, L.; Ogden, D.A. New Engl. J. Med. 1968, 279, 1349-1356.
15. Berlyne, G.M.; Ben-Ari, J.; Kushelevsky, A.; Idelman, A.; Galinsky, D.; Hirsch, M.; Shainkin, R.; Yagil, R.; Zlotnik, M. Quar. J. Med. 1975, 175, 505-621.
16. Lloyd, H.M. Medicine 1968, 47, 53-71.
17. Fujimoto, Y.; Obara, T. Surg. Clin. N. Amer. 1987, 67, 343-357.
18. Murray, T.M.; Peacock, M.; Powell, D.; Monchik, J.M.; Potts, J.T. Jr. Clin. Endocrinol. 1972, 1, 235-246.
19. Brown, E.M. J Clin. Endocrinol. Metab. 1983, 56, 572-581.
20. Reiss, E.; Slatopolsky, E. Endocrinology; DeGroot, L.J., Ed.; Philadelphia, 1979; Vol. 2, Chap. 60.
21. Breslau, N.A. Am. J. Med. Sci. 1987, 294, 120-131.
22. Feinfeld, D.A.; Sherwood, L.M. Kidney International 1988, 33, 1049-1058.
23. Lopez-Hilker, S.; Galceran, T.; Chan, Y-L.; Rapp, N.; Martin, K.J.; Slatopolsky E. J. Clin. Invest. 1986, 78, 1097-1102.
24. Mahaffey, J.E.; Potts, J.T. Jr. Endocrinology; DeGroot, L.J., Ed.; Philadelphia, 1979; Vol. 2, Chap. 59.
25. Parfitt, A.M. Miner Electrolyte Metab. 1982, 8, 92-112.
26. Mundy, G.R. Bone 1987, 8, S9-S16.
27. Simpson, E.L.; Mundy, G.R.; D'Souza, S..M.; Ibbotson, K.J.; Bockman, R.; Jacobs, J.W. N. Engl. J Med. 1983, 309, 325-330.

28. Strewler, G.J.; Williams, R.D.; Nissenson, R.A. J. Clin. Invest. 1983, 71, 769-774.
29. Stewart, A.F.; Insogna, K.L.; Goltzman, D.; Broadus, A.E. Proc. Natl. Acad. 1983, 80, 1454-1458.
30. Rodan, S.B.; Insogna, K.L.; Vignery, A.M.-C.; Stewart, A.F.; Broadus, A.E.; D'Souza, S.; Bertolini, D.R.; Mundy, G.R.; Rodan, G.A. J. Clin. Invest. 1983, 72, 1511-1515.
31. Suva, L.J.; Winslow, G.A.; Wettenhall, R.E.H.; Hammonds, R.G.; Moseley, J.M.; Diefenbach-Jagger, H.; Rodda, C.P.; Kemp, B.E.; Rodriguez, H.; Chen, E.Y.; Hudson, P.J.; Martin, T.J.; Wood, W.I. Science 1987, 237, 893-896.
32. Stewart, A.F.; Wu, T.; Goumas, D.; Burtis, W.J.; Broadus, A.E. Biochem. Biophys. Res. Comm. 1987, 146, 672-678.
33. Strewler, G.J.; Stern, P.H.; Jacobs, J.W.; Eveloff, J.; Klein, R.F.; Leung, S.C.; Rosenblatt, M.; Nissenson, R.A. J. Clin. Invest. 1987, 80, 1803-1807.
34. Thiede, M.A.; Strewler, G.J.; Nissenson, R.A.; Rosenblatt, M.; Rodan, G.A. Proc. Natl. Acad. Sci. 1988, 85, 4605-4609.
35. Thiede, M.A.; Rodan, G.A. Science 1988, 242, 278-280.
36. Merendino, J.J. Jr.; Insogna, K.L.; Milstone, L.M.; Broadus, A.E.; Stewart, A.F. Science 1986, 231, 388-390.
37. Loveridge, N.; Caple, I.W.; Rodda, C.; Martin, T.J.; Care, A.D. Quart. J. Expt. Phys. 1988, 73, 781-784.
38. Rodda, C.P.; Kubota, M.; Heath, J.A.; Ebeling, P.R.; Moseley, J.M.; Care, A.D.; Caple, I.W.; Martin, T.J. J. Endocr. 1988, 117, 261-271.
39. Horiuchi, N.; Caulfield, M.P.; Fisher, J.E.; Goldman, M.E.; McKee, R.L.; Reagan, J.E.; Levy, J.J.; Nutt, R.F.; Rodan, S.B.; Schofield, T.L.; Clemens, T.L., Rosenblatt, M. Science 1987, 238, 1566-1570.
40. Kemp, B.F.; Moseley, J.M.; Rodda, C.P.; Ebeling, P.R.; Wettenhall, R.E.H.; Stapleton, D.; Diefenbach-Jagger, H.; Ure, F.; Michelangeli, V.P.; Simmons, H.A.; Raisz, L.G.; Martin, T.J. Science 1987, 238, 1568-1570.
41. Rodan, S.B.; Noda, M.; Wesolowski, G.; Rosenblatt, M.; Rodan, G.A. J. Clin. Invest. 1988, 81, 924-927.
42. Thompson, D.D.; Seedor, J.G.; Fisher, J.E.; Rosenblatt, M.; Rodan, G.A. Proc. Natl. Acad. Sci. 1988, 85, 5673-5677.
43. McKee, R.L.; Goldman, M.E.; Caulfield, M.P.; DeHaven, P.A.; Levy, J.J.; Nutt, R.F.; Rosenblatt, M. Endocrinology 1988, 122, 3008-3010.
44. Nissenson, R.A.; Diep, D.; Strewler, G.J. J. Biol. Chem. 1988, 263, 12866-12871.
45. Juppner, H.; Abou-Samra, A-B.; Uneno, S.; Gu, W-X.; Potts, J.T. Jr.; Segre, G.V. J. Biol. Chem. 1988, 263, 8557-8560.
46. Saphier, P.W.; Stamp, T.C.B.; Kelsey, C.R.; Loveridge, N. Bone and Mineral 1987, 3, 75-83.
47. Riggs, L.B.; Melton, J.L. III. New Engl. J. Med. 1986, 314, 1676-1686.
48. Avioli, L.V. Geriatrics 1986, 41, 30-37.

49. Heath, H. III; Hodgson, S.F.; Kennedy, M.A. New Engl. J. Med. 1980, 302, 189-193.
50. Mundy, G.R.; Cove, D.H.; Fisken, R. The Lancet 1980, 1317-1320.
51. Dent, D.M.; Miller, J.L.; Klaff, L.; Barron, J. Postgrad. Med. J. 1987, 63, 745-750.
52. Christensson, T.; Hellstrom, K.; Wengle, B.; Alveryd, A.; Wikland, B. Acta. Med. Scand. 1976, 200, 131-137.
53. Evans, R.A. Drugs 1986, 31, 64-74.
54. Zawada, E.T. Jr.; Lee, D.B.N.; Kleeman, C.R. Postgrad. Med. 1979, 66, 91-100.
55. Burt, M.E.; Brennan, M.F. Arch. Surg. 1980, 115, 704-707.
56. Kaye, P.M.; Oliver, J.J. The Lancet 1985, 512.
57. Powell, D.; Singer, F.R.; Murray, T.M.; Minkin, C.; Potts, J.T. Jr. New Engl. J. Med. 1973, 289, 176-181.
58. Sherwood, L.M.; O'Riordan, J.L.; Aurbach, G.D.; Potts, J.T. Jr. J. Clin. Endocrinol. Metab. 1967, 27, 140-146.
59. Malmaeus, J. Scand. J. Urol. Nephrol. Suppl. 70, Uppsala, Sweden 1983.
60. Delmonico, F.L.; Wang, C.A.; Rubin, N.T.; Fang, L.S.; Herrin, J.T.; Cosimi, A.B. Ann. Surg. 1984, 200, 644-647.
61. Johnson, J.W.; Hattner, R.S.; Hampers, C.L.; Bernstein, D.S.; Merrill, J.P.; Sherwood, L.M. Hemodialysis 1972, 21, 18-29.
62. Johnson, W.J.; Goldsmith, R.S.; Arnaud, C.D. Med. Clin. N. Am. 1972, 56, 961-975.
63. David, D.S.; Sakai, S.; Brennan, B.L.; Riggio, R.A.; Cheigh, J.; Stenzel, K.H.; Rubin, A.L.; Sherwood, L.M. N. Engl. J. Med. 1973, 289, 398-401.
64. Pletka, P.G.; Strom, T.B.; Hampers, C.L.; Griffiths, H.; Wilson, R.E.; Bernstein, D.S.; Sherwood, L.M.; Merrill, J.P. Nephron. 1976, 17, 371-381.
65. Diethelm, A.G.; Edwards, R.P. Whelchel, J.D. Surg. Gynecol. Obstet. 1982, 154, 481-490.
66. Garvin,P.J.; Castaneda, M.; Linderer, R.; Dickhans, M. Arch. Surg. 1985, 120, 578-583.
67. Mohamadi, M.; Bivins, L.; Becker, K.L. Clin. Pharmacol. Ther. 1979, 26, 390-394.
68. Christensson, T.; Hellstrom, K.; Wengle, B. Arch. Intern. Med. 1977, 137, 1138-1142.
69. Drinka, P.J.; Nolten, W.E. J. Am. Geriatrics Soc. 1984, 32, 405-407.
70. Field, M.J.; Lawrence, J.R. Med. J. Aus. 1986, 144, 641-644.
71. Grant, C.S.; van Heerden, J.A.; Charboneau, J.W.; James, E.M.; Reading, C.C. World J. Surg. 1986, 10, 555-565.
72. Thompson, N.W. Br. J. Surg. 1988, 75, 97-98.
73. Hamilton, R.; Greenberg, B.M.; Gefter, W.; Kressel, H.; Spritzer, C. Am. J. Sur. 1988, 155, 370-373.
74. Mallette, L.E. Am. J. Med. Sci. 1987, 293, 239-249.
75. Mallette, L.E. Annals. Int. Med. 1982, 97 622-623.
76. Purnell, D.C. Mayo Clin. Proc. 1981, 56, 473-478.

77. Zawada, E.T. Jr.; Lee, D.B.N.; Kleeman, C.R. Postgrad. Med. 1979, 66, 105-111.
78. Neer, R.M.; Potts, J.T. Jr. Endocrinology; DeGroot, L.J., Ed.; Philadelphia, 1979; Vol. 2, Chap. 57.
79. Rosenblatt, M. N. Engl. J. Med. 1986, 315, 1004-1013.
80. Segre, G.V.; Rosenblatt, M.; Reiner, B.L.; Mahaffey, J.E.; Potts, J.T. Jr. J. Biol. Chem. 1979, 254, 6980-6986.
81. Nissenson, R.A.; Teitelbaum, A.P.; Arnaud, C.D. Methods in Enzym. 1985, 109, 48-56.
82. McKee, M.D. Murray, T.M. Endocrinology 1985, 117, 1930-1939.
83. Rizzoli, R.E.; Murray, T.M.; Marx, S.J.; Aurbach, G.D. Endocrinology 1983, 112, 1303-1312.
84. Goldman, M.E.; Chorev, M.; Reagan, J.E.; Nutt, R.F.; Levy, J.J.; Rosenblatt, M. Endocrinology 1988, 123, 1468-1475.
85. DiBella, F.P., Arnaud, C.D. Brewer, H.B. Jr. Endocrinology 1976, 99, 429-436.
86. Teitelbaum, A.P.; Pliam, N.B.; Silve, C.; Abbott, S.R.; Nissenson, R.A.; Arnaud, C.D. In Regulation of Phosphate and Mineral Metabolism; Massry, S.G.; Letteri, J.M.; Ritz, E., Eds.; Plenum Press: New York, 1982; pp 535-548.
87. Horiuchi, N.; Holic, M.F.; Potts, J.T. Jr.; Rosenblatt, M. Science 1983, 220, 1053-1055.
88. Horiuchi, N.; Rosenblatt, M. Am. J. Physiol. 1987, 253, E187-E192.
89. Gray, D.A.; Parsons, J.A.; Potts, J.T. Jr.; Rosenblatt, M.; Stevenson, R.W. Br. J. Pharmac. 1982, 76, 259-263.
90. Segre, G.V.; Rosenblatt, M.; Tully, G.L. III, Laugharn, J.; Reit, B.; Potts, J.T. Jr. Endocrinology 1985, 116, 1024-1029.
91. Parsons, J.A.; Rafferty, B.; Gray, D.; Reit, B.; Keutmann, H.T.; Tregear, G.W.; Potts, J.T. Jr. In Calcium-Regulating Hormones, Talmage, R.V.; Owen, M.; Parsons, J.A. eds., Excerpta Medica, Amsterdam 1975; p. 33.
92. Rosenblatt, M.; Goltzman, D.; Keutmann, H.T.; Tregear, G.W.; Potts, J.T. Jr. J. Biol. Chem. 1976, 251, 159-164.
93. Rosenblatt, M. Potts, J.T., Jr. Endocrine Res. Comm. 1977, 4, 115-133.
94. Tregear, G.W.; Van Rietschoten, J.; Greene, E.; Keutmann, H.T.; Niall, H.D.; Reit, B.; Parsons, J.A.; Potts, J.T., Jr. Endocrinology 1973, 93, 1349-1353.
95. Goltzman, D.; Callahan, E.N.; Tregear, G.W.; Potts, J.T., Jr. Endocrinology 1978, 103, 1352-1360.
96. Mahaffey, J.E.; Rosenblatt, M.; Shepard, G.L.; Potts, J.T., Jr. J. Biol. Chem. 1979, 254, 6496-6498.
97. Goldring, S.R.; Mahaffey, J.E.; Rosenblatt, M.; Dayer, J.M.; Potts, J.T., Jr.; Krane, S.M. J. Clin. Endocrinol. Metab. 1979, 48, 655-659.
98. Rosenblatt, M.; Segre, G.V.; Tyler, G.A.; Shepard, G.L.; Nussbaum, S.R.; Potts, J.T., Jr. Endocrinology 1980, 107, 545-550.
99. McGowan, J.A.; Chen, T.C.; Fragola, J.; Puschett, J.B. Science 1983, 219, 67-69.

100. Doppelt, S.H.; Neer, R.M.; Nussbaum, S.R.; Federico, P.; Potts, J.T., Jr.; Rosenblatt, M. Proc. Natl. Acad. Sci. 1986, 83, 7557-7560.
101. Nussbaum, S.R.; Bendetti, N.V.; Fasman, G.D.; Potts, J.T., Jr.; Rosenblatt, M. J. Prot. Chem. 1985, 4, 391-406.
102. Chorev, M.; Goldman, M.E.; Caporale, L.H.; Levy, J.J.; Reagan, J.E.; DeHaven, P.; Gay, T.; Nutt, R.F.; Rosenblatt, M. In Peptide Chem; Shiba, T.; Sakakibara, S., Eds.; Osaka, 1987, pp. 621-626.
103. Caporale, L.H.; Chorev, M.; Levy, J.J.; Goldman, M.E.; DeHaven, P.A.; Gay, C.T.; Reagan, J.E.; Rosenblatt, M.; Nutt, R.F. In Peptides; Chemistry and Biology, Marshall, G.R., Ed. ESCOM, Leiden, 1988, pp. 449-451.
104. Rivier, J.; Brown, M; Vale, W. Biochem. Biophys. Res. Comm. 1975, 65, 746-751.
105. Arison, B.H.; Hirschmann, R.; Veber, D.F. Bioorganic Chem. 1978, 7, 447-451.
106. Folkers, K.; Horig, J.; Rosell, S; Bjorkroth, U. Acta. Physiol. Scand. 1981, 111, 505-506.
107. Regoli, D.; Escher, E.; Mizrahi, J. Pharmacology 1984, 28, 301-320.
108. Goldman, M.E.; McKee, R.L.; Caulfield, M.P.; Reagan, J.E.; Levy, J.J.; Gay, C.T.; DeHaven, P.A.; Rosenblatt, M.; Chorev, M. Endocrinology 1988, 123, 2597-2599.
109. Martinez, J.; Bali, J-P.; Rodriguez, M.; Castro, B.; Magous, R.; Laur, J.; Lignon, M-F. J. Med. Chem. 1985, 28, 1874-1879.
110. Ondetti, M.A. Circulation 1988, 77, I74-I78.
111. Patchett, A.A.; Harris, E.; Tristram, E.W.; Wyvratt, M.J.; Wu, M.T.; Taub, D.; Peterson, E.R.; Ikeler, T.J.; ten Broeke, J.; Payne, L.G.; Ondeyka, D.L.; Thorsett, E.D.; Greenlee, W.J.; Lohr, N.S.; Hoffsommer, R.D., Jr.; Joshua, H.; Ruyle, W.V.; Rothrock, J.W.; Aster, S.D.; Maycock, A.L.; Robinson, F.M.; Hirschmann, R.; Sweet, C.S.; Ulm, E.H.; Gross, D.M.; Vassil, T.C.; Stone, C.A. Nature 1980, 288, 280-283.
112. Wu, M.T.; Douglas, A.W.; Ondeyka, D.L.; Payne, L.G.; Ikeler, T.J.; Joshua, H.; Patchett, A.A. J. Pharm. Sci. 1986, 74, 352-354.
113. Chang, R.S.L.; Lotti, V.J.; Monaghan, R.L.; Birnbaum, J.; Stapley, E.O.; Goetz, M.A.; Albers-Schonberg, G.; Patchett, A.A.; Liesch, J.M.; Hensens, O.D.; Springer, J.P. Science 1985, 230, 177-179.
114. Chang, R.S.L.; Lotti, V.J. Proc. Natl. Acad. Sci. 1986, 83, 4923-4926.
115. Evans, B.E.; Bock, M.G.; Rittle, K.E.; DiPardo, R.M.; Whitter, W.L.; Veber, D.F.; Anderson, P.S.; Freidinger, R.M. Proc. Natl. Acad. Sci. 1986, 83, 4918-4922.
116. Bock, M.G.; DiPardo, R.M.; Evans, B.E.; Rittle, K.E.; Whitter, W.L.; Veber, D.F.; Anderson, P.S.; Freidinger, R.M. J. Med. Chem. 1989, 32, 16-23.

RECEIVED June 14, 1989

Chapter 18

Interaction of Phenylisopropylamines with Central 5-HT2 Receptors

Analysis by Quantitative Structure–Activity Relationships

Richard A. Glennon and Mark R. Seggel

Department of Medicinal Chemistry, School of Pharmacy, Medical College of Virginia, Virginia Commonwealth University, Richmond, VA 23298–0581

>A QSAR investigation of 27 4-substituted derivatives of 1-(2,5-dimethoxyphenyl)-2-aminopropane (i.e., 2,5-DMA) reveals that the lipophilic character of the 4-substituent is a primary determining factor for 5-HT2 receptor affinity. The length (size/shape ?) of the substituent may also be important. Previous studies have shown that certain 2,5-DMAs act as 5-HT2 agonists whereas preliminary data suggest others may act as antagonists (or at least as mixed agonist-antagonists). Intrinsic activity may be related to electronic as well as lipophilic properties of the 4-substituent.

The discovery of multiple populations of central serotonin receptors (i.e., 5-HT1, 5-HT2, 5-HT3) has ushered in a new era in 5-HT research and has prompted a search for site-selective agents. Recent work from our laboratories has shown that phenalkylamine derivatives bind with varying degrees of affinity and/or selectivity at 5-HT2 sites (1-3). Because there was evidence that such agents might constitute the first class of 5-HT2-selective agonists, we investigated structure-affinity relationships (SAFIR) for 5-HT2 binding (2,3). It was determined that a primary amine and an α-methyl group (though not necessary) result in optimal affinity. Of various aromatic substituents investigated, optimal, though modest, affinity was associated with a 2,5-dimethoxy substitution pattern (i.e., with 2,5-dimethoxy analogs; 2,5-DMAs). It quickly became evident that 4-substitutents play a significant role in modulating the affinity of the 2,5-DMAs for 5-HT2 sites. For example, introduction of a 4-bromo group, to afford DOB, resulted in a greater than 100-fold increase in affinity; the Ki for the parent 2,5-DMA (1) and for DOB (7) = 5,200 and 41 nM, respectively)(2). In order to determine the role of the 4-substituents, we conducted a Hansch analysis on a series of 13 2,5-DMAs for which we had already obtained binding data. The structures of these agents varied only with respect to the 4-position functionality; in the initial series, R4 = H, OMe, OEt, NO2, F, Br, I, Me, Et, n-Pr, n-Bu, t-Bu and n-amyl. A relating equation (Eq 1) suggested that the lipophilicity and the electronic nature of the 4-position substituent contributes to binding (3). For the R4 substituents in question, the Hammett sigma constant σ_p ranged in value from 0.78 to -0.20 and the pi value, π, ranged from 2.67 to -0.28. A Craig plot (π vs σ_p) for these 13 substituents revealed the rather narrow range of values employed and suggested that the electronic term in Eq 1 might reflect some bias in the data set. Thus, it was of interest to prepare and characterize additional agents with a wider range of values, and then to reevaluate the relative contributions of these (and other) parameters. This process took on an iterative character in that as several new agents were evaluated and new regression models developed, the new results were used to direct the design of additional compounds. The new set of agents ultimately included the original 13 compounds plus 14 additional compounds synthesized in our laboratories; radioligand binding data (Ki values) for all 27 compounds are shown in Table I. We have recently submitted a manuscript (that includes the synthesis, binding data, and preliminary QSAR results) for publication (Journal of Medicinal Chemistry), but we take this opportunity to describe our progress and the reasoning underlying these studies.

TABLE I. Affinities (Ki values) of 2,5-DMA Derivatives for Central 5-HT$_2$ Serotonin Binding Sites

	R	Ki (nM)		R	Ki (nM)
1	H	5,200	15	Amyl	7
2	OMe	1,250	16	Hexyl	2.5
3	OEt	2,200	17	Octyl	3
4	NO$_2$	300	18	CH$_2$-C$_6$H$_5$	7
5	F	1,100	19	(CH$_2$)$_3$-C$_6$H$_5$	10
6	Cl	218	20	NH$_2$	26,000
7	Br	41	21	CN	2,400
8	I	19	22	COOCH$_2$CH$_2$CH$_3$	2,460
9	Methyl	100	23	COO(CH$_2$)$_3$CH$_3$	1,530
10	Ethyl	100	24	COCH$_2$CH$_3$	735
11	Propyl	69	25	CONH(CH$_2$)$_2$CH$_3$	7,550
12	iPropyl	76	26	OH	>50,000
13	Butyl	58	27	COOH	>50,000
14	tButyl	19			

[a]Affinities (Ki values) for [^{3}H]ketanserin-labeled 5-HT$_2$ sites. These Ki values have been reported (1-3). For comparative purposes, the Ki value for ketanserin is 1.2 nM. Several derivatives were also examined at [3H]DOB-labeled 5-HT2 sites (13); Ki values are as follows: **1**, 268 nM; **2**, 81 nM; **3**, 113 nM; **7**, 0.79 nM; **8**, 0.8 nM; **9**, 8 nM; **10**, 1.5 nM; **11**, 0.9 nM; **13**, 1.7 nM.

QSAR Analysis

Evaluation of Equation 1. Eq 1 suggests, for the original set of 13 compounds, that lipophilic electron withdrawing groups contribute to affinity (3). For this reason, we prepared and evaluated the esters 22 and 23, the amide 25, the amine 20, and the phenol 26. The phenol was not expected to bind with significant affinity and, indeed, a Ki value could not be determined (Ki > 50,000 nM); likewise, the amine 20 was anticipated to possess a low affinity and this too was found to be the case (Ki = 26,000 nM) (Table I). However, the lipophilic electron withdrawing esters 22 and 23, and the amide 25, also displayed a low affinity and were poorly predicted by Eq 1. Observed and calculated log 1/Ki values for these four compounds are: 20 4.59, 3.95; 22 5.82, 7.53; 23 5.61, 8.06; 25 5.12, 6.22.

Search for New Relationships. If the electronic term is deleted from the regression model (Eq 2), the affinities of 20 (4.59, 5.10), 22 (5.82, 6.92), 23 (5.61, 7.36) and 25 (5.12, 5.92) are predicted with somewhat more (though still less than satisfactory) accuracy. Eq 3 (r = 0.764) shows the relationship between π and affinity for all 17 compounds; inclusion of a sigma term (equation not shown) had no effect (r = 0.764). Further regression analysis was conducted on the expanded set of compounds using the SAS General Linear Models (GLM) procedure. Parameters examined included: π, π^2, σ_p, σ_m, R (resonance effect), F (field effect), the steric parameters E_s, MR (molar refractivity)(4), and MW (molecular weight). Shape was accounted for by use of the B1 parameter (minimum width) as calculated by Verloop and co-workers (5, 6) using the STERIMOL program. The STERIMOL parameters B5 and L (5,6), representative of substituent length, were also included. The SAS RSQUARE procedure was used to identify all possible one- and two-variable equations with r $\geq$ 0.85. No one-variable equation adequately accounted for affinity, and there were no relating equations with a regression coefficient greater than that for Eq 3. Several statistically valid two-variable equations were identified. Eq 4 (r = 0.871; Table II) suggests that the length of the 4-substituent might be important. Thus, although lipophilicity still seems to play a major role, the electronic contribution of these substituents appears to be less important than originally considered, and length may be an additional contributing factor. Nevertheless, the affinities of the esters 22 and 23 were still rather poorly predicted.

One possibility that can not be overlooked is that the esters might undergo hydrolysis under the conditions of the radioligand binding assay. As a consequence, we prepared and evaluated their hydrolysis product (i.e., acid 27). Indeed, the acid, as might have been predicted, displayed little affinity for 5-HT2 sites (Ki > 50,000 nM). Even if there was only a limited degree of hydrolysis, the affinities of 22, 23 and, perhaps, 25 might appear to be artificially low.

To further characterize the influence of lipophilicity and substituent length on affinity, five additional derivatives were prepared: the 4-chloro analog 6, the isopropyl, hexyl, and benzyl derivatives 12, 16, and 18, respectively, and the non-hydrolyzable propionyl derivative 24. The propionyl group of 24 is only slightly lipophilic but is, nevertheless, electron withdrawing. The affinities of

TABLE II. Relating Equations Generated in the Present Investigation

Equation[a]	LOG (1/Ki) =	r	n	SD	F
1.	5.95(0.18) + 0.95(0.15) π + 1.26 (0.49) σ_p	0.894	13	0.44	19
2.	6.07(0.22) + 0.79(0.17) π	0.816	13	0.58	22
3.	5.83(0.22) + 0.80(0.17) π	0.764	17	0.70	21
4.	7.01(0.41) + 1.10(0.17) π - 0.31 (0.10) L	0.871	17	0.55	22
5.	5.86(0.18) + 0.84(0.13) π	0.825	22	0.63	43
6.	6.90(0.34) + 1.11(0.13) π - 0.28 (0.08) L	0.894	22	0.51	38
7.	5.93(0.16) + 0.73(0.09) π	0.858	25	0.61	60
8.	6.96(0.31) + 1.06(0.12) π - 0.28 (0.07) L	0.911	25	0.49	53
9.	4.63(0.63) + 0.70(0.09) π + 0.82 (0.39) B1	0.878	25	0.57	37
10.	6.90(0.28) + 1.04(0.11) π - 0.32 (0.08) K2	0.912	25	0.47	57
11.	6.51(0.24) + 0.98(0.11) π - 0.37 (0.12) B5/B1	0.896	25	0.52	45
12.	6.01(0.14) + 0.75(0.08) π	0.903	23	0.50	93
13.	6.08(0.22) + 0.77(0.22) π	0.725	13	0.55	12
14.	5.93(0.19) + 0.93(0.19) π + 1.22 (0.48) σ_p	0.842	13	0.45	12
15.	3.81(0.48) + 0.50(0.13) π + 1.56 (0.42) B1	0.927	13	0.31	30
16.	3.44(0.68) + 2.02(0.42) B1	0.821	13	0.45	23
17.	3.10(0.52) + 2.40(0.34) B1 - 1.28 (0.41) F	0.914	13	0.33	25

[a]Equation 1 has been previously reported (3).

the five new compounds are predicted reasonably well by Eq 3 and 4 (see Table III). The SAS RSQUARE procedure was used (n = 22) to identify all possible new one-, two-, and three-variable relating equations with r ≥ 0.85. Once again, the importance of the lipophilicity term is underscored by Eq 5 (r = 0.825); inclusion of the sigma term (equation not shown) had no effect (r = 0.825). The correlation with π (Eq 5) is somewhat impoved by the presence of the STERIMOL parameter L (Eq 6; r = 0.894). Three additional equations were identified where r > 0.85: these include the lipophilicity term in combination either with molar refractivity (MR), or the STERIMOL parameters B1 and B5. However, both MR and B5, for the substituents in question, are significantly correlated with L, and inclusion of the B1 term with π was not statistically warranted. These equations continue to demonstrate the importance of the lipophilicity term and, to a lesser extent, a role for a shape/length term.

To further explore the length and lipophilicity factors, the next group of compounds evaluated included the octyl derivative 17 and the phenylpropyl derivative 19. The nitrile 21 was also evaluated. As expected, on the basis of their lipophilicity, the octyl and phenylpropyl derivatives constitute two of the highest affinity agents for 5-HT2 sites. Affinities predicted by original Eq 3, 4, and 6 are shown in Table III. At this point, the regression analyses were repeated using all 25 compounds. Only a single one-variable equation was identified with r ≥ 0.85 (i.e., correlation with π, Eq 7; r = 0.858). Addition of a sigma term did not improve the correlation of pi with

TABLE III. Observed and Predicted Affinities for Several Selected Compounds Using Equations 3, 4, and 6[a]

Agent	R	Observed	LOG (1/Ki) Calculated from:		
			Eq 3	Eq 4	Eq 6
6	Chloro	6.67	6.39	6.71	6.72
12	Isopropyl	7.12	7.04	7.43	7.47
16	Hexyl	8.60	8.38	8.01	8.20
18	Benzyl	8.15	7.43	7.80	7.86
24	Propionyl	6.13	5.87	5.57	5.62
17	Octyl	8.52	9.24	8.57	8.84
19	Phenylpro	8.00	8.34	8.05	8.23
21	Cyano	5.62	5.37	5.07	5.10

[a]See Table II for equations. None of the agents shown in the table were used in deriving Eq 3 or 4. Compounds 6, 12, 16, 18, and 24, but not 17, 19, or 21, were used in deriving Eq 6.

affinity ($r = 0.850$). Of the ten "best" two-variable equations, all contained the π term, and the ten "best" three-variable equations all contained the π and L terms. The most significant correlations with affinity are those that included π in combination with L (Eq 8; $r = 0.911$), B1 (Eq 9, $r = 0.878$), B5 ($r = 0.891$), and MR ($r = 0.876$); however, see comments above regarding use of the B1 and MR terms. Nevertheless, for the 25 parameters involved, there is now a significant internal correlation between π and L ($r = 0.74$), π and B5 ($r = 0.77$) and π and MR (0.81), but not between π and B1 ($r = 0.11$) (Table IV).

TABLE IV. Cross-correlation Matrix for Some of the Substituent Parameters Used in Generating Equations 7-11 (where n=25)

	π	σ_m	σ_p	F	R	MR	L	B1	B5	π^2	Es
π	1.000										
σ_m	-0.509	1.000									
σ_p	-0.033	0.920	1.000								
F	-0.573	0.980	0.823	1.000							
R	0.149	0.419	0.738	0.229	1.000						
MR	0.806	-0.349	-0.148	-0.432	0.262	1.000					
L	0.735	-0.195	-0.040	-0.265	0.254	0.847	1.000				
B1	0.108	0.108	0.177	0.170	0.082	0.079	-0.041	1.000			
B5	0.765	0.765	-0.352	-0.152	-0.434	0.252	0.978	0.862	1.000		
π^2	0.873	-0.492	-0.371	-0.528	0.024	0.783	0.779	-0.067	0.759	1.000	
Es	-0.330	-0.042	-0.209	-0.048	-0.398	-0.565	-0.432	-0.506	-0.540	-0.187	1.00

Though lipophilicity seems to be the single most important factor in determining affinity, a descriptor of size and/or shape continues to surface. Using all 25 compounds, a regression analysis was conducted employing the Kier and Hall shape and connectivity indexes. These included the simple connectivity indexes X1, X2, XP3, XP4, XP5, XP6, XC3, XC4, and the valence indexes XV1, XV2, XVP3, XVP4, XVP5, XVP6, XVC3, and XVPC4 calculated for the 4-position substituents. The shape indexes included K1, K2, K3, KA1, KA2, and KA3. These shape and connectivity indexes were calculated using the program MOLCONN2 (7,8). In no case did the regression coefficient exceed 0.5. This analysis was repeated using these indexes in combination with π and the most significant correlation identified is represented by Eq 10 (r = 0.912). The appearance of the second-order shape index K2 indicates that affinity increases as the spatial density of the 4-substituent increases (i.e., as K2 decreases). The Verloop ratio B5/B1 is another indicator of spatial density (and, for the substituents examined, K2 is highly correlated with B5/B1; r = 0.95), and a correlation between affinity and π in combination with this ratio might be expected. This was found to be the case (Eq 11; r = 0.896). The observed and predicted values for all 25 compounds using Eq 10 and 11 are shown in Table V. (See Figure 1 for a Craig plot of π vs σ_p.)

Overall, it seems as though affinity is related to the lipophilicity and, possibly, the shape of the 4-substituent...a conclusion drawn early on in these studies. However, the affinities of the two esters 22 and 23 are consistently over-predicted. Because of the possibility of hydrolysis, as mentioned above, a correlation between affinity and π was attempted in the absense of the two esters (Eq 12). Eq 12 relates the affinity of all 23 compounds with π (r = 0.903), and Table V shows that the equation fairly accurately predicts affinity at both ends of the affinity scale. The amino derivative 20 is predicted to be of lowest affinity, and the hexyl and octyl derivatives 16 and 17 are correctly predicted to be amongst the highest. Additional studies on newer derivatives will be necessary to further determine the role of shape/length of substituents on affinity. It is entirely possible that an optimal length has already been reached (e.g. compare the affinities of 16 and 17), and/or, that the folding back of the longer alkyl chains may complicate QSAR interpretations. Nevertheless, the effect of branched substituents and benzyl- or phenylpropyl-substituted derivatives remains to be examined.

<u>Evaluation of a Subset</u>. One final issue needs to be addressed: do all of the agents in Table I belong to a common set or do they constitute several subsets of agents? DOM (9) serves as a training drug in drug discrimination studies with rats (9). Furthermore, agents to which the DOM-stimulus generalizes (e.g. 1-5,7-11, 13) possess a high affinity for 5-HT2 sites; in fact, there is a significant correlation between the potencies of these agents in the drug discrimination paradigm and their affinity for 5-HT2 receptors (2). These results suggest that certain of the agents are capable of producing a common (agonist) effect and that the effect is related to their affinity for 5-HT2 sites. Interestingly, neither the tertiary butyl derivative 14 nor the amyl derivative 15 though they possess greater affinities than DOM, produce DOM-like stimulus effects in rats (10). We have recently determined that the DOM-stimulus generalizes to the chloro analog 6 (Ki = 218 nM) (Figure 2) and the isopropyl derivative 12 (Ki

= 76 nM) (Figure 2), but not to the benzyl derivative 18 (Ki = 7 nM) (Figure 3) nor, as anticipated, to the acid 27 (Ki > 50,000 nM) (Figure 2). Another persistent problem is the lack of most of the equations to account for the nitro analog 4. Compound 4 appears to be an agonist, it produces stimulus effects similar to those of DOM (10), it possesses a reasonably high affinity for 5-HT2 sites (Ki = 300 nM)(1), and yet the nitro group is not lipophilic. To examine the possibility that the agents in Table I might represent two subsets of compounds, a correlation between π and affinity was attempted only for those compounds with known agonist activity in the drug discrimination paradigm (i.e., compounds 1-13). Interestingly, this correlation (Eq 13; r = 0.725) is not as significant as Eqs 2, 5, 7 or 12. However, inclusion of a sigma (Eq 14; r = 0.842) or B1 term (Eq 15; r = 0.927) improves the correlation. In fact, there is a sig-

TABLE V. Observed and Predicted Affinities for Agents Using Selected Relating Equations[a]

	R	Obs[a]	Eq 8	Eq 9	Eq 10	Eq 11	Eq 12	Eq 14	Eq 15	Eq 17
1	H	5.28	6.39	5.46	6.18	6.14	6.00	5.93	5.37	5.50
2	OMe	5.78	5.83	5.73	5.87	5.64	5.99	5.58	5.90	6.00
3	OEt	5.65	6.04	6.01	6.02	5.96	6.29	5.99	6.10	6.06
4	NO_2	6.52	5.71	5.84	5.56	5.70	5.80	6.61	6.32	6.32
5	F	5.95	6.38	5.85	6.28	6.27	6.11	6.13	5.98	5.79
6	Cl	6.67	6.74	6.62	6.88	6.84	6.54	6.87	6.97	6.89
7	Br	7.39	6.81	6.84	7.03	6.98	6.66	7.01	7.27	7.21
8	I	7.72	6.97	7.19	7.30	7.24	6.85	7.19	7.71	7.74
9	Methyl	7.00	6.76	6.28	6.72	6.56	6.43	6.24	6.46	6.80
10	Ethyl	7.00	6.90	6.60	6.95	6.74	6.78	6.70	6.68	6.81
11	Propyl	7.16	7.24	6.97	7.24	7.18	7.17	7.22	6.95	6.82
12	iPropyl	7.12	7.44	7.27	7.44	7.39	7.15	7.29	7.53	7.72
13	Butyl	7.24	7.51	7.38	7.57	7.49	7.61	7.72	7.24	6.82
14	tButyl	7.72	7.92	8.16	8.00	8.00	7.50	-	-	-
15	Amyl	8.15	7.86	7.75	7.85	7.93	8.02	-	-	-
16	Hexyl	8.60	8.08	8.13	8.12	8.21	8.42	-	-	-
17	Octyl	8.52	8.66	8.88	8.65	8.92	9.23	-	-	-
18	$CH_2-C_6H_5$	8.15	7.81	7.29	7.36	7.01	7.52	-	-	-
19	Phenpro	8.00	8.11	8.09	8.05	7.79	8.38	-	-	-
20	NH_2	4.59	4.89	4.89	4.87	4.76	5.08	-	-	-
21	CN	5.62	5.19	5.56	5.30	5.58	5.58	-	-	-
22	COOnPr	5.82	6.22	6.73	6.19	6.47	-	-	-	-
23	COOnBu	5.61	6.46	7.12	6.49	6.78	-	-	-	-
24	$COCH_2CH_3$	6.13	6.22	6.02	5.66	5.78	5.58	-	-	-
25	CONHnPr	5.12	4.90	5.85	4.88	5.14	5.87	-	-	-

[a]Log (1/Ki) using affinity (Ki) data from Table I.

18. GLENNON & SEGGEL *Phenylisopropylamine Interaction with 5-HT2* 271

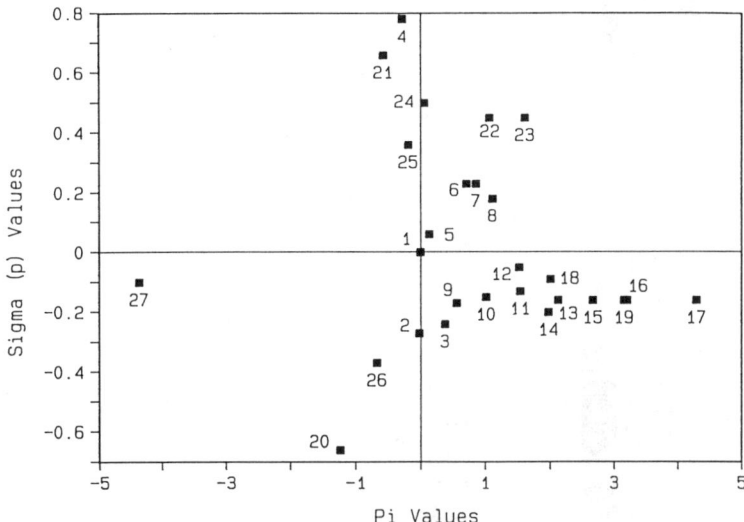

Figure 1. A Craig plot of π versus σ_p values for all 27 substituents included in the present investigation. Note that skewing to the $+\pi$ quadrant resulted as a consequence of our finding that increased lipophilicity enhances affinity, thus leading to the synthesis of a greater number of such derivatives.

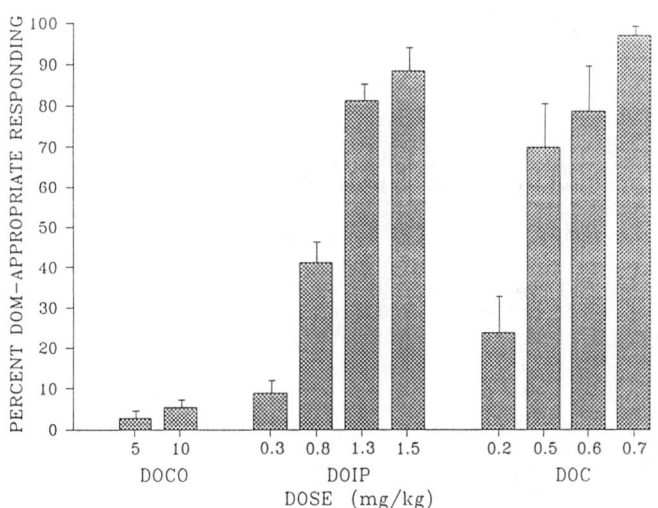

Figure 2. Results of stimulus generalization studies with the 4-COOH (DOCO; **27**), 4-isopropyl (DOIP; **12**), and 4-chloro (DOC; **6**) derivatives of 2,5-DMA in DOM-trained animals. (Drugs were administered via the ip route 15 min prior to testing; figure may not show all doses evaluated.)

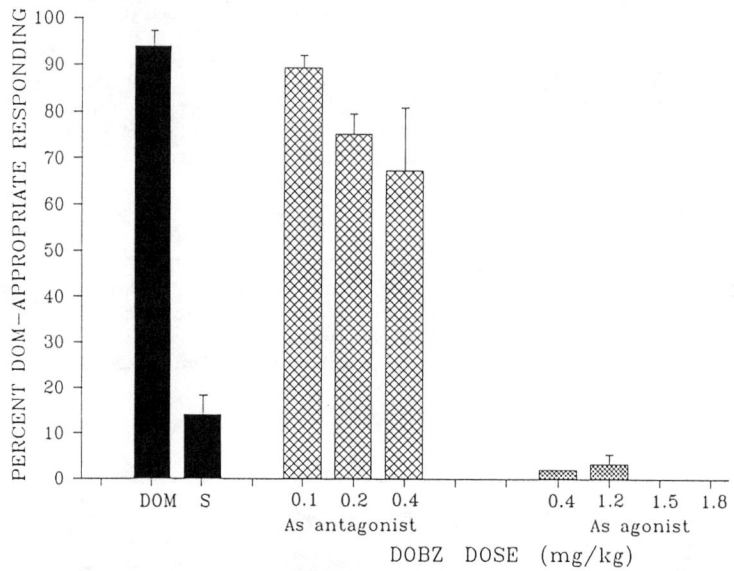

Figure 3. Results of stimulus antagonism and stimulus generalization studies with DOBZ (7) in DOM-trained rats. (DOBZ was administered 10 min prior to DOM in the antagonism studies, and in the absence of DOM in the agonism studies. Doses of DOBZ greater than 0.4 mg/kg in the antagonism studies resulted in disruption of behavior. In the agonism studies, 1.5 and 1.8 mg/kg of DOBZ produced disruption of behavior. DOM = the effect of 1 mg/kg of DOM, and S = the effect of saline.)

nificant correlation with B1 alone (Eq 16; r = 0.821) and this relationship is improved upon addition of an electronic (F) term (Eq 17; r = 0.914). The results suggest that the agonists may indeed be behaving differently. These equations (Table V) seem to better account for the activity of the agonists and, in particular, the affinity of the nitro derivative.

Discussion

The present study investigated the QSAR for the binding of 27 4-substituted 2,5-DMA analogs at 5-HT2 receptors. A preliminary study on 13 agents suggested that the lipophilic and electron withdrawing character of the 4-substituents played a major role in determining affinity (3). However, because the choice of substituents reflected a rather narrow range of parameters, fourteen additional analogs were examined. Analysis of the extended series reveals that lipophilicity is a dominant factor in determining affinity (e.g. see Eq 7 and Eq 12) but that the shape (size/length ?) of the substituent may also play a role (see Eq 8-11). The electronic contribution of the 4-substituent is not as evident as it was in the original series of 13 agents. However, if an analysis is conducted only on those agents known to possess agonist properties, the electronic term surfaces once again (e.g. see Eq 14 and Eq 17). Nevertheless, lipophilicity and shape still appear to play a role. (It might be noted for the original series of agents investigated in the preliminary studies, 11 of the 13 were agonists; this might account for the initial observation that an electronic term is of significance.)

These results suggest that the agents in Table I consist of at least two distinct subsets (from a QSAR point of view). And, indeed, there seems to be supporting evidence for this suggestion. While these QSAR studies were in progress, we demonstrated that tritiated DOB (i.e., the tritiated analog of 7) labels what appears to be the high-affinity state of the 5-HT2 receptors (11). Furthermore, antagonists display a similar affinity for 5-HT2 sites regardless of whether a tritiated antagonist (e.g. ketanserin) or a tritiated agonist (e.g. DOB) is used in the radioligand binding studies. Agonists, on the other hand, display a significantly greater affinity when tritiated DOB rather than tritiated ketanserin is used to label the receptors (12). That is, agonists display a higher affinity for agonist-labeled sites than for antagonist-labeled sites. For example, the affinities (Ki values) of DOB (7), DOI (8), 2,4,5-TMA (2) and 2,5-DMA (1) for [^{3}H]DOB-labeled sites is 0.8, 0.7, 81, and 268 nM, respectively (13). Thus, if the agents in Table I represent a mixture of agonists and antagonists, the affinities of the antagonists would be exaggerated relative to those of the agonists. The QSAR differences observed when the agonist subset was examined alone might simply reflect this fact. Although binding data from [3H]DOB-labeled 5-HT2 sites were only available for nine 2,5-DMA derivatives (see footnote to Table I), we anticipated that a QSAR analysis might provide results comparable to those shown by Eq 13-17. Indeed, a search of all possible one- and two-variable relating equations revealed that the only significant one-variable equations were those containing a π term (r=0.79, n=9) and a B1 term (r=0.80) (Eq 18 and 21, respectively; Table VI). And, amongst the most significant two-variable terms were those containing π in combination with either σ_p

TABLE VI. Relating Equations Generated Using [^{3}H]DOB Binding Data

Equation[a]	LOG (1/Ki) =	r	SD	F
18.	7.20(0.36) + 1.16(0.34) π	0.794	0.67	12
19.	7.42(0.33) + 1.09(0.29) π + 2.16(1.16) σ_p	0.875	0.58	10
20.	4.84(0.69) + 0.80(0.23) π + 1.73(0.48) Bl	0.939	0.41	22
21.	4.37(1.09) + 2.47(0.69) Bl	0.801	0.66	13
22.	2.93(0.71) + 3.66(0.49) Bl - 3.20(0.79) F	0.951	0.37	28

[a]Binding data for only nine agents were available (see footnote to Table I).

(Eq 19, r=0.88) or Bl (Eq 20, r=0.94), or Bl in combination with F (Eq 22, r=0.95). Thus, here also, the same terms identified by Eq 13-17 seem significant in Eq 18-22. Due to the lack of a larger data set, a more detailed analysis of [3H]DOB binding data was not thought to be warranted at this time. However, these preliminary results tend to support our suggestion that data obtained using this new radioligand would be useful for subsequent studies.

A number of the 2,5-DMA derivatives are thought to be 5-HT2 agonists. The above discussion would seem to argue, however, that certain of the 2,5-DMA derivatives might constitute 5-HT2 antagonists (or perhaps mixed agonists-antagonists). As such, the affinities of some of these agents rival that of the classical 5-HT2 antagonist ketanserin (Ki = 1.2 nM). Preliminary results on several of these agents reveal that, indeed, antagonist properties are evident. For example, the 4-benzyl derivative (i.e., DOBZ, 18) and the 4-amyl derivative (i.e., DOAM, 15) were examined as antagonists of the DOM stimulus. Figures 3 and 4 show that neither agent was capable of acting as an antagonist of the DOM-stimulus at the doses evaluated; doses higher than those shown in the figures (i.e., > 0.4 mg/kg for DOBZ, and > 0.2 mg/kg of DOAM) in combination with the training dose of DOM (1 mg/kg) produced disruption of behavior and, thus, antagonism could not be evaluated. On the other hand, the 4-tertiary butyl derivative (i.e., DOTB, 14) does appear to produce some antagonism (14). A study was conducted where doses of DOTB (14) were administered either in combination with 1 mg/kg of DOM or, in the control studies, with 1 mL/kg of 0.9% saline vehicle. Figure 5 shows that DOTB attenuates the stimulus effects of DOM (to approximately 35% DOM-appropriate responding). However, at high doses, DOTB in combination with saline results in an increasing degree of DOM-like responding; higher doses produce disruption of behavior. These results would suggest that at relatively low doses, DOTB behaves as a 5-HT2 antagonist, but that at somewhat higher doses it has some agonist properties. Is DOTB acting as a mixed agonist-antagonist? This is currently being further pursued. Nevertheless, studies with isolated rat aorta (which supposedly possesses peripheral 5-HT2 receptors) suggest this to be the case (Roth, Suba, Seggel & Glennon, unpublished data).

This is the first QSAR study of the affinity of phenylisopropyl-

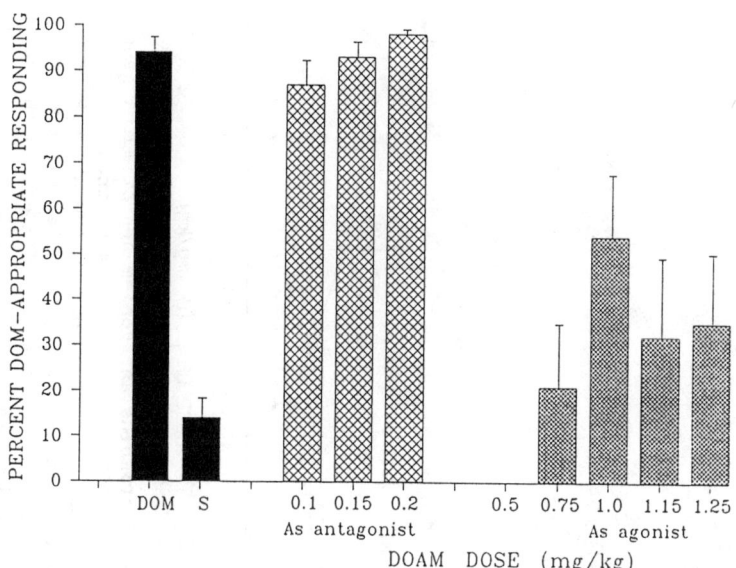

Figure 4. Results of stimulus antagonism and stimulus generalization studies with DOAM (15) in DOM-trained rats. (DOAM was administered 10 min prior to DOM in the antagonism studies, and in the absence of DOM in the agonism studies. Doses of DOAM greater than 0.2 and 1.25 mg/kg in the antagonism and agonism studies, respectively, resulted in disruption of behavior. In the agonism studies, 0.5 mg/kg of DOAM produced 0% DOM-appropriate responding.)

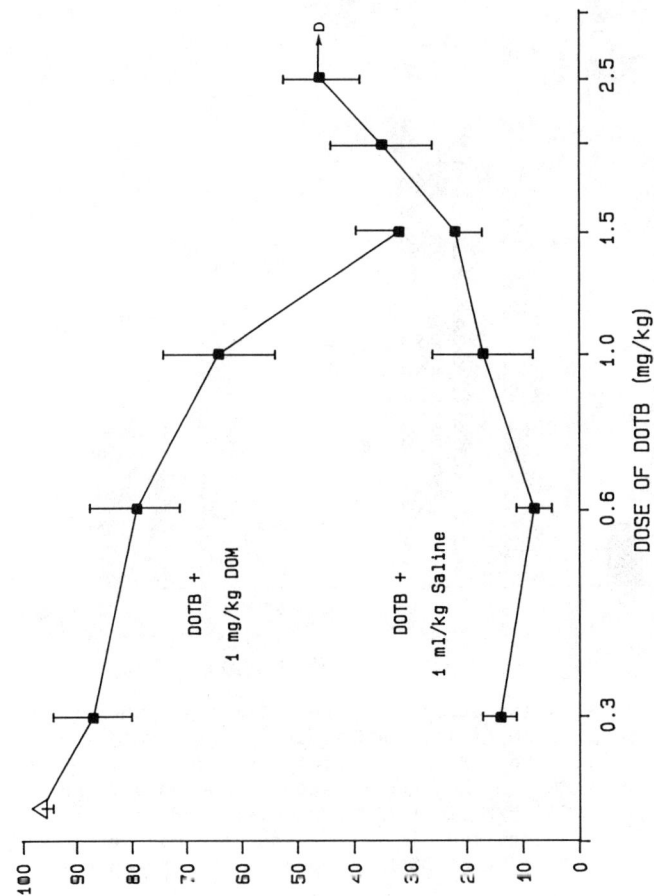

Figure 5. Results of stimulus antagonism and stimulus generalization studies with DOTB (14) in DOM-trained rats. Open triangle represents the effect of DOM in the absence of DOTB; solid squares represent the effect of DOTB administered 1 min prior to 1 mg/kg of DOM (upper curve) or, in the control studies, 1 min prior to 1 mL/kg of saline (lower curve). D = disruption of responding. [A preliminary account of this work has been reported (14).]

amine analogs for central 5-HT2 sites. It would appear that associated with the receptor is a hydrophobic region capable of interacting with 4-position substituents of the phenylisopropylamines. This region seems to be responsible for modulating the affinity, but not necessarily the intrinsic activity, of the 2,5-DMAs. 2,5-DMAs with hydrophilic groups at this position (e.g.-OH, -NH_2) display little to no affinity, whereas agents with lipophilic groups display high affinity (but may lack intrinsic activity). Indolealkylamines (e.g. tryptamine analogs) also bind at 5-HT2 sites and we have attempted to explain the similarities in the modes of binding of these two classes of agents at 5-HT2 receptors (15). There is evidence that polar substituents at the 7-position of tryptamines decrease their affinity (Table VII) (15). In contrast, lipophilic substituents seem to enhance affinity. A 1-methyl group also somewhat enhances affinity (see compound 31, Table VII) suggesting that the hydrophobic region might accomodate both the 1- and 7-position of the tryptamines. This has led us to prepare several 1-substituted and 7-substituted tryptamine derivatives for purpose of comparison. Indeed, both the 1-amyl and 7-amyl derivatives bind with greater affinity than their unsubstituted counterparts (16). Recently, Cohen and co-workers (17) have also found that 1-substituted indolealkylamines display a higher affinity for 5-HT2 receptors than their unsubstituted derivatives.

Though this is the first QSAR study on 5-HT2 binding, several other studies have been conducted on other aspects of phenylisopropylamine pharmacology that may have a direct bearing on the present results. For example, many phenylisopropylamine derivatives are hallucinogenic in humans and we have demonstrated that a significant

TABLE VII. Affinities of Several Indolealkylamines for 5-HT_2 Sites

	R	R'	5-HT_2 Affinity Ki Value (nM)[a]
28	H	H	1,200
29	OH	H	>10,000
30	Br	H	170
31	H	CH_3	400

[a] Data from reference 15. Tritiated ketanserin was used as radioligand.

correlation exists between the human potencies of these agents and both their discrimination derived potencies and their affinities for 5-HT2 receptors (2). In fact, we have proposed that the hallucinogenic effect of the phenylisopropylamines is mediated via a 5-HT2 agonist interaction (2). In 1975, Barfknecht et al (18) reported that that the octanol-water partition coefficients of phenylisopropylamines is an important, though not necessarily exclusive, determinant of their hallucinogenic potency. Using a slightly larger data set, including a series of rearranged isomers, we reported that the overall lipophilicity of these agents might be important, but that lipophilicity was probably primarily a reflection of the lipophilic character of the 4-position substituents (19). We also discussed the importance of electronic effects (i.e., ionization potentials) particularly for those agents lacking a significantly lipophilic 4-substituent (19). Shulgin and Dyer (20) had also demonstrated a relationship between the hallucinogenic potency of a small set of 2,5-DMA analogues and the lipophilic character of their 4-position substituents. Since that time, there have been a number of SAR and QSAR studies on hallucinogenic agents [see Gupta et al (21) for a reveiw], and several of these have made mention of the lipophilic character of the 4-position substituents. Finally, this bings us once again to the nitro compound 4. Early on, we considerered the lack of lipophilic character and potential significance of electronic terms in explaining the activity of this agent (10). We even speculated that weak lipophilic character might be overshadowed by electronic effects of the ring (10). Likewise, Gomez-Jeria and co-workers (22) have confirmed the activity of the nitro analog and have also argued for the importance of an electronic term in determining activity. Of course, other factors may need to be considered; for example, this particular agent may bind in a different manner or the nitro group may influence the orientation of the adjacent methoxy group resulting in a "better" fit for the molecule than expected.

Several other studies have examined the SAR and QSAR of these and various other phenylisopropylamine analogues. Included among these investigations are their interactions at different peripheral serotonin receptors (e.g. isolated rat fundus, sheep umbilical artery); however, the nature of the relationship between the examined 5-HT receptors and 5-HT2 receptors remains, for the most part, unknown. A number of these studies have been mentioned in the review by Gupta and co-workers (21).

Summary

On the basis of the present studies, it appears that the affinity of the 2,5-DMAs for 5-HT2 receptors can be accounted for, primarily, by the lipophilicity of the 4-position substituent. Other factors, particularly those dealing with length or shape, may also play a role. It is entirely possible that the electronic nature of the 4-position substituents is also involved in affinity (at least for a small subset of 2,5-DMA analogs) but this has been more difficult to demonstrate; electronic factors may also be involved in intrinsic activity, particularly as it relates to altering the electron rich character of the aromatic nucleus of the phenylisopropylamines. The present studies also have indirect ramifications with regard to the agonist versus antagonist activity of phenylisopropylamines, and, additionally, may help explain the nature of the interaction of indolealkylamines with these same receptors. With the recent demonstration that agonists display a greater affinity for tritiated agonist-labeled versus tritiated antagonist-labeled sites (12), more realistic QSAR results might be derived using binding data from 5-HT2 sites labeled with [^{3}H]DOB as radioligand. Such studies are already underway.

Acknowledgments: These studies were supported, in part, by US PHS grants DA 01642 and NS 23520. We also wish to express our appreciation to Bryan Misenheimer and Betsy Mack for their assistance in obtaining the drug discrimination data and to Dr. L. B. Kier for helpful discussions regarding the use and interpretation of the connectivity and shape indexes.

Literature Cited

1. Shannon, M.; Battaglia, G.; Glennon, R. A.; Titeler, M. Eur. J. Pharmacol. 1984, 102, 23-29.
2. Glennon, R. A.; McKenney, J. D.; Titeler, M. Life Sci. 1984, 35, 2505-2511.
3. Seggel, M. Youssif, M.; Titeler, M.; Lyon, R. A.; Glennon, R. A. Va. J. Sci. 1986, 37, 122.
4. Hansch, C.; Leo, A. Substituent Constants for Correlation Analysis in Chemistry and Biology; John Wiley and Sons: New York, 1979
5. Verloop, A.; Tipker, J. In QSAR in Drug Design and Technology; Hadzi, D.; Jerman-Blazic, B., Eds.; Elsevier Science: Amsterdam, 1987; pp 97-121.
6. Verloop, A. The STERIMOL Approach to Drug Design; Marcel Dekker: New York, 1987.
7. Hall, L. H. MOLCONN2: A Program for Molecular Topology Analysis; Quincy, MA, 1987.
8. Kier, L. B. Med. Res. Rev. 1987, 7, 417-440.
9. Glennon, R. A. In Transduction Mechanisms of Drug Stimuli; Colpaert, F. C.; Balster, R., Eds.; Springer-Verlag: Berlin, 1988, pp 16-31.
10. Glennon, R. A.; Young, R.; Benington, F.; Morin, R. D. J. Med. Chem. 1982, 25, 1163-1168.
11. Titeler, M.; Herrick, K.; Lyon, R. A; McKenney, J. D.; Glennon, R. A. Eur. J. Pharmacol. 1985, 117, 145-146.
12. Titeler, M.; Lyon, R. A.; Davis, K. H.; Glennon, R. A. Biochem. Pharmacol. 1987, 36, 3265-3271.
13. Titeler, M.; Lyon, R. A.; Glennon, R. A. Psychopharmacology 1988, 94, 213-216.
14. Glennon, R. A. In Clandestinely Produced Drugs, Analogues, and Precursors: Problems and Solutions; Klein, M.; McClain, H.; Sapienza, F.; Khan, I., U.S. Government Printing Office: Washington, DC, 1989.
15. Lyon, R. A.; Titeler, M.; Seggel, M. R.; Glennon, R. A. Eur. J. Pharmacol. 1988, 145, 291-297.
16. Chaurasia, C. S.; Glennon, R. A. Va. J. Sci. 1988, 39, 165.

17. Marzoni, G.; Garbrecht, W. L.; Fludzinski, P.; Cohen, M. L. J. Med. Chem. 1987, 30, 1823-1826.
18. Barfknecht, C. F.; Nichols, D. E.; Dunn, W. J. III J. Med. Chem. 1975, 18, 208-210.
19. Domelsmith, L. N.; Eaton, T. A.; Houk, K. N.; Anderson, G. M.; Glennon, R. A.; Shulgin, A. T.; Castagnoli, N.; Kollman, P. A. J. Med. Chem. 1981, 24, 1414-1421.
20. Shulgin, A. T.; Dyer, D. C. J. Med. Chem. 1975, 18, 1201-1204.
21. Gupta, S. P.; Singh, P.; Bindal, M. C. Chem. Rev. 1983, 83, 633-649.
22. Gomez-Jeria, J. S.; Cassels, B. K.; Saavedra-Aguilar, J. C. Eur. J. Med. Chem. 1987, 22, 433-437.

RECEIVED March 30, 1989

Chapter 19

Analysis of Binding at 4-Aminobutyric Acid Receptor Sites by Structure—Activity Relationships

Philip S. Magee[1] and James W. King[2]

[1]BIOSAR Research Project, Vallejo, CA 94591 and School of Medicine, University of California, San Francisco, CA 94143
[2]U.S. Army Chemical Research, Development and Engineering Center, Aberdeen Proving Ground, Aberdeen, MD 21010—5423

> Series of GABA-ergic compounds were analyzed through the expressed binding response (IC50) in the brain, spinal cord and uptake systems of man and various animals. The technique used was multiple regression of the pIC50 values against the variations in substructural features (1.0/0.0). Nearly all substructural factors made negative contributions relative to the basic GABA structure. Spacing is critical and binding occurs in a sterically restricted lipophilic cleft between the amino and carboxyl sites. The cleft exhibits chiral selection and the most probable binding mechanism is ion-paring.

There are two classes of neural receptors for binding of the inhibitory neurotransmitter, 4-aminobutyric acid (GABA). Of primary interest to this study is the $GABA_A$ receptor which populates both pre- and post-synaptic neural gaps in the CNS. Of lesser interest is the pre-synaptic $GABA_B$ receptor and various non-neural receptor sites. These sites are readily classified by blocking responses to bicuculline (1) or picrotoxinin (2). The post-synaptic $GABA_A$ receptor is a membrane embedded complex mediating the influx and efflux of chloride ion and possessing allosteric binding sites for benzodiazepines, picrotoxinin and some avermectins (3). Reciprocal allosteric modulations among the four classes of receptor sites are easily observed _in vitro_. The receptor has been shown to be a glycoprotein, like most membrane proteins (4). Two and possibly three GABA receptor sites on the complex can be identified kinetically by use of Scatchard plots (5-9). Krogsgaard-Larsen presents evidence for three binding sites (6) which he terms Low, Medium and High (3). A further complication is the presence of an endogeneous protein inhibitor binding to the high affinity sites (10,11). These sites can be exposed for binding studies through a complex washing protocol (9) using the non-ionic detergent, triton X-100 (12). Johnson and co-workers have shown extraordinary increases in GABA binding to rat brain preparations by Triton X-100 extraction (13).

0097—6156/89/0413—0281$06.00/0
© 1989 American Chemical Society

A variety of GABA related chemicals bind to the GABA$_A$ sites to produce both agonist and antagonist responses (3,14,15). Bicuculline (1), picrotoxinin (2), iso-THIP and iso-THAZ appear to be specific GABA receptor antagonists (3,14). Muscimol, THPI, isoguvacine and piperidine-4-sulfonic acid (P4S) are specific GABA agonists while muscimol, THPO, guvacine and nipecotic acid are specific GABA uptake inhibitors (15,17). Muscimol, derived from the mushroom Amanita muscaria, is quite toxic, ten times more potent than GABA as an agonist, and very specific in binding only to high affinity sites (5,18). DeFeudis and co-workers have reported approximately twice as many binding sites for muscimol as for GABA (19). Of the known specific agonists, THPI is the only one capable of penetrating the blood-brain barrier to exert analgesic and anticonvulsant effects (references cited in 6). Clinical responses of GABA agonists and antagonists have been reviewed (20). Of related interest is the report that both anesthetic and convulsant barbituates enhance GABA binding in a dose dependent manner, presumably an allosteric effect (13).

Stereochemistry is clearly involved in the binding of GABA and GABA-ergic compounds. Andrews and Johnston postulate that GABA binds to GABA$_A$ receptors in an extended conformation and to GABA$_B$ receptors in a folded conformation (21). These considerations led Block and King to a detailed conformational study of GABA, muscimol and nipecotic acid based on X-ray crystal data (22). Differences in binding affinities of stereoisomeric GABA-ergic compounds range from small to large (14,17).

A statistical approach to mapping the GABA receptor sites is presented in this study. We attempt to analyze the contributions of key substructures to the measured binding affinities of GABA-ergic compound series. Olsen has noted variations in the rank order potency of GABA analogs between systems (23). By use of multiple regression analysis, we are attempting to quantify these observations in terms of sub-structure contributions (Figure 1).

Research Method. Most studies involve carefully measured IC50's for a substantial number of GABA-ergic compounds on membrane or cellular preparations. Most of the preparations are treated by complex protocols involving extraction with Triton X-100 (9). In many experiments, the GABA receptor sites are first saturated with ^{3}H-GABA, ^{3}H-muscimol, ^{3}H-diazepam or ^{3}H-P4S followed by measurement of the concentration of GABA-ergic compound required to displace 50% of the bound tracer (IC50). Corrections for non-specific binding are made to refine the values. These numbers are converted to pIC50's (log 1/IC50) for regression against the presence or absence of specific substructures.

Indicator variables (I = 1.0/0.0) are used to code the presence or absence of a key substructure. Regression of real numbers (pIC50's) against a matrix of indicator variables is a valid procedure for large sets, as in the Free-Wilson method. However, many of the sets in this study are small (n = 7-10) and it is probable that statistical measures for these sets are only approximate. The overall consistency of substructure dependence in both small and larger sets is considered to validate these measures in a semi-quantitative sense.

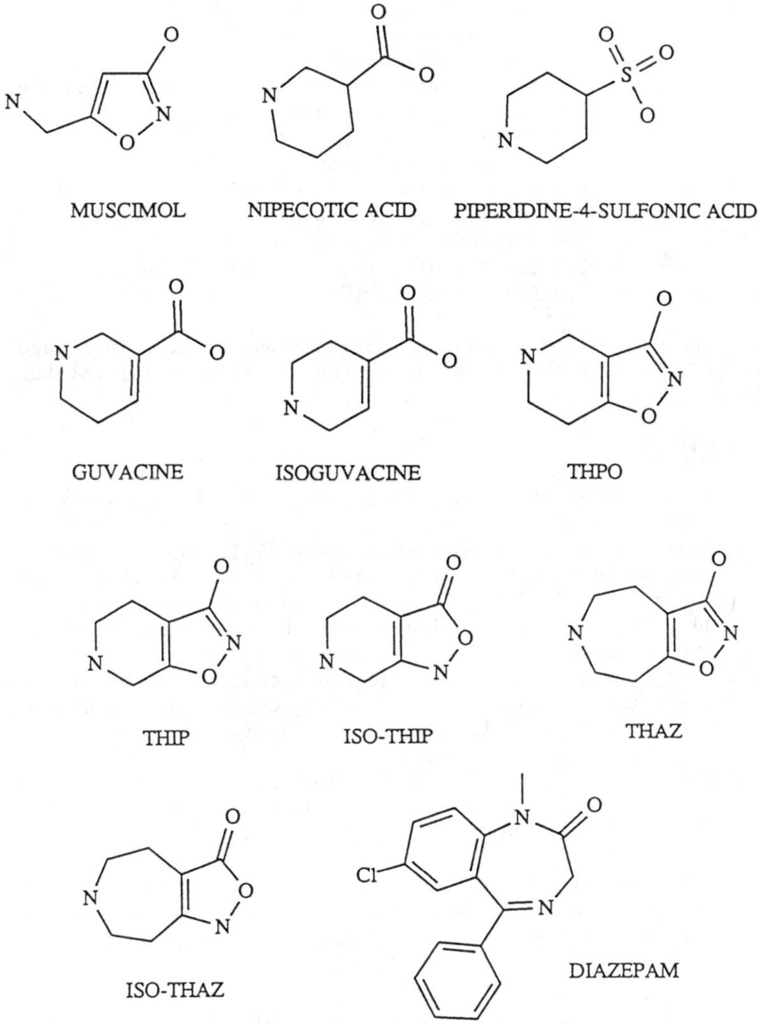

Figure 1. GABA agonists and antagonists.

There are major variations in chemical structure among the sets, making general classification of sub-structures difficult. We have attempted to use indicator variables that describe comparable changes in the same binding region of different sets although the compounds described may not be identical. The following variables are typical.

IHETS =   1.0 for isothiazole ring
IHETO =   1.0 for isoxazole ring
IRNG =   1.0 for aliphatic ring closure
IDB =   1.0 for conjugated double bond or equivalent rigidity
IME =   1.0 for branching methyl group
I2OH, I3OH =   1.0 for 2- or 3-OH groups
ISO3 =   1.0 for SO_3H replacement of COOH
INSUB =   1.0 for alkyl-substituted amino
ILNG, ISL, ISHT =   1.0 for long and short spacing (2- or 4-CH_2's)
INH2 =   1.0 for additional NH_2 group
IR =   1.0 for R-configuration in chiral analogs
ING =   1.0 for unusual or "bad" features

Areas of study are divided into 1. Brain Studies, 2. Spinal Cord Studies, 3. Non-Competitive Binding Studies, 4. Cellular Uptake Studies.

Brain Studies

<u>Human</u>. Displacement of ^{3}H-GABA from 7 different regions of human brain tissue by GABA and 8 other GABA-ergic compounds provides a powerful overview of the GABA binding site (24). Despite several-fold variations in the IC50's for these regions, all pIC50's are colinear (r = 0.938-0.983). The Substantia Nigra correlates lowest (r = 0.938-0.966), while the richest GABA brain region, Cerebellar Cortex, correlates highest with the others (r = 0.946-0.983). Moreover, all provide equivalent SAR equations containing the same factors with similar loadings. The following expression for binding at the Caudate Nucleus is typical of the 9 equations.

pIC50(CN) = -1.35 IRNG - 0.74 IHETS + 0.88 IDB - 2.93 IME + 7.76
    T =   6.18   2.28   4.02   10.77
    n = 9   r = 0.985   s = 0.251   F = 32.56

In a related study containing hydroxylated and sulfonic acid analogs, we again find excellent additivity of substructural features (25). Data are for inhibition of ^{3}H-GABA binding to human cerebellar membranes.

pIC50 = 0.813 IRNG - 1.20 I3OH - 0.770 ISO3 - 3.16 ISHT - 4.25 INSUB + 7.78
    T = 2.15   2.60   2.16   6.43   10.52
    n = 11   r = 0.981   s = 0.522   F = 25.98

Other studies on human cerebellar membranes deal with closely related analogs differing mostly by methyl substitution in the 2,3- and 4-position of GABA (26,27). The SAR equations show nearly perfect additivity of the positional methyl effects, all with strong negative coefficients (-1.55 to -2.51).

Mouse and Rat. Displacement ^{3}H-GABA from a mouse brain membrane preparation shows additive responses of substructures similar to those observed in the human brain (28).

pIC50 = -1.94 ISL - 0.98 ISO3 - 1.05 IRNG - 2.71 INH2 - 1.67 IOH + 7.07
T =   6.18   2.12   2.71   6.58   2.70
n = 14   r = 0.962   s = 0.536   F = 19.70

Binding studies in rat brain preparations involve a broad diversity of compounds and reveal a variety of interesting effects. Inhibition of ^{3}H-GABA binding by Olsen and co-workers shows the criticality of the 4-carbon spacing of GABA as well as sensitivity to methyl substitution alpha to the amino group (29).

pIC50 = -2.37 ISL - 1.05 I4ME + 6.72
T =   7.16   3.49
n = 14   r = 0.909   s = 0.481   F = 26.32

Another study confirms the spacing criticality and shows different negative contributions of R- and S-configured methyl groups in the 4-position (30). Different negative effects for 2- and 3-OH groups are also apparent. Data are for displacement of ^{3}H-GABA in rat brain receptor sites.

pIC50 = -1.67 I4ME(R) - 2.14 I4ME(S) - 1.53 ILNG - 1.69 I2OH - 0.84 I3OH + 6.61
T =   3.40   3.75   3.25   3.13   2.10
n = 13   r = 0.890   S = 0.576   F = 5.33

Bovine. A single study of bovine cortex membranes by Krogsgaard-Larsen and co-workers provides some comparison with human, mouse and rat preparations (31). The selection of compounds does not reveal much detail, but reaffirms the importance of optimal spacing. Data are for competitive displacement of ^{3}H-GABA and ^{3}H-P4S. The pIC50's are colinear (r = 0.963).

pIC50(GABA) =   0.67 IHETO - 1.98 ISL + 6.90
T =   2.68   7.92
n = 13   r = 0.931   s = 0.377   F = 32.73

pIC50(P4S) =   0.43 IDB - 1.99 ISL + 6.88
T =   1.16   4.46
     (not significant)
n = 14   r = 0.820   s = 0.680   F = 11.26

Spinal Cord Studies

Cat. Krogsgaard-Larsen and co-workers present a data set composed principally of heterocyclic and cyclic GABA-ergic analogs (31). Data are IC50's for binding to receptor sites in cat spinal cord. Negative effects for the isothiazole and alicyclic rings are clearly defined.

pIC50 = -1.35 IHETS - 1.21 IRNG - 2.67 ING + 2.01
T =   2.44   3.00   4.81
n =   14   r = 0.867   s = 0.710   F = 10.08

Rat. Equimolar potencies for depolarization of dorsal(DR) and ventral(VR) root fibers of the rat spinal cord have been measured for a varied set of GABA-ergic compounds (32). There are too many features in the data set(6) for high correlation of non-IC50 data, but the dominant features are easily extracted. LogVR and LogDR are colinear (r = 0.891).

LogDR = -0.72 IRNG - 1.27 I3OH + 0.306
T =   2.76   2.70
n =   12   r = 0.748   s = 0.429   F = 5.72

LogVR = -0.73 IRNG - 1.15 I3OH + 0.260
T =   2.70   2.35
n =   12   r = 0.724   s = 0.447   F = 4.95

Non-Competitive Binding Studies

Binding at the GABA complex receptor site has an allosteric effect on the binding of ^{3}H-diazepam at its receptor site. EC50's of diazepam binding have been measured for a varied set of cyclic and acyclic GABA-ergic compounds in rat forebrain membranes (33). The correlation shows negative contributions by methyl groups alpha to amino and especially those in the R-configuration. These two effects are confounded in the analysis in that R-Me groups are counted twice. The other major effect is the very weak response of cyclic vs. acyclic structures (IRNG).

pEC50 = -2.54 IRNG - 1.73 IME - 1.60 IME(R) + 5.48
T =   6.97   3.94   3.65
n =   11   r = 0.952   s = 0.506   F = 22.75

In another study of nine optical isomers of substituted GABA, crotonate and muscimol analogs, IC50's were measured for competitive binding against ^{3}H-GABA, ^{3}H-THPI and ^{3}H-P4S (34). These were compared with the non-competitive (stimulated) EC50 binding of ^{3}H-diazepam. All pIC50's and the pEC50 were colinear (r = 0.905-0.993), despite two orders of magnitude in the concentration response. Correlations are weak (r = 0.571-0.696) for this set (n = 9), but the major negative factor is the double-bond rigidity of the crotonates and muscimols (coefficients = -0.77 to -1.46).

Brain and Cellular Uptake Studies

Extrasynaptic receptors in the supportive glial cells of the CNS are responsible for the uptake (synaptic gap clearance) and transport of GABA to terminate the neurotransmission. These receptor sites differ markedly from the GABA$_A$ sites in SAR response to GABA analogs. Other uptake sites within the neuronal structure appear to differ from those in the glial cells according to comparative studies by Schousboe and co-workers (35). Data are for acyclic and cyclic GABA analogs with uptake measured in cultured astrocytes and mouse brain minislices. The pIC50's show only moderate colinearity (r = 0.780), though the regression equations show similar factors.

pIC50(mouse brain) =   -1.10 I4ME + 0.649 IR - 0.585 IRNG + 4.16
                T =   4.18   2.18   2.22
                n =   12   r = 0.856   s = 0.416   F = 7.31

pIC50(astrocytes) =   -1.21 I4ME + 0.743 IR + 3.98
               T =   2.84   1.53(weak)
               n =   12   r = 0.699   s = 0.700   F = 4.29(weak)

Ring structures and 4-methyl groups are negative factors while the R-configuration is favored for both cyclic and acyclic analogs. The correlations are weak to modest but have the merit of being the only analyzable sets known to us. Structural composition in five other data sets was such that no clear deductions were possible (28, 36-39).

Summary of Substructural Binding Effects

Brain and Spinal Cord (CNS). In the binding of ^{3}H-GABA to human cerebellar membranes, ring-closed analogs show enhanced pIC50's (coefficient = 0.813). This is an exception as all other contributions are negative. The presence of a conjugated double bond is positive in the Caudate Nucleus for ^{3}H-GABA binding (IDB = 0.88) as is the oxazoline hetero-ring in bovine cortex membranes (IHET = 0.67). These small positive effects are opposed by generally large negative binding effects in all other factors.

IRNG =	-0.72 to -1.35(4)	(#) of coefficients
IHETS =	-0.74 to -1.35(2)	
IME, I4ME =	-1.05 to -2.93(4)	
INSUB =	-4.25	
I2OH, I3OH =	-0.84 to -1.69(6)	
INH2 =	-2.71	
ISO3 =	-0.77 to -0.93(2)	
ISHT, ILNG, ISL =	-1.53 to -3.16(7)	

Non-Competitive Binding

IRNG =	-2.54
IME, IME(R) =	-1.60 to -1.73(2)

Cellular Uptake

IRNG =	-0.585
I4ME =	-1.10 to -1.21(2)
IR =	0.649 - 0.743(2) - (R-configuration is favored)

Conclusions. The factors outlined in the Summary of Substructural Effects lead to a number of binding site conclusions.

The receptor site is clearly asymmetric. Chirality is important, both by simple inspection of cases and supported by discrete substructural contributions. The R-configuration for both cyclic and acyclic analogs is favored at uptake sites. If we assume the amino and carboxyl groups to bind specifically, then the cavity between these sites is most likely a flexible, narrow gap. Congeners are permitted to bind (flexibility), but all substituents from the 2- to 4-position reduce the binding energy (steric obstruction). This is true even for small H-bonding groups like NH_2 and OH. There are no H-bonding sites in the gap, hence these groups exert a simple steric effect in the same sense as a CH_3-group. The valley between the amino and carboxyl binding sites is clearly hydrophobic. It is probable that the binding of muscimol is special and occurs with the ring in a vertical, rather than flat orientation. The relative weakness of thiomuscimol binding suggests a difference in acidity rather than a steric effect for a marginally larger ring.

The fact that sulfonic analogs bind with respectable affinity suggests two features about the binding mechanism. First, there are steric requirements for the acid group (1-position) similar to those for the 2- to 4-positions. The binding site for the acid group also lies in a restricted cavity. Second, zwitterionic binding is assured for all analogs as the sulfonic derivatives are 100.00% zwitterionic with no measurable equilibria to neutral species. This would suggest that binding is by ion-paring with no activation energy, much like acetylcholine. Diffusion-controlled ion-pairing may be the only process fast enough for a neural response.

Another factor relating to the binding mechanism is the amino-carboxylate spacing, a factor that is strongly negative whenever the spacing is too short $[(CH_2)_2]$ or too long $[(CH_2)_4]$. It is important to note that binding is _not_ prevented in these cases, but only weakened. This fact, and the many colinearities observed, supports an ion-pairing mechanism that depends only on a distance function. Ion pairs are intact out to 5 Å in non-polar solvents such as benzene. It is reasonable then to assume that such pairs are intact at 2 Å distances on a lipophilic enzyme surface. This degree of latitude would explain the positive but weaker binding of all poorly spaced analogs.

There is nothing in this study that provides a lead to higher GABA-ergic activity as virtually all substructural factors are negative in binding energy. The message may simply be that there is no point in searching for better analogs. Rather, the direction of research should be toward degradable pro-GABA or -muscimol analogs that efficiently load the drug into the CNS.

Acknowledgment. This study was supported by the Chemical Systems Laboratory (Aberdeen Proving Ground, MD) under Contract No. DAAD05-86-M-Q973. We wish to thank Professor John H. Block of Oregon State University for his excellent GABA-related literature search under a subcontract.

Literature Cited

1. Bicuculline, Merck Index, Tenth Edition, No. 1214, p 171.
2. Picrotoxinin, Merck Index, Tenth Edition, No. 7297, p 1069.
3. Krogsgaard-Larsen, P. J. Med. Chem. 1981, 24, 1377.
4. Gavish, M.; Snyder, S. H. Nature 1980, 287, 651.
5. Frere, R. C.; MacDonald, R. L.; Young, A. B. Brain Res. 1982, 244, 145.
6. Falch, E; Krogsgaard-Larsen, P. J. Neurochem. 1982, 38, 1123.
7. Jordan, C. C.; Matus, A. I.; Piotrowski, W.; Wilkinson, D. J. Neurochem. 1982, 39, 52.
8. Napias, C.; Bergman, M. O.; Van Ness, P. C.; Greenlee, D. V.; Olsen, R. W. Life Sci. 1980, 27, 1001.
9. Gardner, C. R.; Klein, J.; Grove, J. Eur. J. Pharmacol. 1981, 75, 83.
10. Greenlee, D. V.; Van Ness, P. C.; Olsen, R. W. Life Sci. 1978, 22, 1653.
11. Toffano, G.; Guidotti, A; Costa, E. Proc. Natl. Acad. Sci. USA 1978, 75, 4024.
12. Triton X-100(Octoxynol), Merck Index, Tenth Edition, No. 6601, p 971.
13. Johnston, G. A. R.; Skerritt, J. H.; Willow, M. In Problems in GABA Research, Okada, Y.; Roberts, E., Eds.; Excerpta Medica, Amsterdam-Oxford-Princeton, 1982, 293-301.
14. Krogsgaard-Larsen, P; Falch, E.; Peet, M. J.; Leah, J. D.; Curtis, D. R. in CNS Receptors: From Molecular Pharmacology to Behavior, Mandel, P.; DeFeudis, F. V., Eds.; Volume 37, Raven Press, New York, 1983, 1-13.
15. Krogsgaard-Larsen, P. In Glutamine, Glutamate and GABA in the Central Nervous System, Hertz, L.; Kvamme, E.; McGeer, E. G.; Schousboe, A. Eds.; Alan R. Liss, Inc., New York, 1983, 537-557.
16. Krogsgaard-Larsen, P.; Hjeds, H.; Curtis, D. R.; Lodge, D.; Johnston, G. A. R. J. Neurochem. 1979, 32, 1717.
17. Krogsgaard-Larsen, P. Mol. Cell Biochem. 1980, 31, 105.
18. Ferrero, P.; Guidotti, A.; Costa, E. Proc. Natl. Acad. Sci. USA 1984, 81, 2247.
19. DeFeudis, F. V.; Ossola, L.; Mandel, P. Biochem. Pharmacol. 1979, 28, 2687.
20. Enna, S. J. In The GABA Receptors, Enna, S. J. Ed.; The Humana Press, Clifton, New Jersey, 1983, Chapter 1.
21. Andrews, P. R.; Johnston, G. A. R. Biochem. Pharmacol. 1979, 28, 2697.
22. Block, J. H.; King, J. W. In Proceedings of the 1986 Scientific Conference on Chemical Defense Research, Vol. 2, 1051.
23. Olsen, R. W. J. Neurochem. 1981, 37, 1.
24. Enna, S. J.; Ferkany, J. W.; Krogsgaard-Larsen, P. In GABA-Neuro-Transmitters; Krogsgaard-Larsen, P.; Scheel-Kruger, J.; Kofod, H., Eds.; Academic Press, New York, 1979, 191-200.
25. Breckenridge, R. J.; Nicholson, S. H.; Nicol, A. J.; Suckling, C. J.; Leigh, B.; Iversen, L. J. Neurochem 1981, 37, 837.
26. Iversen, L. L.; Spokes, E.; Bird, E. In Neurotransmitters, Volume 2; Simon, P., Ed.; Pergamon Press, Oxford, New York, 1979, 3-10.

27. Nicholson, S. H.; Suckling, C. J.; Iversen, L. L. J. Neurochem. 1979, 32, 249.
28. Roberts, E. In Neurotransmitters, Volume 2; Simon, P., Ed.; Pergamon Press, Oxford, New York, 1979, 43-65.
29. Olsen, R.W.; Ticku, M. K.; Van Ness, P. In GABA-Neurotransmitters; Krogsgaard-Larsen, P.; Scheel-Kruger, J.; Kofod, H., Eds.; Academic Press, New York, 1979, 165-178.
30. Honore, T.; Hjeds, H.; Krogsgaard-Larsen, P.; Christiansen, T.R. Eur. J. Med. Chem. 1978, 13, 429.
31. Krogsgaard-Larsen, P.; Jacobsen, P.; Falch, E. In The GABA Receptors; Enna, S. J., Ed.; The Humana Press, Clifton, New Jersey, 1984, 149-176.
32. Allan, R. D.; Evans, R. H.; Johnston, G. A. R. Br. J. Pharm. 1980, 70, 609.
33. Braestrup, C.; Nielsen, M.; Krogsgaard-Larsen, P.; Falch, E. In Receptors for Neurotransmitters and Peptide Hormones; Pepeu, G.; Kuhar, M. J.; Enna, S. J., Eds.; Raven Press, New York, 1980, 301-312.
34. Krogsgaard-Larsen, P.; Falch, E.; Jacobsen, P. In Actions and Interactions of GABA and Benzodiazepines; Bowery, N. G., Ed.; Raven Press, New York, 1984, 109-132.
35. Schousboe, A.; Hertz, L.; Larsson, O. M.; Krogsgaard-Larsen, P. In GABA Neurotransmission, Brain Research Bulletin, Vol. 5, Suppl. 2, 1980, 403-409.
36. Larsson, O. M.; Krogsgaard-Larsen, P.; Schousboe, A. J. Neurochem. 1980, 34, 970.
37. Schousboe, A.; Larsson, O. M.; Hertz, L.; Krogsgaard-Larsen, P. In Amino Acid Neurotransmitters; DeFeudis, F. V.; Mandel, P., Eds.; Raven Press, New York, 1981, 135-141.
38. Johnston, G. A. R.; Allan, R. D.; Kennedy, S. M. E.; Twitchin, B. In GABA-Neurotransmitters; Krogsgaard-Larsen, P.; Scheel-Kruger, J.; Kofod, H., Eds.; Academic Press, New York, 1979, 149-164.
39. Brehm, L.; Krogsgaard-Larsen, P.; Jacobsen, P. In GABA-Neurotransmitters; Krogsgaard-Larsen, P.; Scheel-Kruger, J.; Kofod, H., Eds.; Academic Press, New York, 1979, 247-262.

RECEIVED June 14, 1989

Chapter 20

Quantitative Structure–Activity Relationships for the Cytotoxicity of Substituted Aniline Mustards in Tissue Culture

William A. Denny[1], William R. Wilson[2], and Brian D. Palmer[1]

[1]Cancer Research Laboratory and Section of Oncology, School of Medicine, University of Auckland, Private Bag, Auckland, New Zealand
[2]Department of Pathology, School of Medicine, University of Auckland, Private Bag, Auckland, New Zealand

> The stabilities and aerobic cytotoxicities (growth inhibition and effects on clonogenicity) of substituted aniline mustards against mammalian tumor cells have been determined in tissue culture. Both properties are dependent entirely on substituent electronic effects, and exactly parallel the alkylating ability of the mustards. The halflife and growth inhibition and clonogenicity data are all well fitted by the σ or σ^- electronic parameters. The equations can be used to compute the maximum possible hypoxia selectivity of nitro-substituted aniline mustards. The 4-NO_2 compound does show hypoxia-selectivity (although at a much lower level than predicted), suggesting this approach to the design of hypoxia-selective antitumor agents is valid.

Aniline mustard alkylating agents (1) are an important class of antitumor drugs, with chlorambucil 1 and melphalan 2 still in clinical use more than 20 years after their introduction (2,3). They are useful components of many multi-drug regimens. In common with virtually all cytotoxic antitumor drugs, they have little intrinsic selectivity for cancer cells, and their therapeutic effect is due largely to their cytokinetic selectivity for rapidly-dividing cells. Considerable work aimed at more selective targeting of such compounds has not been successful, resulting in a decline of interest in their further development (4). However, increasing knowledge in tumor biology has recently revived interest in aniline mustards, which have been employed in designs of

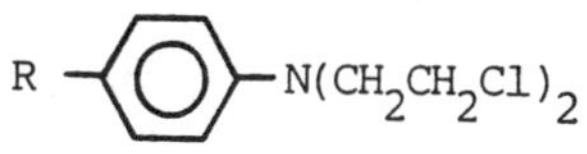

R—⟨O⟩—$N(CH_2CH_2Cl)_2$

1   R = $HOOC(CH_2)_3-$

2   R = $HOOCCH(NH_2)CH_2-$

drugs aimed at selective killing of the chronically oxygen-starved (hypoxic) cells which are known to exist in the interior of most solid tumors (5), and DNA-targeted alkylating agents (6).

Chemical Reactivity of Aniline Mustards

It is well-established that the reactivity of aromatic nitrogen mustards in hydrolysis reactions correlates positively with the degree of electron release to the nitrogen (7). The rate constants for alkaline hydrolysis of a series of aniline mustards correlate well with σ, as shown by Equation 1 (equation 23 of ref. 8).

$$\log k = -1.84(\pm 0.40)\,\sigma - 4.02 \qquad (1)$$
$$n = 11 \quad r = 0.96 \quad s = 0.1$$

The large rho value suggests that the reaction proceeds via the aziridinium ion, the formation of which is the rate-determining step, with its stability being very dependent on the electron density at the nitrogen. Use of the $\bar{\sigma}$ parameter gives a significantly poorer fit (r = 0.68)(8). QSAR studies of the rate of alkylation of 4-nitrobenzylpyridine (NBP) by substituted aniline mustards (compounds XI-1 to XI-9 of ref. 8) also show a similar dependence, resulting in Equation 2.

$$\log(Krel) = -2.16(\pm 0.92)\,\sigma - 2.18 \qquad (2)$$
$$n = 9 \quad r = 0.87 \quad s = 0.53$$

However, for this data set a better fit to the data is obtained by using the σ- parameter (Equation 3)

$$\log(Krel) = -2.10(\pm 0.36)\,\sigma^- - 1.71 \qquad (3)$$
$$n = 9 \quad r = 0.97 \quad s = 0.24$$

This indicates that there is an upper limit to rate of alkylation by the more unstable compounds (with the strongest electron-donor substituents). The better fit of the data to σ- then reflects the nature of this parameter, which reaches a minimum value of -0.16 and then remains essentially constant for all strongly electron-donating groups (9).

Finally, several QSAR studies (8,10,11) have shown that the acute toxicity and antitumor potency of substituted aniline mustards in vivo are also dominated by the electronic properties of substituent groups, although additional factors such as drug lipophilicity play a role in these systems. For example, the potencies (dose of drug required to provide an increase in lifespan of 25%) of a series of compounds against the B-16 melanoma in mice were best fitted by Equation 4 (equation 8 of ref. 11).

$$\log(1/C) = -1.84(\pm 0.44)\,\sigma - 0.17(\pm 0.11)\,\pi - 0.14(\pm 0.06)\,\pi^2 + 4.2 \qquad (4)$$
$$n = 21 \quad r = 0.93 \quad s = 0.32$$

This similar dependence of in vivo potency and mustard reactivity on substituent electronic properties (cf Equations 1 and 4) has been cited (8) as evidence that the biological activity of these compounds is directly due to the rate at which they alkylate cellular DNA. However, in complex in vivo systems the possible roles of drug transport and metabolism as well as DNA repair have also to be considered. Surprisingly, there is virtually no quantitative data available on the cytotoxicity of substituted aniline mustards in mammalian cell culture systems, where the effects of substituent electronic effects on cytotoxicity and stability can be examined in the absence of these factors, and where the use of defined mutant cell lines can provide much information about the mode of drug action.

We report here a QSAR study of the in vitro cytotoxicity of simple substituted aniline mustards, as part of a programme (5) directed towards the rational design of nitrophenyl mustards as hypoxia-selective agents. In order to study the effects of substituent groups, a carefully-selected set of derivatives 3-15 were synthesized (12) for evaluation. The substituent groups were placed at the 3- and 4-positions only, to avoid the steric effects which are known to complicate relationships for ortho substituents (9), and were chosen to have varying lipophilicities but approximately similar size, to minimize possible variations caused by non-specific binding to serum proteins in the culture medium (a property which has been correlated (13) with molecular size). However, the main aim was to provide a series where the electronic properties of the substituent groups spanned the widest possible range, since this property is considered to be the primary modulator of biological activity among the aniline mustards (8). The substituents chosen had σ values ranging from 0.72 (4-SO_2Me) to -0.66 (4-NH_2) in addition to the nitro group (4-NO_2, 0.78), with the other substituents evenly spanning this range. The 3-substituted compounds were included not so much to evaluate positional effects, but for their differing σ values.

The stability and cytotoxicity of the compounds were evaluated against the Chinese hamster cell line AA8, and a derived mutant line (UV4) which lacks the ability to form the incision step of the normal excision repair pathway and is thus hypersensitive to alkylating agents which cause bulky DNA lesions or cross-links (14).

Metabolic Stability of Substituted Aniline Mustards

Preliminary experiments demonstrated that the cytotoxicities of the mustards 3-15 were, as expected, very much greater than that of the corresponding diols which are the end products of hydrolysis. The differential was particularly pronounced with the UV4 cell line, which is hypersensitive to alkylating agents (14). Thus the stability of the mustards against hydrolysis to the the corresponding diols could be conveniently monitored under physiological conditions by bioassay against UV4 cells. Drugs were incubated in Alpha culture medium at 37°C, pH 7.2, in the presence of fetal calf serum (10% v/v). At various times residual cytotoxic activity was assayed by titration of medium against log-phase UV4 cells in 96-well microplate cultures as described previously (15). The fraction of cytotoxic activity remaining was calculated by comparing the IC_{50} (concentration for inhibition of growth to 50% of controls) with that for fresh drug determined concurrently.

Figure 1 shows the results of experiments where two compounds (the unsubstituted mustard 9 and the 4-aminomustard 15) were incubated in culture medium at 37°C either with or without the addition of lethally-irradiated UV4 cells (although such cells will not divide, they retain full metabolic capacity over the period of the experiments). The amount of bioactivity remaining in the solutions was measured at intervals by the methods described above. The presence of metabolically-viable cells makes no difference to the rate of loss of bioactivity, indicating that the stability of these compounds in culture medium is dependent entirely on chemical processes (presumably hydrolysis of the mustard).

The stability of the substituted chloromustards in culture medium at 37°C (in the absence of cells) was then similarly determined, and the results (expressed as half-lives, T1/2) are recorded in Table I and Figure 2. These data (excluding that for compounds 3 and 5, which could not be accurately determined) can be fitted to the expression

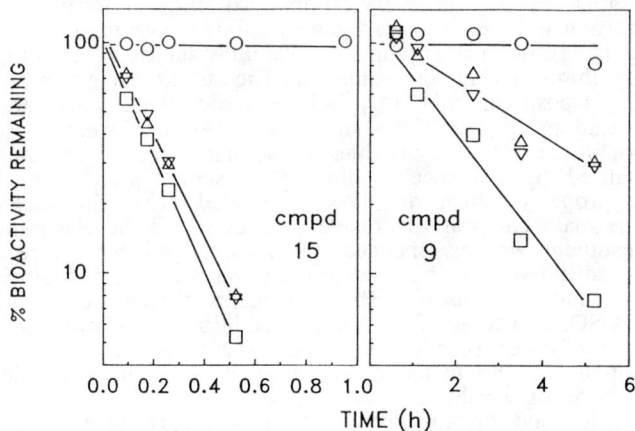

Fig. 1 Rate of inactivation of compounds (9) and (15), assessed by bioassay against UV4. Compounds were incubated at 10 μM (compound 15) and 60 μM (compound 9) at 37°C in : medium (□), medium + 10% fetal calf serum (△), or medium + 10% FCS + lethally-irradiated UV4 feeder cells (▽). Stock solutions in 0.01 M HCl (15) or acetone (9) were held at 0°C and similarly assayed at intervals (○). Percentage bioactivity remaining was calculated as
IC_{50} (stock solution/IC_{50} test solution).

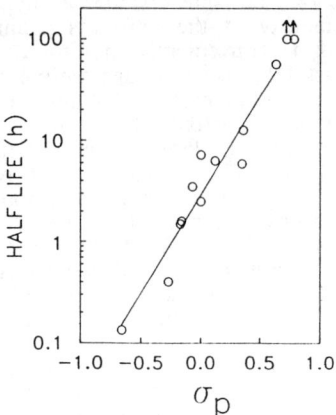

Fig. 2 Biological halflife in culture medium assessed by bioassay of culture supernatant against UV4; relationship to substituent electronic parameter.

Table I. Physicochemical and Biological Data for Substituted Aniline Mustards

$$X \underset{4}{\overset{3}{\diagdown}} \!\!\!\bigcirc\!\!\! - N(CH_2CH_2Cl)_2$$

No.	X	σ	σ^-	T1/2 (h)	IC_{50} (μM)	HF^a	ratio^b	CT_{10} ($\mu M/h$)
3	4-NO$_2$	0.78	1.24	>100	15.8	>5^c	3.2	750
4	3-NO$_2$	0.71	0.71	24^c	96	>1	1.5	
5	4-SO$_2$Me	0.72	0.98	>100	174	3.5	0.9	1120
6	3-SO$_2$Me	0.63	0.63	56.6	76	52	1.0	
7	4-CONMe$_2$	0.36	0.63	12.6	14.6	60	0.95	30
8	3-CONMe$_2$	0.35	0.35	5.9	7.6	>20^d	1.0	
9	H	0	0	2.5	1.2	52	1.1	3
10	4-SMe	0	0	7.2	1.6	ND	0.7	6.3
11	4-Me	-0.17	-0.15	1.5	0.4	46	0.8	2.9
12	3-Me	-0.07	-0.07	3.5	1.2	50	0.9	
13	4-OMe	-0.27	-0.16	0.4	0.45	42	0.8	0.2
14	3-OMe	0.12	0.12	6.3	2.7	42	1.0	
15	4-NH$_2$	-0.66	-0.15	1.6	0.07	25	1.0	0.04

^{a} HF (hypersensitivity factor) = IC_{50}(AA8)/IC_{50}(UV4).
^{b} ratio (hypoxic selectivity in UV4) = IC_{50}(aerobic)/IC_{50}(hypoxic).
^{c} kinetics not first order : value is estimate of first half-life.
^{d} AA8 IC_{50} not reached due to insolubility.

$$\log(T1/2) = 1.94(\pm 0.34)\,\sigma + 0.46 \quad (5)$$
$$n = 11 \quad r = 0.96 \quad s = 0.24$$

The slope of the line (1.99) shows there is a very similar dependence of halflife in culture medium on substituent electronic properties (σ) to that seen by others (7,8) for the dependence of the chemical hydrolysis of substituted aniline mustards (Equation 1 above).

The good fits and the similar magnitude of the slopes of Equations 1 and 5 suggests that the primary determinant of the half-life of the compounds of Table I in culture medium is the rate of hydrolysis of the mustard, and that this is entirely controlled by the degree of electron release. These results also show that several of the derivatives have half-lives of only 1-2h (or even less for the amino compound 15). Thus the stability of these compounds in culture has to be considered when determining their cytotoxicity, for unless very short exposures are used there will be appreciable drug breakdown during the experiment, which will lead to underestimation of the biological potency of the more unstable compounds.

Cytotoxicity of Substituted Aniline Mustards

The cytotoxicities of the compounds were determined against both AA8 (data not shown) and UV4 cells, and the results are given in Table I. The consistently large hypersensitivity shown by the latter cell line (typically 50-fold) suggested that the mechanism of cytotoxicity for all members of the series is bifunctional alkylation resulting in DNA cross-linking. The much higher potency against UV4 allowed a full dataset to be collected, even for compounds which were too insoluble for evaluation against AA8. Two cytotoxicity assays were used. In the first, log-phase cultures of AA8 or UV4 cells were exposed to drugs in 96-well microplates for 1 hr. Drug was then removed by washing the adherent cells with fresh medium, and cultures were grown for a further 72-75hrs before determination of cell numbers and determination of IC_{50} values as above.

In the second assay, stirred suspensions of late log-phase cells were incubated with drug at 10^6 cells/mL, and samples were withdrawn at intervals to assay cell killing by clonogenic assay as described previously (16). This involves diluting the withdrawn cell samples into fresh medium and centrifuging to remove drug. The cells are then plated out on Petri dishes and incubated for 96h, after which time each originally-viable cell will have grown into a colony detectable by a simple dye staining assay.

The measure chosen to reflect cytotoxic patency in the clonogenic assay was the product of the drug concentration multiplied by the time of exposure required at that concentration to reduce the surviving fraction to 10% (CT_{10}; μM-hr), and was determined by interpolation as a representative measure of cytotoxic potency. The assay has the advantage of providing access to early drug/cell contact times, yielding information on initial rates of cell killing by unstable drugs.

Growth Inhibition Assay. These results are shown in Table I and Figure 3a. There is a clear relationship between log IC_{50} values and substituent electronic parameters, except for the 4-nitro compound 3. This compound appears to bind avidly to plastic surfaces, which may reduce the available concentration. The rest of the data is fitted well by Equation 6.

$$\log (IC_{50}) = 2.46(\pm 0.40) \sigma + 0.21 \tag{6}$$
$$n = 12 \quad r = 0.97 \quad s = 0.34$$

The main outlier in this equation is the least stable compound, the 4-amino derivative 15 which has a halflife of only 0.13h (Table I), so that its cytotoxicity is underestimated even with a 1h exposure.

The coefficient of dependency of cytotoxicity upon σ is large (2.96), but similar in magnitude to that observed previously (8) for the dependence of the alkylating ability of substituted aniline chloromustards as measured by their relative rates of alkylation of 4-nitrobenzylpyridine (Equation 2). It is interesting to note that the whole dataset (including the unstable 4-amino compound 15), is equally well fit by the σ- parameter (compare Equations 6 and 7).

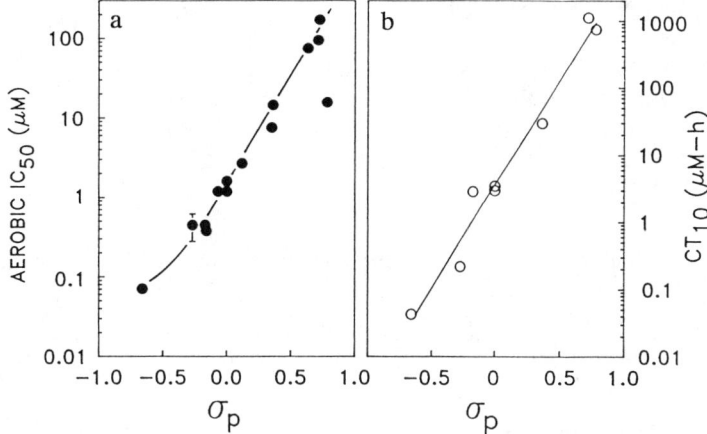

Fig. 3. (a) Growth inhibitory potencies against aerobic UV4 cells after a 1-hr exposure; relationship to substituent electronic parameter. (b) CT_{10} values determined by clonogenic assay; relationship to substituent electronic parameter.

$$\log(IC_{50}) = 2.50(\pm 0.42)\, \sigma^- - 0.02 \quad (7)$$
$$n = 12 \quad r = 0.96 \quad s = 0.22$$

This probably because the cytotoxicities of the more unstable compounds (with the most negative σ values) are underestimated, due to rapid inactivation of drug by hydrolysis during the 1h exposure, and these effects are fortuitously compensated for by the fact that the σ⁻ parameter scale terminates at ca. -0.16 for all highly electron-donating substituents.

Clonogenic Assay. A subset of the compounds (the 4-substituted derivatives) were also assayed for cytotoxicity using the stirred suspension culture assay (Table I and Figure 3b). This assay, by measuring initial rates of cell killing, has the advantage of not underestimating the cytotoxicity of unstable compounds of short half-life. This fact is reflected in the much greater range of CT_{10} values seen (26,000-fold), and by the fact that the data is better fitted by an equation in σ (equation 8) than in σ- (equation 9), in direct contrast with the IC_{50} data.

$$\log(CT_{10}) = 3.07(\pm 0.23)\, \sigma + 0.60 \quad (8)$$
$$n = 8 \quad r = 0.99$$

$$\log(CT_{10}) = 2.56(\pm 0.27)\, \sigma^- + 0.07 \quad (9)$$
$$n = 8 \quad r = 0.92$$

The fact that the cytotoxicity of the substituted aniline chloromustards (measured in both the growth inhibition and clonogenic assays; Equations 6 and 8) and their ability to alkylate 4-nitrobenzylpyridine (Equation 2) show a very similar dependence on electronic parameters is further evidence, together with their large hypersensitivity factors for UV4 versus AA8, that these compounds express their cytotoxicity primarily through alkylation of DNA.

Differential Cytotoxicity of 4-NO_2 and 4-NH_2 Aniline Mustards

This analysis suggests that an important criterion for the design of hypoxia-selective agents, that of a large differential between the cytotoxicities of parent and reduced drug forms (5), can be adequately met by the substituted aniline chloromustards. Earlier work analysing in vivo data (summarized in ref.5) suggested that differential potencies of between 200-400 fold could be obtained between the 4-nitro mustard and its potential 6-electron reduction product, the 4-amino compound. For UV4 cells in culture, the measured IC_{50}s of these two compounds differ by 220-fold (Table I). However, the cytotoxicity of the 4-amino compound (15) is understated and that of the 4-nitro compound (3) may be overstated as mentioned above. The theoretical differential cytotoxicity between the compounds (at least against UV-4 cells, computed by substitution of their σ values into Equation 8) is 26,000-fold in the clonogenic assay.

The observation that the 6-nitro compounds (3 and 4) have broadly expected IC_{50} values from their σ values in Figures 3a and 3b indicates that, in aerobic UV4 cultures, the cytotoxic potency of these compounds is controlled by the intact nitro group (a reason for the low value of the 4-nitro compound 3 has been discussed above). This suggests that there is little net metabolic nitroreduction in aerobic cells, either because the redox potentials of the nitro groups in these molecules are too low, or there is efficient back-oxidation of the initially-formed nitro radical anion by molecular oxygen.

Conclusions

The above results show that both the biological half-lives and the cytotoxic potencies of the substituted aniline chloromustards are dominated by the electronic properties of the substituent group. The half-lives of the compounds vary by over three orders of magnitude, with electron-withdrawing groups stabilizing the compounds by deactivating the mustard. Loss of parent drug in culture medium does not depend on the presence of cells, and presumably reflects spontaneous hydrolysis of the mustard to the much less toxic diol. In the growth inhibition assay the cytotoxicities of the compounds are similarly dependent on substituent electronic effects. The cytotoxicities assay of the more unstable compounds determined by the growth inhibition may be underestimated due to competing hydrolysis over the time course of the experiments, which would account for the σ^- parameter (which reaches a minimum value for all strongly electron-donating groups (9)) providing a better fit to this data than the σ parameter. The fact that cytotoxicity values (CT_{10} values) determined by the clonogenic assay, where drug instability is no longer a factor, are fitted much better by the σ than the σ^- parameter (equations 8 and 9) is consistent with this hypothesis.

The equations describing both the cytotoxicities and the alkylating abilities of substituted aniline chloromustards have similar slopes, suggesting that these compounds express their cytotoxicity primarily by alkylation. It is evident that the critical biological alkylation site is DNA, and that cytotoxicity is due to formation of DNA interstrand crosslinks, since the UV4 mutant defective in repair of these lesions (14) shows marked hypersensitivity to the aniline mustards.

While the majority of the aniline mustards showed no hypoxia-selective cytotoxicity in the growth inhibition assay (Table I), the corresponding 4-NO_2 and 3-NO_2 compounds (3 and 4) were more toxic to UV4 cells under hypoxic conditions, with the 4-NO_2 derivative showing a toxicity ratio of 3.2 (Table I). This is the first direct evidence for the suggestions made by several authors (5,17,18) that compounds such as (3) might act as hypoxia-selective agents via activation of an alkylating function by cellular reduction of the nitro group.

The result is encouraging, since it suggests this approach to the design of hypoxia-selective compounds (the use of cellular nitro-reduction to activate an alkylating moiety) has potential. However, the ratio of 3.2-fold is small compared to the difference in aerobic toxicities between the 4-NO_2 compound (3) and its potential end-stage metabolite, the 4-NH_2 compound (15) actually measured (220-fold) or computed by Equation 6 (3500-fold). Although this ratio must be seen as the theoretical maximum for the hypoxia-selective toxicity of the corresponding nitro compound (it is unlikely that there will be complete reduction to the 6-electron product), the hypoxic selectivity shown by the nitro compounds (3 and 4) are very small fractions of this maximum.

The main reason for this is almost certainly the low nitro group reduction potentials of these compounds. The 4-NO_2 derivative (3) has a reduction potential $E(1)$ of about -500 mV (12), which is likely to be too low for efficient enzymic nitroreduction, and that of the 3-NO_2 compound ($E(1)$ = -470 mV) is similar. Future work in this area is therefore being directed towards the development of more soluble derivatives with higher nitro group reduction potentials.

Acknowledgments

This work was supported by the Auckland Division of the Cancer Society of New Zealand, and by the Medical Research Council of New Zealand. The authors would like to thank Susan Pullen and Robert Lambert for performing the cytotoxicity experiments, and Margaret Snow for preparing the manuscript.

Literature Cited
1. Bratzel, R.P.; Goodridge, J.H.; Huntress, W.T. Cancer Chemother. Repts., 1963, 26, 1.
2. Houghton, J.A.; Cook, R.L.; Lutz, P.J.; Houghton, P.J. Cancer Treatment Repts., 1985, 69, 91.
3. Greig, N.H.; Sweeney, D.J.; Rapoport, S.I. Cancer Chemother. Pharmacol., 1988, 21, 1.
4. Wilman, D.E.V.; Connors, T.A. In Molecular Aspects of AntiCancer Drug Action; Neidle, S., Waring, M.J., Eds.; MacMillan: London, 1983; p 233.
5. Denny, W.A.; Wilson, W.R. J. Med. Chem., 1986, 29, 879.
6. Koyama, M.; Kelly, T.R.; Wanatabe, K.A. J. Med. Chem., 1988, 31, 283.
7. Everett, J.L.; Ross, W.C.J. J. Chem. Soc., 1953, 2386.
8. Panthananickal, A.; Hansch, C.; Leo, A.; Quinn, F.R. J. Med. Chem., 1978, 21, 16.
9. Hansch, C.; Leo, A.J. Substituent Constants for Correlation Analysis in Chemistry and Biology; Wiley: New York, 1979; p 3.
10. Lien, E.J.; Tong, G.L. Cancer Chemotherapy Rept., 1973, 57, 251.
11. Panthananickal, A.; Hansch, C.; Leo, A. J. Med. Chem., 1979, 22, 1267.
12. Denny, W.A.; Wilson, W.R.; Palmer, B.D. Manuscript submitted to J. Med. Chem.
13. Hansch, C. In Drug Design; Ariens, E.J., Ed.; Academic: London, 1971; Vol. 1, p 271.
14. Hoy, C.A,; Thompson, L.H.; Mooney, C.L.; Salazar, E.P. Cancer Res., 1985, 45, 1737.
15. Wilson, W.R.; Denny, W.A.; Finlay, G.F.; Pullen, S.M.; Baguley. B.C. Manuscript submitted to Br. J. Cancer.
16. Wilson, W.R.; Denny, W.A.; Twigden, S.J.; Baguley, B.C.; Probert, J.C. Br. J. Cancer, 1984, 49, 215.
17. Alston, T.A.; Porter, D.J.T.; Bright, H.J. Acc. Chem. Res., 1983, 16, 418.
18. Connors, T.A. In Structure-activity Relationships of Antitumor Agents; Reinhoudt, D.N.; Connors, T.A.; Pinedo, HM.; van der Poll, K.W., Eds.; Nijhoff: The Hague, 1983; p 47.

RECEIVED March 21, 1989

Chapter 21

Quantitative Structure–Activity Relationships of Antibacterial Compounds Based on the Nalidixic Acid Structure

John H. Block[1], Yupei Yu[1], James W. King[2], and Arie Verloop[3]

[1]College of Pharmacy, Oregon State University, Corvallis, OR 97331–3507
[2]U.S. Army Chemical Research, Development and Engineering Center, Aberdeen Proving Ground, MD 21010–5423
[3]1412 AR Naarden, Netherlands

> Over 120 compounds based on the nalidixic acid structure were subjected to QSAR analysis against one Gram-positive and two Gram-negative bacteria. Somewhat inconsistent results were obtained, probably due to two responses being required before significant inhibition of bacterial growth is observed. A successful compound must first penetrate the bacterial cell wall and then fit the receptor on the DNA gyrase. Nevertheless, the results do show that, within limits, several substituents can be placed on the quinoline nitrogen, the choice of substituents at position 6 is limited mostly to fluorine and related small moieties, while a wide variety of basic nitrogen containing substituents can be placed at position 7.

Since the introduction in 1963 of nalidixic acid (Table I), 1-ethyl-1,4-dihydro-7-methyl-4-oxo-1,8-naphthyridine-3-carboxylic acid, as a systemic Gram-negative antibacterial agent, a large number of analogues have been synthesized and evaluated, some of which have come onto the international market.(1,2) A comprehensive review has outlined the synthetic methods, microbiology and structure activity relationships of those derivatives reported prior to 1977.(2)

Over two decades of research have yielded new 6-fluoroquinolones whose activities far surpass that of nalidixic acid. The most significant changes made in the quinolone nucleus, addition of 6-fluorine and 7-piperazine, have provided the fluoroquinolones with activity against Gram-negative bacteria comparable to that of the major classes of antibiotics.(3,4,5) An evolution of structural modifications of nalidixic acid has resulted in increased potency/spectrum such that the newest agents have excellent activity against Gram-negative bacteria (including *Ps. aeruginosa*),

Table I. Selected Fluoroquinolone Antibacterial Agents

	X_6	X_8	R_1	R_3	R_6	R_7
Nalidixic Acid	C	N	C_2H_5	COOH	H	CH_3
Norfloxacin	C	C	C_2H_5	COOH	F	$N(CH_2CH_2)_2NH$
Pefloxacin	C	C	C_2H_5	COOH	F	$N(CH_2CH_2)_2NCH_3$
Enoxacin	C	N	C_2H_5	COOH	F	$N(CH_2CH_2)_2NH$
Ciprofloxacin	C	C	$CH(CH_2CH_2)$	COOH	F	$N(CH_2CH_2)_2NH$
Difloxacin	C	C	$C(CHCH)_2CF$	COOH	F	$N(CH_2CH_2)_2NCH_3$
Oxolinic Acid	C	C	C_2H_5	COOH	$-OCH_2O-$	
Pipemidic Acid	N	N	C_2H_5	COOH		$N(CH_2CH_2)_2NH$

increased activity against Gram-positive bacteria, and, in some instances, better activity against anaerobes.(6,7) This increased potency coupled with better biodistribution properties has broadened the therapeutic potential of quinolones for parenteral and oral treatment of systemic infections other than urinary tract infections.(6) Relative to the first commercially introduced fluoroquinolone, norfloxacin (Table I), subsequent analogues have shown greater oral absorption (pefloxacin, enoxacin), increased serum half-life (pefloxacin), overall increased *in vitro* potency (ciprofloxacin) and an increased spectrum to include anaerobic bacteria (difloxacin).(5,7)

Nalidixic acid and its analogues act by inhibition of bacterial DNA synthesis by inhibiting a bacterial DNA gyrase classified as a type II topoisomerase.(7,8,9,10) Resistance and cross resistance are seen with nalidixic acid and the fluoroquinolones. It seems to be related to decreased cellular permeability and has been considered plasmid independent. (4,5,11,12) More recently, reports have appeared of genes carrying the resistance traits normally found on plasmids.(13,14)

A large number of compounds have been synthesized and tested.(3,6,15,16) A summary of the structure activity relationships contains the following generalizations. Steric bulk originally had been considered important for the N-substituents at position 1.(17) The ethyl analogues are generally more potent than those compounds having smaller or larger substituents. Variants with similar bulk as that of N-ethyl include N-methoxy, N-cyclopropyl and N-vinyl all of which produced active compounds. Also good activity is obtained with substituted phenyl rings attached to the nitrogen at position 1.(18) At this point it is difficult to conclude whether steric, electronic and/or hydrophobic effects best account for the observed activity at N-1 position.

It also has been found that the substituents other than piperazine at position 7 are active against bacteria and the DNA gyrase target enzyme. (19) The key point seems to be a basic nitrogen approximately in the same position as that for the piperazine moiety. Several active compounds with substituted pyrrolidinyl groups at position 7 seem to reinforce this hypothesis.

Quantitative Structure Activity Relationships. With the large number of compounds that has been synthesized and tested, it is surprising that several quantitative structure activity relationship (QSAR) analyses have not been attempted as a means of correlating those structural attributes and physicochemical parameters that significantly affect potency. At least one detailed QSAR study has been reported.(20,21) Initially the pertinent physicochemical parameters were obtained for each position where substituents were varied. Then the study was extended to multiple substituted analogues. The parameters evaluated included the original STERIMOL parameters $L(1)$ for N-substituents at position 1; $L(8)$ and $B4(8)$ at position 8; steric influence at position 6, $Es(6)$; hydrophobicity at position 7, $\pi(7)$; and lipophilicity of the whole molecule, log P. Parabolic relationships were evaluated using the squared terms. Indicator variables included I(7) at position 7 for the presence or absence of a substituent at position 7 and I(7NCO) for the presence or absence of carbonyl functions on the 7-N-heterocyclic substituents, i.e. N'-acylpiperazinyl and 4-carbamoylpiperdinyl. Summed terms of the π-constant at positions 6, 7, and 8, [$\Sigma\pi(6,7,8)$] and its squared value and the summed field-inductive electronic effect, [$\Sigma F(6,7,8)$], also were included. Six of the published equations are shown in Table II.

The first three equations were developed to show the specific contribution of bulk (Eq. 1), lipophilicity (Eq. 2), and length of the substituents (Eq. 3). The 21 compounds used in equation 1 vary at positions 6, 7 and 8. The 22 compounds studied in equation 2 vary only at position 7 having a fluorine at position 6 and N-ethyl at position 1. Finally, equation 3 was developed from eight compounds in which the substituent at positions 1 were varied with a fluorine at position 6 and unsubstituted piperazine at position 7. Equations 4 through 6 show equivalent models based on a sample of 71 compounds. Using more recent biological data, a subset of the compounds from this earlier study was subjected to a QSAR analysis with the results reported later in this chapter. Further, as also will be reported in this paper, the problem of developing more than one statistically equivalent model is common to the quinolone antiinfectives.

Regarding equations 4 - 6, a correlation matrix shows high colinearity between $\pi(7)$ and $\Sigma\pi(6,7,8)$ or Log P of 0.94 and 0.92, respectively, for these 71 compounds. Examination of the data indicates that most of the variance in $\Sigma\pi(6,7,8)$ and Log P is due to $\pi(7)$. Even though equations 4 - 6 are nearly equivalent statistically using $\pi(7)$, $\Sigma\pi(6,7,8)$ or Log P, the π-constant is not really an additive parameter when more than one substituent is present on an aromatic ring.(22)

Purpose. The purpose of the research reported in this chapter was to apply the linear free energy relationship (LFER) or Hansch model and *de novo* or Free-Wilson model to a series of quinolone antibacterial agents. In addition to the hydrophobicity, a goal was to observe the effects, if any, of using different measures of steric influence. These included molar refraction and the newer STERIMOL parameters. Fortunately, there have been a series of articles reporting the antibacterial activity against one Gram-positive (*S. aureus*) and two Gram-negative (*E. coli* and *Ps. aeruginosa*) bacteria by a series of 6,7-disubstituted quinoline and 1,8-naphthyridine 3-carboxylic acids. Consistent biological data is available on over 120 compounds.(23,24,25)

EXPERIMENTAL

Parameters. Log P (F, for *fragment,* in the data tables) and molar refraction (MR) for the substituents were calculated from the molecular fragments and correction terms using CLOGP 3.3.(26) For the substituents at positions 6 and 7 of ring B, each substituted ring B was treated as a distinct compound.

Table II. QSAR Models of a Series of Nalidixic Acid Analogues (Independent Variable: Log 1/MIC where MIC is the minimum inhibitory concentration)

Eq.	L(1)	[L(1)]²	B4(8)	[B4(8)]²	Es(6)	[Es(6)]²	π(7)	[π(7)]²	Log P	[Log P]²	I(7)	I(7NCO)	Σπ(6,7,8)	[Σπ(6,7,8)]²	ΣF(6,7,8)	Constant
1.			+3.770	-1.024	-4.210	-3.236					+1.358					+1.251
	n = 21 s = 0.205		r = 0.978	F = 67.50												
2.							-0.675	-0.244				-0.705				+5.987
	n = 22 s = 0.242		r = 0.943	F = 47.97												
3.	4.102	-0.492														+1.999
	n = 8 s = 0.126		r = 0.955	F = 25.78												
4.	3.532	-0.423	+2.961	-0.868	-3.163	-2.499	-0.633	+0.223			-1.036	-0.774			-0.686	-5.030
	n = 71 s = 0.285		r = 0.961	F = 64.18												
5.	3.036	-0.362	+3.724	-1.023	-3.345	-2.499					+0.986	-0.734	-0.485	-0.205	-0.681	-4.571
	n = 71 s = 0.274		r = 0.964	F = 70.22												
6.	2.528	-0.294	+3.557	-0.985	-3.316	-2.497			-0.370	-0.188	+0.956	-0.792			-0.665	-3.343
	n = 71 s = 0.286		r = 0.961	F = 64.07												

Sigma-rho corrections were assigned equally to the substituents at positions 6 and 7. Revised STERIMOL (L, B1 and B5) parameters were calculated following established methods.(27,28) The data matrices consist of lipophilicity (F), molar refractivity (MR), STERIMOL (L, B1, B5) parameters and *de novo* indicator variables for substituents in positions 1, 6 and 7. The contribution to the partition coefficient by the substituents at positions 6 and 7 were summed, $\Sigma F(6,7)$, as were the molar refractivity contributions, $\Sigma MR(6,7)$. These two sets of summed variables paralleled a similar approach used by Koga in his QSAR analysis of a set of quinoline derivatives.(23) At the same time, the limitations that apply to summed π values, particularly where there is a significant σ-ρ interaction were kept in mind.(22,29,30) The latter is represented as API in the data tables and its significance was evaluated in some of the regression models. In order to check for parabolic relationships, squared terms for the partition coefficient, molar refraction, and STERIMOL parameters were evaluated.

A mixed model using the physicochemical parameters and Free-Wilson indicator variables also was examined. The *in vitro* activity, reported as minimum inhibitory concentration (μg/ml) against *S. aureus, E. coli* and *Ps. aeruginosa*, were converted to molar concentration for the QSAR studies. Multiple linear regression was performed on a CYBER 170/720-2 using the Statistical Interactive Programming System (SIPS) developed at Oregon State University.

Two approaches were used in developing the models. Initially, a forward stepwise procedure was used in which the next most significant variable would enter the model and any variable already in the model that became insignificant (usually p > 0.05) would be dropped. The final model would be checked by adding all of the variables and dropping the insignificant ones starting with the least significant variable. A second approach was to force in a variable initially and then build a model. This was done to evaluate hydrophobicity or steric influence. In each the case, the final models were checked for consistency by omitting five or six randomly selected compounds and examining the consistency of the regression coefficients. This procedure was repeated three times for each model. A number of statistics were obtained including s, the standard error, r, the correlation coefficient, r^2, the percentage of data variance accounted for by the model, F, a test of the overall significance of the model, and t values for the individual regression coefficients in the equation. The antimicrobial activities calculated from the models were compared with the observed biological activities.

<u>Set A.</u> The structures of the 6,7-disubstituted 1-alkyl-1,4-dihydro-4-oxoquinoline-3-carboxylic acids and their *in vitro* antibacterial activity against Gram-positive (*S. aureus* 209P) and Gram-negative bacteria (*E. coli* NIHJ JC-2 and *Ps. aeruginosa* V-1) are shown in Table III.(23) The data matrices for Set A are found in Tables IV - VI.

Table III. Set A: 6,7-Disubstituted 1-Alkyl-1-dihydro-4-oxo-quinoline-3-carboxylic Acids

No.	R_1	R_6	R_7	S. aureus 209P	MIC μg/ml[a] (23) E. coli NIHJ JC-2	Ps. aeruginosa V-1
A18	CH_2CH_3	F	Cl	12.5	1.56	100
A32	CH_2CH_3	F	CH_3	6.25	0.3	50
A33	CH_2CH_3	F	NH_2	>100	3.13	>100
A34	CH_2CH_3	F	$N(CH_2CH_2)_2NH$	0.39	0.05	0.39
A36	CH_2CH_3	F	$N(CH_2CH_2)_2NCH_3$	0.39	0.10	1.56
A37	CH_2CH_3	Cl	$N(CH_2CH_2)_2NH$	1.56	0.20	3.13
A38	CH_2CH_3	Cl	$N(CH_2CH_2)_2NCH_3$	1.56	0.78	25
A39	CH_2CH_3	Br	$N(CH_2CH_2)_2NH$	3.13	0.39	12.5
A40	CH_2CH_3	Br	$N(CH_2CH_2)_2NCH_3$	1.56	0.39	100
A41	CH_2CH_3	CH_3	$N(CH_2CH_2)_2NH$	3.13	0.39	6.25
A42	CH_2CH_3	SCH_3	$N(CH_2CH_2)_2NH$	25	0.78	12.5
A43	CH_2CH_3	$COCH_3$	$N(CH_2CH_2)_2NH$	100	100	>100
A44	CH_2CH_3	CN	$N(CH_2CH_2)_2NH$	12.5	0.39	6.25

Continued on next page

Table III. Set A: continued

No.	R_1	R_6	R_7	MIC µg/ml[a] (23)		
				S. aureus 209P	E. coli NIHJ JC-2	Ps. aeruginosa V-1
A45	CH_2CH_3	NO_2	$N(CH_2CH_2)_2NH$	25	0.78	12.5
A46[b]	CH_2CH_3	H	$N(CH_2CH_2)_2NH$	3.13	0.39	6.25
A47[c]	CH_2CH_3	H	$N(CH_2CH_2)_2NH$	12.5	0.78	1.56
A48	CH_3	F	$N(CH_2CH_2)_2NH$	6.25	0.39	1.56
A49	CH_2CH_2F	F	$N(CH_2CH_2)_2NH$	1.56	0.10	0.78
A50	CH_2CH_2F	F	$N(CH_2CH_2)_2NCH_3$	0.39	0.10	3.13
A51	CH_2CH_2OH	F	$N(CH_2CH_2)_2NH$	1.56	0.39	3.13
A52	$CH_2CH_2CH_3$	F	$N(CH_2CH_2)_2NH$	1.56	0.20	3.13
A53	$CH_2CH=CH_2$	F	$N(CH_2CH_2)_2NH$	3.13	0.20	1.56
A54	$CH_2C_6H_5$	F	$N(CH_2CH_2)_2NH$	1.56	0.78	1.56
A55	CH_2CH_3	F	$N(CH_3)_2$	0.78	0.39	50
A56	CH_2CH_3	F	$N(CH_2)_4$	0.20	0.39	12.5
A57	CH_2CH_3	F	$N(CH_2)_5$	0.78	1.56	50
A58	CH_2CH_3	F	$N(CH_2CH_2)_2O$	0.78	0.20	12.5
A59	CH_2CH_3	F	$N(CH_2CH_2)_2CHOH$	0.39	0.20	12.5
A60	CH_2CH_3	F	$N(CH_2CH_2)_2CHCONH_2$	1.56	1.56	100
A61	CH_2CH_3	F	$N(CH_2CH_2)_2CHN(CH_3)_2$	0.39	0.10	3.13
A62	CH_2CH_3	F	3-oxo-1-piperazinyl	3.13	0.39	12.5

ID	R	X	Substituent			
A63	CH_2CH_3	F	$NHCH_2CH_2NH_2$	>100	6.25	50
A64	CH_2CH_3	$N(CH_2CH_2)_2NH$	Cl	>100	>100	>100
A67	$CH=CH_2$	F	$N(CH_2CH_2)_2NH$	3.13	0.10	0.39
A68	CH_2CH_3	F	$N(CH_2CH_2)_2NCH_2CH_3$	0.39	0.10	3.13
A69	CH_2CH_3	F	$N(CH_2CH_2)_2NCH_2CH_2OH$	0.78	0.10	6.25
A70	CH_2CH_3	F	$N(CH_2CH_2)_2NCH_2CH=CH_2$	0.39	0.39	6.25
A71	CH_2CH_3	F	$N(CH_2CH_2)_2NC_6H_5$	0.39	0.78	50
A72	CH_2CH_3	F	$N(CH_2CH_2)_2NCH_2C_6H_5\text{-}p\text{-}NO_2$	1.65	6.25	>100
A73	CH_2CH_3	F	$N(CH_2CH_2)_2NCHO$	1.56	0.39	6.25
A74	CH_2CH_3	F	$N(CH_2CH_2)_2NCOCH_3$	0.78	1.56	25
A75	CH_2CH_3	F	$N(CH_2CH_2)_2NCOC_6H_5$	1.56	3.13	25
A76	$CH=CH_2$	F	$N(CH_2CH_2)_2NCH_3$	1.56	0.10	3.13
A78	CH_2CH_3	F	$NHCH_2CH_2NHCOCH_3$	>100	25	>100
A79	CH_2CH_3	F	$N(CH_2CH_2)_2NCH_2C_6H_4\text{-}p\text{-}NH_2$	0.39	0.39	12.5
A80[d]	CH_2CH_3	F	$N(CH_2CH_2)_2NH$	>100	>100	>100
A81[e]	CH_2CH_3	F	$N(CH_2CH_2)_2NH$	100	12.5	50
A82[f]	CH_2CH_3	F	$N(CH_2CH_2)_2NH$	50	12.5	50
A1	Nalidixic Acid		See Table I.	>100	3.13	100
A3	CH_2CH_3	H	$N(CH_2CH_2)_2NH$	12.5	0.78	3.13
A83	Pipemidic Acid		See Table I.	25	1.56	12.5
A84	Oxolinic Acid		See Table I.	3.13	0.10	25

[a]Converted to μmoles for the statistical analyses (see Tables IV - VI); [b]Chlorine in position 8; [c]Fluorine in position 8; [d]H in position 3 in place of the carboxylic acid; [e]Methyl ester of A34; [f]Ethyl ester of A34.

Table IV. Physicochemical Parameters and Indicator Variables in Position 1 of Set A (Table III)

NO.	F(1)[a]	MR(1)	L(1)	B1(1)	B5(1)	IE(1)[b]	IEF(1)[c]	IEO(1)[d]	IA(1)[e]	IDM(1)[f]	IV(1)[g]	IP(1)[h]	IM(1)[i]
A18	1.405	1.0163	4.11	1.52	3.17	1	0	0	0	0	0	0	0
A32	1.405	1.0163	4.11	1.52	3.17	1	0	0	0	0	0	0	0
A33	1.405	1.0163	4.11	1.52	3.17	1	0	0	0	0	0	0	0
A34	1.405	1.0163	4.11	1.52	3.17	1	0	0	0	0	0	0	0
A36	1.405	1.0163	4.11	1.52	3.17	1	0	0	0	0	0	0	0
A37	1.405	1.0163	4.11	1.52	3.17	1	0	0	0	0	0	0	0
A38	1.405	1.0163	4.11	1.52	3.17	1	0	0	0	0	0	0	0
A39	1.405	1.0163	4.11	1.52	3.17	1	0	0	0	0	0	0	0
A40	1.405	1.0163	4.11	1.52	3.17	1	0	0	0	0	0	0	0
A41	1.405	1.0163	4.11	1.52	3.17	1	0	0	0	0	0	0	0
A42	1.405	1.0163	4.11	1.52	3.17	1	0	0	0	0	0	0	0
A43	1.405	1.0163	4.11	1.52	3.17	1	0	0	0	0	0	0	0
A44	1.405	1.0163	4.11	1.52	3.17	1	0	0	0	0	0	0	0
A45	1.405	1.0163	4.11	1.52	3.17	1	0	0	0	0	0	0	0
A48	0.876	0.5525	2.07	1.52	2.04	0	0	0	0	0	0	0	0
A49	1.128	1.0318	4.70	1.52	3.17	0	1	0	0	0	0	0	0
A50	1.128	1.0318	4.70	1.52	3.38	0	1	0	0	0	0	0	0
A51	0.245	1.1694	4.79	1.52	3.49	0	0	1	0	0	0	0	0
A52	1.934	1.4001	4.92	1.52	3.78	0	0	0	1	0	0	0	0
A53	1.390	1.4547	5.11	1.52	6.02	0	0	0	0	1	0	0	0
A54	2.444	3.0637	4.62	1.52	3.17	0	0	0	0	0	0	0	0
A55	1.405	1.0163	4.11	1.52	3.17	1	0	0	0	0	0	0	0
A56	1.405	1.0163	4.11	1.52	3.17	1	0	0	0	0	0	0	0
A57	1.405	1.0163	4.11	1.52	3.17	1	0	0	0	0	0	0	0
A58	1.405	1.0163	4.11	1.52	3.17	1	0	0	0	0	0	0	0
A59	1.405	1.0163	4.11	1.52	3.17	1	0	0	0	0	0	0	0
A60	1.405	1.0163	4.11	1.52	3.17	1	0	0	0	0	0	0	0
A61	1.405	1.0163	4.11	1.52	3.17	1	0	0	0	0	0	0	0
A62	1.405	1.0163	4.11	1.52	3.17	1	0	0	0	0	0	0	0
A63	1.405	1.0163	4.11	1.60	3.09	1	0	0	0	0	0	0	0
A67	0.861	0.9909	4.29	1.52	3.17	0	0	0	0	0	0	0	0
A68	1.405	1.0163	4.11	1.52	3.17	1	0	0	0	0	0	0	0
A69	1.405	1.0163	4.11	1.52	3.17	1	0	0	0	0	0	0	0
A70	1.405	1.0163	4.11	1.52	3.17	1	0	0	0	0	0	0	0
A71	1.405	1.0163	4.11	1.52	3.17	1	0	0	0	0	0	0	0
A72	1.405	1.0163	4.11	1.52	3.17	1	0	0	0	0	0	0	0
A73	1.405	1.0163	4.11	1.52	3.17	1	0	0	0	0	0	0	0
A74	1.405	1.0163	4.11	1.52	3.17	1	0	0	0	0	0	0	0
A75	1.405	1.0163	4.11	1.52	3.17	0	0	0	0	0	0	0	0
A76	0.861	0.9909	4.29	1.60	3.09	1	0	0	0	0	1	0	0
A78	1.405	1.0163	4.11	1.52	3.17	1	0	0	0	0	0	0	0
A79	1.405	1.0163	4.11	1.52	3.17	1	0	0	0	0	0	0	0
A3	1.405	1.0163	4.11	1.52	3.17	1	0	0	0	0	0	0	0

[a] Calculated lipophilicity of the substituent; [b] C_2H_5; [c] CH_2CH_2F; [d] CH_2CH_2OH; [e] $CH_2CH_2=CH_2$; [f] $CH_2C_6H_5$; [g] $CH=CH_2$; [h] $n-C_3H_7$; [i] CH_3

Table V. Physicochemical Parameters and Indicator Variables in Position 6 of Set A (Table III)

NO.	F(6)[a]	MR(6)	L(6)	B1(6)	B5(6)	IF(6)[b]	IBR(6)[c]	IS(6)[d]	ICO(6)[e]	ICN(6)[e]	INO(6)[g]	ICL(6)[h]	IM(6)[i]	API[j]
A18	0.370	0.1042	2.65	1.35	1.35	1	0	0	0	0	0	0	0	0.000
A32	0.370	0.1042	2.65	1.35	1.35	1	0	0	0	0	0	0	0	0.000
A33	0.370	0.1042	2.65	1.35	1.35	1	0	0	0	0	0	0	0	0.302
A34	0.370	0.1042	2.65	1.35	1.35	1	0	0	0	0	0	0	0	0.171
A36	0.370	0.1042	2.65	1.35	1.35	1	0	0	0	0	0	0	0	0.171
A37	0.940	0.5801	3.52	1.80	1.80	0	0	0	0	0	0	0	0	0.171
A38	0.940	0.5801	3.52	1.80	1.80	0	0	0	0	0	0	0	0	0.171
A39	1.090	0.8657	3.82	1.95	1.95	0	1	0	0	0	0	0	0	0.171
A40	1.090	0.8657	3.82	1.95	1.95	0	1	0	0	0	0	0	0	0.171
A41	0.876	0.5525	2.87	1.52	2.04	0	0	0	0	0	0	1	0	0.000
A42	0.786	1.3500	4.30	1.70	3.26	0	0	0	0	0	0	0	0	0.000
A43	-0.334	0.0520	4.06	1.60	3.13	0	0	1	0	0	0	0	0	0.311
A44	-0.340	0.5664	4.23	1.60	1.60	0	0	1	0	0	0	0	0	0.397
A45	-0.030	0.8142	3.44	1.70	2.44	0	0	0	1	0	1	0	0	0.366
A48	0.370	0.1042	2.65	1.35	1.35	1	0	0	0	1	0	0	0	0.171
A49	0.370	0.1042	2.65	1.35	1.35	1	0	0	0	0	0	0	0	0.171
A50	0.370	0.1042	2.65	1.35	1.35	1	0	0	0	0	0	0	0	0.171
A51	0.370	0.1042	2.65	1.35	1.35	1	0	0	0	0	0	0	0	0.171
A52	0.370	0.1042	2.65	1.35	1.35	1	0	0	0	0	0	0	0	0.171
A53	0.370	0.1042	2.65	1.35	1.35	1	0	0	0	0	0	0	0	0.171
A54	0.370	0.1042	2.65	1.35	1.35	1	0	0	0	0	0	0	0	0.171
A55	0.370	0.1042	2.65	1.35	1.35	1	0	0	0	0	0	0	0	0.171
A56	0.370	0.1042	2.65	1.35	1.35	1	0	0	0	0	0	0	0	0.171
A57	0.370	0.1042	2.65	1.35	1.35	1	0	0	0	0	0	0	0	0.171
A58	0.370	0.1042	2.65	1.35	1.35	1	0	0	0	0	0	0	0	0.171
A59	0.370	0.1042	2.65	1.35	1.35	1	0	0	0	0	0	0	0	0.171
A60	0.370	0.1042	2.65	1.35	1.35	1	0	0	0	0	0	0	0	0.171
A61	0.370	0.1042	2.65	1.35	1.35	1	0	0	0	0	0	0	0	0.171
A62	0.370	0.1042	2.65	1.35	1.35	1	0	0	0	0	0	0	0	0.171
A63	0.370	0.1042	2.65	1.35	1.35	1	0	0	0	0	0	0	0	0.302
A67	0.370	0.1042	2.65	1.35	1.35	1	0	0	0	0	0	0	0	0.171
A68	0.370	0.1042	2.65	1.35	1.35	1	0	0	0	0	0	0	0	0.171
A69	0.370	0.1042	2.65	1.35	1.35	1	0	0	0	0	0	0	0	0.171
A70	0.370	0.1042	2.65	1.35	1.35	1	0	0	0	0	0	0	0	0.171
A71	0.370	0.1042	2.65	1.35	1.35	1	0	0	0	0	0	0	0	0.171
A72	0.370	0.1042	2.65	1.35	1.35	1	0	0	0	0	0	0	0	0.171
A73	0.370	0.1042	2.65	1.35	1.35	1	0	0	0	0	0	0	0	0.171
A74	0.370	0.1042	2.65	1.35	1.35	1	0	0	0	0	0	0	0	0.171
A75	0.370	0.1042	2.65	1.35	1.35	1	0	0	0	0	0	0	0	0.171
A76	0.370	0.1042	2.65	1.35	1.35	1	0	0	0	0	0	0	0	0.171
A78	0.370	0.1042	2.65	1.35	1.35	1	0	0	0	0	0	0	0	0.302
A79	0.370	0.1042	2.65	1.35	1.35	1	0	0	0	0	0	0	0	0.171
A3	0.227	0.0887	2.06	1.00	1.00	0	0	0	0	0	0	0	0	0.000

[a]Calculated lipophilicity of the substituent; [b]Fluorine; [c]Bromine; [d]SCH_3; [e]$COCH_3$; [f]CN; [g]NO_2; [h]Chlorine; [i]CH_3; [j]$\sigma-\rho$ electronic potential interactions between Position 6 and 7.

Table VI. Physicochemical Parameters and Indicator Variables in Position 7 of Set A (Table III)

NO.	F(7)[a]	MR(7)	RI1(7)[b]	RI2(7)[c]	RI3(7)[d]	RI4(7)[e]	INCO(7)[f]
A1B	0.940	0.5801	0	0	0	0	0
A32	0.876	0.5525	0	0	0	0	0
A33	-1.000	0.4574	0	0	0	0	0
A34	-0.100	2.5039	1	0	0	0	0
A36	0.756	2.9677	1	0	0	0	0
A37	-0.100	2.5039	1	0	0	0	0
A38	0.756	2.9677	1	0	0	0	0
A39	-0.100	2.5039	1	0	0	0	0
A40	0.756	2.9677	1	0	0	0	0
A41	-0.100	2.5039	1	0	0	0	0
A42	-0.100	2.5039	1	0	0	0	0
A43	-0.100	2.5039	1	0	0	0	0
A44	-0.100	2.5039	1	0	0	0	0
A45	-0.100	2.5039	1	0	0	0	0
A48	-0.100	2.5039	1	0	0	0	0
A49	-0.100	2.5039	1	0	0	0	0
A50	0.756	2.9677	1	0	0	0	0
A51	-0.100	2.5039	1	0	0	0	0
A52	-0.100	2.5039	1	0	0	0	0
A53	-0.100	2.5039	1	0	0	0	0
A54	-0.100	2.5039	1	0	0	0	0
A55	0.422	1.3850	0	0	0	0	0
A56	1.216	2.1352	0	1	0	0	0
A57	1.775	2.5990	0	0	1	0	0
A58	0.132	2.2883	0	0	0	1	0
A59	-0.312	2.7521	0	0	1	0	0
A60	-0.782	3.4672	0	0	1	0	1
A61	0.500	3.8953	0	0	1	0	0
A62	-0.432	2.5396	0	0	0	0	1
A63	-0.964	1.7537	0	0	0	0	0
A67	-0.100	2.5039	1	0	0	0	0
A68	1.205	3.4315	1	0	0	0	0
A69	0.097	3.5846	1	0	0	0	0
A70	1.110	3.8699	1	0	0	0	0
A71	2.698	4.3333	1	0	0	0	0
A72	2.341	6.2044	1	0	0	0	0
A73	-0.464	3.0034	1	0	0	0	1
A74	0.048	2.9072	1	0	0	0	1
A75	1.347	3.9235	1	0	0	0	1
A76	0.756	2.9677	1	0	0	0	0
A78	-1.074	2.7170	0	0	0	0	0
A79	1.310	5.8476	1	0	0	0	0
A3	-0.100	2.5039	1	0	0	0	0

[a]Calculated lipophilicity of the substituent; [b-e]ring indicators for [b]-N(CH$_2$CH$_2$)$_2$NH; [c]-N(CH$_2$)$_4$; [d]-N(CH$_2$)$_5$; [e]-N(CH$_2$CH$_2$)$_2$O; [f]An indicator of an amide Nitrogen for 7-N-heterocyclic substituents

Set B. The structures and *in vitro* antibacterial activity of 1,6,7-trisubstituted-1,4-dihydro-4-oxo-1,8-naphthyridine-3-carboxylic acids against *S. aureus* 209 PJC-1, *E. coli* NIHJ JC-2, and *Ps. aeruginosa* Tsuchijima are shown in Table VII.(24) Minimum inhibitory concentrations of **B3A - B3C** and nalidixic acid (**BNA**) which are unsubstituted at position 6 are included for comparison. The 18 compounds represented by **BnA**, **BnB**, and **BnC** where n = 3, 15, 18, and 22 - 24 constitute a set where ethyl is constant at position 1 and the substituent at position 7 is varied consistently (A = pyrrolidinyl, B = piperazinyl, and C - N-methylpiperazinyl) for three different substituents at position 7. This set permits a comparison of both the LFER and *de novo* approaches. The physicochemical parameters and indicator variables of this set of compounds for the substituents at positions 1, 6 and 7 listed in Tables VIII - X.

Set C. The antibacterial activities for a series of 1,4-dihydro-4-oxo-1,8-naphthyridine-3-carboxylic acids with substituted azetidinyl (**C28 - C32**), pyrrolidinyl (**C33 - C48**), and piperidinyl (**C49 - C56**) rings at position 7, fluorine at position 6, and ethyl (series A), vinyl (series B) or 2-fluoroethyl (series C) at position 1 are listed in Table XI. The antibacterial activity was measured against the same microorganisms as in Set B.(25) The data for enoxacin (**C2**) are included for comparison. The physicochemical parameters and indicator variables of this set of compounds for each of the substituents at positions 1 and 7 (fluorine only at position 6) are listed in Tables XII - XIII.

RESULTS AND DISCUSSION

Set A. The biological results in this series of 1-ethyl-7-piperazinyl compounds **A34, A37, A39** and **A41 - A45** which vary only at position 6 indicated that fluorine was the preferable substituent. The substitution of the hydrogen of the piperazine NH group (**A34**) by an alkyl (**A36, A68 - A72, A79**) or acyl (**A73 - A75**) group reduced the activity against Gram-negative bacteria, particularly *Ps. aeruginosa*. The replacement of the 1-ethyl group in **A34** by 2-fluoroethyl (**A49**) and vinyl (**A67**) groups resulted in almost equal activity against Gram-negative bacteria while substitution at position 1 by more (**A51 - A54**) or less (**A48**) sterically hindered groups decreased activity. Esters **A81** and **A82** did not show any significant activity which is consistent with previous reports that the esters are prodrugs which first must be hydrolyzed to the free acid by the infected host.(31,32)

The regression analysis was performed on a modified data set. Nine compounds were not included: **A46** and **A47** - the only two compounds having substituents at position 8; **A64** - the only compound with the piperazine ring at position 6 and chlorine at position 7(an interchange of the normal substitution pattern for positions 6 and 7); **A80** - no carboxyl moiety at position 3; **A81 - A82** - the prodrug esters; **A1, A83** and **A84** - lacking the quinoline ring system. The development of the LFER models for

Table VII. Set B: 1,6,7-Trisubstituted 1,4-Dihydro-4-oxo-1,8-naphthyridine-3-carboxylic Acids

[Structure: 1,4-dihydro-4-oxo-1,8-naphthyridine-3-carboxylic acid core with R_6, R_7, and R_1 substituents]

No.	R_1	R_6	R_7	MIC µg/ml[a] (24) S. aureus 209P JC-1	E. coli NIHJ JC-2	Ps. aeruginosa Tsuchijima
B3A	CH_2CH_3	H	$N(CH_2)_4$	12.5	25	>100
B3B	CH_2CH_3	H	$N(CH_2CH_2)_2NH$	25	6.25	25
B3C	CH_2CH_3	H	$N(CH_2CH_2)_2NCH_3$	25	6.25	25
B15A	CH_2CH_3	Cl	$N(CH_2)_4$	12.5	>100	>100
B15B	CH_2CH_3	Cl	$N(CH_2CH_2)_2NH$	3.13	0.78	6.25
B15C	CH_2CH_3	Cl	$N(CH_2CH_2)_2NCH_3$	6.25	1.56	12.5
B18A	CH_2CH_3	CN	$N(CH_2)_4$	3.13	12.5	25
B18B	CH_2CH_3	CN	$N(CH_2CH_2)_2NH$	6.25	1.56	6.25
B18C	CH_2CH_3	CN	$N(CH_2CH_2)_2NCH_3$	12.5	1.56	12.5
B22A	CH_2CH_3	NO_2	$N(CH_2)_4$	25	100	>100
B22B	CH_2CH_3	NO_2	$N(CH_2CH_2)_2NH$	6.25	6.25	25
B22C	CH_2CH_3	NO_2	$N(CH_2CH_2)_2NCH_3$	12.5	3.13	50
B23A	CH_2CH_3	NH_2	$N(CH_2)_4$	>100	>100	>100
B23B	CH_2CH_3	NH_2	$N(CH_2CH_2)_2NH$	>100	3.13	6.25
B23C	CH_2CH_3	NH_2	$N(CH_2CH_2)_2NCH_3$	25	1.56	12.5
B24A	CH_2CH_3	F	$N(CH_2)_4$	0.39	1.56	3.13
B24B	CH_2CH_3	F	$N(CH_2CH_2)_2NH$	0.78	0.2	0.78
B24C	CH_2CH_3	F	$N(CH_2CH_2)_2NCH_3$	1.56	0.39	1.56
B27A	CH_2CH_3	F	NH_2	>100	1.56	50
B27B	CH_2CH_3	F	$NHCH_2CH_2NH_2$	25	6.25	50
B27C	CH_2CH_3	F	3-oxo-1-piperazinyl	6.25	0.78	12.5
B27D	CH_2CH_3	F	$N(CH_2)_5$	0.78	6.25	12.5
B27E	CH_2CH_3	F	$N(CH_2CH_2)_2O$	1.56	3.13	6.25
B27F	CH_2CH_3	F	$N(CH_2CH_2)_2S$	1.56	3.13	12.5
B27G	CH_2CH_3	F	homopiperazinyl	1.56	0.78	1.56
B27H	CH_2CH_3	F	1-azepinyl	3.13	6.25	50
B27I	CH_2CH_3	F	$N(CH_2)_7$	12.5	25	>100
B27J	CH_2CH_3	F	$N(CH_2CH_2)_2NC_6H_5$	6.25	12.5	>100
B27K	CH_2CH_3	F	$N(CH_2CH_2)_2NCh_2C_6H_5$	0.78	6.25	12.5
B27L	CH_2CH_3	F	$N(CH_2CH_2)_2NCH_2CH_3$	0.78	0.78	3.13
B28A	CH_2CH_3	F	$N(CH_2CH_2)_2NCH_2CH_2CH_3$	1.56	1.56	12.5
B28B	CH_2CH_3	F	$N(CH_2CH_2)_2NCH_2CH_2CH_2CH_3$	3.13	3.13	25
B28C	CH_2CH_3	F	$N(CH_2CH_2)_2NCH_2(CH_3)CH_2CH_3$	1.56	3.13	25
B29	CH_2CH_3	F	$N(CH_2CH_2)_2NCHO$	3.13	1.56	12.5
B30	CH_2CH_3	F	$N(CH_2CH_2)_2NCOCH_3$	3.13	3.13	50
B36	$CH=CH_2$	F	$N(CH_2CH_2)_2NH$	1.56	0.1	0.2
B37	$CH=CH_2$	F	$N(CH_2CH_2)_2NCH_3$	3.13	0.2	0.78
B38	CH_2CH_2F	F	$N(CH_2CH_2)_2NH$	0.39	0.2	0.78
B39	CH_2CH_2F	F	$N(CH_2CH_2)_2NCH_3$	0.78	0.2	1.56
B40	CHF_2	F	$N(CH_2CH_2)_2NH$	6.25	0.78	6.25
BNA[b]	CH_2CH_3	H	CH_3	50	1.56	50
PPA[c]	CH_2CH_3	X_6 = N	$N(CH_2CH_2)_2NH$	6.25	1.56	6.25

[a]Converted to µmoles/ml for the statistical analysis (See Tables 8 - 10); [b]Nalidixic Acid; [c]Pipemidic Acid.

Table VIII. Physicochemical Parameters and Indicator Variables in Position 1 of Set B (Table VII)

NO.	F(1)[a]	MB(1)	L(1)	BI(1)	B5(1)	IE(1)[b]	IV(1)[c]	IEF(1)[d]	ICF(1)[e]
B3A	1.405	1.0163	4.11	1.52	3.17	1	0	0	0
B3B	1.405	1.0163	4.11	1.52	3.17	1	0	0	0
B3C	1.405	1.0163	4.11	1.52	3.17	1	0	0	0
B15A	1.405	1.0163	4.11	1.52	3.17	1	0	0	0
B15B	1.405	1.0163	4.11	1.52	3.17	1	0	0	0
B15C	1.405	1.0163	4.11	1.52	3.17	1	0	0	0
B22A	1.405	1.0163	4.11	1.52	3.17	1	0	0	0
B22B	1.405	1.0163	4.11	1.52	3.17	1	0	0	0
B22C	1.405	1.0163	4.11	1.52	3.17	1	0	0	0
B23A	1.405	1.0163	4.11	1.52	3.17	1	0	0	0
B23B	1.405	1.0163	4.11	1.52	3.17	1	0	0	0
B23C	1.405	1.0163	4.11	1.52	3.17	1	0	0	0
B24A	1.405	1.0163	4.11	1.52	3.17	1	0	0	0
B24B	1.405	1.0163	4.11	1.52	3.17	1	0	0	0
B24C	1.405	1.0163	4.11	1.52	3.17	1	0	0	0
B27A	1.405	1.0163	4.11	1.52	3.17	1	0	0	0
B27B	1.405	1.0163	4.11	1.52	3.17	1	0	0	0
B27C	1.405	1.0163	4.11	1.52	3.17	1	0	0	0
B27D	1.405	1.0163	4.11	1.52	3.17	1	0	0	0
B27E	1.405	1.0163	4.11	1.52	3.17	1	0	0	0
B27F	1.405	1.0163	4.11	1.52	3.17	1	0	0	0
B27G	1.405	1.0163	4.11	1.52	3.17	1	0	0	0
B27H	1.405	1.0163	4.11	1.52	3.17	1	0	0	0
B27I	1.405	1.0163	4.11	1.52	3.17	1	0	0	0
B27J	1.405	1.0163	4.11	1.52	3.17	1	0	0	0
B27K	1.405	1.0163	4.11	1.52	3.17	1	0	0	0
B27L	1.405	1.0163	4.11	1.52	3.17	1	0	0	0
B28A	1.405	1.0163	4.11	1.52	3.17	1	0	0	0
B28B	1.405	1.0163	4.11	1.52	3.17	1	0	0	0
B28C	1.405	1.0163	4.11	1.52	3.17	1	0	0	0
B29	1.405	1.0163	4.11	1.52	3.17	1	0	0	0
B30	1.405	1.0163	4.11	1.52	3.17	1	0	0	0
B36	0.061	0.9909	4.29	1.60	3.09	0	1	0	0
B37	0.061	0.9909	4.29	1.60	3.09	0	1	0	0
B38	1.128	1.0310	4.70	1.52	3.17	0	0	0	0
B39	1.128	1.0310	4.70	1.52	3.17	0	0	0	0
B40	1.322	0.5835	3.30	1.71	2.61	0	0	1	1
B11A	1.405	1.0163	4.11	1.52	3.17	1	0	0	0

[a]Calculated lipophilicity of the substituent; [b]C_2H_5; [c]$CH=CH_2$; [d]CH_2CH_2F; [e]CHF_2

Table IX. Physicochemical Parameters and Indicator Variables in Position 6 of Set B (Table VII)

NO.	F(6)[a]	MR(6)	L(6)	B1(6)	B5(6)	IF(6)[b]	ICN(6)[c]	INO(6)[d]	ICl(6)[e]	INH(6)[f]	API[g]
B3A	0.277	0.0887	2.06	1.00	1.00	0	0	0	0	0	0.512
B3B	0.277	0.0887	2.06	1.00	1.00	0	0	0	0	0	0.512
B3C	0.227	0.0887	2.06	1.00	1.00	0	0	0	0	0	0.512
B15A	0.940	0.5001	3.52	1.80	1.80	0	0	0	1	0	0.627
B15B	0.940	0.5001	3.52	1.80	1.80	0	0	0	1	0	0.627
B15C	0.940	0.5001	3.52	1.80	1.80	0	0	0	1	0	0.627
B18A	-0.340	0.5664	4.23	1.60	1.60	0	1	0	0	0	0.779
B18B	-0.340	0.5664	4.23	1.60	1.60	0	1	0	0	0	0.779
B18C	-0.340	0.5664	4.23	1.60	1.60	0	1	0	0	0	0.779
B22A	-0.030	0.8142	3.44	2.44	2.44	0	0	1	0	0	0.758
B22B	-0.030	0.8142	3.44	2.44	2.44	0	0	1	0	0	0.758
B22C	-0.030	0.8142	3.44	2.44	2.44	0	0	1	0	0	0.758
B23A	-1.000	0.4574	2.78	1.97	1.97	0	0	0	0	1	1.420
B23B	-1.000	0.4574	2.78	1.97	1.97	0	0	0	0	1	1.420
B23C	-1.000	0.4574	2.78	1.97	1.97	0	0	0	0	1	1.420
B24A	0.370	0.1042	2.65	1.35	1.35	1	0	0	0	0	0.627
B24B	0.370	0.1042	2.65	1.35	1.35	1	0	0	0	0	0.627
B24C	0.370	0.1042	2.65	1.35	1.35	1	0	0	0	0	0.627
B27A	0.370	0.1042	2.65	1.35	1.35	1	0	0	0	0	1.088
B27B	0.370	0.1042	2.65	1.35	1.35	1	0	0	0	0	1.088
B27C	0.370	0.1042	2.65	1.35	1.35	1	0	0	0	0	0.027
B27D	0.370	0.1042	2.65	1.35	1.35	1	0	0	0	0	0.027
B27E	0.370	0.1042	2.65	1.35	1.35	1	0	0	0	0	0.027
B27F	0.370	0.1042	2.65	1.35	1.35	1	0	0	0	0	0.027
B27G	0.370	0.1042	2.65	1.35	1.35	1	0	0	0	0	0.027
B27H	0.370	0.1042	2.65	1.35	1.35	1	0	0	0	0	0.027
B27I	0.370	0.1042	2.65	1.35	1.35	1	0	0	0	0	0.027
B27J	0.370	0.1042	2.65	1.35	1.35	1	0	0	0	0	0.027
B27K	0.370	0.1042	2.65	1.35	1.35	1	0	0	0	0	0.027
B27L	0.370	0.1042	2.65	1.35	1.35	1	0	0	0	0	0.027
B28A	0.370	0.1042	2.65	1.35	1.35	1	0	0	0	0	0.027
B28B	0.370	0.1042	2.65	1.35	1.35	1	0	0	0	0	0.027
B28C	0.370	0.1042	2.65	1.35	1.35	1	0	0	0	0	0.027
B29	0.370	0.1042	2.65	1.35	1.35	1	0	0	0	0	0.027
B30	0.370	0.1042	2.65	1.35	1.35	1	0	0	0	0	0.027
B36	0.370	0.1042	2.65	1.35	1.35	1	0	0	0	0	0.027
B37	0.370	0.1042	2.65	1.35	1.35	1	0	0	0	0	0.027
B38	0.370	0.1042	2.65	1.35	1.35	1	0	0	0	0	0.027
B39	0.370	0.1042	2.65	1.35	1.35	1	0	0	0	0	0.027
B40	0.370	0.1042	2.65	1.35	1.35	1	0	0	0	0	0.027
BNA	0.227	0.0887	2.06	1.00	1.00	0	0	0	0	0	0.000

[a]Calculated lipophilicity of the substituent; [b]Fluorine; [c]CN; [d]NO_2; [e]Chlorine; [f]NH_2; [g]σ-ρ electronic potential interactions between position 6 and 7

Table X. Physicochemical Parameters and Indicator Variables in Position 7 of Set B (Table VII)

NO.	F(7)[a]	MR(7)	RI1(7)[b]	RI2(7)[c]	RI3(7)[d]	RI4(7)[e]	INCO(7)[f]	ICH3(7)[g]	IRH1(7)[h]
B3A	1.216	2.1352	0	1	0	0	0	0	0
B3B	-0.100	2.5039	1	0	0	0	0	0	1
B3C	0.756	2.9677	1	0	0	0	0	1	0
B15A	1.216	2.1352	0	1	0	0	0	0	0
B15B	-0.100	2.5039	1	0	0	0	0	0	1
B15C	0.756	2.9677	1	0	0	0	0	1	0
B18A	1.216	2.1352	0	1	0	0	0	0	0
B18B	-0.100	2.5039	1	0	0	0	0	0	1
B18C	0.756	2.9677	1	0	0	0	0	1	0
B22A	1.216	2.1352	0	1	0	0	0	0	0
B22B	-0.100	2.5039	1	0	0	0	0	0	1
B22C	0.756	2.9677	1	0	0	0	0	1	0
B23A	1.216	2.1352	0	1	0	0	0	0	0
B23B	-0.100	2.5039	1	0	0	0	0	0	1
B23C	0.756	2.9677	1	0	0	0	0	1	0
B24A	1.216	2.1352	0	1	0	0	0	0	0
B24B	-0.100	2.5039	1	0	0	0	0	0	1
B24C	0.756	2.9677	1	0	0	0	0	1	0
B27A	-1.000	0.4574	0	0	0	0	0	0	0
B27B	-0.964	1.7537	0	0	0	0	0	0	0
B27C	-0.432	2.5396	0	0	0	0	1	0	0
B27D	1.775	2.5990	0	0	1	0	0	0	0
B27E	0.132	2.2883	0	0	0	1	0	0	0
B27F	0.852	2.9415	0	0	0	0	0	0	0
B27G	-0.003	2.9677	0	0	0	0	0	0	1
B27H	1.902	2.9866	0	0	0	0	0	0	0
B27I	2.893	3.5266	0	0	0	0	0	0	0
B27J	2.669	5.0151	1	0	0	0	0	0	0
B27K	2.538	5.4789	1	0	0	0	0	0	0
B27L	1.205	3.4315	1	0	0	0	0	0	0
B28A	1.654	3.8953	1	0	0	0	0	0	0
B28B	2.103	4.3591	1	0	0	0	0	0	0
B28C	1.883	4.3591	1	0	0	0	0	0	0
B29	0.464	3.0034	1	0	0	0	1	0	0
B30	0.040	3.4672	1	0	0	0	1	0	0
B36	-0.100	2.5039	1	0	0	0	0	0	1
B37	0.756	2.9677	1	0	0	0	0	1	0
B38	-0.100	2.5039	1	0	0	0	0	0	1
B39	0.756	2.9677	1	0	0	0	0	1	0
B40	-0.100	2.5039	1	0	0	0	0	0	1
BNA	0.876	0.5525	0	0	0	0	0	0	0

[a]Calculated lipophilicity of the substituent; [b-e]ring indicator of [b]-$N(CH_2CH_2)_2NH$; [c]-$N(CH_2)_4$; [d]-$N(CH_2)_5$; [e]-$N(CH_2CH_2)_2O$; [f]indicator for N-methyl in the piperazinyl ring; [h]indicator for N-H in the piperazinyl ring

Table XI. Set C: 1,7-Disubstituted 6-Fluoro-1,4-dihydro-4-oxo-1,8-naphthyridine-3-carboxylic Acids

C28 - C32

C33 - C48

C49 - C56

No.	R_1	R	R'	S. aureus 209P JC-1	E. coli NIHJ JC-2	Ps. aeruginosa Tsuchijima
C2	CH_2CH_3	Enoxacin	See Table I	0.78	0.2	0.78
C28A	CH_2CH_3	H_2N		0.78	0.1	0.78
C28B	$CH=CH_2$	H_2N		1.56	0.1	0.39
C29A	CH_2CH_3	$O(CH_2CH_2)_2N$		1.56	6.25	50
C30A	CH_2CH_3	HO		0.78	0.78	3.13
C31A	CH_2CH_3	CH_3O		0.78	3.13	6.25
C32A	CH_2CH_3	CH_3CH_2O		0.78	3.13	25
C33A	CH_2CH_3	H_2N	H	0.2	0.1	0.39
C33B	$CH=CH_2$	H_2N	H	0.2	0.025	0.2
C33C	CH_2CH_2F	H_2N	H	0.39	0.1	0.39
C34A	CH_2CH_3	CH_3NH	H	0.39	0.2	1.56
C34B	$CH=CH_2$	CH_3NH	H	0.78	0.1	0.78
C34C	CH_2CH_2F	CH_3NH	H	0.78	0.2	0.78
C35A	CH_2CH_3	CH_3CH_2NH	H	0.78	0.78	3.13
C36A	CH_2CH_3	CF_3CH_2NH	H	0.78	1.56	50
C36B	$CH=CH_2$	CF_3CH_2NH	H	1.56	6.25	100
C37A	CH_2CH_3	$CH_3CH_2CH_2NH$	H	25	6.25	>100
C38A	CH_2CH_3	$(CH_3)_2N$	H	1.56	0.78	6.25
C38B	$CH=CH_2$	$(CH_3)_2N$	H	3.13	0.39	3.13
C38C	CH_2CH_2F	$(CH_3)_2N$	H	0.78	0.2	6.25
C39A	CH_2CH_3	$(CHO)NH$	H	0.78	0.78	3.13
C39B	$CH=CH_2$	$(CHO)NH$	H	0.78	0.39	3.13
C40A	CH_2CH_3	CH_3CONH	H	0.78	6.25	6.25
C40B	$CH=CH_2$	CH_3CONH	H	1.56	0.78	12.5
C40C	CH_2CH_2F	CH_3CONH	H	1.56	1.56	25
C41A	CH_2CH_3	CF_3CONH	H	0.78	1.56	6.25
C42A	CH_2CH_3	$(CH_3)(CH_3CO)N$	H	0.78	6.25	25
C42B	$CH=CH_2$	$(CH_3)(CH_3CO)N$	H	1.56	3.13	25
C42C	CH_2CH_2F	$(CH_3)(CH_3CO)N$	H	0.39	1.56	25
C43A	CH_2CH_3	NH_2NH	H	0.39	1.56	12.5
C44A	CH_2CH_3	NH_2CONH	H	6.25	3.13	25
C45A	CH_2CH_3	NH_2	HO	1.56	1.56	3.13
C46A	CH_2CH_3	HO	H	0.39	0.78	0.78
C47A	CH_2CH_3	HOOC	H	0.39	1.56	3.13
C48A	CH_2CH_3	Cl	H	0.2	0.39	12.5
C49A	CH_2CH_3	H	NH_2	0.78	0.78	6.25
C50A	CH_2CH_3	NH_2	H	0.2	0.2	1.56
C51A	CH_2CH_3	C_6H_5CONH	H	1.56	2.5	>100
C52A	CH_2CH_3	NH_2CH_2	H	0.39	1.56	12.3
C53A	CH_2CH_3	$CH_3CONHCH_2$	H	0.78	1.56	25
C54A	CH_2CH_3	H	NH_2CO	6.25	2.5	50
C55A	CH_2CH_3	H	HO	1.56	3.13	25
C56A	CH_2CH_3	HO	H	0.78	3.13	6.25

a. Converted to μmole/ml for the statistical analysis (see Tables XII-XIII).

Table XII. Physicochemical Parameters and Indicator Variables in Position 1 of Set C (Table XI)

NO.	F(1)[a]	MR(1)	L(1)	BI(1)	B5(1)	1E(1)[b]	IEF(1)[c]	IV(1)[d]
C28A	1.405	1.0163	4.11	1.52	3.17	1	0	0
C28B	0.861	0.9901	4.29	1.60	3.09	0	0	1
C29A	1.405	1.0163	4.11	1.52	3.17	1	0	0
C30A	1.405	1.0163	4.11	1.52	3.17	1	0	0
C31A	1.405	1.0163	4.11	1.52	3.17	1	0	0
C32A	1.405	1.0163	4.11	1.52	3.17	1	0	0
C33A	1.405	1.0163	4.11	1.52	3.17	1	0	0
C33B	0.861	0.9909	4.29	1.60	3.09	0	0	1
C33C	1.120	1.0163	4.70	1.52	3.17	0	1	0
C34A	1.405	1.0163	4.11	1.52	3.17	1	0	0
C34B	0.861	0.9909	4.29	1.60	3.09	0	0	1
C34C	1.128	1.0310	4.70	1.52	3.17	0	1	0
C35A	1.405	1.0163	4.11	1.52	3.17	1	0	0
C36A	1.405	1.0163	4.11	1.52	3.17	1	0	0
C36B	0.861	0.9909	4.29	1.60	3.09	0	0	1
C37A	1.405	1.0163	4.11	1.52	3.17	1	0	0
C38A	1.405	1.0163	4.11	1.52	3.17	1	0	0
C38B	0.861	0.9909	4.29	1.60	3.09	0	0	1
C38C	1.120	1.0310	4.70	1.52	3.17	0	1	0
C39A	1.405	1.0163	4.11	1.52	3.17	1	0	0
C39B	0.861	0.9909	4.29	1.60	3.09	0	0	1
C40A	1.405	1.0163	4.11	1.52	3.17	1	0	0
C40B	0.861	0.9909	4.29	1.60	3.09	0	0	1
C40C	1.128	1.0318	4.70	1.52	3.17	0	1	0
C41A	1.405	1.0163	4.11	1.52	3.17	1	0	0
C42A	1.405	1.0163	4.11	1.52	3.17	1	0	0
C42B	0.861	0.9909	4.29	1.60	3.09	0	0	1
C42C	1.128	1.0318	4.70	1.52	3.17	0	1	0
C43A	1.405	1.0163	4.11	1.52	3.17	1	0	0
C44A	1.405	1.0163	4.11	1.52	3.17	1	0	0
C45A	1.405	1.0163	4.11	1.52	3.17	1	0	0
C46A	1.405	1.0163	4.11	1.52	3.17	1	0	0
C47A	1.405	1.0163	4.11	1.52	3.17	1	0	0
C48A	1.405	1.0163	4.11	1.52	3.17	1	0	0
C49A	1.405	1.0163	4.11	1.52	3.17	1	0	0
C50A	1.405	1.0163	4.11	1.52	3.17	1	0	0
C51A	1.405	1.0163	4.11	1.52	3.17	1	0	0
C52A	1.405	1.0163	4.11	1.52	3.17	1	0	0
C53A	1.405	1.0163	4.11	1.52	3.17	1	0	0
C54A	1.405	1.0163	4.11	1.52	3.17	1	0	0
C55A	1.405	1.0163	4.11	1.52	3.17	1	0	0
C56A	1.405	1.0163	4.11	1.52	3.17	1	0	0

[a]Calculated lipophilicity of the substituents; [b]C_2H_5; [c]CH_2CH_2F; [d]$CH=CH_2$

Table XIII. Physicochemical Parameters and Indicator Variables in Position 7 of Set C (Table XI)

NO.	F(7)[a]	MR(7)	FR(7)[b]	MR(7)[c]	LR(7)[d]	B1R(7)[e]	B5R(7)[f]	R11(7)[g]	R12(7)[h]	R13(7)[i]	PROXI(7)[j]	INCO(7)[k]
C28A	-0.194	1.7601	-1.7600	0.4574	2.78	1.35	1.97	0	0	0	1.136	0
C28B	-0.194	1.7601	-1.7600	0.4574	2.78	1.35	1.97	0	1	0	1.136	0
C29A	0.310	3.8710	0.2060	2.2883	5.20	1.35	3.42	0	1	0	0.934	0
C30A	-0.248	1.8245	-1.8600	0.2418	2.74	1.35	1.93	0	1	0	1.182	0
C31A	0.410	2.7883	-1.2840	0.7066	3.98	1.35	3.07	0	1	0	1.264	0
C32A	0.939	2.7521	-0.7550	1.1694	4.80	1.35	3.30	0	1	0	1.264	0
C33A	-0.203	2.5039	-1.7600	0.4574	2.78	1.35	1.97	0	0	0	0.568	0
C33B	-0.203	2.5039	-1.7600	0.4574	2.78	1.35	1.97	0	1	0	0.568	0
C33C	-0.203	2.5039	-1.7600	0.4574	2.78	1.35	1.97	1	0	0	0.568	0
C34A	0.003	2.9677	-1.6140	0.9212	3.53	1.35	3.08	0	0	0	0.708	0
C34B	0.003	2.9677	-1.6140	0.9212	3.53	1.35	3.08	0	1	0	0.708	0
C34C	0.003	2.9677	-1.6140	0.9212	3.53	1.35	3.08	1	0	0	0.708	0
C35A	0.612	3.4315	-1.0050	1.3050	4.03	1.35	3.42	0	0	0	0.708	0
C36A	1.371	2.0460	-0.3260	1.0464	5.26	1.35	4.00	1	0	0	0.708	0
C36B	1.371	2.0460	-0.3260	1.0464	5.26	1.35	4.00	1	0	0	0.708	0
C37A	1.141	3.0953	-0.5560	1.0400	6.07	1.35	4.47	1	0	0	0.708	0
C38A	0.656	3.4315	-1.0480	1.3850	3.53	1.35	3.08	1	0	0	0.715	0
C38B	0.656	3.4315	-1.0480	1.3850	3.53	1.35	3.08	1	0	0	0.715	0
C38C	0.656	3.4315	-1.0480	1.3850	3.53	1.35	3.08	1	0	0	0.715	0
C39A	-0.172	3.0034	-1.7200	0.9569	4.22	1.35	3.61	1	0	0	0.559	1
C39B	-0.172	3.0034	-1.7200	0.9569	4.22	1.35	3.61	1	0	0	0.559	1
C40A	-0.348	3.4672	-2.5889	1.4207	5.09	1.35	3.61	1	0	0	0.837	1
C40B	-0.348	3.4672	-2.5889	1.4207	5.09	1.35	3.61	1	0	0	0.837	1
C40C	-0.348	3.4672	-2.5889	1.4207	5.09	1.35	3.61	1	0	0	0.837	1
C41A	0.761	3.5137	-1.0650	1.4672	5.02	1.79	3.61	1	0	0	0.940	1
C42A	-0.121	3.9310	-2.0580	1.8845	4.77	1.35	3.71	1	0	0	0.940	1
C42B	-0.121	3.9310	-2.0580	1.8845	4.77	1.35	3.71	1	0	0	0.940	1
C42C	-0.121	3.9310	-2.0580	1.8845	4.77	1.35	3.71	1	0	0	0.940	1
C43A	-0.511	2.8726	-2.1600	0.8261	3.47	1.35	2.97	1	0	0	0.660	0
C44A	-0.696	3.3721	-2.4000	1.3256	5.06	1.35	3.61	1	0	0	0.715	1
C45A	-0.872	2.0570	-1.7600	0.4574	2.70	1.35	1.97	1	0	0	0.981	1
C46A	-0.280	2.2883	-1.8600	0.7413	2.74	1.35	1.93	1	0	0	0.591	1
C47A	0.105	2.7870	-1.3600	0.2410	3.53	1.60	2.36	1	0	0	0.476	1
C48A	1.399	2.6266	0.0600	0.5801	3.52	1.80	1.80	1	0	0	0.350	1
C49A	0.356	2.9677	0.2270	0.0887	2.06	1.00	1.00	0	0	1	0.000	0
C50A	-0.212	2.9677	-1.7600	0.4574	2.78	1.35	1.97	0	0	1	0.000	0
C51A	1.313	6.2445	-0.2350	3.4601	8.30	1.53	3.84	0	0	1	0.000	1
C52A	0.407	4.3315	-1.1410	0.9212	4.02	1.52	3.05	0	0	1	0.000	0
C53A	-0.007	4.3940	-1.5550	1.0845	5.67	1.52	4.75	0	0	1	0.000	1
C54A	-0.083	3.4672	0.2270	0.0887	2.06	1.00	1.00	0	0	1	0.000	1
C55A	0.279	2.7521	0.2270	0.0887	2.06	1.00	1.00	0	0	1	0.000	1
C56A	-0.312	2.7521	-1.0600	0.2418	2.74	1.35	1.93	0	0	0	0.000	0

[a]Calculated lipophilicity of the substituents; [b]F-Calculated lipophilicity; [c]MR, [d]L, [e]$B1$, [f]$B5$ of R-substituents; [g]ring indicator for $g-N(CH_2)_4$; [h]$-N(CH_2)_3$; [i]$-N(CH_2)_5$; [j]Proximity effect between ring and R-substituent; [k]An indicator of an amide nitrogen for 7-N-heterocyclic substituents

S. aureus and *Ps. aeruginosa* are shown in Tables XIV and XV, respectively. (For *Ps. aeruginosa*, n = 41 because two compounds were deleted: A34 - considered too active relative to the other compounds (see below); A43 - ICO(6) which entered into an earlier model occurs only in the one compound.) A significant LFER model could not be obtained for the *E. coli* data (r = 0.73, $F_{6,35}$ = 7.34). The observed and calculated activities, residuals and standardized residuals for *S. aureus* and *Ps. aeruginosa* based on the statistically accepted models are shown in Table XVI (Eq. 6) and Table XVII (Eq. 7), respectively. The calculated log 1/MIC for the omitted, highly active A34 against *Ps. aeruginosa* was very close to the observed value, 1.64 vs. 1.74, respectively. The correlation matrix of the entire data set will be found in Table XVIII. The two outliers (standardized residual >2.0) from equation 6 (Table XIV) were deleted (A59, A78) and the regression repeated producing a model very similar to that of equation 6 but with another two outliers, A3 and A63. Because it was difficult to rationalize deleting the first two outliers, it was decided to stay with equation 6 (Table XIV). A similar phenomenon was seen with the development of equation 7 (Table XV). There were three outliers (A37, A74, A40) which were dropped. The resulting model was similar to equation 7, but two additional outliers (A61, A67) appeared. Since there was no valid reason for deleting these two compounds, equation 7 was selected as the model for Set A acting against *Ps. aeruginosa*.

Equation 6 (Table XIV) indicates that lipophilicity and molar refractivity of the substituents at position 6 are important determinants of activity against *S. aureus*. There is a parabolic relationship seen with these same descriptors for the substituents at position 7. Comparison of equation 6 for *S. aureus* with equation 7 (Table XV) for *Ps. aeruginosa* indicates a different QSAR. An ethyl substituent at position 1, minimum width (B1) of the substituent at position 6, and the appearance of a piperazinyl ring in position 7 all appear in equation 7. The parabolic relationship of lipophilicity and MR seen in equation 6 for substituents at position 7 is found also in equation 7 indicating that there are optimum lipophilicity and molar refraction ranges for the 7-substituent. At the same time, it must be noted that many of the values for the parameters listed in Tables IV and V show clustering which can bias the results.

Therefore, a subset of 24 compounds (A18, A32 - A34, A36, A55 - A63, A68 - A75, A78 -A79) containing only a fluorine at position 6 and ethyl at position 1 was selected in order to better understand just what descriptors were important for activity at position 7. In contrast with the entire data set of 41 - 43 compounds, LFER models for all three bacterial test systems were derived from the subset as shown in Tables XIX, XX and XXI. Equation 9 (Table XIX) for *S. aureus* indicates that only the presence of an amide nitrogen, INCO(7), and σ-ρ electronic interactions, API, are important determinants. An amide nitrogen at position 7 reduces activity, and there is a parabolic relationship of σ-ρ electronic interactions between positions 6 and 7. The nonlinear result for the latter is probably due to the

Table XIV. LFER Model Development for Set A (Table III) Against S. aureus

Eq.	Intercept	F(6)	MR(6)	F(7)	$F(7)^2$	MR(7)	$MR(7)^2$	r^2	F	d.f.
1.	1.977 (±0.111)			0.546 (±0.127)				0.312	18.581	(1,41)
2.	2.081 (±0.110)			0.916 (±0.178)	-0.284 (±0.103)			0.422	14.581	(2,40)
3.	1.344 (±0.261)			0.806 (±0.166)	-0.347 (±0.096)	0.296 (±0.097)		0.534	14.879	(3,39)
4.	1.577 (±0.241)		-0.880 (±0.258)	0.822 (±0.147)	-0.398 (±0.086)	0.304 (±0.086)		0.643	17.127	(4,38)
5.	0.766 (±0.338)		-0.966 (±0.234)	0.846 (±0.133)	-0.323 (±0.082)	0.918 (±0.899)	-0.106 (±0.034)	0.717	18.754	(5,37)
6.	0.549 (±0.319)	0.699 (±0.242)	-1.116 (±0.220)	0.812 (±0.119)	-0.314 (±0.073)	0.899 (±0.194)	-0.103 (±0.031)	0.771	20.131	(6,36)

n = 43   s = 0.419

See Tables IV - VI for a listing of the independent variables.

Table XV. LFER Model Development for Set A (Table III) Against *P. aeruginosa*

Eq.	Intercept	IE(1)	B1(6)	F(7)	F(7)	MR(7)	MR(7)²	RI1(7)	r²	F	d.f.
1.	1.521 (±0.117)			-0.096 (±0.131)					0.014	0.543	(1,39)
2.	1.658 (±0.109)				-0.362 (±0.100)				0.268	6.965	(2,38)
3.	2.306 (±0.243)	-0.895 (±0.191)		0.338 (±0.138)	-0.284 (±0.082)				0.542	14.586	(3,37)
4.	1.759 (±0.243)	-0.879 (±0.175)		0.259 (±0.130)	-0.332 (±0.077)	0.215 (±0.075)			0.627	15.115	(4,36)
5.	1.053 (±0.341)	-0.818 (±0.162)		0.279 (±0.119)	-0.274 (±0.073)	0.702 (±0.191)	-0.084 (±0.031)		0.693	15.729	(5,35)
6.	1.004 (±0.333)	-0.705 (±0.171)		0.232 (±0.120)	-0.244 (±0.074)	0.580 (±0.199)	-0.074 (±0.030)	0.303 (±0.178)	0.717	14.323	(6,34)
7.	2.082 (±0.176)	-0.524 (±0.176)	-0.904 (±0.371)	0.239 (±0.112)	-0.259 (±0.069)	0.584 (±0.186)	-0.081 (±0.028)	0.479 (±0.181)	0.760	14.964	(7,33)

n = 41   s = 0.375

See Tables IV - VI for a listing of the independent variables.

Table XVI. Comparison of Observed and Calculated MIC's from Eq. 6 (Table XIV)

No.	Observed	Calculated	Residual	Standard. Residual[a]
A18	1.333	1.663	-0.330	-0.849
A32	1.600	1.626	-0.026	-0.066
A33	0.097	-0.045	0.142	0.366
A34	2.914	2.212	0.701	1.807
A36	2.932	2.887	0.045	0.116
A37	2.333	2.080	0.253	0.651
A38	2.351	2.754	-0.403	-1.039
A39	2.084	1.866	0.218	0.562
A40	2.402	2.540	-0.138	-0.356
A41	2.003	2.066	-0.063	-0.163
A42	1.143	1.103	0.039	0.101
A43	0.535	0.663	-0.128	-0.331
A44	1.417	1.201	0.216	0.557
A45	1.141	1.141	0.000	0.000
A48	1.688	2.212	-0.524	-1.350
A49	2.334	2.212	0.122	0.314
A50	2.955	2.887	0.068	0.175
A51	2.332	2.212	0.119	0.307
A52	2.330	2.212	0.117	0.302
A53	2.024	2.212	-0.188	-0.485
A54	2.388	2.212	0.176	0.453
A55	2.551	2.025	0.526	1.356
A56	3.182	2.664	0.518	1.335
A57	2.611	2.783	-0.173	-0.445
A58	2.613	2.311	0.302	0.778
A59	2.932	2.102	0.830	2.138
A60	2.365	1.745	0.620	1.597
A61	2.937	2.959	0.007	0.019
A62	2.027	1.901	0.126	0.324
A63	0.166	0.877	-0.711	-1.832
A67	2.006	2.212	-0.207	-0.533
A68	2.951	3.086	-0.136	-0.349
A69	2.668	2.667	0.000	0.000
A70	2.963	3.143	-0.181	-0.466
A71	3.020	2.559	0.461	1.189
A72	2.463	2.490	-0.026	-0.068
A73	2.347	2.019	0.328	0.845
A74	2.666	2.473	0.192	0.496
A75	2.433	3.158	-0.725	-1.868
A76	2.327	2.887	-0.560	-1.442
A78	0.224	1.140	-0.916	-2.360
A79	3.011	2.956	0.055	0.141
A3	1.382	2.130	-0.748	-1.927

Table XVII. Comparison of Observed and Calculated MIC's from Eq. 7 (Table XV)

No.	Observed	Calculated	Residual	Standard. Residual
A18	0.431	0.646	-0.215	-0.632
A32	0.697	0.647	0.050	0.147
A33	0.097	0.090	0.007	0.019
A36	2.330	1.871	0.459	1.348
A37	2.030	1.339	0.691	2.029
A38	1.146	1.464	-0.318	-0.935
A39	1.483	1.204	0.279	0.820
A40	0.595	1.329	-0.733	-2.155
A41	1.703	1.592	0.111	0.326
A42	1.444	1.430	0.014	0.041
A44	1.717	1.520	0.197	0.578
A45	1.442	1.430	0.013	0.038
A48	2.292	2.270	0.021	0.062
A49	2.636	2.270	0.366	1.075
A50	2.050	2.395	-0.346	-1.015
A51	2.030	2.270	-0.241	-0.707
A52	2.027	2.270	-0.244	-0.715
A53	2.327	2.270	0.057	0.166
A54	2.388	2.270	0.118	0.346
A55	0.745	1.047	-0.302	-0.887
A56	1.386	1.124	0.262	0.769
A57	0.804	0.919	-0.115	-0.337
A58	1.408	1.278	0.130	0.382
A59	1.427	1.233	0.194	0.571
A60	0.558	1.045	-0.487	-1.432
A61	2.062	1.440	0.622	1.828
A62	1.426	1.148	0.278	0.817
A63	0.767	0.643	0.124	0.365
A67	2.910	2.270	0.640	1.879
A68	2.045	1.781	0.264	0.775
A69	1.764	1.892	-0.127	-0.374
A70	1.759	1.812	-0.053	-0.154
A71	0.914	0.590	0.323	0.950
A72	0.356	0.466	-0.111	-0.325
A73	1.745	1.675	0.070	0.205
A74	1.159	1.842	-0.683	-2.006
A75	1.228	1.715	-0.487	-1.431
A76	2.024	2.395	-0.371	-1.090
A78	0.224	0.772	-0.548	-1.611
A79	1.504	1.333	0.171	0.503
A3	1.983	2.062	-0.079	-0.233

Table XVIII. Correlation Matrix of the Variables Used in the Analyses of the S. aureus and P. aeruginosa Test Systems (Tables XIV, XV)

	IE(1)	IF(6)	F(6)	MR(6)	B1(6)	F(7)	$F(7)^2$	MR(7)	$MR(7)^2$	RI1(7)	SA	EC	PA
IE(1)	1.000												
IF(6)	-0.283	1.000											
F(6)	0.067	-0.238	1.000										
MR(6)	0.244	-0.861	0.243	1.000									
B1(6)	0.205	-0.725	0.491	0.834	1.000								
F(7)	0.130	0.158	0.049	-0.137	-0.058	1.000							
$F(7)^2$	0.229	0.250	-0.265	-0.215	-0.159	0.752	1.000						
MR(7)	0.079	0.089	0.015	-0.078	-0.041	0.497	0.497	1.000					
$MR(7)^2$	0.142	0.156	-0.008	-0.135	-0.094	0.565	0.595	0.947	1.000				
RI1(7)	-0.339	-0.362	0.086	0.312	0.263	0.166	-0.027	0.421	0.307	1.000			
SA	-0.077	0.320	0.216	-0.334	-0.119	0.558	0.202	0.500	0.407	0.203	1.000		
EC	-0.328	0.162	0.258	-0.222	-0.041	0.111	-0.247	0.087	-0.011	0.260	0.676	1.000	
PA	-0.449	0.103	0.014	-0.173	-0.152	-0.119	-0.354	0.069	-0.071	0.465	0.414	0.733	1.000

See Tables IV - VI for a listing of the independent variables.

Table XIX. LFER Model Development for a Subset of Series A (Table III; Ethyl at Position 1; Fluorine at Position 6) Against S. aureus

Eq.	Intercept	F(7)	F(7)²	MR(7)	MR(7)²	INCO(7)	API	API²	r²	F	d.f.
1.	2.050 (±0.179)	0.500 (±0.159)							0.310	9.875	(1,22)
2.	2.275 (±0.176)	0.928 (±0.206)	-0.340 (±0.121)						0.498	10.404	(2,21)
3.	1.530 (±0.281)	0.818 (±0.177)	-0.412 (±0.105)	0.306 (±0.230)					0.622	13.067	(3,20)
4.	0.676 (±0.418)	0.866 (±0.148)	-0.351 (±0.089)	0.983 (±0.230)	-0.116 (±0.037)				0.778	16.613	(4,19)
5.	0.906 (±0.418)	0.698 (±0.216)	-0.283 (±0.109)	1.079 (±0.246)	-0.126 (±0.038)		-2.357 (±2.228)		0.791	13.631	(5,18)
6.	1.158 (±0.229)	0.338 (±0.129)	-0.114 (±0.064)	0.185 (±0.188)	-0.026 (±0.025)		1.158 (±3.442)	-71.952 (±10.741)	0.943	46.519	(6,17)
7.	1.272 (±0.194)	0.317 (±0.123)	-0.113 (±0.060)				1.272 (±1.962)	-71.715 (±6.844)	0.939	73.199	(4,19)
8.	1.397 (±0.167)	0.126 (±0.120)	-0.054 (±0.054)			-0.406 (±0.133)	23.611 (±1.771)	-89.709 (±6.751)	0.960	86.027	(5,18)
9.	1.467 (±0.146)					-0.474 (±0.108)	24.167 (±1.618)	-94.327 (±4.822)	0.957	149.696	(3,20)

n = 24   s = 0.207

See Tables IV-VI for a listing of the independent variables.

Table XX. LFER Model Development for a Subset of Series A (Table III; Ethyl at Position 1; Fluorine at Position 6) Against *E. coli*

Eq.	Intercept	F(7)	$F(7)^2$	INCO(7)	API	API^2	RI1(7)	RI3(7)	r^2	F	d.f.
1.	2.693 (±0.156)	0.096 (±0.139)							0.021	6.954	(1,22)
2.	2.916 (±0.141)	0.538 (±0.165)	-0.352 (±0.097)						0.398	6.954	(2,21)
3.	2.916 (±0.164)	0.460 (±0.170)	-0.347 (±0.095)						0.456	5.586	(3,20)
4.	2.650 (±0.188)	0.441 (±0.167)	-0.345 (±0.093)				0.355 (±0.243)		0.505	4.836	(4,19)
5.	2.949 (±0.335)	0.222 (±0.263)	-0.252 (±0.119)				0.504 (±0.262)	0.434 (±0.319)	0.535	4.141	(5,18)
6.	0.333 (±0.337)	0.120 (±0.272)	-0.156 (±0.119)	-0.705 (±0.291)		-8.391 (±7.812)	0.489 (±0.261)	0.353 (±0.319)	0.654	5.367	(6,17)
7.	2.887 (±0.287)	0.341 (±0.326)	-0.068 (±0.095)	-0.947 (±0.233)	13.180 (±3.676)	-16.959 (±7.779)	0.699 (±0.247)	0.398 (±0.291)	0.808	9.605	(7,16)
8.	2.879 (±0.279)	0.326 (±0.207)	-0.075 (±0.090)	-0.940 (±0.226)	12.584 (±3.676)	-61.119 (±13.687)	0.296 (±0.219)	-0.082 (±0.261)	0.807	11.828	(6,17)
9.	2.954 (±0.262) n = 24	-0.477 (±0.102) s = 0.346		-1.010 (±0.208)	13.014 (±3.001)	-58.824 (±11.283)	0.336 (±0.177)		0.799	14.283	(5,18)
10.	2.822 (±0.280) n = 24	-0.398 (±0.102) s = 0.373		-0.921 (±0.219)	15.041 (±3.052)	-63.601 (±9.670)	0.353 (±0.174)		0.753	14.532	(4,19)

See Tables IV - VI for a listing of the independent variables.

Table XXI. LFER Model Development for a Subset of Series A
(Table III; Ethyl at Position 1; Fluorine at Position 6)
Against *P. aeruginosa*

Eq.	Intercept	F(7)	F(7)^2	MR(7)	r^2	F	d.f.
1.	0.766 (±0.288)			0.137 (±0.088)	0.103	2.407	(1,21)
2.	-0.081 (±0.375)			0.828 (±0.243)	0.380	6.130	(2,20)
3.	-0.073 (±0.353)		-0.149 (±0.069)	0.743 (±0.227)	0.501	6.355	(3,19)
4.	0.131 (±0.277)	0.760 (±0.114)	-0.302 (±0.069)	0.760 (±0.178)	0.709	10.971	(4,18)
	n = 23	s = 0.375					

See Tables IV - VI for a listing of the independent variables.

distribution of the σ-ρ interaction in this subset. Two compounds have a σ-ρ interaction equal to 0.0, 19 compounds have an interaction equal to 0.171, and three have an interaction term equal to 0.302, the latter all being inactive. For this analysis, an MIC value of 100 was assumed for the inactive compounds.

This same subset in the *E. coli* test system (Table XXI, Eq. 9) indicates that an amide nitrogen and lipophilicity in position 7 are negative factors, but that the presence of a piperazinyl ring enhances activity. But the significance of the latter coefficient is slighter greater than 0.05 (P = 0.0578). Dropping this indicator variable results in a model (Eq. 10) that is statistically less significant as measured by the correlation coefficient, r, and standard error, s.

The same parabolic relationship for σ-ρ electronic interactions is seen with *E. coli*, but it may be biased for the reasons already discussed. In contrast, the LFER model for *Ps. aeruginosa* (Table XXII, Eq. 4) indicates that only lipophilicity and molar refractivity of the substituents at position 7 are important determinants of activity. In the initial analysis for this subset using the *Ps. aeruginosa* test system, A34 (norfloxacin) was an outlier. Because it is the only compound showing such high activity, it was deleted. The latter's calculated activity was 1.96 versus the observed 1.64.

The observed and calculated activities, residuals and standardized residuals of this subset are shown in Table XXIII (Eq. 9, Table XIX; Eq. 9, Table XX; Eq. 4, Table XXI). The correlation matrix for this subset is found in Table XXIV.

The stability of the regression coefficients found in equation 6 (Table XXII, and equation 7 (Table XVI), was checked by omitting compounds selected by a random number generator. For the models derived from 43 observations (compounds), six randomly selected compounds were omitted each of three times giving equations 1 - 3 (Table XXII) which should be compared to equation 6 (Table XIV). For the models derived from 41 observations, five randomly selected compounds were omitted each of three times giving equations 4 - 6 (Table XXII) which should be compared to equation 7 (Table XVI). Similar procedures were carried out for the subset of 24 compounds. Three compounds were deleted each of three times. In all cases, similar models were obtained although there was some noise in the coefficients.

Many of the compounds listed in Table III were the same as those evaluated by Koga in his earlier QSAR study.(21) This provided a means of comparing the two LFER models using somewhat differently defined parameters. The 36 compounds in common that were tested against *E. coli* were A3, A18, A32, A34, A36 - A42, A44 - A45, A48, A51 - A54, A56 - A62, A67 - A76, A79. No statistically valid model could be obtained on this subset using the independent variables in this study. When using the variables from the earlier QSAR study (Es(6), Es(6)^2, $\Sigma\pi$(6,7,8), $\Sigma\pi$(6,7,8)^2, and I(7NCO) to develop the model, an r = 0.812 was obtained which explains approximately 66 percent of the variance. Thus the inability to

Table XXII. Results from Random Sample Analyses on Set A
(See Eq. 6, Table XIV; Eq. 7, Table XVI)

Eq.	Intercept	F(6)	MR(6)	F(7)	$F(7)^2$	MR(7)	$MR(7)^2$		r^2	F	d.f.
Table XIV											
1.	0.531 (±0.353) n = 37	0.676 (±0.345)	-0.019 (±0.295) s = 0.452	0.812 (±0.133)	-0.311 (±0.081)	0.901 (±0.209)	-0.103 (±0.034)		0.751	15.108	(6,30)
2.	0.779 (±0.328) n = 37	0.655 (±0.210)	-1.237 (±0.194) s = 0.363	0.821 (±0.121)	-0.307 (±0.070)	0.850 (±0.206)	-0.099 (±0.034)		0.820	22.795	(6,30)
3.	0.197 (±0.479) n = 37	0.640 (±0.236)	-1.155 (±0.217) s = 0.406	0.860 (±0.138)	-0.329 (±0.078)	1.145 (±0.275)	-0.136 (±0.039)		0.775	17.175	(6,30)
		IE(1)	B1(6)					RI1(7)			
Table XVI											
4.	1.463 (±0.583) n = 36	-0.490 (±0.186)	-0.501 (±0.400) s = 0.365	0.276 (±0.111)	-0.266 (±0.068)	0.607 (±0.183)	-0.084 (±0.028)	0.448 (±0.185)	0.772	13.602	(7,28)
5.	2.270 (±0.580) n = 36	-0.558 (±0.189)	-0.951 (±0.382) s = 0.379	0.315 (±0.128)	-0.227 (±0.075)	0.517 (±0.226)	-0.069 (±0.036)	0.495 (±0.185)	0.745	11.688	(7,28)
6.	2.214 (±0.695) n = 36	-0.564 (±0.217)	-0.956 (±0.478) s = 0.392	0.298 (±0.140)	-0.273 (±0.078)	0.614 (±0.243)	-0.088 (±0.035)	0.432 (±0.229)	0.725	10.578	(7,28)

Table XXIII. Comparison of Observed and Calculated MIC's from Subsets of Series A (Table III)

No.	Eq. 9 (Table XIX) S. aureus				Eq. 9 (Table XX) E. coli				Eq. 4 (Table XXI) Ps. aeruginosa			
	Observed	Calculated	Residual	Standard Residual	Observed	Calculated	Residual	Standard Residual	Observed	Calculated	Residual	Standard Residual
A18	1.333	1.467	-0.133	-0.691	2.237	2.505	-0.268	-0.876	0.431	0.660	-0.229	-0.676
A32	1.600	1.467	0.133	0.691	2.804	2.536	0.268	0.876	0.697	0.650	0.046	0.137
A33	0.097	0.162	-0.065	-0.338	1.903	1.561	0.342	1.117	0.097	-0.255	0.352	1.037
A34	2.914	2.841	0.072	0.376	3.804	3.720	0.084	0.274				
A36	2.932	2.841	0.091	0.470	3.523	3.312	0.211	0.690	2.330	1.698	0.631	1.863
A55	2.551	2.841	-0.290	-1.502	2.854	3.118	-0.264	-0.864	0.745	1.123	-0.378	-1.117
A56	3.182	2.841	0.341	1.765	2.893	2.739	0.154	0.502	1.386	1.379	0.007	0.022
A57	2.611	2.841	-0.230	-1.193	2.309	2.472	-0.163	-0.534	0.804	1.250	-0.446	-1.315
A58	2.613	2.841	-0.229	-1.184	3.204	3.257	-0.053	-0.172	1.408	1.428	-0.020	-0.059
A59	2.932	2.841	0.091	0.470	3.223	3.469	-0.246	-0.804	1.427	1.354	0.073	0.215
A60	2.365	2.367	-0.003	-0.015	2.365	2.683	-0.319	-1.041	0.558	1.132	-0.575	-1.695
A61	2.967	2.841	0.125	0.650	3.558	3.081	0.476	1.556	2.062	1.798	0.264	0.778
A62	2.027	2.367	-0.340	-1.764	2.932	2.516	0.416	1.358	1.426	1.222	0.204	0.602
A63	0.166	0.162	0.003	0.018	1.672	1.544	0.128	0.417	0.767	0.498	0.269	0.795
A68	2.951	2.841	0.110	0.568	3.541	3.097	0.443	1.448	2.045	1.691	0.354	1.044
A69	2.668	2.841	-0.174	-0.900	3.561	3.626	-0.066	-0.214	1.764	1.687	0.077	0.228
A70	2.963	2.841	0.121	0.629	2.963	3.143	-0.180	-0.588	1.759	1.751	0.008	0.024
A71	3.020	2.841	0.179	0.929	2.719	2.384	0.335	1.093	0.914	0.572	0.342	1.009
A72	2.463	2.841	-0.378	-1.957	1.860	2.555	-0.695	-2.269	0.356	0.543	-0.187	-0.552
A73	2.347	2.367	-0.021	-0.106	2.951	2.884	0.067	0.218	1.745	1.311	0.433	1.279
A74	2.666	2.367	0.298	1.545	2.365	2.640	-0.275	-0.899	1.159	1.567	-0.407	-1.202
A75	2.433	2.367	0.066	0.340	2.131	2.019	0.111	0.364	1.228	1.674	-0.446	-1.316
A78	0.224	0.162	0.062	0.321	1.127	1.597	-0.469	-1.533	0.224	0.713	-0.489	-1.443
A79	3.011	2.841	0.170	0.880	3.011	3.047	-0.036	-0.118	1.504	1.389	0.116	0.341

Table XXIV. Correlation Matrix for a Subset of Series A (Table III; Ethyl at Position 1; Fluorine at Position 6; Tables XIX, XX, XXI)

	F(7)	F(7)²	MR(7)	MR(7)²	INCO(7)	RI1(7)	API	API²	SA	EC	PA
F(7)	1.000										
F(7)²	0.739	1.000									
MR(7)	0.471	0.478	1.000								
MR(7)²	0.533	0.553	0.950	1.000							
INCO(7)	-0.273	-0.209	0.096	-0.008	1.000						
RI1(7)	0.411	-0.202	0.631	0.580	0.146	1.000					
API	-0.444	0.020	0.133	0.050	-0.016	-0.029	1.000				
API²	-0.543	-0.201	-0.126	-0.134	-0.119	-0.214	0.918	1.000			
SA	0.557	0.119	0.528	0.404	0.042	0.470	-0.374	-0.683	1.000		
EC	1.046	-0.306	0.119	0.010	-0.139	0.301	-0.355	-0.555	0.736	1.000	
PA	-0.012	-0.302	0.154	0.016	-0.053	0.450	-0.075	-0.254	0.527	0.761	1.000

See Tables IV - VI for a listing of the parameters.

obtain a satisfactory model for the *E. coli* system holds for both the subset of 36 compounds using two different sets of independent variables and the larger set of 43 compounds.

Set B. In the series of pyrrolidinyl compounds (Table VII: B15A, B18A, B22A, B23A, B24A) the fluoro and cyano groups cause an increase in activity against all the bacteria tested whereas other substituents at position 6 result in a loss of activity, particularly against the Gram-negative bacteria. With respect to the piperazinyl and N-methylpiperazinyl derivatives (series B and C of compounds B15, B18, B22, B23, B24), introduction of a substituent at position 6 tends to enhance the activity against both Gram-positive and Gram-negative organisms. The replacement of hydrogen by halogen, especially fluorine, at position 6 significantly enhances antibacterial activity. Indeed, this one change overall is more significant in this series than are changes in the substituents at position 7.

A QSAR analysis was performed on this set of compounds (Table VII). In contrast with Set A (Table III), it was not possible to construct LFER models with which one would have much confidence because more than one statistically valid equation would be obtained. Examination of the correlation matrices (Tables XXV - XXVI) shows some problems with colinearity. For the *S. aureus* data, four equivalent equations were obtained. Pairs of collinear variables included INH(6) and σ-ρ (r = 0.795), INH(6) and $(\sigma$-$\rho)^2$ (r = 0.901), INO(6) and MR(6)^2 (r = 0.794). The colinearity between INH(6) and the σ-ρ terms occurs because when there is no amine (INH = 0) the σ-ρ terms cluster around 0.63. This leads to a line connecting two clusters of points. Adding to the doubts concerning the *S. aureus* data was the observation that deletion of outliers produced models with different independent variables. In other words, the regression equations produced from the *S. aureus* data with the compounds in Set B seem to be dependent on the particular set of agents (observations). Similar colinearity in the variables that entered the models derived from the *E. coli* data produced three statistically significant equations. Indeed, the tests of significance would oscillate as variables entered and left the models. Lastly, the final equations for the *Ps. aeruginosa* data were composed mainly of indicator variables.

A drawback to any retrospective QSAR study is that the investigator is limited to the available data sets. In an attempt to glean some information from this large group of antibacterial agents, a subset was selected with position 6 held constant with a fluorine substituent. This resulted in 25 compounds (B24A - B24C, B27A - B27L, B28A - B28C, B29 - B30, B36 - B40) in which only the substituents at positions 1 and 7 are varied. Even with a high correlation between IV(1) and L(1), -0.880; L(1) and MR(1), 0.742; and F(7) and MR(7), 0.766, seen in this 25 compound subset, the development of the QSAR model (Table XXVII, Eq. 1) derived from the *S. aureus* data appeared to be normal. Further this equation

Table XXV. Correlation Matrix of the Variables Used in the Analyses of the *S. aureus* Test System with Compounds in Set B

	IEF(1)	IF(6)	INH(6)	INO(6)	F(6)	L(6)	MR(6)	MR(6)²	F(7)	MR(7)	MR(7)²	BPI	BPI²	SA	EC	PA
IEF(1)	1.000															
IF(6)	0.181	1.000														
INH(6)	-0.064	-0.351	1.000													
INO(6)	-0.064	-0.351	-0.079	1.000												
F(6)	0.080	0.443	-0.788	-0.160	1.000											
L(6)	-0.080	-0.444	-0.031	0.317	-0.194	1.000										
MR(6)	-0.137	-0.757	0.245	0.663	-0.395	0.798	1.000									
MR(6)²	-0.128	-0.705	0.131	0.794	-0.297	0.740	0.980	1.000								
F(7)	-0.098	0.081	-0.033	-0.033	0.040	-0.049	-0.072	-0.066	1.000							
MR(7)	-0.015	0.333	-0.079	-0.079	0.117	-0.043	-0.168	-0.158	0.604	1.000						
MR(7)²	-0.045	0.351	-0.104	-0.104	0.140	-0.103	-0.223	-0.209	0.636	0.958	1.000					
BPI	-0.069	-0.201	0.795	0.059	-0.693	0.238	0.357	0.265	-0.205	-0.094	-0.124	1.000				
BPI²	-0.083	-0.286	0.901	0.009	-0.765	0.126	0.319	0.218	-0.194	-0.190	-0.186	0.961	1.000			
SA	0.313	0.590	-0.510	-0.141	0.492	-0.012	-0.330	-0.287	0.215	0.480	0.392	-0.468	-0.586	1.000		
EC	0.349	0.482	-0.208	-0.231	0.260	-0.203	-0.388	-0.371	-0.368	0.071	0.031	-0.217	-0.232	0.500	1.000	
PA	0.339	0.393	-0.103	-0.240	0.131	-0.037	-0.261	-0.273	-0.348	0.079	-0.014	-0.107	-0.155	0.601	0.859	1.000

See Tables VIII - X for a listing of the independent variables.

Table XXVI. Correlation Matrix of the Variables Used in the Analyses of *E. coli* and *Ps. aeruginosa* Test Systems with Compounds in Set B

	IV(1)	MR(1)	IF(6)	ICN(6)	INH(6)	L(6)	F(7)	F(7)²	INCO(7)	RI1(7)	ICH₃(7)	IRH1(7)	BPI	BPI²	SA	EC	PA
IV(1)	1.000																
MR(1)	-0.049	1.000															
IF(6)	0.181	-0.132	1.000														
ICN(6)	-0.064	0.046	-0.351	1.000													
INH(6)	-0.064	0.046	-0.351	-0.079	1.000												
L(6)	-0.080	0.059	-0.444	0.734	-0.031	1.000											
F(7)	-0.098	0.143	0.081	-0.033	-0.033	-0.049	1.000										
F(7)²	-0.128	0.115	0.292	-0.103	-0.103	-0.132	0.847	1.000									
INCO(7)	-0.064	0.046	0.225	-0.079	-0.079	-0.099	-0.304	-0.181	1.000								
RI1(7)	0.181	-0.132	-0.025	0.033	0.033	0.078	-0.084	-0.123	0.033	1.000							
ICH₃(7)	0.174	0.072	-0.237	0.098	0.098	0.147	0.011	-0.210	-0.138	0.394	1.000						
IRH1(7)	0.135	-0.282	-0.128	0.059	0.059	0.095	-0.499	-0.402	-0.160	0.338	-0.280	1.000					
BPI	-0.069	0.051	-0.201	0.083	0.083	0.238	-0.205	-0.103	-0.086	-0.003	0.083	0.042	1.000				
BPI²	-0.083	-0.061	-0.286	-0.029	0.901	0.126	-0.194	-0.124	-0.103	-0.044	0.075	0.027	0.961	1.000			
SA	0.115	0.021	0.590	-0.033	-0.510	-0.012	0.215	0.108	0.056	0.228	-0.047	0.053	-0.468	-0.586	1.000		
EC	0.386	-0.134	0.482	-0.137	-0.208	-0.203	0.368	-0.351	0.104	0.489	0.296	0.296	-0.217	-0.232	0.500	1.000	
PA	0.466	-0.097	0.393	0.149	-0.103	-0.348	-0.416	-0.050	0.448	0.211	0.466	0.601	-0.106	-0.155	0.601	0.859	1.000

See Tables VIII - X for a listing of the independent variables.

Table XXVII. Equations Derived from Subsets of the B Series of Compounds Found in Table VII

S. aureus (LFER model)
1. Log 1/MIC = 0.952[IV(1)] + 0.808[F(1)] - 0.497[F(7)] + 0.422[MR(7)] +
   ($\pm$0.204)   ($\pm$0.343)   ($\pm$0.088)   ($\pm$0.093)

   1.144[RI2(7)] - 0.915[INCO(7)] - 3.953[σ-ρ] - 1.118
   ($\pm$0.253)   ($\pm$0.174)   ($\pm$0.468)   ($\pm$1.137)
   n = 24   s = 0.214   r = 0.958 $F_{7,16}$ = 25.25

S. aureus (de novo model)
2. Log 1/MIC = 1.019[IF(6)] - 1.018[INH(6)] + 1.518
   ($\pm$0.171)   ($\pm$0.232)   ($\pm$0.104)
   n = 22   s = 0.359   r = 0.893 $F_{2,19}$ = 37.30

E. coli (de novo model)
3. Log 1/MIC = 1.278[IF(6)] - 1.262[RI2(7)] + 1.992
   ($\pm$0.181)   ($\pm$0.191)   ($\pm$0.119)
   n = 22   s = 0.389   r = 0.928 $F_{2,19}$ = 58.53

Ps. aeruginosa (de novo model)
4. Log 1/MIC = 1.423[IF(6)] + 0.479[ICN(6)] - 0.282[ICH_3(7)] - 1.016[RI2(7)] +
   ($\pm$0.130)   ($\pm$0.173)   ($\pm$0.134)   ($\pm$0.147)

   1.385
   ($\pm$0.111)
   n = 22   s = 0.267   r = 0.962 $F_{4,17}$ = 53.19

See Tables VIII - X for a listing of the independent variables.

appeared no matter which variables were considered initially or when all variables were entered and the least significant were deleted.

Compound B27D was deleted because the $(CH_2)_5$N-RI3(7) indicator variable which appeared in the initial run is found only on the one compound. It is risky to draw too much from the substituents at position 1 because of the dominance of the ethyl group, but the variety of substituents at position 7 do indicate that hydrophobicity and bulk are important determinants of activity. This model also reinforces the observation that the nitrogen in the heterocyclic rings normally should not be acylated. The predictability of equation 1 (Table XXVII) is found in Table XXVIII. Deletion of outliers only caused additional outliers to appear.

It was not possible to develop unique models using these 25 compounds and the *E. coli* and *Ps. aeruginosa* data. Similar problems with colinearity were encountered. A series of statistically equivalent equations were derived. Further the test of significance oscillated as variables entered and left the model.

Another approach with this data set was attempted. This time, the rules for a *de novo* or Free-Wilson analysis were followed, producing a second subset of 22 compounds (Table VII: B3A - B3C, B15A - B15C, B18A - B18C, B22A - B22C, B23A - B23C, B24A - B24C, B36 - B39). B3B with a 1-ethyl, 6-H and 7-piperazinyl was selected as the reference compound. The indicator variables were vinyl and 2-fluoroethyl at position 1; fluoro, cyano, nitro, chloro and amino at position 6; and pyrrolidyl and N-methylpiperazinyl at position 7. The three *de novo* models (Table XXVII, Eqs. 2 - 4) were derived using both add best and drop worst stepwise procedures. For all three bacterial systems, the 6-fluoro is very significant for activity. The 6-amino group is a negative contributor to activity against *S. aureus*, but has little effect against the other two bacteria species tested. The two Gram-negative bacteria are sensitive to the presence of the pyrrolidinyl ring (RI2(7)) which reduces activity. For *Ps. aeruginosa* only, the 6-cyano group is a positive factor while the N-methyl substituent on the 7-piperazinyl reduces activity. The calculated activities and residuals for the three *de novo* models are listed in Tables XXVIII-XXIX. Even for the two term models (Eqs. 2 - 3) which results in only three calculated values, the equations show good predictability because the experimental results cluster around values of high, intermediate and low antibacterial activities.

The LFER models based on the same 22 compounds as in the *de novo* analysis were derived from all three bacterial systems in an attempt to see if physicochemical parameters would complement the results obtained from the Free-Wilson models. Again, series of statistically equivalent models were obtained. This probably was due to the distribution of values for these parameters.

<u>Set C.</u> There is little correlation between size of the nitrogen containing heterocyclic ring at position 7 and antibacterial activity. The replacement of the piperazinyl group enoxacin (C2, Table XI) by the 3-aminopyrrolidinyl

Table XXVIII. Comparison of Observed and Calculated MIC's from Equation 1, Table XXVII

No.	Observed	Calculated	Residual	Standard. Residual
B24A	2.893	2.893	0.000	0.000
B24B	2.613	2.559	0.054	0.301
B24C	2.331	2.329	0.002	0.009
B27A	0.099	0.320	-0.221	-1.238
B27B	1.071	0.849	0.221	1.238
B27C	1.728	1.824	-0.096	-0.537
B27E	2.313	2.353	-0.039	-0.219
B27F	2.334	2.270	0.064	0.358
B27G	2.331	2.706	-0.376	-2.101
B27H	2.019	1.768	0.252	1.407
B27I	1.444	1.503	-0.059	-0.331
B27J	1.801	2.243	-0.441	-2.468
B27K	2.721	2.503	0.218	1.219
B27L	2.650	2.302	0.348	1.947
B28A	2.366	2.274	0.091	0.510
B28B	2.080	2.247	-0.167	-0.934
B28C	2.382	2.356	0.026	0.144
B29	2.046	2.036	0.010	0.058
B30	2.063	1.977	0.086	0.480
B36	2.309	2.291	0.018	0.102
B37	2.025	2.061	-0.036	-0.199
B38	2.939	2.897	0.042	0.237
B39	2.654	2.667	-0.014	-0.076
B40	1.738	1.720	0.017	0.096

Table XXIX. Comparison of Observed and Calculated Log 1/MIC's from Equations 2 - 4 (Table XXVII)

No.	Equation 2. S. aureus				Equation 3. E. coli				Equation 4. Ps. aeruginosa			
	Observed	Calculated	Residual	Standard Residual	Observed	Calculated	Residual	Standard Residual	Observed	Calculated	Residual	Standard Residual
B3A	1.361	1.518	-0.158	-0.462	1.060	0.730	0.330	0.894	0.157	0.324	-0.167	-0.695
B3B	1.082	1.518	-0.436	-1.277	1.684	1.992	-0.308	-0.833	1.082	1.385	-0.303	-1.261
B3C	1.102	1.518	-0.416	-1.219	1.703	1.992	-0.288	-0.781	1.102	1.104	-0.002	-0.008
B15A	1.410	1.518	-0.108	-0.317	0.206	0.730	-0.523	-1.417	0.206	0.324	-0.118	-0.490
B15B	2.032	1.518	0.513	1.502	2.635	1.992	0.643	1.740	1.730	1.385	0.345	1.435
B15C	1.750	1.518	0.231	0.677	2.352	1.992	0.360	0.974	1.447	1.104	0.344	1.429
B18A	2.000	1.518	0.482	1.410	1.397	0.730	0.667	1.806	1.096	0.803	0.293	1.219
B18B	1.719	1.518	0.201	0.588	1.350	1.992	-0.642	-1.738	1.719	1.865	-0.146	-0.605
B18C	1.435	1.518	-0.083	-0.243	2.340	1.992	0.348	0.943	1.435	1.583	-0.148	-0.614
B22A	1.123	1.518	-0.395	-1.156	0.521	0.730	-0.208	-0.563	0.220	0.324	-0.104	-0.431
B22B	1.745	1.518	0.227	0.663	1.745	1.992	-0.247	-0.669	1.143	1.385	-0.243	-1.009
B22C	1.461	1.518	-0.057	-0.168	2.062	1.992	0.070	0.190	0.857	1.104	-0.247	-1.026
B23A	0.179	0.500	-0.321	-0.940	0.179	0.730	-0.550	-1.490	0.179	0.324	-0.145	-0.602
B23B	0.200	0.500	-0.300	-0.879	2.006	1.992	0.014	0.037	1.706	1.385	0.320	1.332
B23C	1.122	0.500	0.622	1.820	2.327	1.992	0.335	0.907	1.423	1.104	0.319	1.326
B24A	2.893	2.538	0.355	1.040	2.292	2.007	0.284	0.770	1.987	1.747	0.240	0.999
B24B	2.613	2.538	0.075	0.219	3.204	3.270	-0.065	-0.177	2.613	2.808	-0.196	-0.813
B24C	2.331	2.538	-0.207	-0.606	2.932	3.270	-0.338	-0.914	2.331	2.527	-0.196	-0.814
B36	2.309	2.538	-0.229	-0.669	3.503	3.270	0.234	0.632	3.201	2.808	0.393	1.635
B37	2.025	2.538	-0.512	-1.499	3.220	3.270	-0.049	-0.133	2.629	2.527	0.102	0.426
B38	2.939	2.538	0.402	1.176	3.228	3.270	-0.042	-0.113	2.636	2.808	-0.172	-0.714
B39	2.654	2.538	0.116	0.340	3.246	3.270	-0.024	-0.065	2.354	2.527	-0.173	-0.719

moiety (**C33A**) causes an enhancement in activity against all the bacteria tested. Use of a larger ring such as 3- and 4-amino piperidine (**C49A** and **C50A**) results in retention or increase in activity against *S. aureus* and decrease in activity against *Ps. aeruginosa*. On the other hand, replacement with the smaller 3-aminoazetidine group (**C28A**) shows the same level activity as enoxacin.

Addition of substituents to the heterocyclic ring at position 7 can have a pronounced effect on activity. The following substitutions generally reduced the antibacterial response: alkylation (**C34A** - **C37A**) and acylation (**C39A** - **C42A**) of the 3-amino group of **C33A**; replacement of the amino group on **C28A**, **C33A**, **C49A** and **C50A** by a hydroxyl moiety (**C30A**, **C46A**, **C55A** and **C56A**, respectively) and alkylation or formylation of the hydroxyl moiety (**C31A**, **C32A** and **C47A**). Changing the substituent at position 1 can influence the spectrum of antibacterial activity. Replacement of the ethyl (series A) with vinyl (series B) generally enhanced Gram-negative activity and either has little effect or reduces the response against Gram-positive bacteria. The fluoroethyl (series C) tends to reduce the Gram-positive response with little change seen with the Gram-negative microorganisms.

Because of the wide variety of substituents at position 7 for this set of compounds, a QSAR analyses was carried out in the hope that the parameters that determine activity could be better identified. Forty-two compounds were analyzed. Enoxacin (**C2**) was deleted because it was the only example with an unsubstituted piperazinyl ring at position 7. Statistically valid LFER models could not be obtained using the *S. aureus* and *E. coli* data. The models from the *Ps. aeruginosa* data are listed in Table XXX. Equations 1 - 5 were developed by the add best/drop worst procedure. In equation 5 there were two nearly equivalent variables, IE(1) and F(1). Addition of each of these variables individually to equation 5 gave equations 6 and 7, respectively. Both of these equations are nearly equivalent and give nearly identical results in terms of predicting activity. Any combination of two of the three variables causes one of the included variables to lose significance. As seen in Table XXXI, there is a high correlation between IE(1) and F(1) (r = 0.943). Equation 6 with the indicator variable for 1-ethyl was selected since there is not enough variation of substituents at position 1 to give a valid, continuous variable. The comparison of calculated versus experimental log 1/MIC's can be found in Table XXXII.

A structured subset of 25 compounds (Table XI: **C28A** - **C28B**, **C30A**, **C33A** - **C33C**, **C34A** - **C34C**, **C36A** - **C36B**, **C38A** - **C38C**, **C39A** - **C39B**, **C40A** - **C40C**, **C42A** - **C42C**, **C46A**, **C50A**, **C56A**) was selected for a *de novo* analyses. **C50A** was the reference compound containing a 1-ethyl and 7-(N-methylpiperazinyl) substituents. In this subset, the indicator variables were evaluated for the substituents -CH=CH$_2$ and -CH$_2$CH$_2$F at position 1; -OH, -NHCH$_3$, -NHCH$_2$CHF$_3$, -N(CH$_3$)$_2$, -NHCHO, -NHCOCH$_3$, -N(CH$_3$)COCH$_3$ on the rings attached to position 7; and ring indicator variables for the azetidinyl and pyrrolidinyl rings. A *de novo* model could

Table XXX. Development of the LFER Model for Compounds in Set C Against *Ps. aeruginosa*

No.	Intercept	I≡(1)	F(1)	B5R(7)	LR(7)	FR(7)	FR(7)^2	r	F	d.f.
1.	3.11 (±0.287)				-0.348 (±0.068)			0.628	25.944	(1,40)
2.	2.608 (±0.280)				-0.344 (±0.059)	-0.365 (±0.094)		0.750	25.000	(2,39)
3.	2.383 (±0.251)				-0.335 (±0.051)	-1.165 (±0.233)	-0.361 (±0.098)	0.823	26.473	(3.38)
4.	2.338 (±0.242)			-0.331 (±0.165)	-0.122 (±0.117)	-1.431 (±0.260)	-0.443 (±0.103)	0.841	22.518	(4,37)
5.	2.228 (±0.219)			-0.487 (±0.070)		-1.561 (±0.229)	-0.483 (±0.096)	0.837	29.479	(3,28)
6.	2.534 (±0.253)	-0.296 (±0.137)		-0.502 (±0.067)		-1.518 (±0.219)	-0.484 (±0.091)	0.856	25.429	(4,37)
	n = 42		s = 0.392							
7.	3.134 (±0.460)		-0.639 (±0.289)	-0.502 (±0.067)		-1.518 (±0.219)	-0.481 (±0.091)	0.858	25.647	(4,37)
	n = 42		s = 0.391							

See Tables XII - XIII for a listing of the independent variables.

Table XXXI. Correlation Matrix of the Variables Used in the Analyses of the *P. aeruginosa* Test Systems (Tables XXX)

	IE(1)	F(1)	LR(7)	B5R(7)	FR(7)	FR(7)²	MRR(7)	MRR(7)²	RI3(7)	PROXI(7)	SA	EC	PA
IE(1)	1.000												
F(1)	0.943	1.000											
LR(7)	0.010	0.006	1.000										
B5R(7)	-0.150	-0.145	0.868	1.000									
FR(7)	0.285	0.245	0.019	-0.203	1.000								
FR(7)²	-0.263	-0.221	-0.001	0.143	-0.936	1.000							
MRR(7)	-0.102	-0.090	0.929	0.837	0.079	-0.068	1.000						
MRR(7)²	-0.001	0.002	0.857	0.622	0.216	-0.174	0.930	1.000					
RI3(7)	0.325	0.306	-0.106	-0.304	0.371	-0.293	-0.118	-0.099	1.000				
PROXI(7)	-0.232	-0.223	0.123	0.279	-0.367	0.299	0.131	-0.045	-0.842	1.000			
SA	0.015	0.067	-0.250	-0.221	-0.201	0.112	-0.240	-0.226	-0.012	-0.118	1.000		
EC	-0.423	-0.360	-0.488	-0.315	-0.284	0.135	-0.431	-0.468	-0.271	0.880	0.484	1.000	
PA	-0.240	-0.234	-0.627	-0.468	-0.422	0.265	-0.613	-0.516	-0.261	0.139	0.505	0.855	1.000

See Tables XII - XIII for a listing of the variables.

Table XXXII. Comparison of Observed and Calculated MIC's from Eq. 6 (Table XXX)

No.	Observed	Calculated	Residual	Standard. Residual
C28A	2.593	2.428	0.165	0.445
C28B	2.893	2.776	0.117	0.315
C29A	0.876	0.187	0.690	1.855
C30A	1.991	2.426	-0.434	-1.169
C31A	1.710	1.851	-0.141	-0.379
C32A	1.127	1.421	-0.294	-0.791
C33A	2.914	2.428	0.486	1.306
C33B	3.201	2.776	0.426	1.145
C33C	2.939	2.605	0.334	0.899
C34A	2.331	1.886	0.444	1.195
C34B	2.629	2.234	0.395	1.062
C34C	2.654	2.063	0.590	1.588
C35A	2.046	1.600	0.446	1.201
C36A	0.907	0.672	0.234	0.630
C36B	0.602	1.020	-0.418	-1.124
C37A	0.258	0.688	-0.430	-1.156
C38A	1.745	1.752	-0.008	-0.021
C38B	2.043	2.100	-0.057	-0.153
C38C	1.767	1.929	-0.162	-0.437
C39A	2.046	1.611	0.435	1.170
C39B	2.043	1.959	0.084	0.227
C40A	1.762	1.128	0.634	1.704
C40B	1.460	1.476	-0.016	-0.044
C40C	1.182	1.305	-0.124	-0.333
C41A	1.824	1.495	0.329	0.885
C42A	1.177	1.460	-0.282	-0.760
C42B	1.175	1.807	-0.632	-1.700
C42C	1.197	1.637	-0.439	-1.182
C43A	1.428	1.779	-0.350	-0.943
C44A	1.162	1.295	-0.133	-0.359
C45A	2.031	2.428	-0.398	-1.069
C46A	2.614	2.426	0.189	0.507
C47A	2.047	2.226	-0.179	-0.480
C48A	1.434	1.240	0.194	0.522
C49A	1.728	1.365	0.363	0.977
C50A	2.331	2.428	-0.097	-0.262
C51A	0.340	0.639	-0.299	-0.804
C52A	1.445	1.811	-0.366	-0.984
C53A	1.194	1.049	0.145	0.391
C54A	0.860	1.365	-0.505	-1.358
C55A	1.127	1.365	-0.238	-0.640
C56A	1.728	2.426	-0.698	-1.877

not be obtained from the *S. aureus* data. The *de novo* models for *E. coli* (Eq. 1) and *Ps. aeruginosa* (Eq. 2) are listed in Table XXXIII.

For both *E. coli* and *Ps. aeruginosa* only the ethyl substituent at position 1 is an important determinant of activity. The type of ring substituted at position 7 doesn't seem significant for this group of compounds when tested against *E. coli*, while it is with *Ps. aeruginosa*. Substituents on the rings results in decreased antibacterial activity in both test systems. Table XXXIV contains the comparison of calculated and experimental activities for *E. coli* and *Ps. aeruginosa*, respectively.

LFER analyses were carried out with this subset of 25 compounds. A statistically valid model could not be obtained from the *S. aureus* data. Models in the statistical "gray" zone of acceptability were derived from the data of the other two bacteria. In each case, the important physicochemical parameter was molar refraction indicating that steric factors are important for activity in these systems.

One limitation in this subset is the fact that 20 of the 25 compounds have the pyrrolidinyl ring at position 7. Therefore, the analysis was repeated on these 20 compounds (Table XI: C33A - C33C, C34A - C34C, C36A - C36B, C38A - C38C, C39A - C39B, C40A - C40C, C42A - C42C, C46A). The reference compound (C33A) had ethyl in position 1 and the amine for the R-substituent on the pyrrolidinyl ring on position 7. As before, a statistically valid model could not be obtained for *S. aureus*. The de novo models for *E. coli* and *Ps. aeruginosa* were very similar to those found in Table XXXIII including the magnitude of the coefficients and intercept terms.

CONCLUSION

It was the hope that a *unified* QSAR model might have evolved because of the consistency of the test systems, but that was not obtained. Indeed, acceptable LFER models could not be obtained for the *E. coli* in Set B and all three bacteria systems in Set C. There is a consistency seen in the LFER models in that steric parameters, usually represented by molar refraction, and lipophilicity are the best predictors for substituents at position 7. Any interpretation must recognize that this is a complex *in vivo* system in that the assays are being run on intact bacteria.

There is increasing evidence that in order for a compound to be active in this series it has two hurdles to overcome. It first must penetrate the cell wall and then have a favorable free energy of binding for the receptor on DNA gyrase.(33) Lending credence to the importance of penetration are the reports of quinolone resistance factors in bacteria associated with permeability of the drug.(12) Thus in those systems where models could not be obtained, it is reasonable to postulate that there may be a poor correlation between the compounds active against DNA gyrase and those capable of penetrating the cell wall.

Table XXXIII. *de novo* Models Derived from a Subset of Compounds from Set C

E. coli
Log 1/MIC = -1.067[OH] - 1.371[NHCH$_2$CF$_3$] - 0.527[N(CH$_3$)$_2$] - 0.681[NHCHO] - 1.209[NHCOCH$_3$] - 1.393[N(CH$_3$)COCH$_3$] + 3.480
(±0.213) (±0.250) (±0.213) (±0.250) (±0.213) (±0.213) (±0.107)
n = 25   s = 0.319   r = 0.902 $F_{6,18}$ = 13.07

Ps. aeruginosa
Log 1/MIC = -0.577[OH] - 0.524[NHCH$_3$] - 2.307[NHCH$_2$CF$_3$] - 1.210[NH(CH$_3$)$_2$] - 1.017[NHCHO] - 1.594[NHCOCH$_3$] - 1.878[N(CH$_3$)COCH$_3$] +
(±0.128) (±0.139) (±0.157) (±0.139) (±0.157) (±0.139) (±0.139)

0.367[7-Azetidinyl] + 0.744[7-Pyrrolidinyl] + 2.318
(±0.164) (±0.157) (±0.141)
n = 25   s = 0.178   r = 0.980 $F_{9,15}$ = 40.89

Note: All of the substituents in the first equation and substituents 1 - 7 in the second equation are located on the rings attached to position 7.

Table XXXIV. Comparison of Observed and Calculated Log/MIC from the *de novo* analysis (Table XXXIII)

No.	E. coli				Ps. aeruginosa			
	Observed	Calculated	Residual	Standard Residual	Observed	Calculated	Residual	Standard Residual
C28A	3.485	3.480	0.006	0.020	2.593	2.685	-0.091	-0.650
C28B	3.483	3.480	0.003	0.011	2.893	2.685	0.208	1.481
C30A	2.595	2.413	0.182	0.658	1.991	2.108	-0.177	-0.831
C33A	3.504	3.480	0.025	0.089	2.914	3.061	-0.148	-1.051
C33B	4.105	3.480	0.625	2.258	3.201	3.061	0.140	0.997
C33C	3.529	3.480	0.049	0.177	2.939	3.061	-0.122	-0.869
C34A	3.223	3.480	-0.257	-0.930	2.331	2.538	-0.207	-1.474
C34B	3.521	3.480	0.042	0.150	2.629	2.538	0.091	0.649
C34C	3.246	3.480	-0.234	-0.846	2.654	2.538	0.116	0.825
C36A	2.411	2.109	0.302	1.092	0.907	0.754	0.152	1.084
C36B	1.807	2.109	-0.302	-1.092	0.602	0.754	-0.152	-1.084
C38A	2.650	2.953	-0.303	-1.096	1.745	1.852	-0.107	-0.761
C38B	2.947	2.953	-0.006	-0.023	2.043	1.852	0.192	1.364
C38C	3.263	2.953	0.310	1.119	1.767	1.852	-0.085	-0.603
C39A	2.650	2.798	-0.149	-0.537	2.046	2.045	0.001	0.010
C39B	2.947	2.798	0.149	0.537	2.043	2.045	-0.001	-0.010
C40A	1.762	2.271	-0.509	-1.838	1.762	1.468	0.294	2.094
C40B	2.664	2.271	0.393	1.420	1.460	1.468	-0.008	-0.058
C40C	2.386	2.271	0.116	0.418	1.182	1.468	-0.286	-2.036
C42A	1.780	2.086	-0.307	-1.108	1.177	1.183	-0.006	-0.043
C42B	2.077	2.086	-0.009	-0.033	1.175	1.183	-0.008	-0.057
C42C	2.402	2.086	0.316	1.141	1.197	1.183	0.014	0.100
C46A	2.614	2.413	0.201	0.727	2.614	2.485	0.130	0.923
C50A	3.223	3.480	-0.257	-0.930	2.331	2.318	0.013	0.092
C56A	2.030	2.413	-0.383	-1.385	1.728	1.741	-0.013	-0.092

Accordingly it appears that further QSAR investigations should focus separately on those parameters facilitating both cell wall penetration and those required for DNA gyrase inhibition. Then it should be possible to design structures which would maximize both responses.

ACKNOWLEDGMENTS

The authors express their gratitude to Dr. Al Leo, Pomona College Medicinal Chemistry Project, for his guidance in the interpretation of the correction factors used in the CLOGP program.

LITERATURE CITED

1. Lesher, G. Y.; Froelich, E. J.; Gruett, M. D.; Bailey, J. H.; Brundage, R. P. J. Med. Chem. 1962, 5, 1063.

2. Albrecht R. Prog. Drug Res. (Arzneimittel-Forschung) 1977 21, 9.

3. Cornett, J. B.; Wentland, M. P. Ann. Rep. Med. Chem. 1986, 21, 139.

4. Wolfson, J. S.; Hooper, D. C. Antimicrob. Agents Chemother. 1985, 28, 581.

5. Hooper, D. C.; Wolfson, J. S. Antimicrob. Agents Chemother. 1985, 28, 716.

6. Wentland, M. P.; Cornett, J. B. Ann. Rep. Med. Chem. 1985, 20, 145.

7. Liu, L. F. Ann. Rep. Med. Chem. 1986 21, 257.

8. Sugino, A.; Peebles, C. L.; Kreuzer, K. N.; Cozzarelli, N. R. Proc. Natl. Acad. Sci. U.S.A. 1977, 74, 4767.

9. Cozzarelli, N. R. Science 1980, 207, 953.

10. Shen, L. L.; Pernet, A. G. Proc. Natl. Acad. Sci. U.S.A. 1985, 82, 306.

11. Rella, M.; Haas, D. Antimicrob. Agents Chemother. 1982, 22, 242.

12. Inoue, S.; Ohue, T.; Yamagishi, J.; Nakamura, S.; Shimizu, M. Antimicrob. Agents Chemother. 1978, 14, 240.

13. Munshi, M. H.; Haider, K.; Rahaman, M. M.; Sack, D. A.; Ahmed, Z. U.; Morshed, M. G. Lancet 1987, ii, 419.

14. Piddock, L. G. V.; Wijnands, W. J. A.; Wise, R. Lancet 1987, ii, 90.

15. Fernandes, P. B.; Chu, D. T. W. Ann. Rep. Med. Chem. 1987, 22, 117.

16. Fernandes, P. B.; Chu, D. T. W. Ann. Rep. Med. Chem. 1988, 23, 133.

17. Wentland, M. P.; Bailey, D. M.; Cornett, J. B.; Dobson, R. A.; Powles, R. G.; Wagner, R. B. J. Med. Chem. 1984, 27, 1103.

18. Chu, D. T. W.; Fernandes, P. B.; Claiborne, A. K.; Pihuleac, E.; Nordeen, C. W.; Maleczka, R. E., Jr.; Pernet, A. G. J. Med. Chem. 1985, 28, 1558.

19. Domagala, J. M.; Heifetz, C. L.; Mich, T. F.; Nichols, J. B. J. Med. Chem. 1986, 29, 445.

20. Fujita T. In Drug Design: Fact and Fantasy; Jolles G.; Woolridge, K. R. J. H., Eds.; Academic Press: New York, 1984; pp 19-33.

21. Koga H. Kagaku No Ryoiki, Zokan. 1982, 136, 177; Chem. Abstr. 1982, 97, 192622b.

22. Chiou, C. T.; Schmedding, D. W.; Block, J. H.; Manes, M. J. Pharm. Sci. 1982, 71, 1307.

23. Koga, H.; Itoh, A.; Murayama, S.; Suzie, S.; Irikura, T. J. Med. Chem. 1980, 23, 1358.

24. Matsumoto, J.; Miyamoto, T.; Minamida, A.; Nishimura, Y.; Egawa, H.; Nishimura, H. J. Med. Chem. 1984, 27, 292.

25. Egawa, H.; Miyamoto, T.; Minamida, A.; Nishimura, Y.; Okada, H.; Uno, H.; Matsumoto, J. J. Med. Chem. 1984, 27, 1543.

26. CLOGP Program, Release 3.3. Pomona College Medicinal Chemistry Project, Seaver Chemistry Laboratory, Pomona College, Claremont, CA. March 1985.

27. Verloop, A.; Hoogenstraten W.; Tipker J. In Drug Design; E. J. Ariens, E. J., Ed.; Academic Press: New York, 1976; Vol. 7, p 165.

28. Verloop, A. In Pesticide Chemistry; Miyamoto, J.; Kearney, P. C., Eds., Pergamon Press: New York, 1982; pp 339-344.

29. Leo, A. J. Chem. Soc. Perkin II 1983, 825.

30. Leo, A. In Partition Coefficient: Determination and Estimation, Dunn, W. J, III,; Block, J. H.; Pearlman, R. S., Eds., Pergamon: New York, 1986; pp 61-67.

31. Santilli, A. A.; Scotese, A. C.; Yurchenco, J. A. J. Med. Chem. 1975, 18, 1038.

32. Kaminsky, D.; Meltzer, R. I. J. Med. Chem. 1968, 11, 160.

33. Domagala, J. M.; Hanna, L. D.; Heifetz, C. L.; Hutt, M. P.; Mich, T. F.; Sanchez, J. P.; Solomon, M. J. Med. Chem., 1986, 29, 394.

RECEIVED June 21, 1989

TOXICITY MECHANISMS

Toxicity Mechanisms: Introduction

Toxicity of medicinals and agrochemicals to important target organisms is the goal of directed research. Toxicity of the same compounds to non-target humans, animals, fish, useful insects, plants and environmentally important biota is not so easily controlled and may be of public concern. Moreover, the general toxicity of all commonly used chemicals in the household and in every industry is of primary concern in regulating safe handling procedures. As all commercial chemicals fall in the public domain, the task of adequate measurement against the many important species by all modes of ingress is monumentally impossible. The picture is further complicated by problems of runoff, bioconcentration, rates of environmental degradation and by metabolic, chemical and UV intoxification. At the same time, animal research has become increasingly difficult in the face of opposition from the rapidly growing animal rights movement. All of these factors point clearly to the need for developing predictive relations for estimating probable toxicity of new chemicals along with estimates of bioconcentration and clearance from non-target species. Despite important studies over the past decade, this goal remains distant and provides a unique opportunity for the contributions of gifted SAR researchers.

This short section addresses three diverse areas that provide some insight to the nature of SAR research in toxicity mechanisms. In the opening paper, a team from the National Institute of Environmental Health Sciences looks closely at the inhibition of thyroid hormone deiodination by dibenzo-p-dioxins and related polychlorinated biphenyls (PCB's). A second paper by Lipnick of the USEPA examines narcosis in aquatic species and demonstrates the commonality of mechanism in terms of a baseline toxicity concept. The final paper by Magee of BIOSAR and King of Aberdeen Proving Ground (U.S. Army Research) correlates the toxicity of chemical classes to different animals by various modes of administration to reveal both common and diverse mechanisms of death. In proceeding from enzyme-level research through a controlled aquatic environment to the high variability of animal toxicity data, we have attempted to reveal the scope of this growing area of research. The field is wide open for new researchers as the literature contains enormous quantities of untreated toxicity data with a corresponding potential for mechanistic insight.

PHILIP S. MAGEE
BIOSAR Research Project
Vallejo, CA 94591 and
School of Medicine
University of California
San Francisco, CA 94143

Chapter 22

Structurally Specific Interaction of Halogenated Dioxin and Biphenyl Derivatives with Iodothyronine-5'-deiodinase in Rat Liver

U. Rickenbacher[1], S. Jordan, and J. D. McKinney[2]

Laboratory of Molecular Biophysics, National Institute of Environmental Health Sciences, P.O. Box 12233, Research Triangle Park, NC 27709

In in vitro studies, soluble, polar derivatives of polychlorinated biphenyls (PCB) and dibenzo-p-dioxins were shown to inhibit outer (phenolic) ring deiodination of 3,3',5'-triiodothyronine (rT_3) used as substrate for iodothyronine type I deiodinase activity in microsomal fractions of rat liver. Potent inhibition depended on the presence of lateral chlorination (3,5 in biphenyl or 2,3 in dibenzo-p-dioxin derivatives). The most potent PCB ligand exhibited a half-maximal inhibitory concentration similar to the K_m (29 nM) of rT_3. The results are in general agreement with our previous results with human thyroxine binding prealbumin and rat liver nuclear extracts that also show high affinity specific binding of these and related compounds to thyroxine specific binding sites. The functional structural characteristics of these polar PCB and dioxin derivatives involved in binding are in general similar to those found in toxic structures (underivatized) of this type. These relatively metabolically resistant deiodination inhibitor analogs may be useful as selective inhibitors facilitating the further study of biochemical and functional characteristics of protein interactions in thyroid hormone metabolism as well as the study of the possible importance of thyroid hormone antagonism in dioxin and related compound toxicity.

The toxic potency of halogenated aromatic hydrocarbons is dependent on the number and positions of halogen atoms in their molecular structure. Qualitative structure requirements for high toxicity include planarity (or coplanarity) of structure in a shape approximating a rectangle and a sufficient degree of halogenation

[1]Current address: Sandoz Limited 881, CH 4002, Basel, Switzerland
[2]Address correspondence to this author.

approximating a rectangle and a sufficient degree of halogenation concentrated about lateral positions (3,3',4,4',5,5'-positions on the biphenyl nucleus and 2,3,7,8-positions on the dioxin nucleus) of the molecule (1). Treatment of animals with 2,3,7,8-tetrachlorodibenzo-p-dioxin (TCDD), the prototype for certain toxic halogenated aromatic hydrocarbons, produces divergent effects on circulating levels of L-thyroxine (L-T_4) and L-3,5,3'-triiodothyronine (L-T_3) concentrations (2-5). These results along with the lack of information on thyroid hormone modulated responses in these animals have caused uncertainty regarding their functional and biochemical thyroid status (4).

Because of the structural similarities (6) between TCDD and T_4 and the fact that thyroid hormones can modulate TCDD toxicity (7-8), we focused our attention on thyroxine binding proteins as possible sites for critical biochemical interactions that may mediate the toxic responses of these compounds. Initial studies (9-10) involved human thyroxine binding prealbumin (transthyretin or TBPA) (9-10) since it is a major transport protein for T_4 in blood and may be of use as a model for studying the interaction of thyroid hormones with certain nuclear thyroxine receptors (11). Competitive binding studies using TBPA and the thyroxine nuclear receptor extracted from rat liver (12) and polar (soluble) derivatives of dioxin and related polychlorinated biphenyl (PCB) compounds have shown that simple halogenated hydrocarbons which contain only one aromatic ring or linear structures with multiple rings bind with higher affinity than the normal angular diphenylether bridged system characteristic of thyroid hormone analogs. Furthermore, lateral chlorination was common to all compounds which showed high binding activity. In addition, the nuclear receptor showed a remarkably enhanced affinity for laterally substituted compounds that were also planar and highly polarizable. Because these structural properties are also characteristic of highly toxic structures of this type, it was suggested (6) that toxicity may be in part the expression of potent and persistent thyroid hormone agonist activity initiated at the level of the nuclear receptor. In contrast, certain nonreceptor thyroid hormone binding protein interactions may be antagonistic in nature.

Previous workers (13) have shown strong correlations between the binding requirements for TBPA and the inhibitory requirements for rat liver iodothyronine deiodinase activity. The iodothyronine monodeiodinases (ITHD) are the key enzymes of extrathyroidal thyroxine metabolism (14-15). Inactivation of thyroxine's thyromimetic potency occurs by tyrosyl 5-deiodination, producing reverse-T_3 (rT_3). rT_3 (3,3',5'-triiodothyronine) is the most potent of the physiologically occurring inhibitors of ITHD and is thought to be of regulatory importance in modulating deiodinative iodothyronine metabolism (14-15). Previous studies suggest the existence of at least three ITHD isozymes (Types I-III) with different organ distribution and function in T_4 metabolism (14-15). The best characterized isozyme, type I ITHD, prefers rT_3 (K_m^{app} = 50-100 nM) over T_4 (K_m^{app} = 2 µM) as a substrate for deiodination, requires reduced dithiols as cosubstrates, and reacts in a ping-pong mechanism. Type I ITHD is primarily associated with the microsomal fraction of liver, kidney, and the euthyroid brain and is responsible for the production of about 70% of the circulating T_3 in the euthyroid rat (14-15). Thus, the potential inhibitory effects of toxic halogenated aromatic hydrocarbons on the type I ITHD isozyme were of interest since such effects could modulate thyroid hormone metabolism and biochemical thyroid status. Previous studies have shown that such diverse classes of compounds as iodinated contrast media, indicator dyes, anticoagulants, flavonoids, plant extracts, catecholamine (ant) agonists, and thiourea derivatives act as potent inhibitors of type I ITHD (13-15). In this work, we investigated the in vitro deiodinase inhibitory activity of a selected number of hydroxylated PCBs and soluble structural probes for dioxins that had previously been shown to have distinct binding affinities to TBPA (9-10).

Experimental Section

Compounds. $r-T_3$ free acid was purchased from Sigma Chemical Co. (St. Louis, MO). Hydroxylated PCBs were obtained from Ultra Scientific (Hope, RI) with the exception of TCDB-2 (see Table 1 for definition of abbreviations used) which was synthesized previously in our laboratory by adaptation of methods in the PCB synthetic literature (16). TriCDDA and DDA were synthesized in our laboratory as described earlier (9). [^{125}I] $r-T_3$ (specific activity 1000-1400 µCi/µg) was purchased from New England Nuclear Corp. (Boston, MA). If necessary [^{125}I] $r-T_3$ was purified before use by paper electrophoresis (17).

Animals. Male Sprague-Dawley rats (Charles River, MO) weighing 150 to 300g were used. They were fed NIH 31 diet and housed in rooms kept at 21 ± 1°C and 50% relative humidity, with a 12-hr light/dark cycle (lighted from 0700 to 1900 hr). The in vitro enzyme inhibition studies were performed using combined liver microsomal preparations from 6 rats as fed and housed in this manner.

Microsomal preparations were obtained in the following way. Liver was removed, quickly chilled in cold buffer (20 mM tris(hydroxymethyl)aminomethane (Tris), 0.32 M sucrose pH 7.0) and blotted on filter paper. Approximately 3 g were minced and homogenized in 6 volumes of buffer with a motor driven teflon pestle. The homogenate was centrifuged for 15 min at 18,000 x g. The pellet was resuspended in 7 ml buffer and centrifuged as above. The combined supernatant was centrifuged for 1 h at 100,000 x g. The microsomal pellet was mixed with 0.2 M potassium phosphate buffer (pH 7.0) containing 1 mM ethylenediaminetetraacetate (EDTA) and was homogenized in 8 ml of this buffer. An aliquot for the protein determination was saved and the remainder made 1 mM in dithiothreitol (DTT) and frozen in dry ice-acetone. The samples were stored at -70°C for no longer than 45 days. The protein concentration of the DTT-free microsomal preparation was determined (18) using bovine serum albumin as a standard.

rT3 5'-deiodinase assay. rT_3 - 5'-deiodinase activity from liver microsomes was determined by release of [^{125}I$^-$] from

[^{125}I]-rT$_3$ (with modifications) based on the previous method (19). In this assay rT$_3$ is used as substrate instead of T$_4$ because T$_4$ allows alternate (inner ring) deiodination complicating the specific measurement of outer ring deiodination. Direct in vitro inhibition of the 5'-deiodinase by hydroxylated PCBs was analyzed as follows: In a total volume of 200 ul the reaction mixture contained 0.2 M potassium phosphate, 1 mM DTT, 1 mM EDTA, 10 ug microsomal protein and 1 nM [^{125}I] rT$_3$ (pH 7.0). The specific activity was approximately 520,000 cpm/pmol rT$_3$. The PCBs were added in 5 ul methanol yielding 10^{-9} to 10^{-5} M in the reaction mixture. Reaction was started by the addition of the microsomes. Incubation was carried out at 37°C for 5 min. The reaction was terminated by the addition of 300 ul of ice-cold solution of 10 uM L-T$_4$ and 50 uM 6-propyl-2-thiouracil (PTU). Liberated I$^-$ was quantified by separating I$^-$ from rT$_3$ on disposable Sephadex G-25 minicolumns and subsequent counting of [^{125}I] in a Packard Auto Gamma counter (60% efficiency) as described earlier (10). Briefly, after the reaction was stopped a 400 ul aliquot was added to the column and eluted with 1.3 ml 0.2 M potassium phosphate buffer (pH 7.0). This first fraction contained [^{125}I] rT$_3$ eventually bound to proteins (typically < 2% of total substrate concentration). The first fraction was separated from the second fraction (2 ml) containing [^{125}I$^-$]. Free rT$_3$ binds tightly to the gel and does not elute in the volume used. To correct for nonenzymatic deiodination and contamination with [^{125}I$^-$], control incubations were conducted without microsomes or with boiled microsomes yielding the same result (1 to 2% of total counts). The molar amount of I$^-$ liberated was calculated by referring to half of the specific activity for rT$_3$ assuming an equal chance for the 3' or 5' positions to be occupied by [^{125}I] (19). Because of solubility problems the activity of TriCDDA and DDA was measured in a modified procedure: Total volume was 500 ul (opposed to 200 ul) imidazole buffer (pH 7.0) with 5 ug (opposed to 10 ug) microsomal protein, 0.6 nM [^{125}I]rT$_3$ (1 nM total rT$_3$) and 5 mM DDT. The enzyme reaction was much slower under these conditions and linear up to 60 min. An incubation time of 45 min was chosen. At this timepoint again about 20% of the substrate was converted at the highest rate. Reaction was terminated by quickly cooling the samples on ice and adding them directly onto the columns. The time from the termination of the enzyme reaction to the complete elution of the samples took less than 2 min for each sample.

The Lineweaver-Burk (20) plot for the inhibitor interactions was obtained in the following way. The reaction mixture contained 10 ug microsomal protein, 0.05 nM - 0.6 nM [^{125}I]-rT$_3$ [total rT$_3$ concentration was 5 to 60 nM in 200 ul of 0.2 M potassium phosphate buffer (1 mM DTT, 1 mM EDTA, pH 7.0)] alone or in the presence of 1.5 uM TCHB-1, 0.15 uM TCDB-2, 6.25 uM TriCDDA or 0.11 uM TCDB-1. Hydroxylated PCBs and rT$_3$ were added in 5 ul methanol and the dioxin derivatives in 5 ul dimethylsulfoxide (DMSO). There was no difference in the K$_m$ of rT$_3$ (29 nM) between the methanol and the DMSO containing reaction. Incubation time was 5 min at 37°C. Results are the mean of duplicate determinations and representative of two or more independent experiments. Linear regression analysis gave correlation coefficients > 0.9 for all the lines.

Results

Type I deiodinase activity is primarily associated with the microsomal fraction of liver tissue (14-15). The deiodinase activity was assayed using conditions adapted from the literature (19) but using relatively more dilute conditions to enhance the binding activity and facilitate study of the more lipophilic chlorinated aromatic hydrocarbon derivatives. The enzyme reaction was linear with incubation time and protein concentration for all conditions described in the Experimental Section. In addition, the substrate was never depleted by more than about 20%. For example, in the assay used for the hydroxylated PCBs, 10.5 fmols of rT_3 was consumed in 1 min (5.3% of total substrate), 28.2 fmols at 3 min (14.1%), and 47.6 fmols at 6 min (23.8%). The same was true of the procedure used for the dioxin derivatives (data not shown), and under these conditions the reaction was much slower (linear up to 60 min) (Figure 1).

Competitive inhibition studies were done using selected polar hydroxylated PCB and adipamide dioxin derivatives available from previous work with TBPA (9-10). Figures 2 and 3 show the effect of increasing concentrations of various hydroxylated PCBs on the conversion of rT_3 into 3,3'-diiodothyronine (T_2). From these plots, inhibitory concentrations yielding half-maximal inhibition (IC_{50}'s) are estimated (13). A comparison of these values is shown in Table 1. Lateral chlorine substitution (3,5 or *meta*) was common to the two most active PCB compounds (TCDB 1 and 2). The inactivity of TCHB-2 and DDA which lack lateral substituents (3,5 on biphenyl or 2,3 on dioxin nucleus) further supports the importance of lateral substitution for significant binding activity. The result with DDA also argues against the possibility that the adipamide group is a dominant factor in binding. The relatively less polar and more lipophilic compounds, TCHB-1 and TriCDDA showed considerably lower inhibitory activity. However, the similar inhibitory potency of these two compounds would suggest that a lateral hydroxyl group may not be a critical factor in binding.

Table 1

IC_{50} values[a] of chlorinated biphenyl and dioxin derivatives in the 5'-deiodination of rT_3

Inhibitor	Abbrev.	IC_{50} (µM)
3,3',5,5'-tetrachlorodihydroxybiphenyl	TCDB-1	0.07
2,3,5,6-tetrachlorodihydroxybiphenyl	TCDB-2	0.32
3,7,8-trichlorodibenzodioxin adipamide	TriCDDA	4.40
3,5,4'-trichloro-4-hydroxybiphenyl	TCHB-1	7.20
2,4,6-trichloro-4'-hydroxybiphenyl	TCHB-2	NA[b]
dibenzodioxin adipamide	DDA	NA[b]

[a] Half-maximal concentrations estimated from Figs. 2 and 3.
[b] No appreciable activity was observed at the concentration of inhibitor tested.

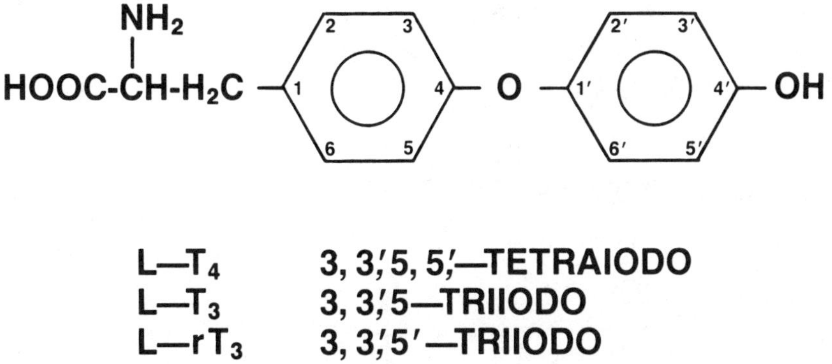

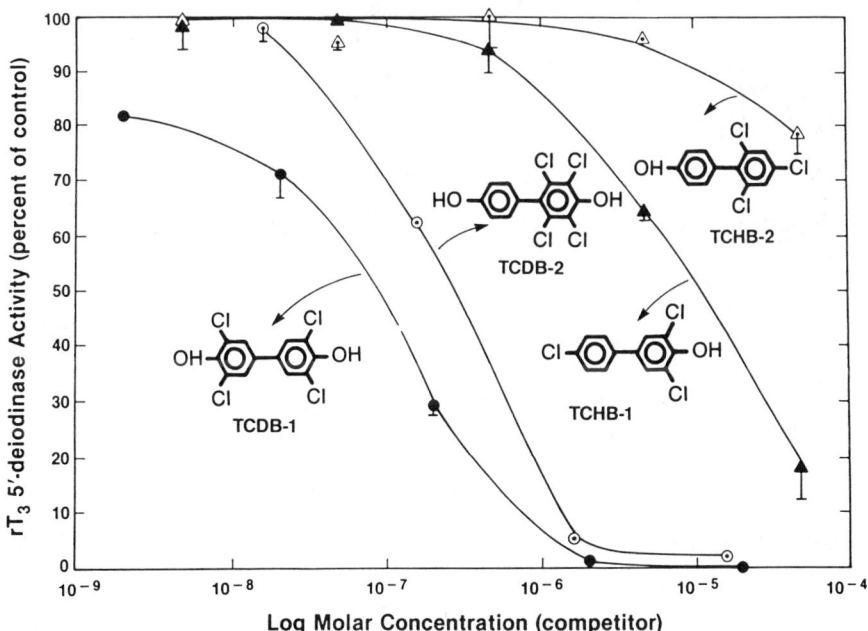

Figure 1. Structure of thyroid hormones and rT_3 used as substrate for 5'-deiodinase type I measurements.

Figure 2. Inhibition of the <u>in vitro</u> conversion of rT_3 into $3,3'-T_2$ by increasing concentrations of various hydroxylated PCBs. Every point is the average of a duplicate determination and representative of two or more independent experiments.

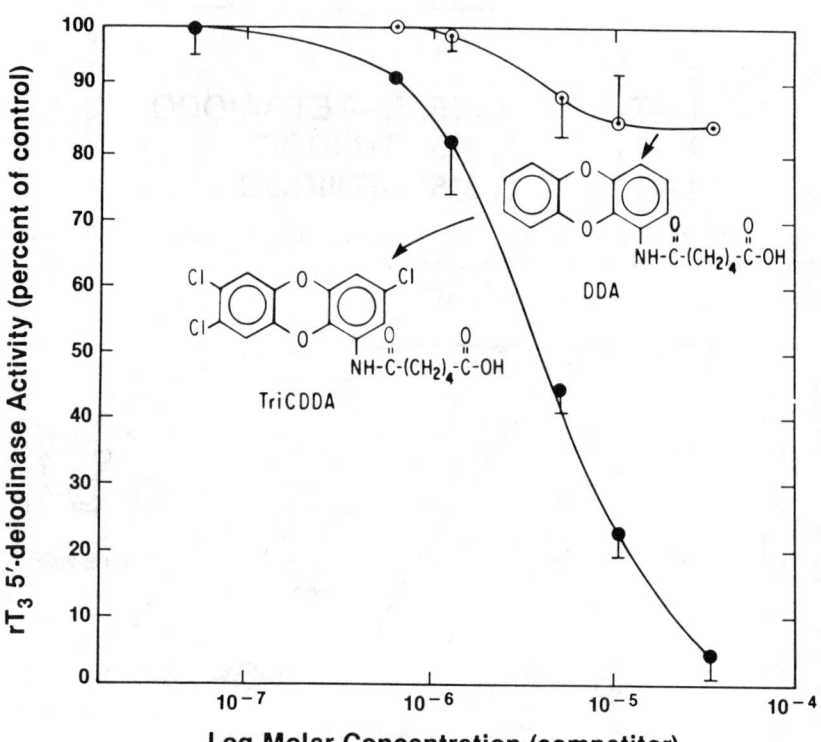

Figure 3. Inhibition of <u>in vitro</u> conversion of rT_3 into $3,3'-T_2$ by increasing concentrations of TriCDDA and DDA. Every point is the average of a duplicate determination and representative of two or more independent experiments.

To verify that the active PCB and dibenzo-p-dioxin derivatives were interacting with the substrate (rT_3) binding site of the enzyme, we investigated the inhibitory mechanism by enzyme kinetic analysis. The results of these experiments are shown in the double reciprocal (Lineweaver-Burk) plot (20) (Fig. 4). A competitive mechanism is strongly suggested by the common intersection point on the vertical axis for the regression lines. The kinetic data derived from this plot for rT_3 are: K_m = 29 nM; V_{max} = 70 pmol of I^-/mg protein/min. The values are similar to those previously reported (13-15) by others for rT_3 using similar methods.

Discussion

Previous studies have shown similar structure-activity relationships for potent inhibitors of type I iodothyronine deiodinase in rat liver and for strong binders to human TBPA (13). The deiodinase active site has been characterized using the hormone-binding site observed in the crystal structure of the T_4-TBPA complex as a model. We have used these structural data as a model for the interaction of PCB and dioxin analogues with their prealbumin-like receptors (9-10). In view of these results, it was anticipated that certain PCB and dioxin derivatives would also be potent inhibitors of type I deiodinase activity. However, in view of the highly lipophilic nature of these chlorinated aromatic hydrocarbons and our previous experience in studying their interaction in crude protein or receptor preparations (12), it was further anticipated that polar derivatives would be technically necessary in order to achieve sufficient solubility in the assay medium and reduce nonspecific binding interactions. In fact, it was not possible to demonstrate deiodinase inhibitory activity for any nonoxygenated PCB compounds in constrast to our previous studies (9-10) involving highly purified TBPA. In crude protein preparations it would be difficult to separate intrinsic binding activity from apparent binding activity that also reflects the poor solubility and varying lipophilic properties of these compounds. Therefore, it is not possible to derive a complete, meaningful quantitative structure-activity relationship for these compounds using the enzyme preparation described.

Interestingly, this problem appears to work in reverse for receptor preparations (e.g., dioxin or Ah receptor) and their assays that are based on the use of a highly lipophilic substrate (e.g., ^{3}H-TCDD) (21-23). Both hydroxylated, halogenated and nonhalogenated aromatic hydrocarbons are in general poor binding ligands for the Ah receptor. This can be explained on the basis of the relatively large dehydration energy costs associated with the binding of such polar, hydrogen bondable hydroxylated compounds (24). The relatively low toxicity of these hydroxylated compounds can be explained in large part on the basis of their more rapid metabolism and elimination as polar conjugates (25). However, we believe that these polar derivatives can serve as useful *in vitro* structural probes for studying the specific binding interactions of the parent compounds with thyroxine binding proteins.

With these problems in mind, we limit our discussion of

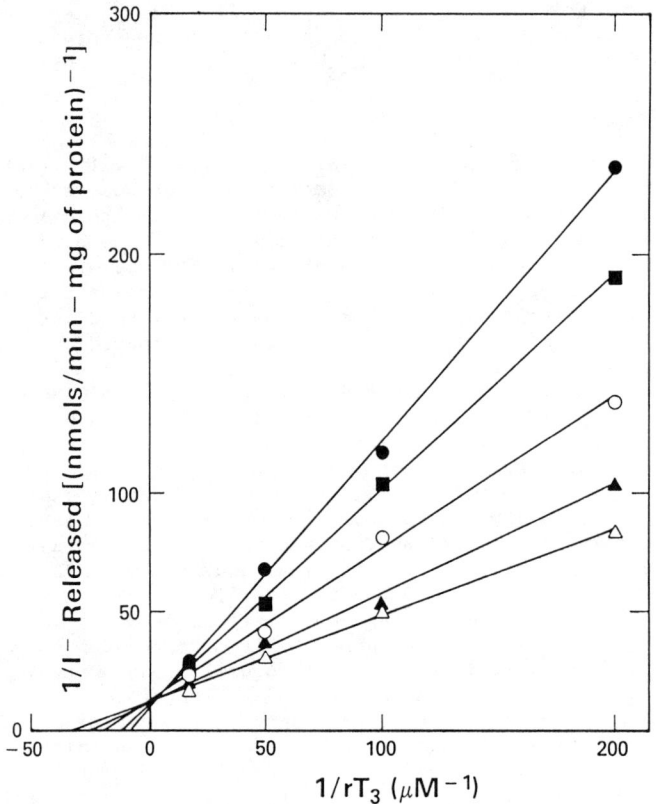

Figure 4. Lineweaver-Burk plot of the conversion of rT_3 into $3,3'-T_2$ in the absence (△) and presence of 1.5 μM 3,5,4'-triCL-4-OH (▲), 0.15 μM 2,3,5,6-tetraCL-4,4'-diOH (○), 6.25 μM TriCDDA (■), 0.11 μM 3,5,3',5'-tetraCL-4,4-diOH (●). For details see "Experimental Section". Every point is the average of a duplicate determination (Range of values was ≤ ± 7.8%) and representative of two or more independent experiments. The lines were fitted by least squares method.

structure-activity relationships primarily to qualitative considerations. The inhibitors studied in this work were selected on the basis of their strong, moderate, or weak binding to TBPA as previously reported (9-10). The deiodinase inhibitory potencies of these compounds are in general similar to their binding activities with TBPA and thyroxine specific nuclear binding sites in rat liver (12). The most potent inhibitor was TCDB-1 which is also the strongest binder to TBPA (10). This compound exhibited an inhibitory potency similar to the K_m of rT_3 supporting its possible use as a selective inhibitor that would facilitate the study of further biochemical and functional characteristics of proteins and reactions in thyroid hormone metabolism (15). TCDB-2 was about 4-5 times less potent than TCDB-1 as an inhibitor and showed about half the affinity to TBPA (10). In contrast, TCHB-1 was about 100 times less potent as an inhibitor (than TCDB-1) but is predicted in the TBPA assay (10) to show similar activity. Likewise TriCDDA is less active as an inhibitor than predicted in the TBPA assay (9). As discussed previously, these latter results are compatible with the less polar, more lipophilic character of these compounds. In view of these results, it is difficult to evaluate the importance of the hydroxyl group for binding (inhibitory) activity, but the TBPA results suggest that polarizable groups of similar size might be able to replace it. This possibility is supported by our molecular and theoretical modeling results with TBPA interactions with PCBs and related compounds (9,10,26).

Iodothyronine type I deiodinase represents another thyroxine specific protein or protein preparation which specifically binds with high affinities certain chlorinated aromatic hydrocarbon derivatives whose functional structural characteristics are in general similar to those found in toxic structures (underivatized) of this type. Although we can not say with certainty that these protein binding interactions mediate some of the toxic effects of these compounds, the studies with the purified protein TBPA which serves as a model for the crude protein preparations of this type suggest that such binding interactions could play an important role. We have not investigated the inhibitory effects of these compounds on other types of deiodination, but all three types of deiodination share common inhibitors (14-15). Such antagonistic effects on thyroid hormone metabolism should be expected to effect biochemical thyroid status, even though treated animals may appear to be functionally euthyroid (4).

In preliminary studies (27), we have demonstrated a dose related depression of specific type I deiodinase activity in livers of TCDD treated rats held 7 days which paralleled a decrease in serum T_4. This result is compatible with the demonstrated dependence of deiodinase activity on serum thyroid hormone levels (14-15), and probably does not reflect direct inhibition, particularly 7 days after treatment. Any competitive inhibitory effect of dioxin on the deiodinase activity appears to be over-shadowed by a serum T_4 related regulation of the enzyme activity or by a biochemical event reflected in the serum T_4 decrease. It has been demonstrated (28) that in hypothyroid rats 5'-deiodinase activity in the liver is depressed. Based on these preliminary in vitro and in vivo results, it is reasonable to anticipate that dioxin would bring about alterations in at least biochemical thyroid status at sublethal doses.

However, complex and anomalous findings have caused uncertainty about the biochemical and functional thyroid status of dioxin treated animals (4). This could be related to a complex interplay of the potential thyroid hormone agonists and antagonists properties of these toxic compounds. Further studies are needed to investigate a possible link between TCDD (and related compounds)-mediated effects and some aspects of thyroid hormone endocrinology.

Acknowledgements

We thank R. Fannin for technical assistance throughout this work, Dr. Michael Kaplan (New England Medical Hospital, Boston) and Dr. R.E. Peterson (Environmental Toxicology Center, University of Wisconsin, Madison) for helpful discussion, and Denise Crawford for performing statistical analyses.

References

1. Poland, A., and Knutson, J.C. Ann. Rev. Pharmacol. 1982, 22, 517-554.
2. Bastomsky, C.H. Endocrinol. 1977, 101, 292-296.
3. Rozman, K., Hazelton, G.A., Klaassen, C.D., Arlotto, M.P., and Parkinson, A. Toxicology 1985, 37, 51-63.
4. Potter, C.L., Moore, R.W., Inhorn, S.L., Hagen, T.C., and Peterson, R.E. Toxicol. Appl. Pharmacol. 1986, 84, 45-55.
5. Rozman, K. In Banbury Report 18: Biological Mechanisms of Dioxin Action; Poland, A., and Kimbrough, R.D., Eds.; Cold Spring Harbor Laboratory, Cold Spring Harbor, New York, 1984; p 345-354.
6. McKinney, J.D., Fawkes, J., Jordan, S., Chae, K., Oatley, S., Coleman, R.E., and Briner, W. Environ. Hlth. Perspect. 1985, 61, 41-53.
7. Rozman, K., Rozman, T., and Greim, H. Toxicol. Appl. Pharmacol. 1984, 72, 372-376.
8. Hong, L.H., McKinney, J.D., and Luster, M.I. Biochemical Pharmacol. 1987, 36(8), 1361-1365.
9. McKinney, J.D., Chae, K., Oatley, S.J., and Blake, C.C.F. J. Med. Chem. 1985, 28, 375-381.
10. Rickenbacher, U., McKinney, J.D., Oatley, S.J., and Blake, C.C.F. (1986) J. Med. Chem. 1986, 29, 641-648.
11. Eberhardt, N.L., Ring, J.C., Latham, K.R., and Baxter, J.D. J. Biol. Chem. 1979, 254(17), 8534-8539.
12. McKinney, J.D., Fannin, R., Jordan, S., Chae, K. Rickenbacher, U., and Pedersen, L. J. Med. Chem. 1987, 30, 79-86.
13. Auf'mkolk, M., Koehrle, J., Hesch, R.-D., and Cody, V.J. J. Biol. Chem. 1986, 261(25), 11623-11630.
14. Kaplan, M.M. Neuroendocrinology 1984, 38, 254-260.
15. Koehrle, J. Brabant, G., and Hesch, R.-D. In Hormone Research Lemarchand-Béraud, T., and Vanhallst, L., Eds; S. Karger Medical and Scientific Publishers, New York, 1987; pp 58-78.

16. Kato, S., McKinney, J.D., and Matthews, H.B. Toxicol. Appl. Phrmacol. 1980, 53, 389-398.
17. Cahnmann, H.J. In The Thyroid and Biogenic Amines Rall and Kopin, Eds; North Holland Publishing Co., Amsterdam, 1972, pp 27-51.
18. Lowry, O.H., Rosenbrough, N.J., Farr, A.L., and Randall, R.J. J. Biol. Chem. 1951, 193, 265-275.
19. Leonard, J.L., and Rosenberg, I.N. Endocrinol. 1980, 107(5), 1376-1383.
20. Lineweaver, H., and Burk, D. J. Am. Chem. Soc. 1934, 56, 658-666.
21. Bandiera, S., Sawyer, T.W., Campbell, M.A., Fujita, T., and Safe, S. Biochem. Pharmacol. 1983, 32(24), 3803-3813.
22. Denomme, M.A., Homonoko, K., Fujita, T., Sawyer, T., and Safe, S. Mol. Pharmacol. 1985, 27(6), 656-661.
23. Denomme, M.A., Homonoko, K., Fujita, T., Sawyer, T., and Safe, S. Chem.-Biol. Interact. 1986, 57(2), 175-187.
24. Long, G., McKinney, J., and Pedersen, L. Quant. Struct.-Act. Relat. 1987, 6, 1-7.
25. Sparling, J., Fung, D., and Safe, S. Biomed. Mass Spectrom. 1980, 7, 13-19.
26. Pedersen, L.G., Darden, T.A., Oatley, S.J., and McKinney, J.D. J. Med. Chem. 1986, 29, 2451-2457.
27. Rickenbacher, U., and McKinney, J.D. Toxicologist 1986, 6, 1239.
28. Larsen, P.R., Silva, J.E., and Kaplan, M.M. Physiological and Endocrine Rev. 1981, 2(1), 87-102.

RECEIVED January 31, 1989

Chapter 23

Base-Line Toxicity Predicted by Quantitative Structure–Activity Relationships as a Probe for Molecular Mechanism of Toxicity

Robert L. Lipnick

Office of Toxic Substances (TS–796), U.S. Environmental Protection Agency, 401 M Street, SW, Washington, DC 20460

> Narcosis represents the most fundamental mechanism of the toxicity of nonelectrolyte organic compounds, and corresponds to *minimum* or *baseline* toxicity. Quantitative structure–activity relationships for chemicals acting by this mechanism for various organisms and routes of exposure provide a valuable probe for determining whether or not a candidate chemical acts via narcosis or by an *electrophile, proelectrophile, cyanogenic*, or other more specific molecular mechanism.

Structure–activity relationships (SAR) and quantitative structure–activity relationships (QSAR) have attracted increased interest by the U.S. Environmental Protection Agency (*1–3*) in the review of new industrial chemicals prior to their manufacture under section 5 of the Toxic Substances Control Act (TSCA) (*4*). There is also a strong impetus in Europe to develop QSAR models to assess the potential toxicity of industrial organic chemicals (*5*). The applicability of such predictive models is dependent upon the availability of measured or predicted values of the parameters used in the QSAR model, and by the validity of the assumption that the mechanism of toxicity of the candidate chemical is the same as that of the compounds used to derive the QSAR, and that they encompass the same range in spanned substituent space (*6*). The relationships that should ideally be satisfied for such models are illustrated in Figure 1. In the absence of more specific effects, nonelectrolyte organic compounds act by a narcosis mechanism, which represents *minimum* or *baseline* toxicity. Baseline toxicity QSAR models can be used as a probe to identify chemicals acting by more specific toxicological mechanisms.

Demonstrating Commonality of Molecular Mechanism

The assumption of the commonality of mechanism is rarely stated in a QSAR study, but can be based implicitly or explicitly on one or more of the following:

This chapter not subject to U.S. copyright
Published 1989 American Chemical Society

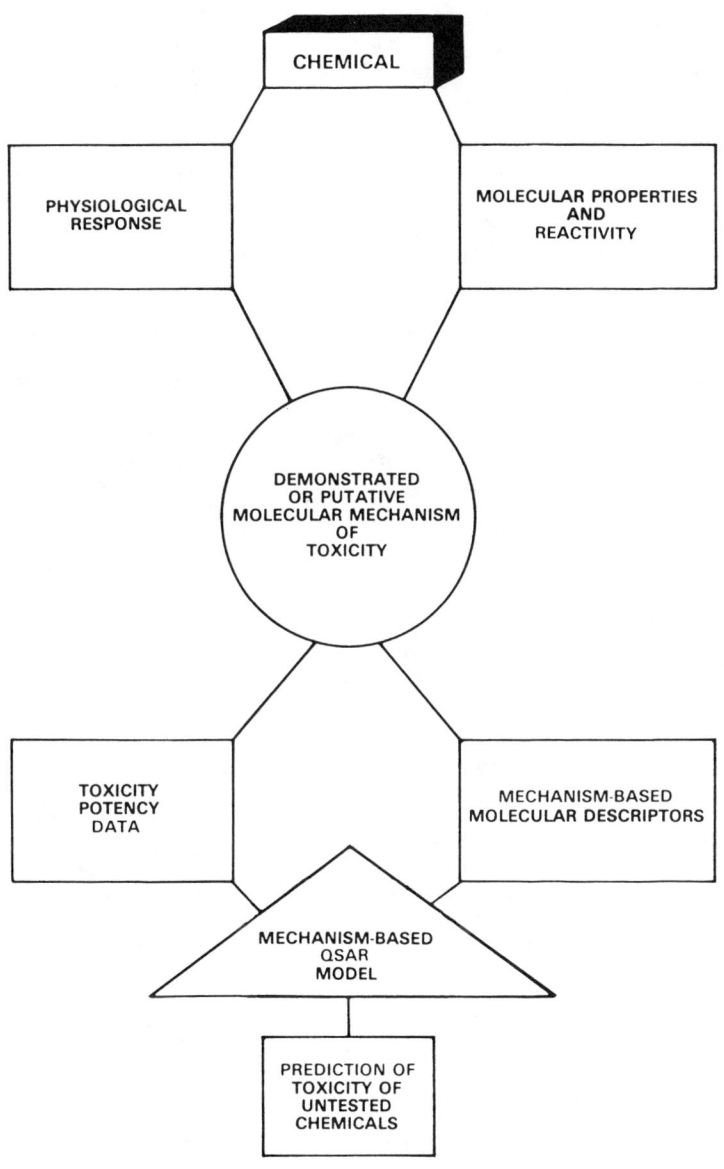

Figure 1. Mechanism–based QSAR model and its relationship to chemical and biological data of the molecules used in its derivation. A quantitative prediction of the toxicity of an untested chemical requires the availability of a QSAR model for related compounds acting by a putative common molecular mechanism and the ability to measure or predict the required mechanism–based molecular descriptors used in the QSAR model.

1. Receptor binding
2. Symptomatology
3. QSAR correlation
4. Additivity of biological activity
5. Chemical properties

Receptor Binding. The first criterion consists of similar types of binding with isolated receptors or other molecular sites of action and understanding of the pathway of cascading biochemical and physiological perturbations leading to the observed end effect. Information from the exact structure of and other receptors enzymes is becoming an extremely powerful tool in the design of new drugs (7).

Symptomatology. A second criterion consists of commonality of syndromes, and biochemical and physiological changes associated with the toxic end effect. Similar symptomatology has provided a traditional criterion for assessing the degree to which a group of compounds may be acting by a common mechanism, and as evidence for a change in mechanism within a series. Chemicals producing a narcosis or anesthetic physiological response are frequently classified on this basis due to the lack of a specific receptor that has been characterized at the molecular level. More recently an attempt has been made to mathematically transform a large number of quantitative symptomatology parameters to a smaller number of orthogonal descriptors in which symptomatology syndromes could be defined through cluster analysis (8-11).

QSAR Correlation. Statistical quality of QSAR correlation can be employed as a third criterion of commonality of mechanism. This approach can prove very meaningful when coupled with a mechanistic interpretation of the role of molecular descriptors used in the correlation, and with the significance of the slope and intercept. The quality of statistical fit and the interpretation of the parameter or parameters used in the correlation can provide a valuable insight into molecular mechanism. Recently, Hansch analysis has been combined with molecular graphics and modeling studies in which the activities of a series of substrates to an enzyme receptor have been related to the hydrophobic, electronic, and steric requirements for reversible binding (12).

Additivity of Biological Activity. Additivity of toxicity is a valuable means for demonstrating commonality of mechanism. In most cases, compounds acting by different mechanisms show toxici-ties that are less than additive (13). It has been demonstrated that the aquatic toxicity of up to 50 nonelectrolyte organic chemicals acting by a narcosis mechanism exhibits strictly additive behavior (14-16).

Chemical Properties. Finally, similarity in chemical properties and reactivity based upon mechanistic and physical organic chemistry can also be used to support a hypothesis of commonality of mechanism (17). By this means, a trained organic chemist can suggest the limitations of a mechanism even for distantly related compounds based upon their chemical properties, thus permitting the identification of chemicals that can potentially act by more specific mechanisms.

Discovery of the Correlation Between Narcosis Potency and Water Solubility and Molecular Weight

Interest has existed for some time regarding what property of a molecule is responsible for the production of a specific type of biological response (18-20). In the case of narcosis or anesthetic response, this was complicated by the finding that narcosis could be produced by a wide variety of different, apparently unrelated compounds, including simple saturated monohydric alcohols. Perhaps the earliest systematic investigation of the mechanistic basis of narcosis was that of Cros at the University of Strasbourg, who in 1863 reported that toxicity of simple alcohols administered to mammals increases with decreasing water solubility, up to a point of maximum potency, beyond which it decreases, until the alcohols become very insoluble and act like fatty substances (21).

Richardson in England (22) in 1869 found that the toxicity of simple monohydric alcohols to mammals and birds increased with increasing molecular weight, and this relationship has been referred to as Richardson's law (23). Rabuteau in France (24) in 1870 concluded that the increase in the toxicity to frogs immersed in solutions of alcohols reflected an increase in chainlength. Both of these scientists were apparently unaware of Cros' earlier, more fundamental discovery. Although Cros' discovery was confirmed several years later by Dujardin-Beaumetz and Audigé (25-27), it was apparently not until 1893 that the relationship between water solubility and narcosis-produced toxicity was investigated more carefully. Georges Houdaille, a doctoral student of Charles Richet at the University of Paris, observed a correlation between water solubility and minimum narcosis concentration to fish and tadpoles for a variety of hypnotic agents (28). This relationship is frequently referred to as Richet's law (29).

Discovery of the Correlation of Narcosis Potency and Partition Coefficient

The discovery of a parameter (olive oil/water partition coefficient) upon which a mechanistic interpretation for narcosis could be based was made independently six years later by Charles Ernest Overton at the University of Zurich (30-31) and by Hans Horst Meyer (32) and his collaborator Fritz Baum (33) at the University of Marburg. Prior to this discovery, Walter Dunzelt, a student of Meyer, attempted to confirm the Houdaille data on the relationship between water solubility and minimum toxic concentration, using tadpoles and small fish (34).

Dunzelt concluded in his 1896 Inaugural Dissertation (34) that although this correlation seemed to be generally correct, it did not hold for two compounds. Bromal hydrate was found to be very soluble in water and produce narcosis in fish at low concentrations. By contrast, methyl urethane was found to produce only a slight narcotic effect, even though it was only slightly soluble in water. This finding of Dunzelt may have provided Meyer with the impetus to search for another chemical property that better correlated with narcotic potency. Meyer further supported the relevance of partition coefficient to the mechanistic basis of narcosis by demonstrating that the temperature dependence of narcosis in tadpoles followed the same trends as the corresponding olive oil/water partition coefficients (35).

Overton's discovery evolved from studies of cell permeability with algae, and later, tadpoles (36-37). Overton provided considerable support two years later for this lipoid theory of narcosis in his classic monograph "Studien über die Narkose," (31) whose findings have recently been reviewed (38). Overton's tadpole data have been employed in several QSAR studies (39-41) (Lipnick, R.L. In *QSAR in Drug Design*; Fauchère, J.L., Ed.; Alan R. Liss: New York, in press.).

More Recent Studies of Narcosis

For the next five decades, the correlation of partition coefficient with narcosis potency continued to be an important area for research, with studies of various compounds performed using whole organisms, organs, cells, and enzymes. This work has been discussed in a number of major reviews (42-51), and in papers from an international conference on this subject held in Paris in 1950 (52).

Despite the considerable number of studies of the mechanism of narcosis, the details regarding its molecular basis have proven very elusive, and the theoretical basis for narcosis or anesthesia still represents a subject of considerable debate. Current research is focused on two major theories, disorganization of membrane lipoid constituents (53-54), and binding to one or more enzymes or proteins (55-56).

Linear QSAR Models for Narcosis Baseline Toxicity

In the mid 1960s Hansch and co-workers employed regression analysis to develop quantitative structure-activity relationships (57). They reported a number of linear QSARs for simple nonelectrolytes acting by a narcosis mechanism (40), in the form,

$$\log (1/C) = A \log P + B \qquad (1)$$

where C is the molar concentration producing a standard biological response, and P is the n-octanol/water partition coefficient.

When an organism is placed in an aqueous solution of a toxicant, pseudo-steady-state partitioning takes place between the toxic site of action and the aqueous phase. For organisms in which air is extracted from the water by means of gills, exchange of toxicant takes place mainly via this route, and the rate of uptake is controlled by the cross-sectional area, and the rate of blood flow across the gills.

Recently, a number of investigators have reported linear QSARs for the toxicity of simple nonelectrolytes to aquatic organisms (58-63). A general representation of such QSARs and their limitations is depicted in Figure 2. These equations are of enormous value to EPA and other regulatory agencies in setting testing priorities for chemicals in commerce and for new industrial chemicals prior to their manufacture.

A corresponding pseudo-steady-state partitioning takes place between the toxicant vapor and this same site of action for gaseous anesthetic

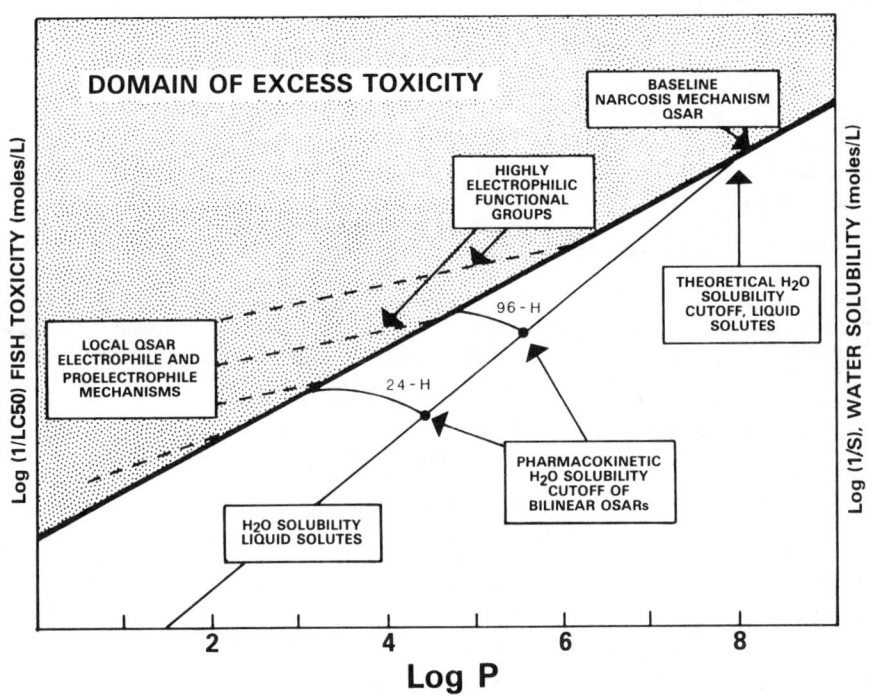

Figure 2. Baseline narcosis aquatic toxicity QSAR model. For nonelectrolytes acting by a narcosis mechanism, no toxicity is observed if the predicted toxic concentration exceeds the water solubility. With decreasing test duration, pseudo-steady-state partitioning is not achieved for very hydrophobic chemicals, and the location of the water solubility cutoff shifts to chemicals having a lower partition coefficient. Compounds acting by more specific mechanisms produce toxicity at lower aqueous concentrations than predicted by narcosis, and fall within the domain of excess toxicity.

agents. In this case, the oil/gas partition coefficient and not the oil/water partition coefficient is the appropriate model parameter and yields a linear correlation as shown in Figure 3. By contrast, only scatter is observed if octanol/water is used as the model parameter as shown in Figure 4 (64).

Model Partitioning Systems

Overton and Meyer both used olive oil as a partitioning system to model the physicochemical properties of the putative membrane lipoid site of action. Although Overton attempted to use melted cholesterol and other substances he thought might serve as a better reference phase, he abandoned this approach due to problems with the formation of inseparable emulsions (31). Collander in Finland (65) experimented with a variety of aqueous organic solvent systems and found that for many simple nonelectrolytes, the values were well-correlated according to the following equation:

$$\log P_1 = a \log P_2 + b \quad (2)$$

where P_1 and P_2 are two such partitioning systems. The Collander equations were investigated more fully by Leo and Hansch. They concluded that solutes able to act as either strong hydrogen-bond donors or acceptors require an additional correction to account for the relative ability of the organic solvents to interact in this fashion with respect to one another (66). Although n-octanol is now commonly used as a model organic phase, it should be kept in mind that it may be a better model for some biological systems than for others. Meyer and Hemmi (67) compared the ability of partition coefficients from different solvent systems to model narcosis data, and they found oleyl alcohol to yield the best results. For complex organic molecules containing hydrogen bonding functional groups, the choice of a particular model system could lead to apparent outliers which may not reflect a true change of mechanism at the molecular site of action.

Estimation of Partition Coefficients

The development of the fragment constant methodology for predicting log P values directly from chemical structure (68-71) and its computerization (72-73) is furthering the use of n-octanol as a standard reference phase. Another approach that has been pursued is the use of retention times determined on a reversed-phase HPLC column to estimate log P, using a set of standard solutes for reference (74-76). These correlations display limitations similar to those in changing from one partitioning system to another if hydrogen-bonding interactions are significant. The interactions of this type of solute with biological substrates have also been modeled using solvatochromic parameters (77).

Species Sensitivity to Narcosis

Generally, the slopes of QSARs for narcosis are close to unity indicating that (1) n-octanol provides a reasonable model for the physicochemical properties of the biological site of action, (2) pseudo-

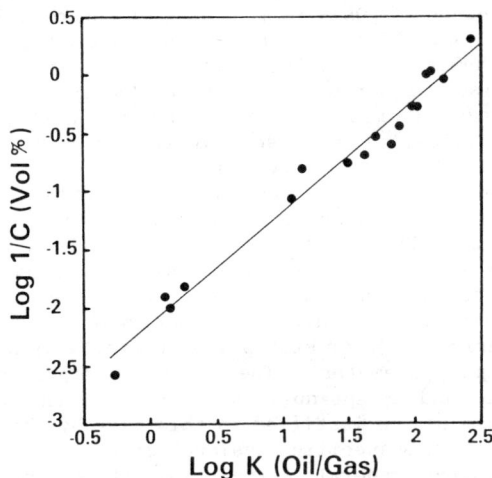

Figure 3. Linear relationship between oil/gas partition coefficient and minimum concentration in air required to produce anesthesia in mice on a log log scale. (Reproduced with permission from Ref. 64. Copyright 1987 Elsevier).

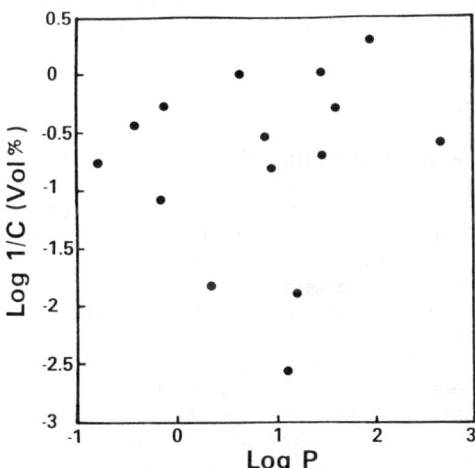

Figure 4. Lack of a relationship between octanol/water partition coefficient and minimum concentration in air required to produce anesthesia in mice on a log log scale. (Reproduced with permission from Ref. 64. Copyright 1987 Elsevier).

steady-state partitioning has been achieved, and (3) any loss of toxicant via metabolism is reasonably constant within the series studied. The intercept in QSAR is a measure of the relative sensitivity of the organism. Both Meyer and Overton postulated that the relative sensitivity of chemicals is related to the percent of lipoid tissue within the organism, higher organisms being the most sensitive due to the high concentration of such lipoids in nerve cells. In comparing relative sensitivities, it is essential that a common physiological endpoint be employed. Thus, QSAR equations developed from data on multi-generational toxicity to algae, protozoa, and bacteria have intercepts that reflect higher sensitivity than simple narcosis (78).

Data from Führner (79) on the narcosis by alcohols to 23 different organisms representing a broad range in taxonomic class were analyzed to examine the Meyer-Overton hypothesis. Two of the organisms in this data set were tested with a sufficient number of compounds to satisfy the Topliss criterion (80) for statistical validity with respect to the number of tests per parameter. The remaining organisms were tested using only ethanol and n-heptanol. Nevertheless, the use of only two points per organism can be justified in this case by the observation of common symptomatology, a specific physicochemical interpretation of the log P parameter, and general association of saturated monohydric alcohols with a narcosis mechanism. The slopes and intercepts derived from these data of Führner are arranged in order of increasing sensitivity in Table I.

There is an evident association of greater sensitivity with increasing phylogenetic development. The partial overlap between invertebrates and vertebrates may reflect differences in the specific biological endpoint chosen by Führner for observation.
The intercepts obtained for fish and tadpoles are similar to those reported in the literature based upon other test data (81-82).

Water solubility cutoff

Pseudo-steady-state partitioning can take place in aquatic organisms immersed in an aqueous solution of toxicant or via exposure to vapor by inhalation through the lungs. The concentration of a toxicant at the narcosis site of action is a function of both its aqueous concentration and partition coefficient. Increasingly lower aqueous concentrations are required to produce a common molar concentration at the site of action with increasing partition coefficient. These relationships are illustrated in Figure 2.

Overton pointed out in his classic monograph "Studien über die Narkose" that a classification existed at that time of substances which narcotize both plants and animals (31,38), and those such as sulfonal that narcotize only animals. Overton provided a much simpler and more reasonable explanation for this apparent dichotomy. The narcotizing concentration of sulfonal to tadpoles is very close to the water solubility of this compound. Overton found that algae required 6-7 times as great a concentration to produce narcosis as tadpoles and other higher aquatic organisms, and therefore require concentrations of sulfonal exceeding its water solubility to produce narcosis.

Table I. Relationship Between Species Sensitivity to Narcosis and Taxonomic Hierarchy[a,b]

Species	Type	Sensitivity (Intercept)	Slope	Narcosis Endpoint
Noctiluca miliaris	Protozoa	0.33	0.96	Tentacle movement
Covoluta roseoffensis	Sea urchin	0.63[c] 0.64[d]	0.98[c] 0.98[d]	Suppress movement
Physa fontinalis	Freshwater snail	0.65	1.04	Locomotion
Tomopteris onisciformis	Worm	0.70	1.00	Suppress swimming
Pecten operculus	Small mussel	0.72	1.01	Muscle contraction on contact
Soles (cutlellus) pellucidus	Mussel	0.72	1.01	Muscle contraction on contact
Asterias rubens	Sea star	0.72	1.01	Locomotion
Spio vulgaris	Worm	0.72	1.01	Locomotion
Gammarus pulex	Arthropod	0.73	1.03	Swimming
Mysis (macromysis) flexuosa	Marine crab	0.73	1.03	Swimming
Aeolis drummondi and rufibranchialis	Marine night snail	0.74	1.06	Locomotion
Amphioxus vulgaris	Acranian	0.75	1.13	Swimming on contact

Continued on next page

Table I. Continued

Species	Type	Sensitivity (Intercept)	Slope	Narcosis Endpoint
Idothea tricuspidata	Arthropod	0.75	1.00	Swimming
Pleuronectes platessa	Fish	0.77	1.20	Swimming
Rana esculenta	Water frog	0.78	1.14	Narcosis
Cyclopterus lumpus	Fish	0.80	1.20	Narcosis
Cydippe pileus	Cleonphones	0.82	0.97	Tentacle paralysis
Rana fusca	Tadpole	0.82	1.16	Narcosis
Triton vulgaris	Water salamander	0.83	1.12	Narcosis
Rana agilis	Tadpole	0.85	1.14	Narcosis
Sepiola rondeletti	Cephalopod	0.88	1.25	Narcosis
Actinia equina	Sea anemone	0.90	1.02	Tentacle paralysis
Phoxinus laevis	Minnow fish	1.00[c]	0.89[o]	Narcosis

[a] Data from Fühner (79). [b] Data for ethanol (log P = −0.235) and heptanol (log P = 2.410) were used to algebraically obtain the slope (A) and the intercept (B) for the equation log $(1/C) = A$ log $P + B$, where C is the limiting narcosis producing concentration in moles/L.
[c] Calculated by regression analysis from data on 5 alcohols.
[d] Calculated from data for ethanol and heptanol. [e] Calculated by regression analysis from data for 10 alcohols.

Limiting water solubility also accounts for the appearance of a cutoff or abrupt change from toxic to nontoxic within a series of compounds (Figure 2). Overton demonstrated, however, that prior to the appearance of such a cutoff, an increase in partition coefficient, successively lower concentrations of toxicant achieve this same effect. This is accompanied by the need for increasing test duration to achieve pseudo-steady-state partitioning between the site of action and the aqueous test solution.

Influence of Melting Point on Water Solubility Cutoff

The water solubility cutoff for an organic compound is a function of the minimum concentration inducing narcosis and the water solubility (Figure 2). The latter is related to both the partition coefficient and the melting point of the compound (83). Melting point is a measure of the enthalpy of fusion and reflects the strength of intermolecular forces present in the crystalline state. These forces are considerably strengthened by the presence of hydrogen bonds. Moreover, compounds containing aliphatic chains and other functional groups offering multiple degrees of conformational freedom and lacking symmetry must lose these degrees of freedom in passing from the liquid to the solid state (84-85). For example, phenanthrene and anthracene are isomers having the same partition coefficient (log P = 4.49). While the former produced narcosis at 0.0112 mmoles/L, the latter showed no effect at saturation, regardless of the duration of exposure (37,40). Phenanthrene possesses a lower degree of symmetry than anthracene, and a considerably lower melting point. The former melts at 98°, and the latter at 218° (37).

Bilinear Relationships and Pharmacokinetic Cutoff

The toxicity of chemicals to aquatic organisms acting by a narcosis mechanism exhibits linear QSAR relationships with tests of sufficient duration to achieve pseudo-steady-state partitioning between the aquarium aqueous donor phase and the site of action within the fish. For shorter test durations, a higher concentration is required to permit the needed internal toxic concentration to be achieved, resulting in a non-linear relationship between toxicity and log P. Such data have been fit by a bilinear model (86). This relationship is illustrated in Figure 2. If the toxic concentration required at this shorter test duration exceeds the water solubility, no toxicity will be observed in a saturated solution, even though it will be seen at lower concentration in tests of longer duration. Thus, pentachlorobenzene was reported to be nontoxic at saturation in a 96-hour test (60), but exhibited a 14-day LC50 of 0.18 mg/L (58). Bilinear relationships are also observed in rat oral LD50 (Figure 5), where there is also a lack of equilibrium partitioning between the site of administration and the site of action (64,87).

Excess Toxicity

Narcosis represents the most fundamental molecular mechanism of the toxicity of nonelectrolyte organic compounds. QSAR correlations derived

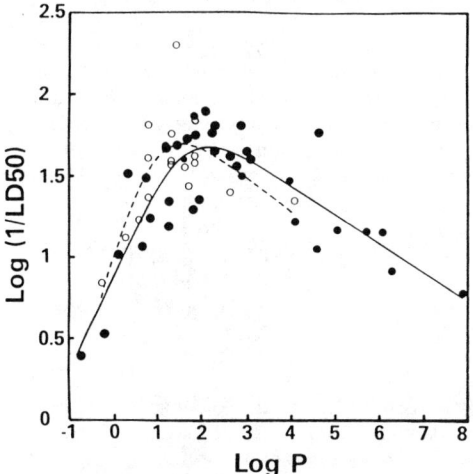

Figure 5. Bilinear relationship between octanol/water partition coefficient and rat oral LD50 of saturated monohydric alcohols and saturated monoketones on a log log scale. (Reproduced with permission from Ref. 64. Copyright 1987 Elsevier).

from data on saturated monohydric alcohols and other classes well-associated with a narcosis mechanism can be used as a probe to examine the toxicological behavior of other nonelectrolyte organic compounds (Figure 2). Mechanistic organic chemistry provides a rational basis for uncovering the molecular mechanism of nonelectrolytes more toxic than predicted. Nonelectrolyte organic compounds acting by a more specific mechanism can be identified by comparing their toxicity with that predicted by a baseline narcosis equation, according to the following equation,

$$T_e = C_{pred}/C_{obs} \qquad (3)$$

where T_e is the excess toxicity parameter (63), and C_{pred} and C_{obs} are the baseline QSAR predicted and observed toxicities, respectively. The excess toxicity of most nonelectrolytes can be rationalized in terms of an electrophile or proelectrophile molecular mechanism, using mechanistic organic chemistry as a basis for predicting such reactivity (81,88-89).

Electrophile Mechanism. Electrophile toxicants undergo direct covalent bond formation with nucleophilic functional groups such as sulfhydryl that are present in enzymes and other critical biological macromolecules. A schematic representation of the biochemical and physiological processes involved are illustrated in Figure 6. Electrophilicity has been associated for some time with biological activity (92). The compounds p-nitrobenzylpyridine (90) and p-nitrothiophenol (91) have been employed as model nucleophiles, and the pseudo-first order reaction rates have been used as parameters to correlate with electrophilic toxicity. Examples of three types of electrophile toxicants, those undergoing direct nucleophilic displacement, Michael-type addition, and Schiff base formation with amino groups, such as ε-amino groups of lysine are shown in Table II.

Within a series of electrophiles bearing a common reactive functional group in a similar electronic environment, excess toxicity decreases with increasing partition coefficient. Thus, acetaldehyde has an excess toxicity of 131, and butyraldehyde, 45 (88). A modification in the electronic environment of a functional group that results in increased electrophilicity can produce an increased excess toxicity, even for a compound with a higher partition coefficient. Thus, pentafluorobenzaldehyde has both a higher log P and greater excess toxicity than trimethoxybenzaldehyde. The former, however, is considerably more reactive due to the presence of five strongly electron withdrawing groups compared to three electron donating groups in the latter (88). There is a similar trend in comparing the Michael-type acceptors acrolein, 2-hydroxyethyl acrylate, and acrylamide. Likewise, allyl bromide and allyl chloride have almost the same log P value, but differ greatly in their excess toxicity due to the greater ability of bromide relative to chloride to act as leaving group (63). Work is in progress in quantifying these differences based upon electronic molecular descriptors.

Proelectrophile Mechanism. Proelectrophiles, which produce electrophiles by means of metabolic transformation, represent a second class

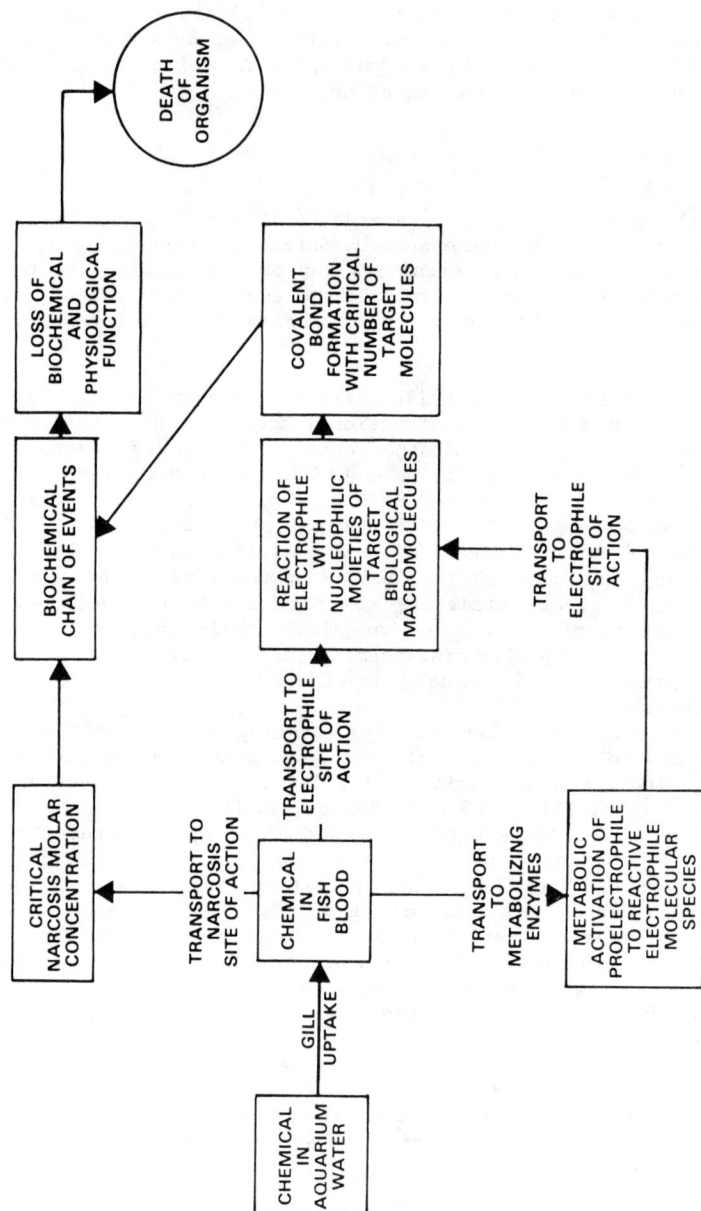

Figure 6. Schematic representation of the physiological and biochemical processes involved in the production of lethality to fish from toxicants acting by narcosis, electrophile, and proelectrophile mechanisms.

of nonelectrolytes showing excess toxicity. Some examples of proelectrophiles are illustrated in Table III. Allyl alcohol undergoes metabolic oxidation by the ubiquitous enzyme alcohol dehydrogenase to acrolein, a Michael-type acceptor electrophile (63), shown in Table II. It has been proposed that the excess toxicity of 1,3-dibromopropane (88) reflects the same mechanism as was proposed for its mutagenicity, i.e., metabolism to the electrophilic four membered sulfonium cation. The excess toxicity of pentaerythritol triallyl ether can be accounted for based upon metabolism via a free radical mechanism to the corresponding hemiacetal, which can readily cleave to yield acrolein (63).

Cyanogenic Mechanism. The acute toxicity of a third class of simple nonelectrolytes can be readily rationalized in terms of a cyanogenic mechanism, involving metabolic release of cyanide. Free cyanide is highly toxic to fish with a TLm of 0.69 mg/L (93). Some examples of cyanogenic toxicants are shown in Table IV. Lactonitrile is functionally similar to a hemiacetal, and can readily lose cyanide by a simple mechanism. In the case of malononitrile and allyl cyanide, free radical metabolic oxidation yields a stable species which on free radical perhydroxylation becomes functionally equivalent to lactonitrile.

Chronicity Ratio. 24-Hr LC50/96-Hr LC50

Aquatic toxicologists frequently report fish toxicity data for 24 and 96-hr durations. The chronicity ratio, R, may be defined as,

$$R = LC50_{24-h}/LC50_{96-h}$$

where $LC50_{24-h}$ and $LC50_{96-h}$ are the 24-h and 96-h toxicity values, respectively (88). Values of this ratio greater than 2 have been employed by aquatic toxicologists as an indicator of the potential need to perform chronic testing on a chemical to assess its potential risk if released into the environment (95). An analysis of data for 37 nonelectrolytes demonstrated a correlation between excess toxicity and a chronicity ratio greater than 2 (88). This finding is consistent with the hypothesis that chronicity can result from covalent binding by an electrophile or proelectrophile toxicant, or by another more specific mechanism such as metabolic release of cyanide. These results are illustrated graphically in Figure 7. All chemicals in this study (88) with a T_e value less than 6.7 have a chronicity ratio (R) of 2 or less, while the majority of those with T_e values greater than 6.7 show chronicity ratios greater than 2. R represents only a single temporal probe for investigating cumulative toxicity. Certain highly reactive chemicals having R values less than 2 may produce cumulative damage within shorter test durations. Conversely, if the rate of metabolic transformation to toxicant is slow or requires a complex series of steps, longer test durations may be required to assess such potential chronicity.

Use of Screening Data to Investigate the Limitations of the Narcosis Mechanism

The use of QSAR for predictive purposes is placed upon a more secure foundation if a rationale can be developed for why the candidate

Table II. Electrophiles and Excess Toxicity[a]

Chemical	Log P	T_e
Nucleophilic Substitution		
Ethylene oxide	−0.792	490
Propylene oxide	−0.273	97
Allyl bromide	1.590	>490
Allyl chloride	1.450	35
Chloroacetonitrile	0.219	1960
α,α-Dichloroxylene	3.866	86
Phenyl disulfide	4.440	21
3,4-Dichloro-1-butene	1.972	11
Michael-type addition		
Acrolein	0.101	>81000
2-Hydroxyethyl acrylate	−0.058	1540
Acrylamide	−0.859	180
Schiff base formation		
Acetaldehyde	−0.224	131
Butyraldehyde	0.834	45
Pentafluorobenzaldehyde	2.449	51
α,α,α-Trifluoro-m-tolualdehyde	2.594	40
Benzaldehyde	1.495	31
2,4,5-Trimethoxybenzaldehyde	1.381	11

[a] Calculations of T_e adapted from refs. 63 and 88, which were based upon linear or bilinear baseline toxicity QSAR models.

Table III. Proelectrophiles and Excess Toxicity[a]

Chemical	Log P	T_e
Allyl alcohol	−0.250	16000
1,3-Dibromopropane	1.987	87
Pentaerythritol triallyl ether	−1.599	18000

[a] Calculations of T_e adapted from refs. 63 and 88, which were based upon linear or bilinear baseline toxicity QSAR models.

Table IV. Cyanogenic Nonelectrolytes and Excess Toxicity

Chemical	Log P	T_e
Lactonitrile[a]	-0.850	23800[c]
Malononitrile[b]	-1.198	88700
Allyl cyanide	0.120	16

[a] Data from 93. [b] Data from 94. [c] Average of three 96-h LC50 values.

chemical is likely to act by the same mechanism as those chemicals used to derive the equation. In the case of nonreactive nonelectrolytes, the identification of related compounds that have been reported to produce narcosis, anesthesia, or act as hypnotics (96) can be valuable in establishing the likelihood of a baseline narcosis mechanism. Fish toxicity screening data have proven especially useful in assessing the limitations of the narcosis QSAR baseline toxicity model to these organisms (97). Data retrieved from four such studies (98-101) were used to investigate the utility and limitations of the narcosis model for alcohols containing no other heteroatom functional groups. Although the data for most of these compounds proved to be consistent with the model, primary and secondary propargylic alcohols were found to show excess toxicity.

Acetylenic Alcohols. The finding of a more specific mechanism for primary and secondary propargylic alcohols was confirmed in a subsequent study on the fathead minnow in which 96-h LC50 values were determined for a series of acetylenic alcohols. All primary and secondary propargylic alcohols were found to produce excess toxicity, consistent with a proelectrophile mechanism involving transformation by alcohol dehydrogenase in the fish to the corresponding α,β-unsaturated propargylic aldehyde or ketone electrophilic toxicant. Furthermore, the toxicities of tertiary propargylic alcohols, which cannot undergo this metabolic transformation, were very close to the QSAR predictions. Primary homoproparglic alcohols were also found to produce excess toxicity. A mechanism involving tautomerization of the aldehyde metabolite to the conjugated allene was proposed. This allene can act as a Michael-type acceptor electrophile (Veith, G.D.; Lipnick, R.L.; Russom, C.L. *Xenobiotica*, in press).

Local QSAR Models. Fish toxicity screening data have also been employed to investigate the limitations of "local" QSAR models (Figure 2) for phenols (102) and anilines (103) using these same data. Most of these data are from studies conducted or sponsored by the U.S. Fish and Wildlife Service (USFWS). The Service has screened thousands of chemicals for effects using a wide range of organisms and routes of

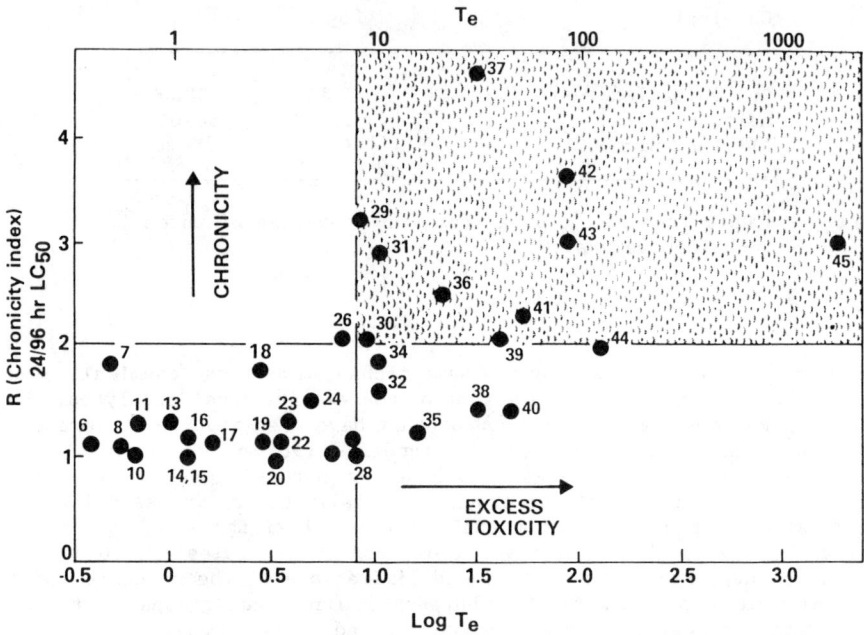

Figure 7. Relationship between excess toxicity (T_e) and chronicity ratio (R) for the toxicity of 37 nonelectrolyte organic compounds to the fathead minnow as follows (by increasing T_e): 2,4,5-Trimethyloxazole (5), N,N-Dimethyl-p-toluidine (6), 4-Methyloxazole (7), m-Bromobenzamide (8), 2-Fluorotoluene (9), n-Butyl sulfide (10), 1,2-Dichloropropane (11), 4,7-Dithiadecane (12), Naphthalene (13), 1,3-Dichloropropane (14), 4-Benzoylpyridine (15), 1,1,1,3,3,3-Hexafluoropropan-2-ol (16), t-Butyl disulfide (17), Butan-2-one oxime (18), s-Trioxane (19), Furan (20), 2,4-Dichlorobenzamide (21), Pyrrole (22), Deca-1,9-diene (23), 2-Cyanopyridine (24), 2-Adamantanone (25), Hexanal (26), Adamantane (27), 1,4-Dicyanobutane (28), Chloromethylstyrene (29), 2,6-Diphenylpyridine (30), 3,4-Dichlorobut-1-ene (31), 2,4,5-Trimethoxybenzaldehyde (32), 4-Nitrobenzamide (33), Pentane-2,4-dione (34), Allyl cyanide (35), Phenyl disulfide (36), Benzaldehyde (37), 2,5-Diphenylfuran (38), α,α,α-Trifluoro-m-tolualdehyde (39), Butanal (40), Pentafluorobenzaldehyde (41), α,α-Dichloro-p-xylene (42), 1,3-Dibromopropane (43), Acetaldehyde (44), and Chloroacetonitrile (45). (Figure adapted from data in Tables 2 and 3 in Ref. 88).

administration (104). In the early 1950s, Wood (98) tested 3,400 chemicals for toxicity to 3-4 species of freshwater fishes, followed by work of Hollis and Lennon on these same organisms (99), both at the Service's Laboratory in Kearneysville, WV. The chemicals were subsequently transferred to another USFWS laboratory in Michigan and used as part of a study to search for a selective lampricidal agent (100). The most toxic chemicals were then sent to the USFWS Laboratory in Galveston, Texas, and were tested in search of an agent selectively toxic to the red tide dinoflagellate (105); and subsequently to the University of Idaho, where they were tested for selective toxicity to various fish species (101). The use of chemicals as piscicidal agents has been reviewed in a Russian publication (106).

Conclusion

QSAR models can be a valuable means of predicting the toxicity of untested nonelectrolyte organic chemicals. The models need to be derived from a series of chemicals acting by a common molecular mechanism and encompass an adequate domain of spanned substituent space in their physical and chemical properties. The acute toxicity of many classes of nonelectrolytes is consistent with a narcosis or baseline toxicity mechanism. The ability to apply such models for predictive purposes also requires information to suggest that the candidate chemical acts by the same mechanism.

Acknowledgments

The author is grateful to Dr. James H. Gilford, Chief, Environmental Effects Branch, U.S. Environmental Protection Agency for his encouragement in the preparation of this article. The author wishes to thank Elsevier Publishing Company for kindly permitting the reproduction of Figures from *QSAR in Drug Design and Toxicology* (64).

Literature Cited

1. Lipnick, R.L. *Environ. Toxicol. Chem.* **1985**, *4*, 255-257.
2. Veith, G.D. *State-of-the-Art of Structure Activity Methods Development*. U.S. Environmental Protection Agency, Environmental Research Laboratory, Duluth, MN. 1981; EPA-560/81-029; NTIS PB 81-187-239.
3. Auer, C.M.; Gould, D.H. *J. Envir. Sci. Hlth.* **1987**, *C5(1)*, 29-71.
4. *Toxic Substances Control Act*, Public Law 94-469, October 11, 1976.
5. *Structure-Activity Relationships in Toxicology and Ecotoxicology: An Assessment.* European Chemical Industry Ecology and Toxicology Center, Brussels, February 24, 1986, Monograph No. 8, 86 pages; ISSN 0773-6347.
6. Hansch, C. In *Biological Activity and Chemical Structure*; Keverling Buisman, J.A., Ed.; Elsevier: Amsterdam, 1977; pp. 47-61.
7. Ariëns, E.J. In *Innovative Approaches in Drug Research*; Harms, A.F., Ed.; Elsevier: Amsterdam, 1986; pp. 9-22.
8. Drummond, R.A.; Russom, C.L.; Geiger, D.L.; DeFoe, D.L. In *Aquatic Toxicology and Environmental Fate: Ninth Volume*, Peston, T.M.; Purdy, R. Eds.; STP 921, American Society for Testing and Materials: Philadelphia, 1986; pp. 415-435.

9. McKim, J.M.; Schmieder, P.K.; Niemi, G.J.; Carlson, R.W.; Henry, T.R. *Environ. Toxicol. Chem.* **1987**, *6*, 295–312.
10. McKim, J.M.; Schmieder, P.K.; Niemi, G.J.; Carlson, R.W.; Henry, T.R. *Environ. Toxicol. Chem.* **1987**, *6*, 313–328.
11. McKim, J.M.; Bradbury, S.P.; Niemi, G.J. *Environ. Health Persp.* **1987** *71*, 171–186.
12. Hansch, C.; Klein, T.E. *Acc. Chem. Res.* **1986**, *19*, 392.
13. Broderius, S.; Kahl, H. *Toxicology* **1985**, *6*, 307–322.
14. Könemann, H. *Ecotox. Environ. Safety* **1980**, *4*, 415–421.
15. Könemann, H. *Toxicology* **1981**, *19*, 229–238.
16. Hermens, J.; Canton, H.; Jansen, P.; DeJong, R. *Aquat. Toxicol.* **1984**, *5*, 143–154.
17. Magee, P.S. *Chemtech* **1981**, *11*, 378–384.
18. Fränkel, S. *Die Arzneimittel-Synthese: Auf Grundlage der Beziehungen zwischen Chemischen Aufbau und Wirkung;* 4th edition; Julius Springer: Berlin, 1919; pp. 26–135.
19. W.A. Sexton, *Chemical Constitution and Biological Activity;* 3rd edition; Van Nostrand: Princeton, 1963.
20. Albert, A. *Selective Toxicity: The Physicochemical Basis of Therapy;* 7th edition, Chapman and Hall: London, 1985.
21. Cros, A.F.A. *Action de l'alcool amylique sur l' organisme;* Thesis, Faculté de Médecine de Strasbourg. Strasbourg, 1863.
22. Richardson, B.W. *Medical Times and Gazette (London)* December 18, 1869, *2*, 703–706.
23. Munch, J.C.; Schwartze, E.W. *J. Lab. Clin. Med.* **1925**, *10*, 985–996.
24. Rabuteau. *L'Union Medicale* **1870**, *10 (3rd series)*, 165–173.
25. Dujardin–Beaumetz; Audigé. *C.R. Hebd. Seances Acad. Sci.* **1875**, *81*, 192–194.
26. Dujardin–Beaumetz; Audigé. *C.R. Hebd. Seances Acad. Sci.* **1876**, *83*, 80–81.
27. Dujardin–Beaumetz; Audigé. *Researches Expérimentelles sur la Puissance Toxique des Alcools;* Octave Doin: Paris, 1879.
28. Houdaille, G. *Étude Expérimentelle et Critique sur les Nouveaux Hypnotiques;* Thesis, Faculty de Medécine de Paris, Paris, 1893.
29. Richet, C. *C.R. Soc. Biol. (Paris)* **1893**, *54*, 775–776.
30. Overton, E. *Vierteljahrsschr. Naturforsch. Ges. Zuerich* **1899**, *44*, 88–135.
31. Overton, E., *Studien über die Narkose, zugleich ein Beitrag zur allgemeiner Pharmakologie;* Gustav Fischer: Jena, Germany, 1901.
32. Meyer, H. *Arch. Exp. Pathol. Pharmakol.* **1899**, *42*, 109–118.
33. Baum, F. *Arch. Exp. Pathol. Pharmakol.* **1899**, *42*, 119–137.
34. Dunzelt, W. *Vergleichende Experimentaluntersuchungen über die Stärke der Wirkung einiger Narcotica;* Inaugural Dissertation, Hohen Medicinischen Facultät der Universität Marburg: Marburg, Germany, 1896.
35. Meyer, H. *Arch. Exp. Pathol. Pharmakol.* **1901**, *46*, 338–345.
36. Overton, E. *Vierteljahrsschr. Naturforsch. Ges. Zuerich* **1895**, *40*, 159–201.
37. Overton, E. *Vierteljahrsschr. Naturforsch. Ges. Zuerich* **1896**, *41*, 383–406.
38. Lipnick, R.L. *Trends Pharmacol. Sci.* **1986**, *5*, 161–164.
39. Leo, A.; Hansch, C.; Church, C. *J. Med. Chem.* **1969**, *2*, 766–771.
40. Hansch, C.; Dunn, W.J., III. *J. Pharm. Sci.* **1972**, *61*, 1–19.

41. Lipnick, R.L. In *Aquatic Toxicology and Hazard Assessment: 11th Symposium*; Suter, C.W., II, and Lewis, M., Eds.; STP 1007, American Society for Testing and Materials: Philadelphia, 1989; 468–489.
42. Winterstein, H. *Die Narkose: In Ihrer Bedeutung für die Allgemeine Physiologie*; 2nd edition; Julius Springer: Berlin, 1926.
43. Kochmann-Halle, M. In *Handbuch der Experimentelle Pharmakologie*; Heffter, A. Ed.; Julius Springer: Berlin, 1923; Vol. I, pp. 449–469.
44. Henderson, V.E. *Physiol. Rev.* 1930, *10*, 171–220.
45. McElroy, W.D. *Quart. Rev. Biol.* 1947, *22*, 25–58.
46. Butler, T.C. *Pharmacol. Rev.* 1950, *2*, 121–160.
47. Mullins, L.J. *Chem. Revs.* 1954, *54*, 289–323.
48. Featherstone, R.M.; Muehlbacher, C.A. *Pharmacol. Rev.* 1963, *15*, 97–121.
49. Seeman, P. *Pharmacol. Rev.* 1972, *24*, 583–655.
50. Ueda, I.; Kamaya, H. *Anesth. Analg.* 1984, *63*, 929–945.
51. Abu-Hamdiyyah, M. *Langmuir*, 1986, *2*, 310–315.
52. *Mécanisme de la Narcose*, Centre Nationale de la Recherche Scientifique, Proceedings of an International Colloquium: Paris, April 19–26, 1950 (No. XXVI), Paris, 1951.
53. Miller, K.W.; Smith, B.E. In *A Guide to Molecular Pharmacology-Toxicology*; Featherstone, M., Ed.; Marcel Dekker: New York, 1973; Part II, Chapter 11.
54. Miller, K.W. *Int. Rev. Neurobiol.* 1985, *27*, 1–61.
55. Franks, N.P.; Lieb, W.R. *Nature* 1982, *300*, 487–493.
56. Franks, N.P.; Lieb, W.R. *Trends Pharmacol. Sci.* 1987, *8*, 169.
57. Martin, Y.C. *Quantitative Drug Design*; Marcel Dekker: New York, 1978.
58. Könemann, H. *Toxicology* 1981, *19*, 209–221.
59. Lipnick, R.L.; Dunn, W.J., III. In *Quantitative Approaches to Drug Design*; Dearden, J.D., Ed.; Elsevier: Amsterdam, 1983; pp. 265–266.
60. Veith, G.D.; Call, D.T.; Brooke, L.T. *Can. J. Fish. Aquat. Sci.* 1983, *40*, 743–748.
61. Veith, G.D.; Call, D.T.; Brooke, L.T. In *Aquatic Toxicology and Hazard Assessment: Sixth Symposium*; Bishop, W.E., Ed.; STP 802, American Society for Testing and Materials: Philadelphia, 1983; pp. 90–97.
62. Hermens, J.; Canton, H.; Jansen, P.; DeJong, R. *Aquat. Toxicol.* 1984, *5*, 143–154.
63. Lipnick, R.L.; Watson, K.R.; Strausz, A.K. *Xenobiotica* 1987, *17*, 1011–1025.
64. Lipnick, R.L.; Pritzker, C.S.; Bentley, D.L. In *QSAR in Drug Design and Toxicology*; Hadzi, D. and Jerman-Blazic, B., Eds.; Elsevier: Amsterdam, 1987; pp. 301–306.
65. Leo, A.; Hansch, C.; Elkins, D. *Chem. Revs.* 1971, *71*, 575.
66. Leo, A.; Hansch, C. *J. Org. Chem.* 1971, *30*, 1539.
67. Meyer, K.H.; Hemmi, H. *Biochem. Z.* 1935, *277*, 39.
68. Rekker, R.F. *The Hydrophobic Fragmental Constant*; Elsevier: Amsterdam, 1977.
69. Hansch, C.; Leo, A. *Substituent Constants for Correlation Analysis in Chemistry and Biology*, Wiley-Interscience: New York, 1979.
70. Leo, A. *J. Chem. Soc. Perkin Trans. II* 1983, 825–838.
71. Leo, A. *J. Pharm. Sci.* 1987, *76*, 166–168.

72. Chou, J.T.; Jurs, P.C. *J. Chem. Inf. Comput. Sci.* **1979**, *19*, 172-178.
73. Leo, A.; Weininger, D. *Medchem Software Release 3.33*, Medicinal Chemistry Project, Pomona College, Claremont, CA, 1985.
74. Veith, G.D.; Austin, N.M.; Morris, R.T. *Water Res.* **1978**, *13*, 43-47.
75. Burkhard, L.P.; Kuehl, D.W.; Veith, G.D. *Chemosphere* **1985**, *14*, 1551-1560.
76. Tipker, J.; Groen, C.P.; Van Den Bergh-Swart, J.K.; Van Den Berg, J.H.M. *J. Chromatogr.* **1988**, *452*, 227-239.
77. Kamlet, M.J.; Doherty, R.M.; Taft, R.W.; Abraham, M.H.; Veith, G.D.; Abraham, D.J. *Environ. Sci. Technol.* **1987**, *21*, 149-155.
78. Lipnick, R.L.; Hood, M.T. *Correlation of chemical structure and toxicity of industrial organic compounds to daphnia, algae, bacteria, and protozoa;* Abstracts of papers, Seventh Annual Meeting, Society of Environmental Toxicology and Chemistry, Alexandria, VA, November 2-5, 1986.
79. Führner, H. *Z. Biol.* **1912**, *57*, 465-494.
80. Topliss, J.G.; Costello, R.J. *J. Med. Chem.* **1972**, *15*, 1066-1068.
81. Lipnick, R.L. In *QSAR in Toxicology and Xenobiochemistry;* Tichý, M., Ed.; Elsevier: Amsterdam, 1985; pp. 39-52.
82. Thurston, R.V.; Gilfoil, T.A.; Meyn, E.L.; Zajdel, R.K.; Aoki, T.I.; Veith, G.D. *Water Res.* **1985**, *19*, 1145-1155.
83. Banerjee, S.; Yalkowsky, S.H.; Valvani, S.S. *Environ. Sci. Technol.* **1980**, *14*, 1227-1229.
84. Herbrandson, H.F.; Nachod, F.C. In *Determination of Organic Structures by Physical Methods,* Braude, E.A. and Nachod, F.C., Eds.; Academic: New York, 1955; pp. 3-23.
85. Valvani, S.C.; Yalkowsky, S.H. In *Physical Chemical Properties of Drugs;* Yalkowsky, S.H., Sinkula, A.A. and Valvani, S.C., Eds.; Marcel Dekker: New York, 1980; pp. 201-229.
86. Kubinyi, H. *J. Med Chem.* **1977**, *20*, 625-629.
87. Lipnick, R.L.; Pritzker, C.S.; Bentley, D.L. In *QSAR and Strategies in the Design of Bioactive Compounds;* Seydel, J.K., Ed.; VCH: Weinheim, FRG, 1985, pp. 420-423.
88. Lipnick, R.L. In *Risk Assessment of Chemicals in the Environment;* Richardson, M.L., Ed.; Royal Society of Chemistry: London, 1988; pp. 379-397.
89. Lipnick, R.L. *Environ. Toxicol. Chem.* **1989**, *8*, 1-12.
90. Hermens, J.; Busser, F.; Leeuwanch, P.; Musch, A. *Toxicol. Environ. Chem.* **1985**, *9*, 219-236.
91. Cheh, A.M.; Carlson, R.E. *Anal. Chem.* **1981**, *53*, 1001.
92. Ross, W.C.J. *Biological Alkylating Agents: Fundamental Chemistry and the Design of Compounds for Selective Toxicity;* Butterworths: London, 1962.
93. McKee, J.E.; Wolf, H.W. *Water Quality Criteria;* 2nd ed., The Resources Agency of California, State Water Resources Control Board: Sacramento, CA, 1963.
94. *Acute Toxicities of Organic Chemicals to Fathead Minnows (Pimephales promelas);* Brooke, L.T.; Call, D.J.; Geiger, D.L.; Northcott, C.E., Eds.; Center for Lake Superior Environmental Studies, University of Wisconsin-Superior, vol 1, 1984.

95. Birge, W.J.; Black, J.A. In *Aquatic Toxicology and Hazard Assessment: Eighth Symposium*; Bahner, R.C. and Hansen, D.J., Eds.; STP 891; American Society for Testing and Materials: Philadelphia, 1985; p. 51.
96. Wheeler, K.W. In *Medicinal Chemistry*; Campaigne, E.E. and Hartung, W.H., Eds.; Wiley: New York, 1963; Vol. VI, pp. 1-245.
97. Lipnick, R.L.; Johnson, D.J.; Gilford, J.H; Bickings, C.K.; Newsome, L.D. *Environ. Toxicol. Chem.* 1985, 4, 281-296.
98. Wood, E.M. *The Toxicity of 3400 Chemicals to Fish*; U.S. Fish and Wildlife Service, Kearneysville, WV, 1953; In EPA Report No. 560/6-87-002; NTIS PB 87-200-275.
99. Hollis, E.H.; Lennon, R.E. *The Toxicity of 1085 Chemicals to Fish*; U.S. Fish and Wildlife Service, Kearneysville, WV, 1953; In EPA Report No. 560/6-87-002; NTIS PB 87-200-275.
100. Applegate, V.C.; Howell, J.H.; Hall, A.E., Jr. *The Toxicity of 4,346 Chemicals to Larval Lampreys and Fishes*; Special Scientific Report, Fisheries No. 207, U.S. Fish and Wildlife Service: Washington, DC, 1957.
101. MacPhee, C.; Ruelle, R. *Lethal Effects of 1888 Chemicals upon Four Species of Fish from Western North America*; University of Idaho; Forest, Wildlife, and Range Experiment Station: Moscow, ID, Bulletin No. 3, 1969.
102. Lipnick, R.L.; Bickings, C.K.; Johnson, D.E.; Eastmond, D.A. In *Aquatic Toxicology and Hazard Assessment: Eighth Symposium*; Bahner, R.C.; Hansen, D.J., Eds.; STP 891; American Society for Testing and Materials: Philadelphia, 1985; pp. 153-176.
103. Newsome, L.D.; Johnson, D.E.; Cannon, D.J.; Lipnick, R.L. In *QSAR in Environmental Toxicology-II*; Kaiser, K.L.E., Ed.; D. Reidel: Dordrecht, Netherlands, 1987; pp. 231-250.
104. Walker, C.R.; Menzie, C.M.; Bowles, W.A., Jr. *J. Chem. Inf. Comput. Sci.* 1981, 21, 29-35.
105. Marvin, K.T.; Proctor, R.R., Jr. *Preliminary Results of the Systematic Screening of 4,306 Compounds as 'Red Tide' Toxicants*; Data Report 2; U.S. Fish and Wildlife Service, Bureau of Commercial Fisheries Biological Laboratory, Galveston, TX: Washington, DC, March, 1964.
106. Sudakova, E.V. *Izv. Gos. Nauchno-Issled. Inst. Ozern. Rechn. Rybn. Khoz.* 1977, 121, 97-132; *Chem. Abstr.* 1978, 89, 71897W.

RECEIVED July 14, 1989

Chapter 24

Correlations and Mechanisms of Chemical Toxicity in Animals

Philip S. Magee[1] and James W. King[2]

[1]BIOSAR Research Project, Vallejo, CA 94591 and School of Medicine, University of California, San Francisco, CA 94143
[2]U.S. Army Chemical Research, Development and Engineering Center, Aberdeen Proving Ground, MD 21010–5423

> Toxic response of animals to important chemical classes can be analyzed in terms of mechanistic descriptors, even with data from many different laboratories. We estimate that 8-10 data points are needed to support each descriptor with inter-laboratory data, a doubling of the requirement for single labs. Given the large data sets available from toxicity compilations like Sax and RTECS, we were able to achieve many successful correlations. Differences among animals or between different routes of administration were easily perceived in mechanistic terms. Important conclusions can be drawn in a three-way comparison of chemical class, animals and routes by examining the form of correlation and the mean level of toxicity.

Toxicity data derived from single laboratories on selected animal subjects are known to correlate well with mechanistic descriptors such as logP. Examples are reported for barbiturates on rats, mice and rabbits (1), bicyclophosphates on mice (2), 2,6-dialkylanilines on rats (3), and phenols on mice (4). Optimal dependance on logP is frequently observed.

Data reported from different laboratories may involve different strains, size, sex, age, vehicle and technique on the reported animal. All of these factors make unknown contributions to the variance of inter-laboratory data. Moreover, there is no consistent protocol for reporting toxicity data. Perusal of a recent compilation of pesticides (5) reveals LD50 toxicities reported as: about 50, 400-800, 3362, approximately 335, 318-557, 1.125, 568.9, 2900±800, 1717 (1366-2156) and 1077±78. All levels of confidence from rough estimates to precise knowledge are expressed in these reports. However reported, these data are converted to single LD50 numbers in RTECS (6) and Sax (7), with a bias toward selecting the lowest of several values. These compilations provide the only convenient source of large data sets of diverse chemical structure in most areas of interest. A key objective of this study is to analyze such sets in terms of mechanistic descriptors to assess differences between chemical classes, animals and routes of administration. We hope to achieve a partial analysis of the data without attempting to pick-up all the interlab variance related to animal and experimental factors.

Inter-laboratory toxicity data has been used to develop a general, non-mechanistic model for predicting rat LD50 ($\underline{8}$). Further, the correlatability of RTECS data in mechanistic terms was recently demonstrated by Lipnick in developing a bilinear logP relation for alcohol toxicity to rats (n = 127) ($\underline{9}$). One of our objectives is to uncover more complex mechanistic factors than the frequently reported dependance on logP.

Phenol Toxicity

In order to get a fix on the difference in variance, a set of simple phenols (n = 26) with ip toxicity to albino mice was selected to study the structure-activity variance for single laboratory data ($\underline{4}$). In a comparative study (n = 50), a diverse set of phenols with ip toxicity to "mice" provided the expected variance for inter-laboratory data ($\underline{7}$). Multiple regression was used to analyze all data sets. In the single lab data set, measured logP was found to correlate better than $\Sigma\Pi$, but the latter was acceptable and necessary for comparison with the interlab analysis. Molecular weight correction was found to be important for the single lab analysis and is used throughout this study. Corrections for phenol ionization and for incremental toxicity of nitrophenols are important in the interlab analysis. In both sets, the phenol toxicity depends on a low positive slope of $\Sigma\Pi$ without curvature over a broad structural variation. The lower slope observed for the interlab data may be a simple consequence of the weaker correlation (r = 0.764 vs. r = 0.931) which generally causes reduction in the regression coefficients. The interlab data set (n = 50) ($\underline{7}$) comes from 17 different sources and yet, the many animal and experimental variables do not obscure the correlation with mechanistic descriptors.

Albino mouse, ip toxicity ($\underline{4}$)

Log MW/LD50 = 0.429 $\Sigma\Pi$ - 0.546
T = 12.49
n = 26   r = 0.931   F = 156.0

Mouse, ip toxicity ($\underline{7}$)

Log MW/LD50 = 0.199 $\Sigma\Pi$ - 0.196 $\Sigma\bar{\sigma}$ + 0.438 INO2 - 0.274
T = 5.65   4.47   4.12
n = 50   r = 0.764   F = 21.44

Mouse iv toxicity also correlates with a low positive slope of $\Sigma\Pi$. Corrections are needed for hydroxyphenols (IOH) and nitrophenols (INO2). Both corrections are positive suggesting the operation of intoxication mechanisms, perhaps via quinones for hydroxyphenols. The nitrophenol correction is similar to that observed for ip toxicity (0.438).

Mouse, iv toxicity (7)

Log MW/LD50 = 0.126 ΣΠ + 0.503 IOH + 0.605 INO2 + 0.060
       T =   4.57   5.55   4.86
       n = 30   r = 0.815   F = 17.13

By contrast, mouse, rat and guinea pig oral toxicities correlate only with reactivity factors, $\Sigma\bar{\sigma}$ and/or I26 (or $\upsilon_{2,6}$). The 2,6-factor appears to be more than steric as the indicator variable, I26, often correlates better than $\upsilon_{2,6}$. There is no dependance on lipophilicity.

Mouse, oral toxicity (7)

Log MW/LD50 = 0.425 $\Sigma\bar{\sigma}$ + 0.388 I26 - 0.707
       T =   7.14   3.05
       n = 42   r = 0.854   F = 52.54

Rat, oral toxicity (7)

Log MW/LD50 = 0.396 $\Sigma\bar{\sigma}$ + 0.520 I26 - 0.876
       T =   5.55   4.54
       n = 71   r = 0.729   F = 38.58

Guinea pig oral toxicity (n = 13) correlates only with the electronic effect (0.618 $\Sigma\bar{\sigma}$ - 0.822, r = 0.914), while dermal (rat) and sub-cutaneous (rat, mouse) toxicities all depend on positive slopes of $\Sigma\bar{\sigma}$ or $\upsilon_{2,6}$. Rabbit dermal depends on both.

Rabbit, dermal toxicity (7)

Log MW/LD50 = 0.482 $\Sigma\bar{\sigma}$ + 0.338 $\upsilon_{2,6}$ - 1.10
       T =   5.58   3.39
       n = 19   r = 0.889   F = 30.26

The transition between simple dependance on lipophilicity to exclusive dependance on reactivity factors led us to consider two separate classes of phenol toxicity as discussed in the conclusions. In addition, the clear dependance on steric or "ortho" effects for 2,6-substituents also brings the commonly accepted uncoupling of oxidative phosphorylation into question as a mechanism of death.

Diarylamine Rodenticides

Mouse oral data on diarylamine rodenticides was recently reported by Dreikorn and O'Doherty without any attempt at analysis (13). Set 1 (n = 14) has one 2,4,6-trinitrophenyl ring with substituents varied on the other ring. Set 2 (n = 18) differs in having one 2-CF_3,4,6-dinitrophenyl ring. These are presumed to be uncouplers of oxidative phosphorylation with NH-acidity as an important factor (14). One outlier was deleted from each set.

Mouse Oral (Set 1) (14)

$$\text{Log MW/LD100} = 1.30 \ \Sigma\bar{\sigma} - 0.284$$
$$T = 2.41$$
$$n = 13 \quad r = 0.588 \quad F = 5.81$$

Mouse Oral (Set 2) (14)

$$\text{Log MW/LD100} = 2.20 \ \Sigma\bar{\sigma} + 0.390$$
$$T = 7.82$$
$$n = 17 \quad r = 0.896 \quad F = 61.07$$

Both sets depend exclusively on an electronic effect controlling the population of the toxic anion, an effect analogous to that of phenols in uncoupling oxidative phosphorylation (14). The weaker coefficient of Set 1 relative to Set 2 is probably a consequence of the weaker correlation (r = 0.588 vs. r = 0.896), rather than a real difference in response. The data were extracted from bar graphs in the publication (13) and may have more error than tabulated values.

Aryl N-Methylcarbamate Toxicity

Correlations of insecticidal carbamates against the housefly (AChE pI50) (10) and brown planthopper (AChE pKd, Log1/LC50) (11,12) have revealed complex factors such as electronic and steric effects combined with selective hydrogen-bonding of polar substituents. These problems are complicated by the fact that ortho-, meta- and para-substituents correlate by separate mechanisms in honey-bee toxicity (P. S. Magee, unpublished study, 1986, PM330). It was not surprising then, to observe a similar complexity in a large set of aryl N-methylcarbamates (n = 27) against rats. Both steric (υ_2) and non-steric (I2) effects at the ortho-position were coupled with enhanced toxicity for H-bonding substituents (HB).

Rat, Oral Toxicity (7)

$$\text{Log MW/LD50} = -0.223 \ \Sigma\Pi + 1.35 \ \upsilon_2 - 1.40 \ \text{I2} + 0.438 \ \text{HB} + 0.493$$
$$T = 3.85 \quad 2.10 \quad 2.79 \quad 2.12$$
$$n = 27 \quad r = 0.765 \quad F = 7.78$$

In both mouse (n = 14) and chicken (n = 8) oral toxicities, the only discernible factor is a strong positive dependance on H-bonding groups (1.01 HB, 0.98 HB). None of the mechanistic factors were significant for these small sets. Active site binding is the principal mechanistic step in carbamate toxicity.

Organophosphate Toxicity

Analysis of large sets of oral toxicity data (7) for rats (n = 214), mice (n = 121), chicken (n = 65), guinea pig (n = 46) and rabbit (n = 39) was generally unsuccessful in identifying mechanistic factors. None of the derived equations are considered strong enough to provide a useful predictive model. The extreme diversity of structure within 18 subclasses of phosphates, phosphonates and phosphoramidates is presumed responsible for the failure. Future success will probably depend on factoring these sets into less diverse groups. The major useful output of these studies were the mean values of Log MW/LD50 used for inter-class, inter-animal and inter-route comparisons.

A study intended for comparison with the animal sets suggested the involvement of lipophilicity in phosphate toxicity. Data on honeybee toxicity of phosphates, phosphonates and phosphoramidates were acquired from UC Riverside (15). This data set from a single laboratory provided a measure of our method for a broad range of structural variation. Pi values of the groups around P→O were summed with a correction for P→S to provide a $\Sigma\Pi$ estimated to parallel logP. Seven outliers, all of unusual structure (10%), were eliminated during regression. Indicator variables for a phenolate leaving group (IOAr) and for phosphoramidates (INP) were important descriptors.

Honeybee Topical (Dust) (15)

Log MW/LD50 = $-0.146 \; \Sigma\Pi + 0.811 \; IOAr - 0.521 \; INP + 2.51$
T = 5.84　　　6.68　　　3.02
n = 66　　r = 0.691　　F = 18.85

Phosphates differ from carbamates with phosphorylation as the rate determining step. The involvement of logP ($\Sigma\Pi$) is probably related to transport rather than binding.

Aniline Toxicity

Aniline studies were disappointing, despite the availability of large data sets. We had hoped to compare anilines with phenols point by point, but the correlation quality does not permit it. The equation for rat oral toxicity is significant, though quite weak (100 r^2 = 35.3). It indicates a negative steric effect for 2,6-substituents, opposite to that observed for phenols, and a negative dependance on H-bonding substituents. This suggests involvement of the amino-group in toxicity with other H-bonding groups as disrupting factors.

Rat, Oral Toxicity (7)

Log MW/LD50 = $-0.239 \; \upsilon_{2,6} - 0.337 HB - 0.685$
T = 2.40　　　5.01
n = 53　　r = 0.594　　F = 13.66

Another study on mouse intraperitoneal toxicity is below standard significance and is shown only because it supports the negative steric effect for 2,6-substituents and shows a low dependance on $\Sigma\Pi$ similar to the phenols.

Mouse, ip Toxicity (7)

Log MW/LD50 = $0.191 \; \Sigma\Pi - 0.259 \; \Sigma\bar{\sigma} - 0.501 \; \upsilon_{2,6} - 0.062$
T = 2.26　　　1.84　　　1.84
n = 27　　r = 0.550　　F = 3.33

Despite these weak results, the mean value of Log MW/LD50 was evaluated for two animals (rat,mouse) and two routes (oral,ip) needed to locate anilines on the inter-comparison scales.

Pyridine Toxicity

Unlike anilines, the pyridine sets correlated easily and with great consistency. The consistent factor is a strong, negative dependance on substituent electronic behavior, indicating a nucleophilic toxicity mechanism. Correlations are remarkably high considering the small size of the sets (n = 10-14). The mouse iv set contains no sigma minus groups and depends simply on $\Sigma\sigma$. The rat oral set has a major outlier. Deletion strongly improves the correlation but does not alter the interpretation. Lipophilic dependance in the mouse ip relation suggests that other factors may emerge when larger sets are studied.

Mouse, iv Toxicity (7)

Log MW/LD50 = $-0.775\ \Sigma\sigma + 0.156$
T = 3.49
n = 14   r = 0.710   F = 12.19

Mouse, ip Toxicity (7)

Log MW/LD50 = $0.226\ \Sigma\Pi - 1.14\ \Sigma\bar{\sigma} + 0.288$
T = 2.51   5.01
n = 14   r = 0.881   F = 19.05

Rat, Oral Toxicity (7)

Log MW/LD50 = $-0.618\ \Sigma\bar{\sigma} - 0.999$
T = 2.66
n = 10   r = 0.686   F = 7.08

Log MW/LD50 = $-0.707\ \Sigma\bar{\sigma} - 1.08$
T = 8.45
n = 9   r = 0.955   F = 71.44

Special Inter-Lab/Inter-Animal Factors

Originally, we planned to study the special factors affecting toxicity from multiple laboratories in some detail. Different experimental techniques and animal variables clearly affect the toxicity end-point when carried out by different investigators. Of all the sets studied, phenol toxicity to mice (ip) seemed ideal for analysis because of the set size (n = 50) and simplicity of the class 1 mechanism. The project was abandoned on learning that the 50 data points come from 17 different laboratories. There are too many variables to support a systematic analysis.

However, it was possible to study one of the more interesting animal factors, the differential toxicity of chemicals to male and female of the same strain. Toxicity measures for 51 pesticides were assembled for 78 male/female pairs (65 rat, 10 mouse, 3 other). Of the 78 pairs, a simple count showed M>F (n = 37) and F>M (n = 40), an apparently equal distribution. The bias of higher toxicity to the female is revealed in summing Log MW/LD50 [Σ(F) = 8.05, Σ(M) = 5.60], and also in the slope of regression of M vs. F 0.940).. It is interesting that the expression correlates highly (100 r^2 = 93.0) and passes accurately through the origin after deletion of two extreme outliers. Normal variation in animal toxicity testing is

$$\text{Log MW/LD50(M)} = 0.940 \text{ Log MW/LD50(F)} - 0.014$$
$$T = 33.84$$
$$n = 76 \quad r = 0.969 \quad s = 0.233 \quad F = 1145$$

roughly a factor of 3.0 (log = 0.477) (8), which is nearly identical with 2s (0.466) as defined above. If the residuals are normally distributed, then 5% or 4 cases/76 would be expected to exceed 2s. There are actually 7 cases plus the two outliers or 9/78 (11.5%), which in combination with the female toxicity bias, indicates a real genetic factor.

In other studies, Hollingworth has documented colinear relations (r = 0.785-0.929) for pairs of animals (n = 9) tested against the same series of organophosphates (n = 7-32) (16). Toxicity orders can be deduced from these relations and from similar studies of 88 insecticides by Magee (P. S. Magee, unpublished study, 1976, PM 97). These orders are not general, however, even within the organophosphate class. Uchida and O'Brien determined the LD50 of the insecticide, Dimethoate, on seven animals and related the data accurately (r = 0.974) to the rate of liver degradation in each animal (17,18). This suggests that differential degradation can play a key role in defining animal toxicity orders. Major differences have also been reported in the toxic response to butylated hydroxytoluene (BHT) of five different strains of male mice (21).

It is rare to have data such as Hollingworth reports for most classes of compounds. In the usual data set, each animal is exposed to a different series of phenols, carbamates or phosphates, and colinear regressions are impossible. Moreover, not all data sets provide satisfactory correlations thereby eliminating the intercept as a reliable measure of relative animal toxicity. There is, however, one number that expresses the most probable toxicity of each class, namely the mean value of Log MW/LD50. For a random set of related compounds, this number should be fairly stable for n>30. These values are all taken at face value (no T-test) and there is obvious overlap in some comparisons. However, there is a remarkable degree of consistency in the orders observed to date.

The following inter-animal toxicity orders are based on mean Log MW/LD50 values.

Organophosphates (Oral)

Pigeon, Duck > Quail, Cat > Rat > Dog, Mouse > Gpig > Rabbit
(1.47) (1.43) (1.17) (1.02) (0.630) (0.521) (0.487) (0.366) (0.226)

Carbamates (Oral)   Chicken > Rat > Mouse
                        (0.921) (0.306) (0.227)

Phenols (Oral)   Mouse, Gpig > Rat > Rabbit
                  (-0.499)(-0.485)(-0.720)(-0.974)

Phenols (Dermal, iv, sc, ip)   Mouse > Rat > Rabbit

Anilines, Pyridines (Oral, ip)   Mouse > Rat

Inter-Route and Inter-Class Comparisons

The effect of administration route on drug action is discussed in some detail by Benet (19) and by Rowland (20). Oral administration forces a first-pass route through the liver, subjecting the toxicant to enhanced metabolism. Other routes are weaker metabolically, though in some cases, skin can display up to 80% or more of liver metabolite activity. In the rat, for example, skin is more efficient than liver in degrading aryl carbamates. Our results support this thesis in terms of mean Log MW/LD50 values for phenol toxicity but not for carbamate toxicity.

Phenols (Rat)   iv > ip > sc,   dermal > oral
                (0.373) (-0.157) (-0.429) (-0.469) (-0.720)

Phenols (Mouse)   iv > ip > sc > oral
                  (0.431) (0.048) (-0.130) (-0.499)

Phenols (Rabbit)   dermal > oral
                    (-0.869) (-0.974)

Carbamates (Rat)   oral >> dermal
                    (0.306)   (-0.463)

Inter-Class Comparisons. Classes of toxic compounds are easily ordered by the same measure.

Mouse (Oral)   Diarylamines > Phosphates > Carbamates > Phenols > Anilines
                (1.15)   (0.487)   (0.227)   (-0.499) (-0.723)

Rat (Oral)   Phosphates > Carbamates > Phenols > Anilines > Pyridines
              (0.630)   (0.306)   (-0.720) (-0.875)   (-2.88)

Birds (Oral)   Phosphates > Carbamates
                (1.17-1.47) (0.921)

Data are lacking for diarylamines and phosphates for other routes of administration.

Conclusions

Inter-laboratory toxicity data give structure-activity correlations of sufficient precision to classify mechanism and indicate the mode of death. Stronger correlations would be expected by using indicator variables for gender, age, size, etc., but these are unlikely to enhance the mechanistic description. Good analyses were achieved in all cases of sufficient size (n) and data span. With ten or more data points per descriptor, regression coefficients were uniformly strong and consistent among related sets.

Intraperitoneal and intravenous toxicity of phenols to the mouse depend on small positive slopes of $\Sigma\Pi$ with no observable optimum. This simple behavior (Class 1) cannot be causally defined but suggests absorption-desorption from a lipid pool as the rate-limiting step. All other toxicities explored (oral, dermal, sc) to mouse, rat, guinea pig and chicken correlate with positive slopes of $\Sigma\bar{\sigma}$ and/or 2,6-effects (Class 2). Diarylamines, anilines and pyridines also appear to behave as Class 2 toxicants against mice. These reactivity factors indicate target site expression, consistent with death by irreversible inhibition.

Oral toxicity of aryl N-methylcarbamates to the rat indicate a complexity similar to that observed against the housefly (10) and brown planthopper (11,12). Complex ortho-effects and strong H-bonding for some substituents combine to modify toxicity. The major mechanistic step in carbamate poisoning is strong binding to an AChE site. The parallel factors involved for insects and animals are consistent with death due to AChE inhibition.

Honeybee toxicity was amenable to partial analysis and gave hope that the animal data could be resolved. However, the bee set is far simpler in structural scope than those reported for animal toxicity. Failure to resolve these sets suggests that factoring into subsets will be needed to reduce the complexity of some 18 organophosphate classes.

Mean values of Log MW/LD50 were used to establish orders among animals, routes of administration and toxicant classes. This method seems ideal for comparing non-overlapping structural sets. Moreover, the orders appear useful for classifying mechanism. For example, inter-animal and inter-route orders confirm that phosphates and carbamates kill by a similar mechanism that differs from death by phenol toxicity. This is clearly indicated by reversals in (rat, mouse) and (oral, dermal) orders.

By combining mechanistic insight from regression analysis with toxicity orders from Log MW/LD50, we hope to develop an extensive knowledge base in three dimensions: animals, routes, toxicants. Work in the immediate future will continue with organophosphates and extend to organohalides, organomercurials, conjugated vinyls and nitroaromatics.

Adknowledgement. We wish to thank the Chemical Systems Laboratory (Aberdeen Proving Ground, MD) for generous support through Battelle Columbus Laboratories (Columbus, OH) under Delivery Order No. 1398.

Literature Cited

1. Hansch, C., Steward, A. R., Anderson, S. M. and Bentley, D., J. Med. Chem., 1967, 11, 1.
2. Casida, J. E., Eto, M., Moscioni, A. D., Engel, J. L., Milbrath D. S. and Verkade, J. G., Toxicol. Appl. Pharmacol, 1976, 36, 261.
3. Durden, J. A., J. Med. Chem., 1973, 16, 1316.

4. Biagi, G. L., Gandolfi, O., Guerra, M. C., Barbaro, A. M. and Cantelli-Forti, G. J. Med. Chem. 1975, 18, 868.
5. Spencer, E. Y., Guide to the Chemicals Used in Crop Protection, Publication 1093, Canadian Government Publishing Centre, Ottowa, Canada, 1982.
6. U. S. Department of Health and Human Services. Registry of Toxic Effects of Chemical Substances (RTECS), 1981-1982, 1983.
7. Sax, N. I. Dangerous Properties of Industrial Materials, 6th Edition, Van Nostrand Reinhold, New York 1984.
8. Enslein, K. and Craig, P. N. J. Environ. Path.-Tox. 1978, 2, 115.
9. Lipnick, R. L., Pritzker, C. S. and Bentley, D. L. In QSAR and Strategies in the Design of Bioactive Compounds; Seydel, J. K. Ed.; VCH Verlagsgesellschaft, West Germany, 1985.
10. Goldblum, A., Yoshimoto, M. and Hansch, C. J. Agric. Food Chem. 1981, 29, 277.
11. Nishioka, T., Fujita, T., Kamoshita, K. and Nakajima, M. Pest. Biochem. Physiol. 1977, 7, 107.
12. Kamoshita, K., Ohno, I., Kasamatsu, K., Fujita, T. and Nakajima, M. Pest. Biochem. Physiol. 1979, 11, 104.
13. Dreikorn, B. A. and O'Doherty, G. O. P. CHEMTECH 1985, 15, 424.
14. Hansch, C., Kiehs, K. and Lawrence, G. L. J. Am. Chem. Soc. 1965, 87, 5770.
15. Atkins, E. L. et al. Toxicity of Pesticides and Other Agricultural Chemicals to Honey Bees 1973; Reducing Pesticide Hazards to Honey Bees 1981, Agricultural Extension, University of California, Riverside.
16. Hollingworth, R. M. In Insecticide Biochemistry and Physiology; Wilkinson, C. F., Ed.; Plenum Press, New York, 1978, Chapter 12.
17. Uchida, T., Dauterman, W. C. and O'Brien, R. D. J. Agric. Food Chem. 1964, 12, 48.
18. Uchida, T. and O'Brien, R. D. Toxicol. Appl. Pharm. 1967, 10, 89.
19. Benet, L. Z. J. Pharmacokin. Biopharm. 1978, 6, 559.
20. Rowland, M. J. Pharm. Sci. 1972, 61, 70.
21. Kawano, S., Nakao, T. and Hiraga, K. Toxicol. Appl. Pharmacol. 1981, 61, 475.

RECEIVED June 14, 1989

INDEXES

Author Index

Blair, T. A., 232
Block, John H., 2,231,301
Bordás, Barna, 169
Craig, A. Morrie, 70
Denny, William A., 291
DesJarlais, Renee L., 60
Dixson, J. A., 157
Donoghue, Orla, 123
Doweyko, Arthur M., 82
Draber, W., 215
Franklin, P. H., 232
Glennon, Richard A., 264
Goldman, Mark E., 243
Hammock, Bruce D., 169
Han, Jian Hwa, 123
Henry, Douglas R., 26,58,70
Jordan, S., 354
King, James W., 281,301,390
Kral, R. M., 157
Kuntz, Jr., Irwin D., 60
Kurita, Yasuyuki, 183
Kyomura, Noboru, 136
Lavine, Barry K., 123
Leid, M., 232
Lipnick, Robert L., 366
Magee, Philip S., 37,134,136,147, 281,352,390

Matolcsy, György, 169
McKinney, J. D., 354
Murray, T. F., 232
Nishioka, Takaaki, 105
Oda, Jun'ichi, 105
Ohta, Hiroki, 136
Palmer, Brian D., 291
Pittel, B., 215
Plummer, E. L., 157
Ragsdale, Nancy N., 198
Rickenbacher, U., 354
Rosenblatt, Michael, 243
Seggel, Mark R., 264
Seibel, George L., 60
Siebenaller, J. F., 232
Sisler, Hugh D., 198
Sumi, Kazuo, 105
Székács, András, 169
Takahashi, Yoji, 136
Takayama, Chiyozo, 183
Trebst, A., 215
Tsushima, Kazunori, 183
Verloop, Arie, 301
Ward, Anthony J. I., 123
Wilson, William R., 291
Yu, Yupei, 301

Affiliation Index

Bayer AG, 215
BIOSAR Research Project, 37,134,136,147, 281,352,390
Clarkson University, 123
FMC Corporation, 157
Kyoto University, 105
Louisiana State University, 232
Merck Sharp and Dohme Research Laboratories, 243
Mitsubishi Kasei Corporation, 136
Molecular Design Limited, 26,58,70
National Institute of Environmental Health Sciences, 354
Oregon State University, 2,70,230,232,301

Plant Protection Institute of the Hungarian Academy of Sciences, 169
Ruhr-University Bochum, 215
Sumitomo Chemical Company, Limited, 183
U.S. Army Chemical Research Development and Engineering Center, 281,301,390
U.S. Department of Agriculture, 198
U.S. Environmental Protection Agency, 366
Uniroyal Chemical Co., Inc., 82
University of Auckland, 291
University of California–Davis, 169
University of California–San Francisco, 37,60,134 136,147,281,352,390
University of Maryland, 198
Virginia Commonwealth University, 264

Subject Index

A

A_1 adenosine receptor
appearance in chick atria, 235,236f
coupling to K^+ channels via guanine nucleotide regulatory protein, 237
involvement in negative chronotropic response of adenosine, 233
porcine atrial receptor characterization, 237–241
rank order potency of adenosine analogues, 233
use of radioligands for direct labeling, 233
A_2 adenosine receptor, rank order potency of adenosine analogues, 233
Acetolactate synthase
use of global sequence similarity to find inhibitor, 108
catalytic reactions, 108–109
Acetylcholine
development from biologically active molecules, 8–9
two-dimensional modeling, 152t,153f
Acetylcholine-binding site, physical nature, 151
Acetylcholinesterase inhibition, mechanism for organophosphate and organocarbamate, 147–155
Acetylenic alcohols, mechanism of toxicity, 383
Actinomycins, development from natural products, 3,5–6
Activation of drugs in bound state
binding energies, 39–40
calculation of molecular electrostatic potentials, 41
Cambridge Crystallographic Data File, 40–41
Acyclovir, selective toxicity, 13–14
Adenosine
cardiovascular effects, 233
function, 233
negative chronotropic properties, 233–237
receptor subtypes, 233
Adenosine analogues
effect of receptor antagonist on response, 234–235,236f
inhibition of spontaneous beating rate in chick atria, 234t
Adenosine-receptor-mediated cardiac responses, role of regulation of adenylyl cyclase activity, 234
Adenylyl cyclase, sensitivity to inhibition vs. embryogenesis, 235,237

Agricultural science, relationship with medicinal science, 37
Agrochemical research, development, 134
Aldicarb, two-dimensional modeling, 152t,153f
Amantadine, development, 18,20
Amino acid sequences, availability, 105
4-Aminobutyric acid
agonists, 282
antagonists, 282
binding sites, 281
classes of neural receptors for binding, 218
effect of stereochemistry on binding, 282
4-Aminobutyric acid receptor
classification, 281
kinetic identification, 281
4-Aminobutyric acid receptor site, structure–activity relationship analysis of binding, 281–288
4-Aminobutyric acid receptor site, mapping
bovine brain, 285
brain and cellular uptake, 286–287
cat spinal cord, 285–286
experimental procedure, 282,284
human brain, 284
mouse and rat brains, 285
noncompetitive binding, 286
rat spinal cord, 286
substructural binding effects, 287
Aniline mustard(s)
chemical reactivity, 291–293
cytotoxicity of substituted compounds, 296,297f,298
differential cytotoxicity, 298
metabolic stability of substituted compounds, 293
physicochemical and biological data, 293,294f,295t,296
QSAR of in vitro cytotoxicity, 292–298
structures, 295t
Aniline mustard alkylating agents
structure, 291,293
use as antitumor drugs, 291
Aniline toxicity
mouse intraperitoneal toxicity, 394–395
rat oral toxicity, 394
Antifungal agents, selective toxicity, 13
Antiinfective agents based on nalidixic acid structure
activities and residuals, 338,339–340t
activities for oxoquinoline-3-carboxylic acids, 321,323t
correlation between size of N and activity, 338,341

Antiinfective agents based on nalidixic acid structure—*Continued*
  correlation matrices, 334,335–336t
  correlation matrix for oxoquinoline-3-carboxylic acids, 321,326t
  data matrices for oxoquinoline-3-carboxylic acids, 306,310–312t
  de novo model, 341,345,346t
  descriptors important for activity at position 7, 321,327–329t,330
  development of QSAR model, 334,337t,338
  effect of ethyl substituent on activity, 345,347t
  effect of fluorine substitution on activity, 313,321
  effect of lipophilicity and molar refractivity on activity, 321,331–333t
  effect of substituents on activity, 334
  model development, 306
  parameter calculation, 304,306
  physicochemical parameters and indicator variables for naphthyridine-3-carboxylic acids, 313,315–320
  QSAR models, 341,342–344t
  regression analysis on oxoquinoline-3-carboxylic acids, 313,321,322t,324t
  residuals for oxoquinoline-3-carboxylic acids, 321,325t
  stability of regression coefficient, 330
  structures of azetidinyl-substituted naphthyridine-3-carboxylic acids, 313,318t
  structures of naphthyridine-3-carboxylic acids, 313,314t
  structures of oxoquinoline-3-carboxylic acids, 306,307–309t
Apomorphine, three-dimensional substructure query, 33,34f
Aryl-binding sites, differences between organophosphates and organocarbamates, 154
Aryl N-methylcarbamate(s)
  enzyme inhibition mechanism, 143–144
  enzyme inhibitor data, 141t,142
  resistance mechanism for green rice leafhopper, 137
  resistance vs. enzyme inhibition, 142–143
  structure, 137,141t
  structure–activity relationship, 137
Aryl N-methylcarbamate toxicity, rat oral toxicity, 393
N-Aryl oxadiazolones
  carbamate resistance mechanism of green rice leafhopper, 137

N-Aryl oxadiazolones—*Continued*
  enzyme inhibition data, 138,140t,143
  enzyme inhibition mechanism, 143–144
  structures, 137,138–140t
Atropine, development from natural products, 3–4
Azidothymidine, selective toxicity, 13

B

Bacterial diseases, use of selective toxicity, 15
Ballast position, definition, 168
Benzimidazoles
  mechanism of fungicidal activity, 206–207
  structures of fungicides, 206–207,208f
Benzoylurea insecticides, QSAR models, 163,165f
Binding, view as soft reaction, 44–45
Binding events
  modeling with physicochemical descriptors, 45–46
  modeling with quantum chemical descriptors, 46–47
Binding-site mapping
  methods, 48
  philosophy, 50–51
  statistical docking methods, 48,49–50f
Bioactive molecules
  development by selective toxicity, 10–16
  development from biochemically active molecules, 8–10,12
  development from natural products, 3–9
  developmental approaches, 2
  exploitation of secondary effects, 18,20
  metabolism of xenobiotic compounds, 16–18
  QSAR, 21–22
  receptor mapping, 19–21
  screening, 22–23
Bioactive substances
  definition, 230
  examples of end result, 230
Biochemically active molecules, development of bioactive molecules, 8–10,12
Biochemically specific fungicides
  development, 198
  resistance problems, 198–199
Biological activity, requirement, 61
Biological bases of local similarity of proteins
  amino acid sequence similarity among enzymes, 111,112t
  sequence–function relationships in proteins, 113
  sequence similarity due to molecular evolution, 111

INDEX

Biological bases of local similarity of
proteins—*Continued*
sequence similarity found in functionally
important regions, 111,113
Biological response, influencing
factors, 159
Biphenyl pyrethroids, dihedral angles vs.
activity, 183–184
Bordeaux mixture, development, 18
Bufencarb, two-dimensional modeling,
152t,153f

C

Calcium homeostasis, normal conditions,
243–245
Cambridge Crystallographic Database,
description, 61
Cambridge Crystallographic Data File, source
of valid molecular models, 40
Carboxamides
structure, 207,208f
structure–activity relationship, 209
Carboxin, structure, 207,208f
Cardiac glycosides, development from natural
products, 3–4
Cephalosporins, development from natural
products, 3,5–6
Chemical reactivity of aniline mustards
correlation with degree of electron
release to N, 291–293
effect of substituent electronic
properties, 292
rate constants for alkaline hydrolysis,
291–292
Chemical toxicity to animals
aniline toxicity, 394–395
aryl N-methylcarbamate toxicity, 393
diarylamine rodenticides, 392–393
organophosphate toxicity, 393–394
phenol toxicity, 391–392
pyridine toxicity, 395
special interlaboratory and interanimal
factors, 395–397
Chlorambucil, structure, 291,293
Chromones, π-densities and net charges,
221,222t,223,225
Chronicity ratio
definition, 381
effect of excess toxicity, 381,384f
Cluster analysis
disadvantages, 161
substituent selection, 158
Cocaine, development from natural
products, 3–4
Commonality of molecular mechanism
additivity of biological activity, 368

Commonality of molecular mechanism—
Continued
chemical properties, 368–369
QSAR correlation, 368
receptor binding, 368
symptomatology, 368
Comparative molecular field analysis,
description, 71
Computational chemist, role in design of
molecules, 28
Computational chemistry, role in design and
study of drug and agricultural
chemicals, 26
Computer-aided molecular design
hardware trends, 29–31
likelihood of success at stages in drug
development process, 26,27f
merging of statistical and molecular
modeling, 33,35
motivations for use, 26,28
potential benefit of application at lowest
level of development, 26,27f,28
role of computational chemist, 28
software trends, 31–35
trends, 28–29
Computer techniques to study mechanism and
activity, practical application, 58
Cortisone, development from biologically
active molecules, 10
Crystal lattice energy, description, 40
Crystallographic data of proteins,
availability, 106
α-Cyano-3-phenoxybenzyl
2-(4-chlorophenyl)-2-methylpropionate
conformational analyses of substructures,
184,185f
conformer pairs, 191–192,193t
optimization of geometry, 184,186
segments, 186,187f
shape comparisons using least-squares
fitting method, 186
[^{3}H]Cyclohexyladenosine, radioligand for
A_1 adenosine receptors, 233
Cyclosporin, development from natural
products, 5,7
Cytotoxicity of substituted aniline mustards
clonogenic assay, 298
determination, 296
growth inhibition assay, 296,297f,298

D

DDT, development, 18,20
Degree of matching, calculation, 85,87
Deltamethrin
candidates for active conformers of
substructures, 192,193f

Deltamethrin—*Continued*
  conformational analyses of substructures, 184,185f
  conformational energies, 186,191t
  dihedral angles, 186,191t
  optimization of geometry, 184,186
  segments, 186,187f
  shape comparisons using least-squares fitting method, 186
  superimposition of conformers, 192,194f
Design of highly bioactive drugs, importance of knowledge of active conformers, 183
Design of parathyroid hormone antagonists
  agonist and antagonist properties, 256,257t
  developmental approaches, 255–256,258
  inhibition potency, 256,257t
  methodological considerations, 253–254
  structure–activity relationships of peptide analogues, 254–258
Diarylamine rodenticides, mouse oral toxicity, 392–393
Diflubenzuron, crystal structure, 163,164f
Dihydrofolate reductase analysis of hypothetical active-site lattice (HASL)
  binding predictions, 98,99f
  comparison of fitted orientations for inhibitors, 98,101f
  construction of HASL, 98,100f
  inhibitor structures, 94,97f,98
Dimethyl phosphate, transition-state models of serine hydroxyl displacement of 3,4-dimethylphenolate ion, 153f
DNA sequencing of genes, technological advancement, 105
Drug design, factors influencing application of protein structure and sequence data, 106
Drug development, molecule identification, 61
Drug-metabolizing enzymes, function, 16

E

Electrophilic superdelocalizabilities, correlation of structure–activity relationships, 47
Electrostatic potential, calculation, 63
Energy-minimized structures
  formation methods, 38
  ground-state modeling, 39
  receptor–inhibitor recognition point, 39
Ergosterol, biosynthetic pathway, 199,200f
Ergosterol biosynthesis inhibitors
  mechanism groups, 199
  sterol $\Delta^8 \rightarrow \Delta^7$ isomerization inhibitors, 204t,205f,206

Ergosterol biosynthesis inhibitors—*Continued*
  sterol C-14 demethylation inhibitors, 201,202f,203,205f
Esfenvalerate, comparison to chrysanthemic acid, 183
Experimental descriptors for mathematical modeling
  comparison to quantum chemically derived descriptors, 43–44
  function, 42
  nature, 43
  selection, 42–43

F

Fenitrothion, two-dimensional modeling, 152t,153f
Fenvalerate
  candidates for active conformers of substructures, 192,193f
  conformational analyses of substructures, 184,185f
  conformational energies, 186,188–189t,191
  conformer pairs, 191–196
  dihedral angles, 186,188–189t,191
  optimization of geometry, 184,186
  segments, 186,187f
  shape comparisons using least-squares fitting method, 186
  superimposition of conformers, 192,194–195f
Fiber optic technology, advantages in computer-aided molecular design, 31
6-Fluoroquinolones, activity against Gram-negative bacteria, 301
Function, definition, 106
Fungicide(s)
  benzimidazoles, 206–207,208f
  carboxamides, 207,208f,209
  characterization of target sites, 199
  ergosterol biosynthesis inhibitors, 199–206
  melanin biosynthesis inhibitors, 209,210f,211
  phenylcarbamates, 206–207,208f
Fungicide development
  biochemically specific compounds, 198
  multisite biochemical inhibitors, 198

G

Global sequence similarity of proteins
  use to find inhibitor of acetolactate synthase, 108,109f
  family classification, 107
  subfamily classification, 107

INDEX

Glutathione reductase, homology graphs, 115,116f,117
Glutathione synthetase
  application of local sequence similarity to find inhibitors, 108,109t,110
  sequence segment as part of adenosine-5'-triphosphate-binding site, 110–111
Green rice leafhopper
  mechanism of resistance, 136–144
  resistance to carbamate insecticides, 176

H

Halogenated aromatic hydrocarbon(s), toxic potency, 354
Halogenated aromatic hydrocarbon–iodothyronine 5-deiodinase interaction
  animals, 356
  inhibition of conversion vs. concentration, 358,359–360f
  inhibitory concentrations, 358t
  inhibitory potency, 363
  Lineweaver–Burk plot, 361,362f
  materials, 356
  rT_3 5'-deiodinase assay procedure, 356–357
  structure–activity relationship, 361
Hardware trends in computer-aided molecular design
  fiber optic technology, 31
  optical disk technology, 31
  three-dimensional graphics work stations, 30–31
Histamine, development from biologically active molecules, 8,10,12
Homology graphing
  application to finding lead structures, 118,119f,120
  calculation procedure, 114,116f
  definition, 114
  function, 114
  glutathione reductase, 115,116f,117
  identification of sequence–chemical structure relationships, 117
  influencing factors, 114–115
Houseflies, cholinesterase inhibition, 148
5-HT2 binding
  role of substituents at position 4, 264–265
  structure–affinity relationships, 264
5-HT2 receptor–phenylisopropylamine interaction by quantitative structure–activity relationship analysis
  antagonist evaluation, 272f,274,275–276f
  Craig plot, 269,270f
  effect of length on affinity, 268t

5-HT2 receptor–phenylisopropylamine—*Continued*
  effect of lipophilicity on affinity, 268t,269
  effect of polar vs. lipophilic substituents on affinity, 277t
  effect of size and shape on affinity, 269
  equations, 266,267t
  evaluation of equation 1, 265–266
  evaluation of ester hydrolysis product, 266
  evaluation of subsets, 269,270f,271,272f
  factors influencing affinity, 273
  future research, 279
  observed and predicted affinities, 266–271
  radioligand-binding data, 265t
  search for new relationships, 266–271
  significant one- and two-variable equations, 273,274t
  use of SAS general linear model procedure, 266
Humoral hypercalcemia of malignancy
  description, 248
  peptide identification, 249–250
Hydrocortisone, development from biologically active molecules, 10
Hydroxyquinoline(s)
  comparison of π densities and net charges with those of quinolines, 223
  QSAR, 220t
Hydroxyquinoline inhibition of photosystem II
  comparison of experimentally measured potency with calculated inhibitory potency, 220t
  correlation matrix for regression equations, 220,221t
  displacement of [^{14}C]methibuzin, 218f,219
  effect of tris(hydroxymethyl)aminomethane treatment, 218,219t
  experimental materials, 216
  experimental procedures, 216–217
  inhibitory potency, 217t
  localization of inhibition site, 217,218t
  mechanism, 224
  regression equations for inhibitory potency, 220,221t
Hypercalcemia
  incidence, 250–251
  medical management, 251
  surgical management, 251–252
Hypercalcemia factor
  mediation of actions through stimulation of parathyroid hormone receptor, 249
  sequence homology with parathyroid hormone, 249

Hypermolecule, definition, 48
Hyperparathyroid hypercalcemia
  management, 251–252
  pathophysiology, 245–251
  potential clinical uses of parathyroid
    hormone antagonists, 245
Hyperparathyroidism
  medical management, 251
  surgical management, 251–252
Hypothetical active-site lattice
  comparison of molecular lattices,
    85,87,88f
  creation of molecular lattice,
    83,84f,85t,86f
  definitions, 83,85t
  effect of resolution, 91–96
  illustration of lattices for
    p-aminobenzoic acid, 85,86f
  logic flow chart, 98,102f,103
  methodology, 98,102f,103
  partial pK_i distribution, 87,89,90f,91
  schematic representation of construction,
    83,84f
  study of dihydrofolate reductase, 94–101

I

Insect(s), use of selective toxicity, 15
Insect development disrupter discovery
  biochemical pathway affected by
    benzoylphenylureas, 161,162f,163
  biological evaluation methods,
    163,166,167f
  CNDO2 electron density calculation,
    164f,166
  comparative biological response of
    benzoylphenylureas, 166,167f
  identification of ballast position, 168
  importance of metabolism for biological
    response, 166
  initial sequential simplex optimization
    design set, 163,165f
  multiple linear-regression analysis of
    topical data, 166
Insect juvenile hormone esterase, inhibition
  by trifluoromethyl ketones, 169
Interanimal toxicity orders
  anilines, 397
  carbamates, 396
  organophosphates, 396
  phenols, 397
  pyridines, 397
Interclass comparisons for animal toxicity
  birds, 397
  mouse, 397
  rat, 397

Interroute and interclass comparisons for
  animal toxicity
  carbamates, 397
  phenols, 397
Iodothyronine monodeiodinases
  function, 355
  inhibition, 355–356
Iodothyronine type I deiodinase
  dose vs. activity, 363
  role in protein-binding interactions,
    363–364

L

Lead structure identification by homology
  graphing
  enzyme–reaction data base, 118,120
  example, 120t
  future prospects, 120
  procedure, 118,119f
Leave-one-out approach, use for test of
  significance, 178
Leukemias, treatment agents developed from
  natural products, 5,8–9
Local sequence similarity of proteins
  use to find inhibitors of glutathione
    synthetase, 108,109t,110
  biological bases found between sequences
    of proteins, 111,112t,113
$\log k'$, predictor of $\log P$ for organic
  molecules, 126–130
$\log k_w$, predictor of $\log P$ for organic
  molecules, 129,130f
$\log P$
  prediction by $\log k'$, 126–130
  prediction by $\log k_w$, 129,130f
Lymphomas, treatment agents developed from
  natural products, 5,8–9

M

Mathematical modeling of receptor sites
  experimental descriptors, 42–44
  multiple regression analysis, 42
  need, 41–42
  quantum chemically derived descriptors,
    43–44
Mechanistic descriptors, correlation with
  toxicity data, 390
Medicinal science, relationship with
  agricultural science, 37
Melanin biosynthesis inhibitors
  pathway of secondary metabolism, 209,210f
  structures, 209,210f,211
Melphalan, structure, 291,293

INDEX

Metabolic stability of substituted aniline mustards, effect of chemical processes, 293,294f
Methotrexate, selective toxicity, 11–12
N-Methylcarbamate, transition-state models of serine hydroxyl displacement of 3,4-dimethylphenolate ion, 153f
Micellar liquid chromatography in quantitative structure–activity relationship modeling of organic compounds
  effect of surfactant type on degree of correlation between log k' and log P, 129,130f
  efficiency, 126
  experimental procedure, 124,126,127t
  list of substituted benzenes used, 126,127t
  log k' and log k_w as predictor of log P, 129,130f
  microphotographs of C_{18} stationary phase, 124,125f
  structure of anisotropic phase, 126
  transport properties of organic molecules, 126,128f,129
Model partitioning systems, selection, 372
Molecular electrostatic potentials, calculation, 41
Molecular lattices
  calculation of degree of fit, 85,87
  comparison, 85,87,88f
  creation, 83,84f,85t,86f
  quantitation of molecular overlap, 87,88f
Molecular mechanism, demonstration of commonality, 366,368–369
Molecular shape
  description, 71
  parameterization by spherical harmonics, 71
Molecular shape analysis method, description, 71
Molecular shape descriptors, See Shape descriptors of molecules
Morpholines, sterol $\Delta^8 \rightarrow \Delta^7$ isomerization inhibition, 204t,205f,206
Multiple endocrine neoplasia, description, 246–247
Multisite biochemical inhibitors
  advantages and disadvantages, 198
  examples, 198

N

Nalidixic acid
  inhibition of bacterial DNA synthesis, 303
  QSAR relationships, 301
  structure, 301,302t

Naphthoquinones
  inhibition of photosystem II electron flow, 219t,220
  π densities and net charges, 221,222t,223,225
Narcosis
  bilinear relationships and pharmacokinetic cutoff, 377,378f
  chronicity ratio, 381,384f
  cyanogenic mechanism, 381,383t
  electrophile mechanism, 379,380f,382t
  excess toxicity, 377,379
  focus of current research, 370
  influence of melting point on water solubility cutoff, 377
  proelectrophile mechanism, 379,381,382t
  species sensitivity, 372,374,375–376t
  use of screening data to investigate limitations of narcosis mechanism, 381,383
  water solubility cutoff, 374,377
Narcosis base-line toxicity
  linear QSAR models, 370,371f,372
  oil–gas coefficient vs. minimum concentration in air required to produce anesthesia, 372,273f
Narcosis potency
  correlation with partition coefficient, 369–370
  correlation with water solubility and molecular weight, 369
Natural products, development of bioactive molecules, 3–9
Nonpeptide parathyroid hormone antagonists, future prospects, 258
Norfloxacin
  activity, 303
  structure, 302t,303

O

Octanol–water partition coefficient
  determination by micellar liquid chromatography, 124–130
  determination by shake-flask method, 124
  estimation by reversed-phase liquid chromatography, 124
Optical disk technology, advantages in computer-aided molecular design, 31
Oral antidiabetic agents, development, 18
Organocarbamate inhibitors
  binding mechanism, 153–155
  distance from carbonyl to binding center, 151,152t
  effect of bulk tolerance, 148,150t,151
  effect of molecular structure, 148

Organocarbamate inhibitors—*Continued*
  leaving group requirements, 155
Organophosphate inhibitors
  binding mechanism, 153–155
  distance from phosphoryl to binding
    center, 151,152*t*
  effect of bulk tolerance, 148,149*t*,151
  effect of molecular structure, 148
  leaving group requirements, 155
Organophosphate toxicity, 393–394
Osteoporosis
  description, 250
  effect of parathyroid hormone, 250

P

Parathyroid hormone
  biological activities, 244
  indirect calcium-level-increasing
    actions, 244
  renal actions, 244*t*
  sequence homology with hypercalcemia
    factor, 249
  stimulation of bone resorption, 244
  synthesis, 244–245
Parathyroid hormone antagonists
  design, 253–258
  potential clinical uses, 253
  therapeutic potential, 243–258
Parathyroid hormone receptor, signal
  transduction mechanism for
  stimulation, 245
Partial pK_i distribution
  equal distribution, 89
  estimation of distribution, 89,90*f*,91
Partition coefficients, estimation, 372
Pathophysiology of hyperparathyroid
  hypercalcemia
  causes, 246–247
  correlation with hypertension, 245
  humoral hypercalcemia of malignancy,
    248–250
  incidence of disorders, 250–251
  osteoporosis, 245
  side effects, 245–246
  symptoms, 245
Penicillins, development from natural
  products, 3,5–6
Penicillopepsin, second-generation
  computer-assisted inhibitor design,
  63,64*t*,65*f*,66
Pharmacological screens
  advantages, 22–23
  discovery of biologically active
    compounds, 22
Phenol(s)
  essential chemical element for inhibition,
    225,226*f*

Phenol(s)—*Continued*
  inhibition mechanism, 224
  inhibitors of photosystem II, 215–216
Phenol toxicity
  albino mouse intraperitoneal toxicity, 391
  mouse
    ip toxicity, 391
    iv toxicity, 391–392
    oral toxicity, 392
  rabbit, dermal toxicity, 392
  rat, oral toxicity, 392
Phenoxyacetic acids, development from
  biologically active molecules, 10,12
3-Phenoxybenzyl (*R*)-2-(4-ethoxyphenyl)-
  3,3,3-trifluoropropyl ether
  candidates for active conformers of
    substructures, 192,193*f*
  conformational analyses of substructures,
    184,185*f*
  conformational energies, 186,190*t*,191
  conformer pairs, 192,194–195*t*,196
  dihedral angle, 186,190*t*,191
  optimization of geometry, 184,186
  segments, 186,187*f*
  shape comparisons using least-squares
    fitting method, 186
  superimposition of conformers,
    192,195*f*
Phenylcarbamates
  structure–activity relationships, 207
  structure of fungicides, 206–207,208*f*
Phenylisopropyladenosine, affinity ratios,
  238,241*t*
Phenylisopropylamines
  hallucinogenic affect, 278
  QSAR analysis, 264–278
Photosystem II inhibitors
  binding site, 223
  effect of tris(hydroxymethyl)aminomethane
    treatment on inhibitory potency,
    218,219*t*
  hydroxyquinolines, 216–226
  naphthoquinones, 219*t*,220
  phenol family, 215–216
  urea–triazinone family, 215–216
Porcine atrial A_1 adenosine receptor
  agonist inhibition of specific binding,
    238–240*f*
  characterization, 237–241
  inhibition of adenylyl cyclase activity
    vs. guanosine 5′-triphosphate
    concentration, 238,239*f*
  rank order potency of adenosine analogues,
    237,238*t*
Postsynaptic 4-aminobutyric acid A receptor,
  description, 281
Primary hyperparathyroidism,
  description, 246
Prodrug design, examples, 16–17

INDEX

Proteins
  biological bases of local similarity, 111,112*t*,113
  factors influencing application of structure and sequence data to drug design, 106
  global sequence similarity, 107–108,109*f*
  homology graphing, 113–120
  identification factors, 107
  local sequence similarity, 108,109*t*,110
  methods to derive three-dimensional model from sequence, 106
  testing of similarity, 107
Pyrethrins, development from natural products, 5–6
Pyrethroid discovery
  active-site binding, 159,160*f*,161
  biological response models, 159
  substituent selection, 158,160*f*
  synthesis of 2-biphenylethanol analogues, 161,162*f*
  topical assay testing of compounds, 158–159
Pyridine toxicity
  mouse intraperitoneal and intravenous toxicity, 395
  rat oral toxicity, 395
Pyrrolozidine, structures, 71,72*f*
Pyruvate oxidase, catalytic reactions, 108,109*f*

Q

Quantitative agrochemical design strategies
  advantages, 157–158
  goals, 157–158
  insect development disrupter discovery, 161–168
  pyrethroid discovery, 158–162
Quantitative structure–activity relationship(s) (QSAR)
  antiinfective agents based on nalidixic acid structure, 304–347
  function, 21–22
  nalidixic acid derivatives, 303–304,305*t*
Quantitative structure–activity relationship analysis, 5-HT$_2$ receptor–phenylisopropylamine interaction, 264–278
Quantitative structure–activity relationship model(s)
  acetylenic alcohols, 383
  applicability, 366
  bilinear relationships and pharmacokinetic cutoff, 377,378*f*
  chronicity ratio, 381,384*f*
  cyanogenic mechanism, 381,383*t*

Quantitative structure–activity relationship model(s)—*Continued*
  demonstration of commonality of molecular mechanism, 366,368–369
  discovery of correlation between narcosis potency and partition coefficient, 369
  discovery of correlation between narcosis potency and water solubility and molecular weight, 369
  effect of melting point on water solubility cutoff, 377
  electrophile mechanism, 379,380*f*,382*t*
  estimation of partition coefficients, 372
  excess toxicity, 377,379
  linear models for narcosis base-line toxicity, 370,371*f*,372,373*f*
  local models, 383,385
  mechanism-based model vs. chemical and biological data of molecules used in derivation, 366,367*f*
  model partitioning systems, 372
  proelectrophile mechanism, 379,381,382*t*
  species sensitivity to narcosis, 372,374,375–376*t*
  use of screening data to investigate limitations of narcosis mechanism, 381,383
  water solubility cutoff, 374,377
Quantitative structure–activity relationship modeling of organic compounds, application of micellar liquid chromatography, 124–130
Quantum chemically derived descriptors for mathematical modeling
  comparison to experimental descriptors, 43–44
  function, 44
Quinolones
  comparison of π densities and net charges with those of hydroxyquinolines, 223*t*
  π densities and net charges, 221,222*t*,223,225
2-(2-Quinolyl)cyclohexane phenylhydrazone
  electrostatic potential, 64,65*f*,66
  fit in active site of penicillopepsin, 64,65*f*
  structure, 64

R

Radioligands, labeling of A$_1$ adenosine receptors, 233
Receptor–inhibitor recognition point
  discussion of concept, 39
  ground-state modeling, 39
Receptor mapping
  approaches, 19

Receptor mapping—*Continued*
  definition, 19
  example, 19–21
  limitations, 21
  requirements, 19
Receptor sites in medicine and agrochemistry, need for mathematical modeling, 41–44
Resistance to fungicides, development of counteraction approaches, 198–199
Resistance mechanism of green rice leafhopper
  acetylcholinesterase preparation and assay, 137
  enzyme inhibition of aryl N-methylcarbamates, 141t,142–144
  enzyme inhibition of N-aryl oxadiazolones, 138–140t,143–144
  statistical data analysis, 137,142
  structures of compounds, 137,138–141t
Resolution
  π-aminobenzoic acid partial pK_i estimations, 94,95t
  definition, 91
  effect on estimates of partial pK_i, 94,95f
  effect on number of lattice points, 91,93f
  effect on predictivity, 92,93f,94
  error in partial pK_i values, 94,96f
  testing of predictivity vs. resolution, 91,92t
Reversed-phase liquid chromatography
  comparison of different modes, 126,127t
  estimation of octanol–water partition coefficient, 124

S

Secondary hyperparathyroidism
  description, 247
  reversal mechanism, 247
  sequence of events, 247
Second-generation computer-assisted inhibitor design method
  application to penicillopepsin, 63,64t,65f,66
  description, 61
  design of hydrogen-bonding interactions, 67
  design procedure, 61–63
  evaluation of electrostatic potential, 63
  future research, 67–68
  number of potential interactions for tight ligand binding, 67
  scoring of orientations, 62–63

Second-generation computer-assisted inhibitor design method—*Continued*
  variable description used in docking procedure, 62,64t
Selective toxicity of bioactive molecules
  advantages and disadvantages, 13,15–16
  description, 11
  examples, 11–16
  treatment of bacterial diseases, 15
  use against viruses, 15–16
  use as insecticide, 10
Sequence–chemical structure relationships
  difficulties in searching, 113–114
  identification from homology graphs, 117
Sequential simplex optimization
  description, 161
  insect development disrupter discovery, 161–168
Serotonin receptors, discovery of multiple populations, 264
Shape descriptors of molecules
  binding-moment approach to molecular orientation, 74
  biodata correlation matrix, 72t
  calculated log P vs. acute morbidity of pyrrolizidine alkaloids, 74,75f
  canonical correlation, 76,79,80t
  comparison of structures, 76,79t
  examples, 71
  experimental methodology, 72,74
  experimental procedures, 71–72
  modeled structures, 74,75f
  modification, 47
  problems, 71
  soft independent modeling by class analogy, 74,76,77–78f
  structure, 72,73f
  structure alignment, 74
  superpositions of structures, 74,77f
Significance tests, *See* Test of Significance, 177–178
Similarity of proteins
  amino acid sequence, 107
  protein function, 107
Small molecule–enzyme receptor interaction
  effect of molecule's structure, 70–71
  stages, 70
Soft independent modeling by class analogy of molecular shape
  advantages and disadvantages, 76
  cluster dendrogram, 76,77f
  description, 74,76
  schematic representation, 76,78f
Software trends in computer-aided molecular design
  merging of statistical and molecular modeling, 33,35
  three-dimensional structural data bases, 32–33,34f

INDEX

Software trends in computer-aided molecular design—*Continued*
  trends, 31–32
Spherical harmonics, parameterization of three-dimensional molecular shape, 71
Statistical docking method
  examples, 49–50f
  procedure, 48
Stepwise regression analysis on trifluoromethyl ketones
  correlation between calculated and measured pI_{50} values, 173,175f,176,177t
  correlation between total lipophilicity and inhibitory potency, 170
  error in prediction, 170–171
  interpretation of regression equations, 178,179t,180
  procedure, 171–177
  tests of significance, 170–178
  theory, 170–171
Sterol C-14 demethylation inhibitors
  cytochrome P-450 binding site models, 203
  development of resistance, 203
  interaction mechanism, 201,205f
  structure, 201,202f
  structure–activity relationships, 201,203
Sterol $\Delta^8 \rightarrow \Delta^7$ isomerization, mechanism pathway, 204,205f
Sterol $\Delta^8 \rightarrow \Delta^7$ isomerization inhibitors
  high-energy or transition-state intermediates, 204t
  morpholines, 204
  target enzymes, 204t,206
Structure–activity relationships
  activation in bound state, 39–41
  binding as soft reaction, 44–45
  computer generation, 82
  function, 37–38
  inhibition of green rice leafhopper, 136–144
  insights from energy-minimized structures, 38–39
  methodology, 38
  modeling of binding events with physicochemical descriptors, 45–46
  modeling of binding events with quantum chemical descriptors, 46–47
  need for mathematical modeling, 41–44
  restrictions to statistical approach to analysis, 38
  use of hypothetical active-site lattice, 83–103
meta-Substituted *N*-methyl arylcarbamates, vs. housefly head acetylcholinesterase, 49f
ortho-Substituted aryl *N*-methylcarbamates
  analysis for binding to acetylcholinesterase, 45–46
  vs. housefly head acetylcholinesterase, 49,50f

7-Substituted 5-hydroxytryptamine derivatives, binding to LSD receptor site, 47

T

Test of significance
  F value, 177–178
  leave-one-out approach, 178
Tetracyclines
  development from natural products, 3,5–6
  selective toxicity, 11,13
Three-dimensional graphics work stations
  future developments, 30–31
  high-end systems, 30
  low-end systems, 30
  mid-range systems, 30
Three-dimensional structural data bases
  Brookhaven Protein Databank, 32
  Cambridge Crystallographic Structure Database, 32
  pharmacophore and toxicophore searching, 33
  THOR data base system, 32
  three-dimensional module of MACCS-II, 33,34f
Thyroid hormones, structure, 358,359f
Thyroxine-binding prealbumin, competitive binding studies, 355
Thyroxine-binding protein, sites for interactions that mediate toxic responses, 355
Tocainide, development, 16–17
Toxicity data, correlation with mechanistic descriptors, 390
Toxicity of medicinals and agrochemicals, need for developing predictive relations for estimating probable toxicity, 352
Toxicity testing, normal variations, 396
Toxic potency of halogenated aromatic hydrocarbons, qualitative structure requirements, 354
2-(Trifluoromethyl)hydroxyquinolines, inhibition of photosystem II, 217–226
Trifluoromethyl ketones
  effect of inhibitory potency on binding, 179t,180
  effect of lipophilicity on binding, 179
  effect of molar volume on binding, 178
  effect of *ortho* substituents on binding, 180
  enzyme inhibition assay, 171
  inhibition of hydrolytic enzymes, 169
  stepwise regression analysis, 170–171
  structures, 171,172t,174
  synthesis, 171
3,3',5'-Triiodothyronine, structure, 358,359f
Trimethoprim, selective toxicity, 11

U

Uridine 5'-diphosphate-*N*-acetyl-
  glucosamine, crystal structure,
  163,164*f*
Urea/triazinones
  essential chemical element for inhibition,
    225,226*f*
  inhibition mechanism, 224

V

Viruses, use of selective toxicity, 15–16

X

Xenobiotics, function, 16–18

*Production: Raymond L. Everngam, Jr.
Indexing: Deborah H. Steiner
Acquisition: Cheryl Shanks*

**Elements typeset by Hot Type Ltd., Washington, DC
Printed and bound by Maple Press, York, PA**

Other ACS Books

Chemical Structure Software for Personal Computers
Edited by Daniel E. Meyer, Wendy A. Warr, and Richard A. Love
ACS Professional Reference Book; 107 pp;
clothbound, ISBN 0–8412–1538–3; paperback, ISBN 0–8412–1539–1

Personal Computers for Scientists: A Byte at a Time
By Glenn I. Ouchi
276 pp; clothbound, ISBN 0–8412–1000–4; paperback, ISBN 0–8412–1001–2

Biotechnology and Materials Science: Chemistry for the Future
Edited by Mary L. Good
160 pp; clothbound, ISBN 0–8412–1472–7; paperback, ISBN 0–8412–1473–5

Polymeric Materials: Chemistry for the Future
By Joseph Alper and Gordon L. Nelson
110 pp; clothbound, ISBN 0–8412–1622–3; paperback, ISBN 0–8412–1613–4

The Language of Biotechnology: A Dictionary of Terms
By John M. Walker and Michael Cox
ACS Professional Reference Book; 256 pp;
clothbound, ISBN 0–8412–1489–1; paperback, ISBN 0–8412–1490–5

Cancer: The Outlaw Cell, Second Edition
Edited by Richard E. LaFond
274 pp; clothbound, ISBN 0–8412–1419–0; paperback, ISBN 0–8412–1420–4

Practical Statistics for the Physical Sciences
By Larry L. Havlicek
ACS Professional Reference Book; 198 pp; clothbound; ISBN 0–8412–1453–0

The Basics of Technical Communicating
By B. Edward Cain
ACS Professional Reference Book; 198 pp;
clothbound, ISBN 0–8412–1451–4; paperback, ISBN 0–8412–1452–2

The ACS Style Guide: A Manual for Authors and Editors
Edited by Janet S. Dodd
264 pp; clothbound, ISBN 0–8412–0917–0; paperback, ISBN 0–8412–0943–X

Chemistry and Crime: From Sherlock Holmes to Today's Courtroom
Edited by Samuel M. Gerber
135 pp; clothbound, ISBN 0–8412–0784–4; paperback, ISBN 0–8412–0785–2

For further information and a free catalog of ACS books, contact:
American Chemical Society
Distribution Office, Department 225
1155 16th Street, NW, Washington, DC 20036
Telephone 800–227–5558